Planets Outside the Solar System: Theory and Observations

NATO Science Series

A Series presenting the results of activities sponsored by the NATO Science Committee. The Series is published by IOS Press and Kluwer Academic Publishers, in conjunction with the NATO Scientific Affairs Division.

A. Life Sciences	IOS Press
B. Physics	Kluwer Academic Publishers
C. Mathematical and Physical Sciences	Kluwer Academic Publishers
D. Behavioural and Social Sciences	Kluwer Academic Publishers
E. Applied Sciences	Kluwer Academic Publishers
F. Computer and Systems Sciences	IOS Press
1. Disarmament Technologies	Kluwer Academic Publishers
2. Environmental Security	Kluwer Academic Publishers
3. High Technology	Kluwer Academic Publishers
4. Science and Technology Policy	IOS Press
5. Computer Networking	IOS Press

NATO-PCO-DATA BASE

The NATO Science Series continues the series of books published formerly in the NATO ASI Series. An electronic index to the NATO ASI Series provides full bibliographical references (with keywords and/or abstracts) to more than 50000 contributions from internatonal scientists published in all sections of the NATO ASI Series.
Access to the NATO-PCO-DATA BASE is possible via CD-ROM "NATO-PCO-DATA BASE" with user-friendly retrieval software in English, French and German (WTV GmbH and DATAWARE Technologies Inc. 1989).

The CD-ROM of the NATO ASI Series can be ordered from: PCO, Overijse, Belgium

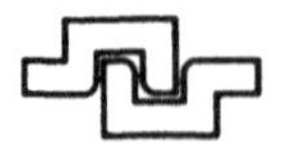

Series C: Mathematical and Physical Sciences – Vol. 532

Planets Outside the Solar System: Theory and Observations

edited by

J.-M. Mariotti †

and

D. Alloin
European Southern Observatory,
Garching, Germany

Kluwer Academic Publishers

Dordrecht / Boston / London

Published in cooperation with NATO Scientific Affairs Division

Proceedings of the NATO Advanced Study Institute on
Planets Outside the Solar System: Theory and Observations
Cargèse, Corsica, France
May 5-15, 1998

A C.I.P. Catalogue record for this book is available from the Library of Congress.

ISBN 0-7923-5708-6

Published by Kluwer Academic Publishers,
P.O. Box 17, 3300 AA Dordrecht, The Netherlands.

Sold and distributed in North, Central and South America
by Kluwer Academic Publishers,
101 Philip Drive, Norwell, MA 02061, U.S.A.

In all other countries, sold and distributed
by Kluwer Academic Publishers,
P.O. Box 322, 3300 AH Dordrecht, The Netherlands.

Printed on acid-free paper

TABLE OF CONTENTS

PREFACE

The question of the existence of other worlds and other living beings has been present in the human quest for knowledge since as far as Epicurus. For centuries this question belonged to the fields of philosophy and theology. The theoretical problem of the formation of the Solar System, and hence of other planetary systems, was tackled only during the 18th century, while the first observational attempts for a detection started less than one hundred years ago.

Direct observation of an extra-solar planetary system is an extraordinarily difficult problem: extra-solar planets are at huge distances, are incredibly faint and are overwhelmed by the bright light of their own stars. With virtually no observational insight to test their models, theoreticians have remained for decades in a difficult position to make substantial progress.

Yet, the field of stellar formation has provided since the 1980s both the theoretical and observational evidences for the formation of discs at the stage of star birth and for debris materials orbiting the very young stellar systems. It was tempting to consider that these left-overs might indeed later agglomerate into planetary systems more or less similar to ours. Then came observational evidences for planets outside the Solar System.

From now on, the interplay between theory and observations can booster the development of a new field of Astrophysics and Planetary Sciences. Therefore, we thought it was quite timely to propose an ASI on the subject. In addition, numerous projects for the detection and observation of extra-solar planets are springing. They rely either on indirect methods (gravitational pull of the planet), or on direct methods (actual detection of the planet emission). Again it was an appropriate time to examine all of these projects.

But of course the field of extra-solar planets does not pertain only to Planetary Sciences and Astrophysics: the questions of the emergence of life on Earth and of the existence of other forms of life in the Universe are major ones. We chose to devote part of the ASI to these subjects.

The ASI was attended by 61 persons. The scientific program was organized in three main parts so as to span the various aspects related to the extra-solar planets search. The first lectures were intended to discuss the formation processes for stars and their accompanying discs and the subsequent formation of planets. Then, a large part of the ASI was devoted to the description of experiments, in progress or in a planning phase, to detect extra-solar planets. Finally, in the third part of the ASI we touched upon questions related to the occurence of life on Earth. The field of exobiology was quite new to many of the participants: it required special attention and numerous additional discussions under-the-olive-tree. Several seminars and posters were contributed by the attendees, enlightening some particular points and presenting new methods or results.

This ASI was organized with the support of NATO through its Scientific Affairs Division. We acknowledge as well funding from the Centre National de la Recherche Scientifique (Division de la Formation Permanente), from the European Southern Observatory and from the National Science Foundation.

The organization of the School was under the responsibility of Georgette Hubert, in close coordination with the staff of the Formation Permanente at CNRS and with the staff of the Institut d'Etudes Scientifiques de Cargèse. I would also like to thank Georgette Hubert, Alessandro Pizzella and Maria-Eugenia Gómez and Paul Williams for their efficient support in the editing of the book.

Thanks to the lively contributions from all lecturers and the enthusiastic participation of the students, the atmosphere of the ASI has been quite warm and pleasant. For obvious reasons, a well-known person in Cargèse, the priest, was quite interested in the subject and joined our discussions, songs and dances during the closing diner. Most participants expressed the wish that other meetings on this topic should be organized in the future: they are welcome to act in this way.

One very sad event happened after the school: Jean-Marie Mariotti, who had been the driving force in the organization of this ASI, passed away on July 28th from a rapidly evolving leucemia. This book has been prepared with having steadily in mind Jean-Marie and the two wonderful weeks we shared with him in Cargèse, thinking and discussing about other planets, other worlds, ways to find them as we are convinced they exist, and other possible forms of life.

Santiago, November 1998 — Danielle Alloin

Ce sont de grandes lignes calmes qui s'en vont
à des bleuissements de vignes improbables.
La terre en plus d'un point mûrit
les violettes de l'orage; et ces fumées de sable
qui s'élèvent au lieu des fleuves morts,
comme des pans de siècles en voyage...

St.John Perse
Extract of Anabase, VII

Jean-Marie MARIOTTI (1955 - 1998)

Jean-Marie Mariotti, astronome à l'Observatoire de Paris et en détachement auprès de l'Observatoire Européen Austral (ESO) à Garching depuis octobre 1997, nous a quittés le 28 juillet 1998, emporté à l'âge de 43 ans et en moins d'un mois par une leucémie foudroyante. Ses derniers jours à l'hôpital universitaire de Munich furent empreints d'un grand courage et d'une extraordinaire sérénité : en furent témoins tous ceux qui purent garder contact avec lui dans ces moments qui allaient se révéler ultimes. Ses cendres reposent dans ce petit cimetière de Trivaux, au pied de cet Observatoire de Meudon dont il aimait le calme encore agreste.

Le 14 septembre, une Journée scientifique, tenue en hommage à Jean-Marie, rassembla à Meudon près d'une centaine de collègues et d'amis, venus souvent de bien loin : les exposés, tournés vers l'avenir, y développèrent la diversité et la fécondité des idées que Jean-Marie avait semées au cours de sa trop brève carrière. Nous allons tenter d'en rendre compte ici. Jean Marie naquit en 1955 près de Paris. Lors de ses études aux lycées Montaigne et Louis-le-Grand, nous savons que l'astronomie l'attirait déjà: il fut de ce petit groupe passionné qui remit en service la coupole de la Sorbonne, rue Saint Jacques. En 1978, il obtient son diplôme d'ingénieur de l'Ecole Supérieure d'Optique à Orsay, cette école fameuse qui a donné à l'astronomie tant de femmes et d'hommes de talent. Conforté dans ses premiers goûts par une rencontre décisive avec Antoine Labeyrie lors d'une conférence de ce dernier, qui venait tout juste de proposer l'interférométrie des tavelures et préparait le couplage interférométrique et historique de deux télescopes disjoints, Jean-Marie décide alors de s'orienter vers l'astronomie et entre à Meudon dans la Formation doctorale Astrophysique et techniques spatiales de l'Université Paris VII. De cette double formation d'opticien et d'astronome, il gardera toujours le goût des élégantes réalisations instrumentales et de leur application immédiate d'authentiques problèmes d'astrophysique.

Sans barguigner, il s'embarque aussitôt vers la haute résolution angulaire en optique, un domaine à haut risque, encore fortement marqué du scepticisme des astronomes quant à ses perspectives à moyen terme. Sa thèse de troisième cycle , préparée à Lyon sous la direction de François Sibille, développe les potentialités de l'interférométrie des tavelures dans le proche infrarouge avec les détecteurs encore rudimentaires de l'époque (un seul pixel !) et les applique à des enveloppes circumstellaires observées à Kitt Peak et à Zelentchuk. Après un bref séjour à Milan chez O.Citterio et P. di Benedetto, il rejoint l'Observatoire de Lyon comme assistant-astronome (un statut assez peu enviable à l'époque) pour y préparer son Doctorat d'Etat qu'il soutiendra en 1987. A Lyon l'auront rejoint Steve Ridgway, puis Christian Perrier de retour de coopération à La Silla, avec lesquels il développera amitié et coopérations durables. Cette thèse établit nombre de résultats fondamentaux pour l'interférométrie multi-télescopes, résultats en partie obtenus au Plateau de Calern sur le premier interféromètre (I2T) et qui servent de référence dans les programmes actuels. A l'automne 1988, il rejoint en tant qu'Astronome-adjoint l'Observatoire de Paris au sein du département de recherche spatiale (DESPA) et sa compétence est assez vite internationalement reconnue. Car c'est aussi en 1988 que débute à l'Observatoire européen austral (ESO) la construction du très grand télescope européen VLT, qui doit comprendre un mode interférométrique de couplage cohérent de ses télescopes : de 1988 à 1992, Jean-Marie est consultant puis président du VLT Interferometry Panel de l'ESO, à une période clef où s'affine, sous la responsabilité de Jacques Beckers, la conception détaillée du VLTI et de son instrumentation. Soucieux de maintenir une cohérence entre des objectifs "sol" à court terme et des objectifs spatiaux plus lointains, il devient en 1996 membre de l'Infrared Interferometry Cornerstone Advisory Group auprès de l'Agence spatiale européenne et en 1997 membre du Planet Finder Advisory Group de la NASA : il va y jouer un rôle clef dans l'émergence du projet de mission européenne IRSI (InfraRed Space Interferometer), plus connue sous le nom de DARWIN. A l'Observatoire de Paris, il crée un petit groupe d'interférométrie où vont se succéder les thèses de Vincent Coudé du Foresto (1994), de Zhao Peiqian (1995), de Guy Perrin (1996), de Frédéric Cassaing (à l'ONERA, en 1997) et où se préparent encore celles de Bertrand Mennesson, de Cyril Ruilier et de Pierre Kervella. Le Programme national Haute résolution angulaire en astrophysique du CNRS/INSU succède à un Groupement de recherche et est créé dès 1991 : Jean-Marie, en tant que directeur-adjoint et aux côtés de Renaud Foy à la tête de son Conseil scientifique, en sera l'animateur incontesté lors des années difficiles de maturation du programme VLTI. A l'automne 1997, choisi par l'ESO pour succéder à Oskar von der Luehe à la tête du programme VLTI, il rejoint Garching, conscient des enjeux et des difficultés de ce programme qui est désormais en directe compétition avec celui de l'interféromètre Keck sur le Mauna Kea.

Jean-Marie avait une profonde compréhension de la cohérence en optique, et de la façon de la traiter dans un interféromètre. En 1984, avec di Benedetto, il publie à partir de données obtenues au Calern une analyse approfondie de la stabilité des trajets optiques dans un interféromètre ; en 1988, avec Ridgway, ils inventent l'analyse spectrale-spatiale dite Double Fourier, déjà pressentie par Jean Gay et qui étend élégamment à un interféromètre spatial les principes classiques de la spectrométrie de Fourier. Ces deux articles, aux côtés de ceux de Fizeau, de Michelson, de Labeyrie et de Shao, sont inclus dans le très sélectif ouvrage de Lawson paru en 1997.

De fait, un troisième article de cette sélection, bien que non signé par le toujours modeste Mariotti, apporta en 1991 une contribution capitale qu'il inspira et rendit effective : cet article développe le concept du filtrage spatial de faisceaux affectés par la turbulence atmosphérique à l'aide d'une fibre optique monomode. Appliqué depuis 1992 par Coudé du Foresto puis par Perrin, ceci conduit à gagner, par rapport aux méthodes antérieures, plus d'un ordre de grandeur sur la détermination des visibilités (amplitude de la composante de Fourier de l'objet échantillonnée par l'interféromètre) des objets astronomiques observés à travers l'atmosphère : la précision relative atteint aujourd'hui quelques dix-millièmes .

L'impact sur les observations stellaires fut immédiat, conduisant à des diamètres et des températures effectives d'une extraordinaire précision. Les interféromètres en construction (VLTI, Keck) ou en fonctionnement se réfèrent presque tous à ce concept. Elaborant sur les performances déjà démontrées ou attendues des fibres monomodes dans le visible et le proche infrarouge (jusqu' une longueur d'onde de 10 micromètres environ), Jean-Marie proposa en 1996, puis en 1998 au Colloque de Kona, une vue futuriste où les six ou sept grands télescopes optiques du site de Mauna Kea pourraient être couplés de façon cohérente avec la même commodité que celle habituelle aux radioastronomes pour le transport des signaux sur des distances kilométriques : cette idée, aujourd'hui reprise par François Roddier, pourrait bien un jour "ancrer" notre CFHT ou son successeur à ce site extraordinaire.

Sa contribution à l'astronomie débuta en 1983 par de très soigneuses observations, en interférométrie des tavelures dans l'infrarouge, d'enveloppes circumstellaires (GL 2591, MWC 349). La maîtrise acquise par Perrier et lui dans cette difficile technique les conduisit à publier en 1987 un article questionnant la réalité de la "première" naine brune observée, VB8B proposée comme compagnon de l'étoile van Biesbroeck 8. Tâche peu agréable mais qui fut rigoureusement confirmée par la suite, et qui marqua le début de l'intérêt de Jean-Marie pour les objets de faible masse. En 1987, la découverte d'un compagnon d'une étoile naine rouge par imagerie infrarouge à haute résolution, en parallèle à celle effectuée par vitesse radiale par le groupe de Michel Mayor à Genève, le conduisit a réaliser le potentiel unique de la combinaison de telles mesures pour obtenir la masse de façon très précise, fixer la limite basse de la séquence principale et recenser les éventuelles naines brunes. Il entrepris alors, avec le regretté Duquennoy (Genève), un travail systématique de recherche dans les deux hémisphères, utilisant rapidement l'optique adaptative : ce travail de longue haleine, commence à produire les résultats escomptes, préfigurant les programmes fondamentaux que les très grands télescopes et le VLTI permettront de mener à terme. Dans l'esprit de Jean-Marie, la question des exo-planètes et celle d'une éventuelle vie sur celles-ci étaient étroitement associées, comme chez beaucoup d'astronomes, à une réflexion fondamentale sur la place de la Terre et de l'homme dans l'univers. Cet ancrage philosophique, qui prenait chez lui une dimension éthique, joua certainement un rôle lors de son implication croissante dans ce sujet : avec Alain Léger, il va développer le concept optique de la mission spatiale Darwin , améliorant une élégante méthode d'annulation du signal stellaire par interférences destructives. La proposition initiale supposait des performances instrumentales extrêmes, mais en quelques années il va proposer de remarquables solutions, rendant la mission bien plus réalisable à court terme : filtrage optique monomode du faisceau et modulation interne du signal. Ce concept de mission, aujourd'hui étudiée par l'Agence spatiale européenne sous le nom d'IRSI (InfraRed Space Interferometer), va rallier l'intérêt des collègues américains et représente aujourd'hui l'une des approches de spectro-imagerie les plus prometteuses pour la détection de signes biotiques.

Quelques semaines avant sa disparition, Jean-Marie, associé à Michel Mayor et d'autres, observait à l'Observatoire de Haute-Provence et ils découvraient une exo-planète de type jovien autour de l'étoile 14 Her (Gl 614), ayant la période la plus longue parmi les 12 exoplanètes découvertes à ce jour par mesure de vitesse radiale. Ultime récompense de trop brèves années. Au-delà de cette annonce, les deux années de mesures intensives auxquelles il participait ont déjà permis de constater qu'une proportion notable des étoiles proches observées étaient perturbées par des compagnons substellaires ou des planètes joviennes.

Bien que peu porté sur le métier d'enseignant, Jean-Marie respecta les obligations faites aux astronomes, et enseigna plusieurs années l'optique aux étudiants du DESS "Optique et matériaux" de Paris VII : ils n'oublieront sans doute pas ses leçons, inspirées parfois du Micromegas de Voltaire ! Mais c'est aussi avec Danielle Alloin qu'il portera le souci d'entraîner une nouvelle génération d'astronomes vers les promesses et les difficultés de la haute résolution angulaire ou de la découverte d'exo-planètes : ils organisèrent à Cargèse trois écoles fameuses, dont les ouvrages publiés sont des références essentielles. Avec son épouse Françoise et leurs deux enfants (Apolline, 6 ans et Octave, 3 ans), nous cultiverons la mémoire de Jean-Marie, ce collègue brillant et discret qui, dans les eaux parfois tumultueuses de la compétition scientifique, sut garder la candeur d'un enfant et le paisible sourire, teinté d'humour, d'un philosophe.

Pierre Léna

Mariotti J.-M., 1981, Thèse de doctorat de 3ème cycle, Université Paris VII, "Interférométrie des tavelures en infrarouge, applications aux enveloppes circumstellaires et aux régions de formation d'étoiles.

Mariotti J.-M., 1987, Thèse de doctorat d'état, Université Claude Bernard Lyon, "Imagerie à la limite de diffraction en infrarouge, méthodes et résultats astrophysiques.

Mariotti J.-M., Di Benedetto P., 1984, "Path length stability of synthetic apertures telescopes", in Very large telescopes, their Instrumentation and Programs, Ulrich M.H. and Kjaer K., ESO, Garching.

Mariotti J.-M., Ridgway S., 1988, Astron.Astrophys. 195, 350.

Lawson P.L., 1997, "Selected Papers on Long Baseline Stellar Interferometry", SPIE Milestone Series.

Coudé du Foresto V., Ridgway S., 1991, "Fluor, a stellar interferometer using single-mode infrared fibers", in High resolution imaging by interferometry II, Beckers J. and Merkle F., ESO, Garching.

Coudé du Foresto V., Ridgway S., Mariotti J.-M. 1997, Astron.Astrophys.Suppl.Ser. 121, 379.

Mariotti J.-M., Coudé du Foresto V., Perrin G., Zhao P., Léna P., 1996, Astron.Astrophys.Suppl.Ser. 116, 381.

Perrier C., Mariotti J.-M., 1987, Ap.J. 312, L27.

Mariotti J.-M., Perrier C., Duquennoy A., Duhoux P., 1990, Astron.Astrophys. 230, 77.

Mariotti J.-M., Perrier C., 1991, "Hunting the Brown dwarf", The Messenger 64, 29.

Léger A., Mariotti J.-M., Mennesson B., Ollivier M., Puget J.-L., Rouan D., Schneider J., 1996, Icarus 123, 249.

Ollivier M., Mariotti J.-M., 1997, Appl.Opt. 36, 5340.

Mayor M., Queloz D., Beuzit J.-L., Mariotti J.-M., Naef D., Perrier C., Sivan J.-P., 1998, in Protostars and Planets IV.

"Diffraction-limited imaging with very large telescopes", 1989, eds. Alloin D. and Mariotti J.-M., ASI vol. C274, Kluwer.

"Adaptive optics for astronomy", "Adaptive optics for astronomy", 1994, eds. Alloin D. and Mariotti J.-M., ASI vol.C423, Kluwer.

"Planets outside the Solar System: theory and observations", 1998, eds. Mariotti J.-M. and Alloin D., ASI in press, Kluwer.

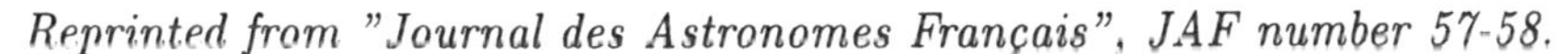
Reprinted from "Journal des Astronomes Français", JAF number 57-58.

LIST OF PARTICIPANTS

ALLOIN D.
European Southern Observatory
Alonso de Cordova 3107 / Vitacura, SANTIAGO 19 - CHILE
dalloin@eso.org

ARTYMOWICZ P.
Stockholm Observatory
13336 SALTSJOBADEN - S
pawel@astro.su.se

BAUDOZ P.
Observatoire de la Côte d'Azur
06304 NICE Cedex 04 - F
baudoz@obs-nice.fr

BELY P.
Space Telescope Science Institute, 3700 San Martin Drive
BALTIMORE, MD 21218 - USA
bely@stsci.edu

BENIT J.
Institut d'Astrophysique Spatiale, Université Paris XI
Bât. 121, 91405 ORSAY Cedex - F
benit@ias.fr

BOCCALETTI A.
Observatoire de Meudon, 5, Place Jules Janssen
92195 MEUDON - F
boccalet@despa.obspm.fr

BURROWS A.
Steward Observatory, University of Arizona
TUCSON, Arizona 85721 - USA
burrows@jupiter.as.arizona.edu

CHARBONNEAU D.
Harvard University, Dept. of Astronomy
60 Garden Street, CAMBRIDGE, MA 02138 - USA
dcharbonneau@cfa.harvard.edu

CHUECA S.
Instituto de Astrofisica de Canarias
38200 LA LAGUNA / Tenerife - E
chueca@iac.es

CLAES P.
ESTEC SCI/SAF
Keplerlaan 1, 2201 AZ NOORDWIJK - NL
pclaes@astro.estec.esa.nl

CLAUDI R. Osservatorio Astronomico di Padova
Vicolo dell'Osservatorio 5, 35122 PADOVA - I
claudi@pd.astro.it

COLAVITA M. JPL, 4800 Oak Grove Drive
PASADENA, California 91109 - USA
M.M.Colavita@jpl.nasa.gov

DAMGOV V. Space Research Institute, 6 Moskovska St., P.O.BOX 799
SOFIA 1000 - BULGARIA
VDAMGOV@BGEARN.ACAD.BG

DECIN G. Instituut voor Sterrenkunde, Celestynenlaan 200 B
3001 LEUVEN - B
greet@ster.kuleuven.ac.be

DESIDERA S. Dipartimento di Astronomia, Universitá di Padova
Vicolo dell'Osservatorio 5, 35122 PADOVA - I
desidera@pd.astro.it

DUTREY A. IRAM, 300 rue de la Piscine, Domaine Universitaire
38406 ST. MARTIN D'HÈRES - F
dutrey@iram.fr

FALKESGAARD J. Astronomisk Observatorium
Juliane Maries Vej 30, 2100 COPENHAGEN - DK
banshee@astro.ku.dk

FERRARI M. European Southern Observatory
Karl-Schwarzschild-Str. 2, 85748 GARCHING b.Muenchen - D
mferrari@eso.org

GOMEZ J.L. Instituto de Astrofisica de Andalucia
Apartado Postal 3004, 18080 GRANADA - E
jlgomez@iaa.es

GRININ V. Astronomical Institute of St. Petersburg University
Petrodvorets 198904, ST.PETERSBURG - RUSSIA
grinin@VG1723.spb.edu

HARWIT M. Cornell University, Dept. of Astronomy
510 Space Science Bldg., ITHACA, NY 14853 - USA
mharwit@ibm.net

HINZ P. Steward Observatory, 933 N. Cherry Ave.
TUCSON, AZ 857 21-0065 - USA
phinz@as.arizona.edu

KALTENEGGER L. c/o Prof. Neger, Petersgasse 16
8010 GRAZ - A
lisa@fubphpc.tu-graz.ac.at

KASTING J. Pennsylvania State University, 525 Davey Lab.
UNIVERSITY PARK, PA 16802 - USA
kasting@essc.psu.edu

KONACKI M. Torun Centre for Astronomy, Nicolaus Copernicus University
Gagarina 11, 87-100 TORUN - PL
kmc@astri.uni.torun.pl

LABEYRIE A. Observatoire de Haute Provence
04870 SAINT MICHEL L'OBSERVATOIRE - F
labeyrie@obs-hp.fr

LANGLOIS M. Steward Observatory, 933 N. Cherry Ave.
TUCSON, AZ 857 21-0065 - USA
maud@as.arizona.edu

LAURENT E. 14, Avenue de Verdun, 91670 LIMOURS - F
lauc@ens.math-info.univ-paris5.fr

LECAVELIER A. CNRS, Institut d'Astrophysique de Paris
98bis Bld. Arago, 75014 PARIS - F
lecaveli@iap.fr

LEGER A. Université Paris XI, Bât. 121
91405 ORSAY Cedex - F
leger@iaslab.ias.fr

LOISEAU S. JPL, MS 306-388, 4800 Oak Grove Drive
PASADENA, CA 91109-8099 - USA
loiseau@huey.jpl.nasa.gov

MARIOTTI J.-M.

MAUREL M-Ch. Institut Jacques Monod, Tour 43
2 Place Jussieu, 75251 PARIS Cedex 05 - F
Marie-Christine.Maurel@ijm.jussieu.fr

MENNESSON B. Observatoire de Paris DESPA
5, Place Jules Janssen, 92195 MEUDON Cedex - F
bertrand@bluenote.obspm.fr

MISCHNA M. Penn State University, 503 Walker Building
UNIVERSITY PARK, PA 16802 - USA
mischna@essc.psu.edu

MORGAN R. Steward Observatory N411, University of Arizona
TUCSON, AZ 85720 - USA
rhonda@as.arizona.edu

MOUTOU C. Observatoire de Haute Provence
04870 ST.MICHEL L'OBSERVATOIRE - F
moutou@obs-hp.fr

NABER R. Kapteyn Institute, P.O.BOX 800
97 AV.GRONINGEN - NL
richard@astro.rug.nl

NAEF D. Observatoire de Genève
1290 SAUVERNY - CH
dominique.naef@obs.unige.ch

OLLIVIER M. Institut d'Astrophysique Spatial, Bât. 121
Université de Paris XI, 91405 ORSAY Cedex - F
ollivier@ias.fr

OLSSON B. Pennsylvania State University, Dept. of Geosciences
436 Deike Bldg., UNIVERSITY PARK, PA 168022713 - USA
olsson@essc.psu.edu

OWEN T. University of Hawaii, Institute for Astronomy
2680 Woodlawn Dr., HONOLULU, HI 96822 - USA
owen@hubble.IFA.Hawaii.Edu

PALACIOS J. Ciudad Universitaria de Cantoblanco
28049 MADRID - E
javier@astro1.ft.uam.es

PAPALOIZOU J. Queen Mary - Westfield College
Mile End Road - LONDON E1 4NS - UK
J.C.B.Papaloizou@qmw.ac.uk

PAVLOV A. Pennsylvania State University, Dept. of Geosciences
436 Deike Bldg., UNIVERSITY PARK, PA 168022713 - USA
pavlov@essc.psu.edu

QUELOZ D. JPL, MS 306-473, 4800 Oak Grove Drive
PASADENA, CA 91109 - USA
didier@huey.jpl.nasa.gov

ROSTOPCHINA A. Crimean Astrophysical Observatory p/o Nauchny
334413 CRIMEA - UKRAINE
arost@crao.crimea.ua

RUILIER C. Observatoire de Paris, DESPA
5, Place Jules Janssen, 92195 MEUDON Cedex - F
ruilier@hplyot.obspm.fr

SABATKE E. Steward Observatory, University of Arizona
933 North Cherry Avenue, TUCSON, AZ 85721-0065 - USA
esabatke@u.arizona.edu

SACKETT P. Kapteyn Institute, University of Groningen
Postbus 800, 9700 AV GRONINGEN - NL
psackett@thales.astro.rug.nl

SCALLY A. University of Cambridge, Institute of Astronomy
Madingley Road, CAMBRIDGE CB3 OHA - UK
aylwyn@ast.cam.ac.uk

SEAGER S. Harvard University, Smithonian Center for Astrophysics
60 Garden Street, CAMBRIDGE, MA 02138 - USA
sseager@cfa.harvard.edu

SCHINDLER T. Penn State University, Dept. of Meteorology
503 Walker Bldg., UNIVERSITY PARK, PA 16802 - USA
tls246@psu.edu

SHAKHOVSKOY D. Crimean Astrophysical Observatory p/o Nauchny
334413 CRIMEA - UKRAINE
dshakh@crao.crimea.ua

SOLANO E. Universidad Europea - CEES
28670 VILLAVICIOSA DE ODON/MADRID - E
esm@vilspa.esa.es

TAMBOVTSEVA L. Central Astronomical Observatory Pulkovo
Pulkovskoe Shosse 65/1, 196140 ST. PETERSBURG - RUSSIA
tamb@pulkovo.spb.su

THOMAS E. Observatoire de la Côte d'Azur
Avenue Copernic, 06130 GRASSE - F
ethomas@obs-azur.fr

VERMAAK P. South African Astronomical Observatory, P.O.BOX 9
7935 CAPE TOWN - SOUTH AFRICA
pierre@saao.ac.za

WILHELM R. DLR Berlin-Adlershof, Rudower Chaussee 5,
12489 BERLIN - D
rainer.wilhelm@dlr.de

WUCHTERL G. Institut fuer Astronomie der Universitaet Wien
Tuerkenschanzstr. 17, 1180 WIEN - A
wuchterl@amok.ast.univie.ac.at

ZUCKER S. School of Physics and Astronomy, Raymond and Beverly Sackler
Faculty of Exact Sciences, Tel-Aviv University
TEL-AVIV 69978 - ISRAEL

Part I : Theory and Modeling

Un largo día se cubrió de agua,
de fuego, de humo, de silencio, de oro,
de plata, de ceniza, de transcurso,
y allí quedó esparcido el largo día:
cayó el árbol intacto y calcinado,
un siglo y otro siglo lo cubrieron
hasta que convertido en ancha piedra
cambió de eternidad y de follaje.

Pablo Neruda
Las Piedras del Cielo

THE UNFINISHED HISTORY OF PLANET SEARCHES

MARTIN HARWIT
511 H Street, SW
Washington, DC
20024-2725, USA
(Also Cornell University)

Abstract

The recent discovery of planets around many nearby stars, implies that life in the Universe may be common. The further discovery of intelligent life on other planets could bring about a great change in our mindset here on Earth. This paper describes how exploratory processes of all kinds, whether they are astronomical observations or explorations by humans, have historically been supported, but also carefully monitored by governments and the military. Given the societal implications of the discovery of intelligent life elsewhere, the time may be ripe to alert governmental bodies that such a finding may be imminent. A set of studies to prepare for this eventuality could then be initiated to permit nations to react with greater insight if and when such a discovery is made.

1. Introduction

Nearly half a century ago, Arthur C. Clarke imagined a historian of the year 3000 looking back on the 20th century and writing:

> To us a thousand years later, the whole story of Mankind before the twentieth century seems like the prelude to some great drama, played on the narrow strip of stage before the curtain has risen and revealed the scenery. For countless generations of men, that tiny, crowded stage --- the planet Earth --- was the whole of creation, and they the only actors. Yet toward the close of the fabulous century, the curtain began slowly, inexorably to rise, and Man realized at last that the Earth was only one of many worlds. . . . (T)he childhood of our race was over and history as we know it began . . .

This was written in 1951, before satellites had been launched into orbit, before adaptive optics was conceived as a reality, and before most of the novel technologies that helped us discover planets around nearby stars were more than glimmers in the imaginations of visionaries [1].

How did we come this far in just fifty years? To answer this question, we must go back a few centuries and retrace our steps.

J.-M. Mariotti and D. Alloin (eds.), Planets Outside the Solar System: Theory and Observations, 3–11.

2. The Centrality of Man

The relationship between the world's great explorers and their governments and military bodies, has always been complex. At times it has been openly hostile, on occasion it has been friendly and rewarding. Almost always, the motivations of the two sides were at odds.

For a millennium after Ptolemy, his Earth-centered world model defined the official view of the Christian Church. It gave mankind fashioned in the image of God, its deserved place at the center of the Universe. Then, in the 13th century, this view began to be questioned on theological grounds. Why should God be restricted to creating only one Earth, when He might have chosen to create many others? In 1440, Bishop Nicolas of Cusa, in his book "Docta Ignorantia" proposed that other stars like the Sun existed, and that the Sun, Moon and stars were populated.

While Nicolas of Cusa was subsequently made a cardinal by the Church, his spiritual successor, Giordano Bruno, on February 17, 1600, was burned at the stake as a heretic. What had changed the mind of the Church in the course of a century and a half was Copernicus's book "De revolutionibus erbium coelestium" of 1543, which dared to displace the Earth and mankind from the center of the Universe. While Nicolas of Cusa had been safe in his theological speculations, the work of Copernicus had made the question of man's place in the Universe uncomfortably verifiable and immediate. The scientist unfortunate enough to point this out was Galileo Galilei.

While the original inventor of the telescope is not known, we know for certain that a spyglass surfaced at an annual fair in Frankfurt in September 1608. Within months Galileo had fashioned a superior instrument of improved design, that magnified some twenty times. In August 1609, he presented this device to the Venetian senate, alerting them to its ability to quickly identify distant ships at sea, a matter of utmost importance in warfare. So pleased was the senate that it at once rewarded Galileo with a lifetime appointment at the University of Padua. His further uses of the spyglass, however, got him in trouble. He discovered the phases of Venus, mountains on the Moon, and moons around other planets, all of which showed the existence of other worlds rather similar to Earth. How could one maintain the centrality of man or God, when Earth was no longer central to the Universe? Galileo, suffered his tribulations precisely because his observations challenged Church dogma on the centrality of man created in the image of God.

3. The Centrality of Governing Bodies

The issue of other worlds had raised practical questions even before Copernicus. When the New World --- America --- was discovered in 1492, and other civilizations were found there, a new problem arose for the Church. Some of the newly encountered civilizations were, by any measure, advanced in the arts, but believed in different gods. This posed a number of problems. Should these populations be considered the descendants of Adam and Eve, like the peoples of the Old World? Were they heathens constituting a challenge to the Christian faith? Might they be humans untouched by original sin? Or were they to be considered a new species of animals, like the apes? This last alternative would permit them to be butchered at will. The issue was ultimately settled by edict in 1537. Forty-five years after the discovery of the Americas, Pope Paul III issued a papal bull declaring the native American capable of conversion to the Catholic faith -- and, therefore, not to be butchered [2].

4. The Helping Military Hand

The friendly terms with the military that scientists had initiated long before Galileo have persisted and flourished. Without military input, astronomical progress in the twentieth century would have been far less impressive. The successes of radar-, radio-, infrared-, x-ray, and gamma-ray astronomy all attest to this.

Radar was one of the most powerful weapons systems developed during World War II. Invented just before the onset of the War and further developed throughout this conflict, radar enabled British intelligence to see the approach of German aircraft long before they could cross the British Channel. Fighter planes could be scrambled in time, and the approaching airplanes encountered. Later, Allied bombers were equipped with radar so that their crews might see their targets through cloud cover and drop their bombs with greater accuracy.

Radar operates at radio wavelengths. On two successive mornings during the war, a most serious alert shook up defense forces all over England. Years later, James Stanley Hey, in charge of trouble-shooting the British wartime radar network, reported on this incident in a 1946 letter to the journal *Nature* [3]:

> It is now possible to disclose that, on one occasion during the War, Army equipments observed solar radiation of the order of [a hundred thousand] times the power expected from the Sun . . . This abnormally high intensity of solar radiation occurred on February 27 and 28, 1942. . .The main evidence that the disturbance was caused by electromagnetic radiations of solar origin was obtained by the bearings and elevations measured independently by the [radar] receiving sets, sited in widely separate parts of Great Britain . . .

At the time of these observations, the observatory at Meudon detected strong solar flares and it soon became evident that solar radio emission was enhanced during periods of solar activity. Shortly after the war, Hey and his coworkers also discovered that the galaxy now known as Cygnus A emits a strong radio flux; Martin Ryle, who would later win a Nobel prize for his astronomical discoveries, used discarded British military radar equipment to set up a radio astronomy research group at Cambridge; A. C. B. Lovell established a similar group at Manchester; and Jan Oort, in the Netherlands, used discarded German radar equipment to start a Dutch radio-astronomy program.

Thus were radar- and radio-astronomy begun. The next twenty-five years would revolutionize astrophysics with discoveries of radio galaxies, quasars, cosmic masers, pulsars, superluminal sources, and the microwave background -- findings often so puzzling that a new, noncommittal name had to be invented for them, names that would identify these phenomena without prejudging what they signified or what physical processes might be at work. In planetary research, radio and radar techniques have also revolutionized our understanding of planetary magnetospheres, the surfaces of Mercury and Venus, and the compositions of asteroids and the moons of other planets.

5. The Immediate Post-War Era

Throughout the post-war years, the defense establishments of the major powers kept improving their surveillance capabilities. As the armed forces in many of these countries outgrew a particular technique and no longer needed it, the military would pass it on to colleagues in astronomy [4]. These instruments provided an often-serendipitous view of the universe through entirely new eyes. These eyes might be blind to ordinary visible light, but they allowed astronomers to sense infrared and ultraviolet radiation, X-rays and gamma rays -- radiations that had never been tapped before.

Over a period of a dozen years starting in the early 1930s, Wernher von Braun and a huge army of technical experts had painstakingly learned how to build the powerful V-2 rockets with which Hitler hoped to turn the tide of war. These missiles could be guided accurately to such targets as London or Antwerp. After the war, the U.S. military captured many of the V-2s and brought them to the United States for testing. As part of the test flights, pioneers like Richard Tousey and Herbert Friedman of the U.S. Naval Research Laboratories were permitted to put useful astronomical payloads on board [5]. They conducted the very first ultraviolet and X-ray observations of the sun. The V-2s became the basis for America's first post-war rocket-astronomical discoveries and also the foundation on which an entire U.S. rocket industry would arise.

6. The Space Race

The United States was not the only beneficiary of German rocket techniques. At the end of the war, the Soviet Union had assembled their own set of German experts and had begun to develop a powerful rocket industry of their own. With the launch of Sputnik in 1957, they exhibited the impressive capabilities their rockets had attained in the dozen years since the war. The accuracies with which the Soviet Union was able to place satellites into Earth orbit showed that their rockets could now reach any place on Earth with great precision and presumably with significant nuclear warheads. Moreover, the high ground of space gave them the ultimate means of surveillance of military installations anywhere on Earth.

To counter this ascendancy, the United States created a crash program to develop both more powerful rockets and more incisive surveillance techniques from space. The Space Race of the 1960s had started! It would culminate in the Moon landings of 1969 and then level off.

As part of the space race, the U.S. Air Force, and probably the air forces of other nations as well, were showing great interest in studying the infrared emission from the Universe. The military forces began to launch sensitive infrared telescopes above the atmosphere, with the aim of potentially intercepting fast approaching enemy missiles. The problem for the Air Force was that it would have to distinguish the faint glow emitted by an incoming projectile from the infrared-radiation of background stars or galaxies. Nobody knew what that might be. My colleagues and I at Cornell University had built and flown the first liquid helium cooled rocket-borne telescopes. These permitted us to publish the results of the first far-infrared astronomical observations from above the atmosphere [6]. Since we were slightly ahead of the U. S. Air Force in the development of these techniques, the Air Force, for a while, decided to fund our research. They were interested in what we would find. Later, by spending considerably larger sums than we had been able to obtain, the Air Force also made far better maps. At that point they no longer needed the rather primitive surveys a small university group could carry out, and instead went on to tackle the problem in style.

7. Military Advances in Astronomy

The Air Force's mapping of the heavens reached a new climax in 1996 and '97 with the completion of a survey that yielded the most exquisitely detailed map of our Milky Way at both mid-infrared and ultraviolet wavelengths. Their mission, the *Midcourse Space Experiment (MSX)*, surveyed the sky for ten months. The maps they constructed are currently being made available to the astronomical community. A recent paper describes one entirely new finding, a population of more than 2,000 extremely low temperature Galactic clouds, opaque even in the mid-infrared. They are seen in absorption against a bright infrared background. Their sizes are similar to those of the cores of giant molecular clouds, but they lack the envelopes that the giant clouds normally exhibit, and may represent a class of "pre-protostellar cores" [7].

Perhaps the most phenomenal discovery by the U.S. military was a tantalizing finding encountered in 1968. So secret was this discovery that no one in the astronomical community was told until five years later, in 1973. Ever since then, it has defied full explanation despite the intense efforts of the most talented theoretical astrophysicists. Here is its history.

At the beginning of the space age, the United States and the Soviet Union had finally agreed to ban atmospheric testing of nuclear bombs. The Atmospheric Test Ban Treaty signed in the Kennedy/Khrushchev era forced nations to carry out their tests underground. The problem the treaty posed for the opposing military establishments was that an enemy could learn a great deal about the yield of a tested weapon by sensing the seismic waves it generated. The United States worried that the Soviet Union might strive to hide such information and seek greater secrecy by exploding their test devices, not underground but at great distances out in space. Were the Soviet military to try that, our military planners argued, we could still gain a great deal of information by sensing the burst of gamma rays released in such highly energetic explosions. This idea culminated in the construction of a set of satellites for the so-called "Vela" project. Several Vela satellites were launched into earth-circling orbits, with the thought that at least one of these instruments would always be positioned to see gamma rays emitted from any direction in space, whether it was on the day or night side of the Earth.

What the Vela satellites detected were gamma ray bursts alright! The first one must have been a tremendous shock to the military. And then others followed, one every few months. Eventually, these bursts revealed themselves not to be Soviet nuclear bombs; they were unimaginably vaster explosions somewhere out in space. Once this was established, and the military realized that the problem was none of their concern, they published their findings in the premier astronomical publication, the Astrophysical Journal Letters [8]. The news stunned the astronomical community, which has wrestled with the problem of understanding these bursts ever since. One recently discovered burst appears to come from a galaxy at red shift z = 3.42 [9]. For a few seconds its emission may have rivaled the luminosity of the entire Universe. What its influence might have been on star formation is anybody's guess. And if life existed in the gamma-burst's host galaxy, it must have seriously suffered from the enormous pulse of ionizing radiation, equivalent to $\sim 10^{30}$ hydrogen bombs.

These are only some of the most striking examples of military technologies adopted by astronomers, but there are many others. Adaptive optics and powerful lasers used with them were developed for solely military purposes, and still are being improved under the military's lead; they have already proved immensely valuable in astronomy, particularly in the search for exoplanets. The widely available 2.4 meter mirror blanks, used also on the Hubble Space Telescope, became popular because the military was using downward viewing devices launched into space from the Space Shuttle; 2.4 meter telescopes were the largest size compatible with the Shuttle cargo bay. The entire computer revolution --- particularly the small computer thrust --- was initiated by the military for a variety of weapons developments [10, 11].

8. Theorists Play the Same Game

Astronomical observers were not alone in working both sides of the curtain of secrecy that partitioned the worlds of academic science and military technology throughout the Cold-War. Theorists may have been even more adapt at playing the game. One of the outstanding contemporary American theorists, Kip Thorne, has reminisced on conversations that he and other American astrophysicists often had with leading Soviet colleagues at the height of the Cold War [11]. Most of the top astrophysicists on both sides had been involved in the development of the arsenals of atomic and hydrogen bombs. And so, at some stage, the discussions on an intricate problem of astrophysics would tend to grind to an embarrassing halt, as both sides realized they were getting uncomfortably close to revealing a military secret --- though each side probably already knew as much as the other. The physics of hydrogen bombs is very similar to the physics of supernova explosions, except that the explosions of the astronomers are incomparably more powerful than those of the military.

Astrophysicists who have straddled the fence between academic and military work read like a *Who's Who* of theorists. On the American side, J. Robert Oppenheimer, father of the atomic bomb, published some of the most incisive early work on neutron stars and black holes. In 1939, he and his students noted the possible birth of these then-hypothetical objects in supernova explosions. That same year, on the eve of the War, Hans Bethe elucidated the release of nuclear energy that makes the sun and stars shine. Later Bethe became the chief theorist of the Manhattan project that built the atomic bomb. Edward Teller, father of the American hydrogen bomb, similarly had made important contributions to astrophysics before the War. John Archibald Wheeler, one of the pioneers of relativity theory and black holes was also involved in nuclear bomb projects. In Germany, the young astrophysicist Carl Friedrich von Weizsäcker worked on the German atomic bomb. In the Soviet Union, Yakov Borisovich Zel'dovich, perhaps the most prolific theoretical astrophysicist of the post-war era, was deeply involved in the design of the Soviet atomic and hydrogen bombs. And the father of the Soviet hydrogen bomb, Andrei Dmitrievich Sakharov, who later became celebrated for his efforts at establishing world peace, also made fundamental contributions to cosmology and our understanding of the origins of the Universe. Accompanying these giants of theoretical astrophysics came an army of less well-known theorists. They have elucidated the origin of the Universe as the most profound explosion of all time. They have also investigated the enormously energetic outbursts we see in the center of galaxies and the smaller but still extremely powerful supernova explosions. These and other outbursts have been of the greatest interest to astronomers -- and to the military, with their professional interests in explosive devices.

9. Controlling Influences

The strong connections between government and exploration of all kinds, ranging from remote-sensing astronomical observations, to actual journeys by robotic or human explorers, is clear. Military instrumentation has time and again redirected astronomical efforts. The early radio-astronomical findings, the discoveries made possible by novel infrared-sensing equipment, the unexpected recordings of cosmic gamma-ray bursts, all induced astronomers to shift their attention and channel their energies to understanding such newly-discovered phenomena. And wherever human explorers have established inroads, governments have soon followed to prescribe the rules of engagement.

Provided that living matter is reasonably common in the Universe, the discovery of exoplanets has brought us a step closer to finding life elsewhere in the Cosmos. Until we actually discover life on other worlds, particularly intelligent life, nobody will pay a great deal of attention. However, once this discovery is made, all subsequent decisions will be taken by governments, most likely in consultation with the military. The wishes of scientists will then be largely ignored, just as they were in the invention of atomic bombs.

Leo Szilard was the first physicist to conceive the idea of a nuclear chain reaction and the possibility of constructing an atomic bomb. On the eve of World War II, in early August of 1939, Szilard persuaded Albert Einstein, to write President Roosevelt, alerting him to the possibility that Germany might build an atomic bomb [13]. This letter ultimately led to the establishment of the Manhattan project and the construction of the first atomic bombs. After the war, however, public officials came to consider Szilard a meddling nuisance for his pleadings for international control of atomic bombs. A somewhat similar fate awaited Oppenheimer. When the creator of the atomic bomb argued against development of the even-more-powerful hydrogen bombs, some years after the War, he was summarily declared a security risk and dismissed as an advisor to the US government.

The intervention of governments on matters of public importance is natural in democratically constituted nations, and should be considered inevitable. Different governments, however, might very well find little common ground. The decisions and actions of governments often are motivated by local concerns and can be unpredictable and unilateral. The recent, unexpected detonation of a family of nuclear devices, by both India and Pakistan shows the extent to which nations act independently, even in the face of overwhelming opposition from other powers.

The recent discovery of planets around several nearby stars implies that life, and perhaps intelligent life, may be common. As specialists we should call these potential discoveries to the attention of a broader political leadership. That way some form of consideration might be given to the eventuality well before a discovery takes place. Likely public reactions to an announcement might, by these means, be anticipated, steps to quiet potential hysteria might be planned, and a more deliberate set of policy alternatives fashioned. So much of science fiction has involved a "Starwars" response to the discovery of intelligence elsewhere in the Universe, that a more reasoned, long-term approach is urgently needed. As scientists we are not sufficiently cognizant of practical politics. Nor are we equipped to formulate the requisite white papers. Any advice we might offer will gain public acceptance only if it stimulates wide debate, particularly among people in public life.

10. Summary

Our discovery of planets around so many nearby stars, implies that life in the Universe may be common. We must recognize that the discovery of intelligent life on other planets could bring about a great change in our mindset here on Earth. This could well be as powerful as the realization, in the late 1940s, that atomic bombs might destroy all of civilization. The finding of intelligent life elsewhere could lead to a similar awakening. Life elsewhere could constitute a threat to survival on Earth. However, like nuclear energy, the existence of life elsewhere might also bring great benefits to humankind. No question can exist that government leaders will decide what steps are to be taken. As scientists our role can be to help our governments, by providing advance notice and time to deliberate appropriate actions *before* these momentous discoveries are made, rather than later, when public hysteria might catch the leadership unprepared and force its hands.

11. Acknowledgments

I am pleased to note that my astronomical research is supported by NASA grant NAG5-3347.

12. References

1. Clarke, A. C. (1951) *The Exploration of Space*, Harper, New York, p195.

2. Guthke, K. S. (1983) *Der Mythos der Neuzeit -- Das Thema der Mehrheit der Welten in der Literatur- und Geistesgeschichte von der koperanischen Wende bis zur Science Fiction,* Francke, Bern, 1983, p52.

3. Hey, J. S. (1946) Solar Radiations in the 4 -- 6 meter Radio Wavelength Band, Nature, **157**, 47.

4. Harwit, M. (1981) *Cosmic Discovery -- The Search, Scope and Heritage of Astronomy*, Basic Books, New York.

5. DeVorkin, D. (1993) *Astronomy With a Vengeance --- How the Military Created the Space Sciences After World War II*, Springer-Verlag, New York.

6. Houck, J. R., Soifer, B. T., Pipher, J. L., and Harwit, M. (1971), Ap. J. Lett. **169**, L31--L34.

7. Egan, M. P. *et al.* (1998) A Population of Cold Cores in the Galactic Plane," Ap. J. Lett, **494** L199--L202.

8. Klebesadel, R. W., Strong, I. B., and Olson, R. A. (1973) Observations of Gamma-Ray Bursts of Cosmic Origin, Ap.J. **182**, L85- L88.

9. Kulkarni, S. R. *et al.* (1998) Identification of a host galaxy at redshift z = 3.42 for the γ-ray burst of 14 December 1997, Nature, **393**, 35 -- 39.

10. Goldstine, H. H. (1972) *The Computer from Pascal to von Neumann*, Princeton.

11. Ceruzzi, P. E. (1989) *Beyond the Limits --- Flight Enters the Computer Age*, MIT Press.

12. Thorne, K. S. (1994) *Black Holes & Time Warps -- Einstein's Outrageous Legacy*, Norton, New York.

13. Einstein, A. (1939) Letter to F. D. Roosevelt. A copy of this letter, dated August 2, 1939, is reprinted in *Leo Szilard: His Version of the Facts*, eds. Spencer R. Weart and Gertrud Weiss Szilard (Cambridge, MIT Press, 1978), p94.

LATEST STAGES OF STAR FORMATION AND CIRCUMSTELLAR ENVIRONMENT OF YOUNG STELLAR OBJECTS

A. DUTREY
Institut de Radio Astronomie Millimétrique
300 rue de la Piscine,
F-38406 Saint-Martin-d'Hères, France

1. Introduction

A few years ago, the scenario of planet formation was dictated by models of the Solar System because of the lack of appropriate observations. Recent discoveries of extra-solar planets are now changing the first ideas we had about the formation of planets. As an example, the discovery of Jupiter-like planets orbiting very closely from their suns (radius <1 AU) was unexpected and has induced new models.

At the early stages of planetary formation, when the star itself forms, instruments like the HST, mm arrays and adaptive optic systems on large optical and near-infrared (NIR, hereafter) telescopes are also slowly but surely changing our ideas about the circumstellar environment of young stellar objects (YSOs). In fact, the knowledge of physical processes leading to star formation has drastically evolved since the last 20 years.

The lecture is devoted to the formation and evolution of the circumstellar environment around TTauri stars which are low-mass-stars ($\sim 0.5 - 2\,M_\odot$), similar to the sun when it was in its pre-main-sequence (PMS) phase. I will first present a "standard" picture of the star formation, including a very short review on theoretical aspects. The main part of the lecture will then focus on recent observational results obtained at high angular resolution ($\leq 2-1''$) from the mm to the optical range. Most of these results have been obtained in the closest star forming region visible from the Northern hemisphere (where are currently located most of the best ground based telescopes), the Taurus-Auriga clouds, located at about ~150 pc (or 1"=150 AU). As such, they are not representative of all star forming regions. The results presented here have been selected from their capability to bring new informations on the field, to give a global view of problems related to star

J.-M. Mariotti and D. Alloin (eds.), Planets Outside the Solar System: Theory and Observations, 13–49.

formation. In this sense, this lecture is not a review and only a few results are shown. For a deeper understanding of the physical processes, observational results or the theoretical aspects of the stellar formation, the reader will be invited to refer to the reviews or articles whose references are given at the beginning of each section.

1.1. OVERVIEW OF PMS STAR EVOLUTION

Stars form in molecular clouds. With typical kinetic temperatures within the range $\sim 10 - 50$ K, molecular clouds are cold and radiate mainly in the mm and submm domain. For this reason, the study of the stellar formation processes have started recently with the current generation of large radiotelescopes and interferometers.

Observations remain however limited. The confusion with the surrounding molecular clouds, the low angular resolution (best single-dishes have a spatial resolution of $\sim 10''$ or 1500 AU at the distance of the closer star-forming regions) and the lack of sensitivity of current mm radiotelescopes and interferometers, strongly limit the observations of all the stages of the star formation. Particularly from the cloud collapse to the embedded proto-star phases, our understanding of these first stages mainly come from theoretical arguments and low resolution data.

Giant molecular complexes similar to the OrionA clouds form stars of all mass range (including OB stars) while dark molecular regions like the Taurus-Auriga clouds or the Ophiuchus region form only low-mass stars ($\sim 0.5 - 2\,M_{\odot}$). Moreover, low-mass star forming regions are not all identical. For example, the Taurus-Auriga clouds have filamentary structures (Cernicharo 1990) and relatively isolated star formation. On the contrary, the ρ Oph cloud is a cluster with a higher density of newly born stars. This can be clearly seen in fig.6 of Motte et al. 1998 where the authors compare the 1.3mm dust distribution in typical areas of both clouds.

1.2. STANDARD SCENARIO FOR STAR FORMATION

Standard scenarii of stellar formation are complicated by the fact that many stars form in binary or multiple systems. Following Shu et al. 1987 (their fig.7), star formation can be divided in four main stages:

1. The cloud core formation (in molecular clouds)
2. The inside-out collapse (pre-stellar collapsing core)
3. The main accretion/ejection phase (angular momentum dissipation)
4. The newly born PMS star and its protoplanetary disk (visible TTauri star)

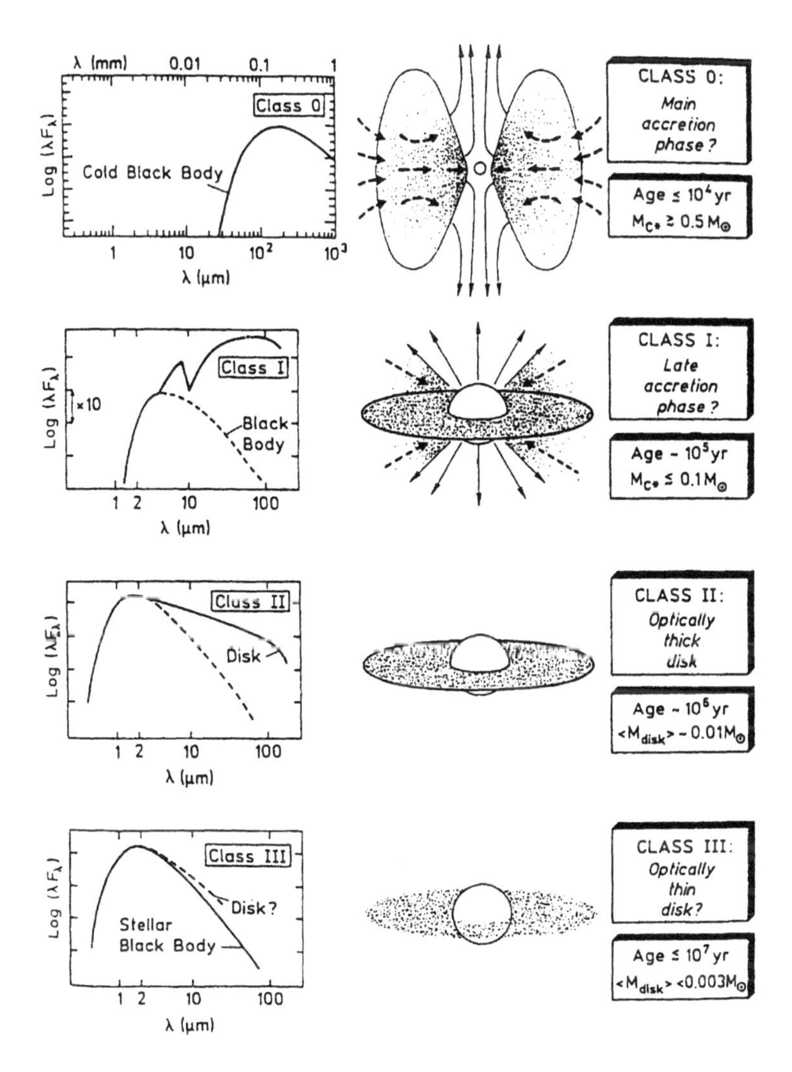

Figure 1. The standard observational classification of PMS phases associated to the evolution of the PMS star. From Ph.André 1994. The corresponding evolution of the PMS star is given in Fig.2

Stages 1) and 2) will be only mentionned, more details can be found in Shu et al. 1991, Mouchovias 1991.

Stages 3) and 4) correspond to four main phases, from the so-called Class 0 to Class III. Fig.1 and fig.2 summarize the properties of these different classes, figure 1 shows the observed standard classification (adapted from Adams et al. 1987 and André et al. 1994) and figure 2 the PMS star evolution in the HR diagram for a star with a solar metallicity (from Siess 1996).

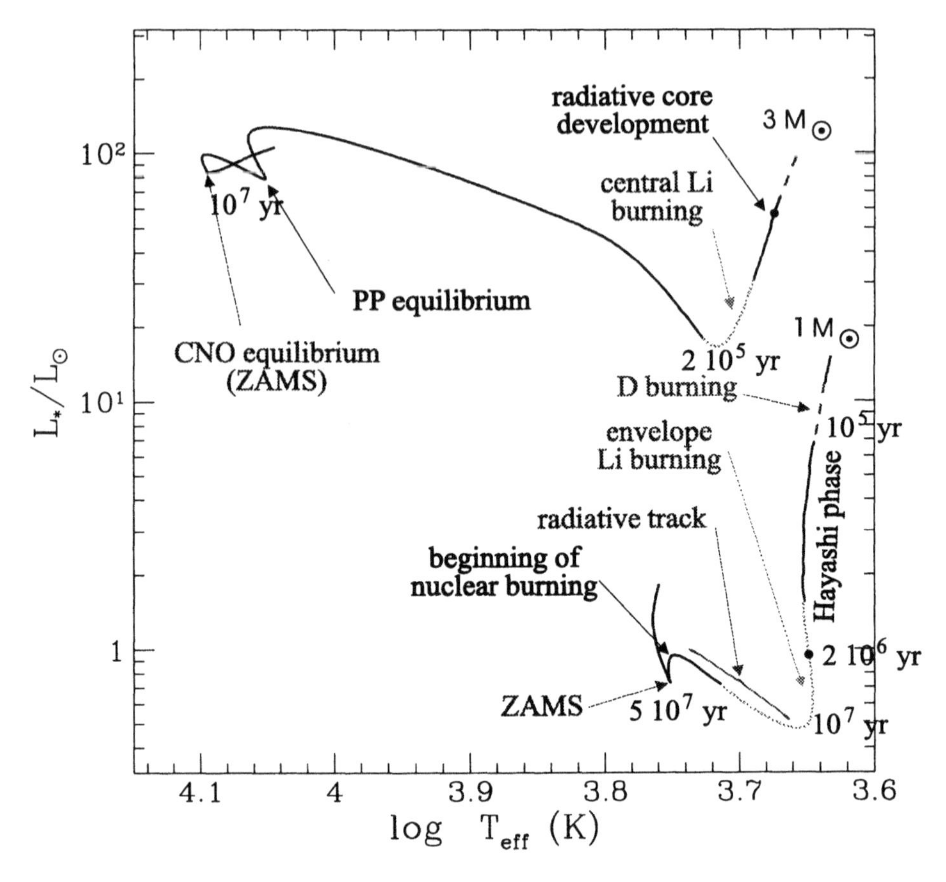

Figure 2. Evolutionary tracks of 1 and 3 $M_\odot$,respectively. Stars have a solar metallicity. From the Ph.D Thesis of L.Siess 1996. Fig.1 gives the evolution the circumstellar material.

1.2.1. *The pre-stellar phase*

Cores slowly form in molecular clouds by loosing their magnetic and turbulent supports through the process of ambipolar diffusion. They are supposed to follow a quasistatic evolution toward a $\sim r^{-2}$ density distribution as expected for a singular isothermal sphere. The formation of pre-stellar cores, similar to those observed in NH_3 (Myers & Benson 1983) takes place in about $\sim 10^5 - 10^6$yr.

As soon as an isothermal sphere is formed in the center of the core, conditions are reached to proceed to the *inside-out collapse*. The material close to the center is strongly accreted on the proto-star and an accreting disk is formed while the outer envelope of gas and dust is infalling. This point is considered as *the zero age* of the young star. The collapsing core is strongly embedded in its parent cloud making observations very difficult and confuse. Clouds and cores are rotating at rate of $\Omega_c \sim 1\text{km}\cdot\text{s}^{-1}\cdot\text{pc}$ (Ho et al. 1977) or more. At the center of the collapsing core, such a rotation rate would create a centrifugal barrier $R_c = \Omega_c.R^2/G.M_*$, where R is the initial core radius (typical values for R are $\sim 10^{17}$ cm or 0.3 pc) and M_* the mass of the central object. Assuming $M_* = 1\,\mathrm{M}_\odot$, the size of the central accretion disk in rotation is $\sim$ 100 AU.

Dissipation of the angular momentum resulting from the cloud rotation is one of the key problem of the star formation.

1.2.2. *The main accretion and ejection phases*

Although details are not understood, angular momentum dissipation proceeds essentially through the existence of a powerful jet expelled from the proto-star. The material is flowing along the poles while the object is still surrounded by a large ($\sim$ 10000 AU) and possibly flattened envelope (Galli & Shu 1993a,b) of infalling material. This corresponds to the Class 0 and Class I phases.

Details about the theory of star formation are beyond the scope of this lecture. More details can be found in a recent book by Hartmann (1998). However, theory indicates that in proto and PMS stars, the dissipation of the angular momentum may proceed through 1) viscous torque (Shakura & Sunyaev 1973, Lynden-Bell & Pringle 1974) or 2) MHD jet torque (Chan & Henriksen 1980, Blandford &Payne 1982).

Viscous accretion disk model In viscous accretion disk models (a subclass of active disks which can be opposed to passive disks reprocessing the stellar light) the angular momentum is removed by the viscous dissipation which also produces the accretion luminosity L_{acc}. For a Keplerian disk which is geometrically thin (H being the scale height), optically thick (at IR wavelengths) and in steady state, the viscosity ν is parametrized by the so-called α parameter $\nu = \alpha \cdot H \cdot C_s$ ($\alpha \leq 1.0$) where C_s is the local sound speed. The surface density at a radius r from the star is linked to the viscosity and to $\dot{M_{acc}}$ by $\Sigma(r) = \frac{\dot{M_{acc}}}{3\pi\nu} \cdot (1 - \sqrt{R_*/r})$ (R_* is the stellar radius). The disk accretion luminosity is not directly related to the viscosity but only to the accretion rate: $L_{acc} = \frac{G \cdot M_* \dot{M_{acc}}}{2 \cdot R_*}$

MHD jet model In such models, the magnetic field lines require a special configuration (Pudritz et al. 1991). B lines are open at large scale and cross over the disk (bipolar configuration) with an opening angle $\theta \lesssim 30^o$ (from the jet axis). The disk material, frozen along the magnetic field lines, is magnetically accelerated as long as it reaches the Alfven surface (at radius r_a from the star) well above the disk plane. This is the Lorentz force, acting on the frozen particules, which is responsible for the MHD acceleration creating the ejection (Ferreira 1997). At the Alfven surface, the magnetic field stops to maintain the jet in rigid co-rotation with the disk, the jet slowly disconnects from the magnetic lines. The ratio of mass loss rate to mass accretion rate is given by $\frac{\dot{M}_{acc}}{\dot{M}_j} \simeq (r_a/r_o)^2$ where r_a/r_o ($>> 1$) is the lever arm, r_o being the footpoint of the magnetic field line. Centrifugal acceleration is however possible until the jet reaches the fast magnetosonic surface. In the main accretion phases (classes 0,I), typical expected values are of order $\dot{M}_j \sim 0.1\dot{M}_{acc}$. The MHD jet torque remains the most efficient process to simultaneously carry out the angular momentum and accrete material on the central object. Note that everything is expected to happen within radius of 0.1-1 AU from the star.

1.2.3. *Connexion between the PMS star and its accretion disk*

A few years ago, star and disk interactions were thought to proceed through a hot turbulent boundary layer (Basri & Bertout 1989). However, observations of hot spots on the stellar surface (Bouvier & Bertout 1989) and the comparison of the rotation speed of CTTs with those of (diskless) WTTs (Bouvier et al. 1993, Edwards et al. 1993) seem to favor models where the interaction between the disk and the star proceeds through magnetic accretion columns or funnel flows. The hot spots are shocks at the basis of the accretion columns (Bertout et al. 1988).

In such models (*e.g.* Shu et al. 1994, Ferreira 1995,1997) the disk magnetic field, still open, reconnects to the bipolar stellar magnetic field at the so-called "X" point at radius r_X (where $B_{disk} = B_{star}$) forming a neutral magnetic line. At r_X, a fraction of the material accreted from the disk is ejected along open stellar magnetic fields producing X winds while the other part is braked by the stellar magnetosphere and falls down along stellar magnetic loops (the accretion columns). With $r_{CO} \leq r_m \leq r_X$ where r_{CO} is the corotation radius (the star magnetosphere rotates at the disk speed) and r_m the magnetospheric radius ($\sim$ inner disk radius), the disk rotates slower than the star and the resulting outward transfer of angular momentum balances the inward flux from the accreting material. The star is braked because the disk "locks" the rotation speed of the star. This mechanism works efficiently as long as the disk magnetic flux is not completely dissipated (Ferreira et al. 1998).

1.3. FROM CLASS 0 TO CLASS III

1.3.1. *Standard observational classification*

The low-mass proto stars and PMS stars can be classified as follows (Adams et al. 1987 and André 1994, see Fig.1)

Class 0 The object is now $\sim 10^4$ years old and is in its main accretion phase. It has a powerful, very collimated outflow and is surrounded by a large envelope of gas and dust. The proto-star, strongly embedded in its parent cloud, is not visible and radiates purely in the mm, submm range with a $T_k \sim 10-30$K. Its *Spectral Energy Distribution* or SED has a spectral index α_{IR} between $20 \leq \lambda \leq 100\mu$m (André 1994) which is positive: $0 \leq \alpha_{IR} \leq 3$. Recent work conducted by Montmerle and collaborators (*e.g.* Casanova et al. 1995) have shown that some of them are strong X-ray emitters. The origin of the X emission is not yet fully understood but seems to be associated to a strong magnetic activity linked to the star.

Class I The SED of the object evolves slowly towards a spectral index: $-2 \leq \alpha_{IR} \leq 0$, and the proto-star is becoming an unresolved infrared (IR) source. The object is still accreting and ejecting an outflow but these activities are slowly decreasing. It is usually surrounded by a less massive envelope (see Bontemps et al. 1996) and an accretion (rotating) disk. At this stage, the object is about $\sim 10^5$ years old.

Class II The main ejection and accretion stage is now over, the envelope is dissipated and the PMS star, still surrounded by a rotating accretion disk (of mass a few $\sim 0.01\,M_\odot$), becomes visible. Compared to a stellar black body, the SED presents an IR excess coming from an optically thick disk at infrared wavelengths with $-3 \leq \alpha_{IR} \leq -2$. The star is now a classical emission line TTauri star (CTTs) which has strong emission from optical lines such as H_α (signature of accretion and ejection close to the star). Note that, such emission lines are subject to variability. A CTTs is defined by an equivalent linewidth of H_α: $W_{H_\alpha} \geq 5$ Å. Such a PMS star is about $\sim 10^6$ years old.

Class III With mass around $\sim 0.003\,M_\odot$ or less, the rotating disk is becoming optically thin and the SED is more and more similar to a stellar black body. The star, still in its PMS phase, is now about a few $\sim 10^7$ years old and does not present strong optical emission lines. With $W_{H_\alpha} \leq 5$ Å, this is a weak line TTauri star (WTTs). Such a star has usually intense X-ray emission associated to magnetic corona activity (Montmerle 1991).

In this general scheme, many points remain unclear and the frontier between the different classes of objects is not strict. For example, there also exist WTTs without IR excess, the NTTs (Naked TTauri stars). A relatively important fraction of WTTs and NTTs have ages of same order than those of CTTs (Stahler & Walter 1993).

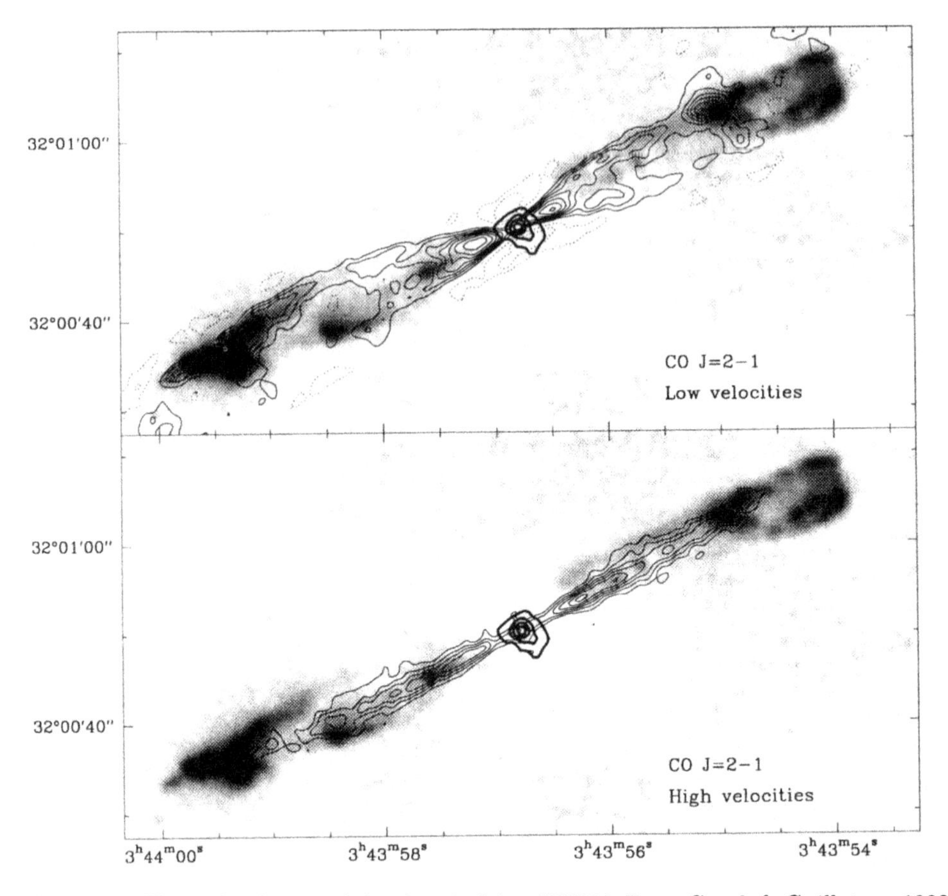

Figure 3. The molecular jet of the class 0 object HH211. From Gueth & Guilloteau 1998. CO J=2-1 observations from the IRAM interferometer superimposed to the vibrationally H_2 emission at 2.12 μm (from McCaughrean et al. 1994). The black contours correspond to the 1.3mm continuum emission.

2. Jets and flows

In 1980, the first bipolar outflow was found in CO J $= 1 \rightarrow 0$ by Snell et al. around L1551. The authors interpreted it as cavities driven by the stellar wind emanating from the L1551 proto-star. A few years later, the Herbig-Haro (hereafter HH) objects were definitely associated to highly collimated outflows (Mundt et al. 1983). With a high velocity jet of $\sim 200\,\mathrm{km.s^{-1}}$, a proto-star of $\sim 10^5$ yr old can generate a flow of size up to ~ 20 pc. In some cases, these giant HH objects associations are indeed observed *e.g.* HH111 (Reipurth et al. 1997, fig.1 and 2). Up to now, we only observe secondary indicators of jets (entrained material). Some of them are closely related to the jet energetics but the nature of the jet itself is not yet fully known and may even evolve with the proto-star and PMS star.

2.1. MASS LOSS IN PROTO AND PMS STARS

Observational evidences of mass loss are found from the cm to optical wavelengths. Details can be found in the reviews of the proceedings from the IAU symposium no.182 and Bachiller 1996.

Fig.3 (from Gueth & Guilloteau 1998) presents a good example of a jet emitted by a Class 0 (HH211). The vibrationally excited H_2 emission (McCaughrean et al. 1994) is shown in grey scale while the CO $J = 2 \rightarrow 1$ emission at high and low velocities is seen in black contours.

The main observational properties of the outflows are the following:

Thermal radio jets are observed at cm wavelengths at $\sim$0.1" (VLA interferometer). They correspond to free-free emission coming from the densest part of ionized winds or jets (see also Rodriguez 1997).

Molecular flows are likely the most spectacular mass loss phenomena. Powerful bipolar outflows are observed in the Class 0 and Class I phases. In the Class 0 phase, they are strongly collimated while they usually exhibit large opening angles in the Class I stage. They are seen in mm/submm molecular lines and NIR vibrationally excited lines of H_2. Recent high resolution images (e.g. HH211 shown in Fig.3) in CO $J = 2 \rightarrow 1$ from mm arrays reveal a high velocity component along the jet axis and a low velocity component corresponding to the walls of the cavity created by the jet travelling in the molecular cloud. Along the jet axis, molecular shock tracers like SiO $v = 0$ $J = 2 \rightarrow 1$ are also observed (Guilloteau et al. 1992). They likely originate from internal shocks generated by velocity variations inside a possible pulsating jet (Raga & Kofman 1992). Molecular flows also exhibit strong terminal bow-shocks resulting from the impact of the jet with the ambient medium (L1448: Dutrey et al. 1997).

HH objects are seen in optical and IR lines (SII, FeII, H_α,OI, H_2...). These objects usually look like aligned series of bright optical knots. They correspond to terminal bow-shocks and internal shocks located along the jet axis (Hartigan et al. 1987). When proper motions are measured, they are in correct agreement with bow-shocks models (Eislöfel & Mündt 1992). Recent studies of the ionization state of HH objects show that such jets are partially ionized, the ionization decreasing with distance to the star and internal shocks (Bacciotti 1997).

Forbidden emission lines (*e.g.* OI at 6300 Å, SII, NII...) are also observed around Classical TTauri stars (Hartigan et al. 1995). The emission is generally blueshifted because the disk hides the receding flow. Recent spectro-imaging technics allow to map micro-jets which are thought to be responsible for such an emission (DG Tau, Lavalley et al. 1998) at a scale of ~ 50 AU. These observations provide direct evidence for two mass loss components: a high velocity wind more likely coming from the inner part of the disk or the star (the jet itself ?) and a low velocity flow which may originate from entrained material in the wake of the jet. Inside the jet, knots associated to internal shocks are also observed.

Permitted emission lines (*e.g.* Balmer lines, HeI, FeII,NaD) are also found around classical TTauri stars (Edwards et al. 1994). They are usually centrally peaked, present large wings, 60-70 % of them have blue absorption features and some of them have inverse P Cygni profiles (indicative of accretion). The complexity of such profiles cannot be reproduced by a simple model of wind or accretion because lines usually exhibit features associated to accretion and ejection motions (Edwards 1997). Models of magnetic accretion funnels seem to explain the redshifted part (and the central peak) of the line profiles while the blueshifted absorbed component are explained by stellar winds.

In conclusion, there are observational evidences of mass loss at all stages of star formation but the outflow activity is clearly decreasing with increasing age (see below). The exact nature and origin (disk?) of the jet or flow or wind is not accurately known and may evolve from Class 0 to Class II. However, there are some common features seen in all classes of objects. Particularly, HH objects, molecular outflows and micro-jets exhibit knots along the jet axis. They are likely related to internal shocks due to time variability in velocity (Raga & Kofman 1992). The nature of such shocks (continuous C or strong dissociative J shocks) is not really known. Recent ISO observations in Far IR lines (Cabrit et al. 1998, Saraceno et al. 1998) suggest that both kind of shocks might exist. Such shocks lead also to a rich and complex molecular chemistry (Bachiller & Pérez-Gutierrez 1997) along the jet axis.

Finally note that the dynamical age deduced from the flow is always an underestimate of the age of the proto-star because the time at which the outflow activity starts and the tangential velocity of the jet are not well known.

2.2. EVOLUTION OF THE MASS-LOSS AND ACCRETION RATES

Several observational results show that outflow and wind activities decrease from the Class 0 to the Class II phases (Calvet 1997). For example, in a recent CO study, Bontemps and coauthors (1996) have shown that on a sample of Class 0 and I objects located mainly in Taurus-Auriga, the CO outflow momentum flux F_{CO} is decreasing with time and with the mass of the envelope $< M_{env} >$. They typically found that F_{CO}(Class 0) $\sim 10 \times F_{CO}$(Class I) while $< M_{env}$(Class 0) $>\sim 10\times < M_{env}$(Class I) $>$. In a limited SII survey performed on 22 TTauri stars in Taurus-Auriga, Gomez et al. 1997 found that the SII emission line is respectively present in 60% of Class I and only 10% of Class II objects.

Simultaneously, the mass accretion rate decreases with time from values of $< \dot{M}_{acc} >\sim 10^{-4}$ $M_{\odot}$/yr in FU Ori objects (a proto-star in a very strong accretion episod, see Hartmann & Kenyon 1996 for a review) to $< \dot{M}_{acc} >\sim 10^{-9}$ $M_{\odot}$/yr for Class II objects (Hartigan 1995, Hartmann 1997). Even if these results suffer from strong uncertainties depending on the various observational methods and tracers used for measuring the accretion and the ejection, there is a clear correlation between mass loss and mass accretion rates. Observational results seem to converge towards $\dot{M}_{acc}/\dot{M}_j \sim 10^{-1}$, at least for the first stages. Following Hartmann (1997) and taking such a value, a class II Tauri star of 0.5 $M_{\odot}$ mass and $\sim 10^6$yr old having a mass accretion rate of $\dot{M}_{acc} \sim 10^{-8}$ $M_{\odot}$/yr would have accreted only about $M_{acc} \sim 0.01\,M_{\odot}$ and ejected about $M_j \sim 0.001\,M_{\odot}$ in its life. This indicates that 90 % of the mass of the TTauri star must be accreted in the early stage of the star formation. It is clear that the next generation of high-angular resolution optical telescopes associated to spectro-imaging capabilities will provide key results about the origin of accretion/ejection phenomena in TTauri stars.

3. Circumstellar material around Class 0 and Class I objects

Only a few objects have been imaged with high enough spatial resolution ($\sim 150 - 300$ AU) to allow detailed investigations of the morphology and kinematics of Class 0 and I sources. Recent results from mm arrays (Guilloteau et al. 1997) show that at a scale of $\sim$ 10000-5000 AU, the dust distribution at 2.7 and/or 1.3mm is definitely more complicated than the spherical envelope assumed by many models of infall.

We now have a few examples where the dust emission appears flattened on scale $3000 - 500$ AU in a direction perpendicular to the jet axis (*e.g.* HH211, L1157). To explain such assymetry, the influence of the magnetic field can be invoked (Galli & Shu 1993a). Infall in a flattened core is also a possibility (Hartmann et al. 1994). Moreover, the large scale envelope can be disturbed by the outflow (*e.g.* L1157: Gueth et al. 1997, their fig.1 or L1527: Motte et al. 1998, their fig.1).

Most of the studies performed so far, have been done either in continuum (mm,submm,IR) or in CO lines. CO lines and continuum observations are complementary and cannot be considered without each other. After a description of the kinematics of such objects, this latter point will be illustrated through two representative examples: the Class 0 object L1157 and the Class I object L1551.

3.1. KINEMATICS OF PROTO-STARS

Studying the kinematics of Class 0 and Class I objects is a difficult task because they are strongly embedded in their parent molecular cloud. Along the line of sight, the CO velocity pattern is usually strongly dominated by the outflow motions (HL Tau, Cabrit et al. 1996). Therefore the main problem is how to disentangle on a given line of sight between 1) outflow motions, 2) possible rotation or 3) infall. It is however possible to analyse some typical velocity configurations.

An infalling envelope in front of a warmer core Because of symmetry, infall motions in a spherical core do not produce projected velocity gradient. A cold optically thick ^{13}CO spherical envelope in infalling radial motions near free-fall velocities in front of an unresolved warmer core of about $\sim$ $1\,M_\odot$ would present an absorption dip (Saito et al. 1997) of linewidth $\sim$ $0.1 - 0.3$ km/s. High spectral resolution is then needed to see the dip while high angular resolution, only provided by an interferometer, is necessary to resolve out the absorption in front of the core. Gueth et al. 1997 have shown (their figure 7) that an interferometer creates an apparent absorption dip whether the envelope is emitting or absorbing. "Real" self-absorption is only confirmed when it is also seen in single-dish spectra.

Therefore, *at small scales*, this is only the comparison of interferometric data (spatial resolution to determine the location of the dip in front of the dense core) with single-dish data (to confirm that the dip is not an instrumental artifact) which allows to conclude on infall motions.

A resolved rotating or infalling disk For flattened structures (disk or toroid) inclined along the line of sight, velocity gradients can be detected. However to decide between rotation and infall, it is still necessary to have a resolved image. This point is illustrated by Dutrey 1996 (her fig.2) who presents simulations of a CO disk in Keplerian rotation and in infall near free-fall velocity. In both cases, the integrated spectra, as seen by a single-dish radiotelescope are rather similar. However, when the disk is resolved, the *velocity gradient is along the major axis for a Keplerian disk* and *along the minor axis for infall motions*. Note that if the disk is seen pole-on, in both cases, there is no velocity gradient.

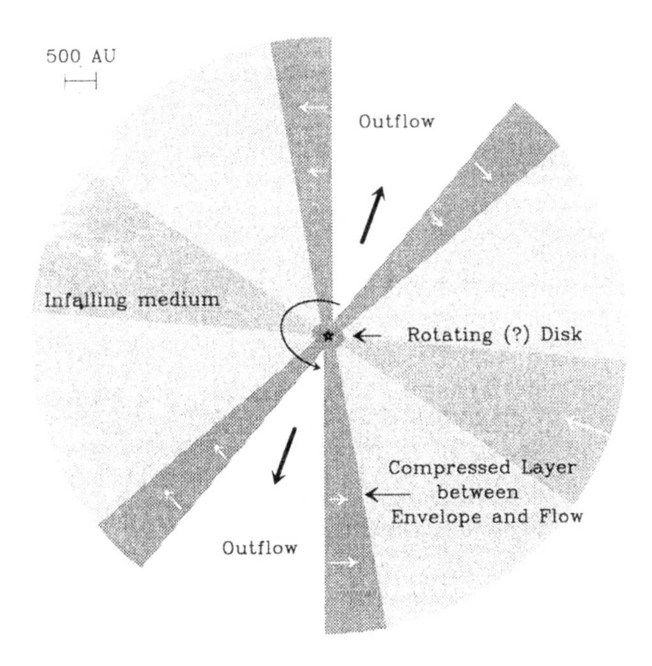

Figure 4. From Gueth et al. 1997. General schematic view of the region around the L1157-mm proto-star as deduced from continuum and CO lines study. The different components are (a) an extended infalling envelope of ~ 3 $M_{\odot}$, possibly including a large flattened component, (b) a biconical cavity excavated by the outflow, (c) a dense layer at the cavity walls of mass ~ 0.65 $M_{\odot}$ and compressed by the outflow, (d) a dense dusty circumstellar structure (possibly a rotating disk) containing about 0.2 $M_{\odot}$.

3.2. EXAMPLES OF CLASS 0 AND CLASS I OBJECTS

3.2.1. *The Class 0 object L1157*

L1157 is located in the Cepheus region at about 440 pc ($1'' = 440$ AU). With a luminosity of 11 $L_{\odot}$, this object presents all the attributes of a Class 0 source including a strong mm flux coming from the thermal dust emission of the protostellar condensation and a powerful molecular flow (Gueth et al. 1996). In a recent study at resolution $\sim 3''$, Gueth et al. 1997 have investigated the kinematics (^{13}CO $J = 1 \rightarrow 0$ and C^{18}O $J = 1 \rightarrow 0$ images) and the morphology of the dust (2.7mm) and gas around the proto-stellar condensation. Their main conclusions, summarized by the model given in fig.4 are the following:

- The dust thermal emission at 2.7mm is resolved in two components: a $\sim$ 500 AU flattened core of ~ 0.2 $M_{\odot}$ (at least), possibly the disk, surrounded by a large envelope of diameter $\sim$ 10000 AU and mass $\sim 3\,M_{\odot}$.
- Inside the envelope, the bipolar outflow has created a biconical cavity of heated and compressed dust and gas.

- CO data show strong indication for infalling motions in front of the central dense core. High resolution images and considerations on the excitation conditions suggest that the infalling medium is confined in a large ($\sim$ 10000 AU) flattened structure.

3.2.2. *The Class I object L1551*

L1551-IRS5 is a Class I object located in the Taurus-Auriga cloud at 150 pc from us and luminosity $\sim 20\,L_\odot$. It exhibits a well developed outflow (Snell et al. 1980), an optical jet (Mundt & Fried 1983) and HH objects (Herbig 1974). Moreover, this object seems to be in a FU Or stage (Stocke et al. 1988). In recent 3mm observations performed with the BIMA interferometer at $\sim 0.5''$ resolution (or 70 AU) Looney et al. 1997 have clearly shown that the small-scale structure of the circumstellar environment of L1551-IRS5 is completely different when looking at $\sim 3''$ and $0.5''$ (their fig.6). They report that the source is a binary system of separation $\sim$ 50 AU and that the best model for the dust distribution has three components:

- A large envelope of size $\sim$ 2200 AU and mass $\sim 0.28\,M_\odot$ (which dominates the dust emission at scales larger than $3''$)
- A flattened structure, possibly a circumbinary disk, of mass $\sim 4 \cdot 10^{-2}\,M_\odot$ and size $\sim$ 1000 AU
- Two point-like sources, presumably the circumstellar disks, separated by $\sim$ 50 AU of masses $\sim 2.4 \cdot 10^{-2} - 9 \cdot 10^{-3}\,M_\odot$.

Moreover, ^{13}CO $J = 1 \rightarrow 0$ observations performed with the Nobeyama array by Ohashi et al. 1997 at a resolution $\sim 4''$ reveal that:

- The ^{13}CO gas has an extended weak component delineating the edges of the CO outflow.
- Most of the ^{13}CO emission arises from a compact flat component of size $\sim 1200 \times 670$ AU which has a position angle perpendicular to the jet axis.
- The compact gas component has a velocity gradient along its minor axis as expected for infalling motions in disk.

In summary, around this proto-star which is more evolved than L1157, we find a less massive envelope and a (circumbinary) disk which may be infalling on the central sources.

3.3. CONCLUSION

Can we consider the global pictures deduced from the previous observations as a general scheme for proto-stars? Of course not, it is clear that differences should exist among low-mass star forming regions, depending for instance of the star formation mode: in cluster or isolated.

L1551-IRS5, considered so far as the proto-type of a class I single object is indeed a close binary system. This fact has been recently confirmed by Rodriguez et al. 1998 who performed observations at 7mm (spatial resolution $\sim 0.05''$) of the thermal dust emission originating from the circumstellar disks of the stars. This discovery is not surprising because about $\sim 60\%$ of MS stars are in fact binaries. Many Class 0 and Class I objects, actually considered (and modelled) as single stars should also be binary or even multiple systems.

It clearly demonstrates that a real understanding of the circumstellar properties of the proto-stars requires high angular resolution. With resolution of $\sim 0.1''$ or even below, the next generation of mm arrays will definitely change the crude picture we have today on YSOs.

Finally, recent models suggest that Class I objects might be in reality Class 0 sources seen pole-on. Their physical properties should be the same but the lower extinction expected for a near pole-on source may explain the differences in the observed properties. It might be true for some Class I objects, however I would like to point out that most of the Class 0 sources exhibit strongly collimated CO outflows (HH211) while it is not the case for Class I objects (the opening angle of the outflow of L1551-IRAS5 is larger than 90° at a few arcseconds from the star). Such a difference is impossible to explain by projection effects.

4. Circumstellar disks surrounding Class II sources

When the star is about 10^6 years old, most of the envelope is expected to be dissipated and the outflow activity is much weaker. The PMS star becomes a Class II object, also radiating at optical wavelengths.

In fig.1, the third panel (from the top) shows the typical SED of a CTTs. There is still a strong IR excess coming from the thermal dust emission of the circumstellar disk. This emission is optically thick in the IR domain, providing informations on the kinetic temperature $T_k(r)$. It becomes partially optically thin at submm wavelengths $\lambda \simeq 500 - 100\mu m$. In the mm range, most of the emission is optically thin (except the inner 30-60 AU), thereby tracing the *mass of the disk*.

Many observational results (*e.g.* Beckwith et al. 1990) show that $T_k(r)$ falls of as $\propto r^{-0.5} - r^{-0.6}$. Simple disk models like passive geometrically thin disks and viscous accretion disks (see §1) both predict that $T_k(r)$ must be proportional to $r^{-0.75}$ and fail to reproduce the observed SEDs. This discrepancy can have several physical origins. Among them, 1) the disk is flaring (as expected in case of hydrostatic equilibrium) and the outer part intercepts more stellar light, resulting in a weaker slope for $T_k(r)$ (Kenyon & Hartmann 1987), 2) the disk is back-warmed by a tenuous envelope (Natta

1993), 3) the outer ($r > 100 - 200$ AU) disk which essentially radiates in the mm/submm range (optically thin emission) is heated by cosmic rays and the interstellar UV field. Beyond a certain radius, $T_k(r)$ is then essentially uniform, with values similar to those encountered in dark clouds $T_k \sim 10 - 15$ K (Dutrey et al. 1996).

Therefore, *the properties of dust disks surrounding CTTs cannot be understood by the interpretation of the SED alone.* Resolved images are required to investigate the physical processes at work. I will present first the properties of dust disks and discuss in a second part the properties of CO disks.

4.1. DUST IN PROTOPLANETARY DISKS

Building planets implies the growth of the interstellar dust grains. Interstellar dust grains are composed of a mixture of amorphous carbons, silicates and graphite (*e.g.* Pollack et al. 1994). Observed properties of dust in circumstellar disks are indeed different than those observed in the interstellar medium.

In interstellar clouds and in star forming regions, dust particles of radius a are efficient absorbers of short wavelength radiations (UV, optical domain with $\lambda < a$). In equilibrium between heating and cooling, they reemit at longer wavelengths (mainly from mm to FIR) a continuous spectrum that closely resembles a thermal spectrum. At the shorter wavelengths, the scattering of the light from external sources dominate the thermal emission from the grains. The limit between the thermal regime and the scattering regime is around $\sim$ 2-4 μm.

For spherical particles, the dust properties at frequency ν can be characterized by the dust absorption coefficient

$$\kappa_\nu = \sigma_\nu/(4/3\pi a^3 \rho Z) = Q_{abs}(\nu)3/(4a\rho Z)(\mathrm{cm}^2/\mathrm{g}) \tag{1}$$

where Z is the gas to dust ratio (supposed equal to 100, the value found in the interstellar medium), ρ the density and σ_ν the radiation cross-section, $\sigma_\nu = Q_{abs}(\nu)\pi a^2$, Q_{abs} being the absorption efficiency. The absorption cross-section cannot exceed the geometrical cross-section of the spherical grains ($Q_{abs} < 1$), and observations at frequency ν are mainly sensitive to grains of size $a \leq \lambda$.

Details about the dust properties can be found in the reviews by Hildebrand 1983, Genzel 1990.

4.1.1. *Thermal regime*

The absorption coefficient is modelled by $\kappa_\nu = \kappa_o \times (\nu/10^{12}\mathrm{Hz})^\beta$. In the mm/submm domain, the dust emission is optically thin and the knowledge of κ_o and β allow the determination of the disk mass.

Continuum mm/submm data come from several radiotelescopes and suffer from different uncertainties. However, several groups (Beckwith 1990, Beckwith & Sargent 1991, Mannings et al. 1994, Dutrey et al. 1996) now agree that the best fits of κ_ν are obtained for $\sim 0.6 \leq \beta \leq 1$ and $\kappa_\nu = 0.1 \times (\nu/10^{12}\mathrm{Hz})^1$ cm^2.g^{-1} (including all material dust+gas, with gas/dust=100). This value of β is smaller than the value of $\beta \simeq 2$ found in molecular clouds and implies that grain growth is occuring. Indeed, current models of dust fail to reproduce the high value of κ_ν and the low β observed in the mm/submm range (Pollack et al. 1994). To produce the large absorption coefficient observed, Eq.1 indicates that the grain mean density ρ should be lowered. Grains in disks should be similar to aggregates (Henning & Stognienko 1996) and may even have fractal structures (Wright 1987). Agladze et al. 1996 have recently performed laboratory measurements of mm absorption spectra of various silicate grains at the low temperatures encountered in the interstellar medium and in disks. They pointed out that κ_ν should be larger than predicted from current models of dust. Even if they represent a promising step in the right direction, these new results do not allow yet to obtain dust properties in good agreement with those observed in protoplanetary disks. They clearly demonstrate that new laboratory measurements and new models are required to understand the grain properties (and grain growth) in protoplanetary disks.

4.1.2. *Scattering regime*

The dust particles scattering the stellar light are responsible for reflexion nebulae seen in all star forming regions. In protoplanetary disks, similar to those surrounding HH30 or GM Aur (see next sub-section), the column densities are high and optical depths of ~ 1 are quickly reached, the optical and NIR mainly probe the disk surface. These data do not allow an estimate of the disk mass without an opacity correction and detailed modellings (single or multiple scattering models, Lazareff et al. 1990, Whitney et al. 1992). The scattering is asymetric, depending of the dust phase function, forward scattering being easier to produce than backward scattering (GG Tau, fig.10, next section). Analyses of the optical and NIR properties of grains suggest that the grains responsible for scattering are relatively small $a \sim 0.05 - 0.8\mu$m, and usually fit the interstellar extinction curve (Burrows et al. 1996, Whitney et al. 1997, Close et al. 1998).

4.1.3. *Polarisation*

The origin of the polarisation is completely different in the optical/NIR and in the mm/submm range. For details, see the review by Bastien 1991.

Optical/NIR λ: The polarisation of the light is due to single or multiple scattering on grains. The polarisation pattern depends on the geometry of the scattering medium and the dust grains properties. For example, the pattern of the polarisation vectors around R Monocerotis/NGC2261 (Close et al. 1997b, fig.3) is centro-symetric, exactly as expected for a reflexion nebula. Other examples of recent NIR/optical polarisation studies in star-forming regions are found in Whitney et al. 1997.

mm/submm λ: The linear polarisation observed in dark clouds and cores is due to alignment of non spherical grains. The extinction is maximum along the (projected) major axis of the grains (dichroic effect). Grains tend to align with their shortest axis parallel to the magnetic field, providing information on its direction. In a recent paper, Akeson & Carlstrom 1997 described and analysed the first interferometric polarimetric measurements at 3mm in two Class 0 objects. These observations are still limited by the resolution ($\sim$ 300AU) of the mm array, but they clearly demonstrate the interest of the technics.

Finally, to properly constrain the dust properties, multi-wavelengths studies will provide a clear improvement. Particularly, one would like to know if the dust can be fitted by a single size distribution explaining the properties from the mm to the optical range and how particles are located within the disk, e.g. do larger particles settle in the disk plane ?

4.1.4. *Imaging dust disks*

Circumstellar dust disks imaged with mm/submm interferometers: Since 1990, mm arrays have started to survey disks around low-mass stars (Ohashi et al. 1991, Koerner et al. 1995, Dutrey et al. 1996, Mundy et al. 1996, Lay et al. 1997). Most of the continuum emission originates from the thermal spectrum of dust and possible free-free emission coming from the jets is usually negligible (e.g. DO Tau, Koerner et al. 1995). Only a few disks have been imaged at a resolution which is high enough to allow a proper estimate of the surface density law of the dust disk $\Sigma(r)$ ($\propto r^{-p}$). As explained in §4.2.2, mm arrays are only sensitive to the bulk of the dust emission because as soon as the dust emission becomes too optically thin (typically around $R \sim 200$ AU) it escapes the threshold of detection of current interferometers. Recent results obtained on partially resolved disks (Dutrey et al. 1996, Mundy et al. 1996) suggest however that the surface density law might be shallower than previously thought with $p \leq 1.2$. Disk masses inferred from mm arrays are usually within the range $10^{-1} - 10^{-3}\,M_\odot$.

Proplyds in the Trapezium Cluster: Proplyds are protoplanetary disks around low-mass stars which are seen in silhouette against the strong HII region associated to the Trapezium Cluster in Orion A (*e.g.* McCaughrean & O'Dell 1996). In a recent mm survey, Lada et al. (1996) have detected 3 proplyds and obtained significant upper limits on several others. They got masses within $\sim 0.007 - 0.016\,M_\odot$ and an upper limit $M_{3\sigma} \sim 0.0017\,M_\odot$. This result suggests that disks around low-mass stars in the Trapezium cluster may have masses of same order as those encountered in low-mass star forming regions.

The HH30 disk: So far, the HH30 observations performed by Burrows et al. 1996 using the HST give one of the most complete picture (kinematics excepted) of the material surrounding a PMS star. The dust disk (their fig.2) is seen edge-on (80°), and appears as a dark lane with an outer radius of about $R_D \sim 250$ AU. Only the flaring surface of the disk is seen in scattered light. The star, obscured by the disk material, is not seen ($A_v \geq 30$ mag) and ejects material perpendicular to the disk plane (jet). Since the disk is seen edge-on the vertical distribution $h(r)$ can be estimated. Even though the results are model dependent, multiple scattering model fit to the data suggest that the disk is pressure supported (dominated by the central star) with the best fit given by $h(r) \sim r^{1.45}$. The authors also estimate the surface density law $\Sigma(r) \sim 1/r$ and the disk mass $\sim 6.10^{-3}\,M_\odot$.

4.2. GAS DISKS INFERRED FROM CO ROTATIONAL LINES

Gas disks around TTauri stars are not different from dust disks. The recent studies of the disk surrounding the CTTs GM Aur performed by Stapelfeld et al. 1997 in V band using the HST (fig.6) and by Dutrey et al. 1998 with the IRAM interferometer in ^{12}CO J=2-1 (fig.5) demonstrate that both dust and gas disks have the same geometry *i.e.* 1) the *same radius* ~ 500 AU, 2) the *same inclination angle* $i \sim 60^o$ and 3) the *same position angle.* This is a clear evidence that both CO observations and dust scattering the stellar light are tracing the *same physical disk.*

Gas disks around TTauri stars are mostly observed in rotational lines of CO because the carbon monoxyde remains the most abundant molecule in disks after H_2. The first rotational levels of CO are very easily populated by collision with H_2 and completely thermalized at the high H_2 densities ($\geq 10^6$cm^{-3}) encountered in disks.

CO observations also allow to study the kinematics which can only be addressed by spectroscopy. In the last years, a fundamental step was made by mm interferometers which have shown that disks around CTTs are indeed in *Keplerian rotation* around their central stars (Koerner et

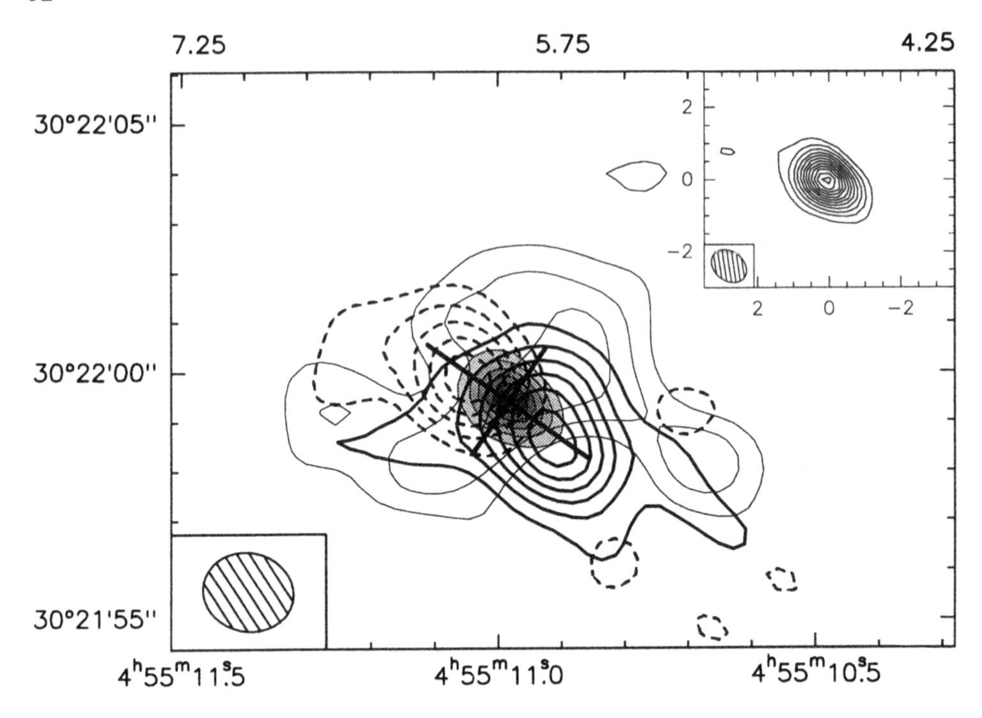

Figure 5. From Dutrey et al. 1998 (their fig.1). The GM Aur disk observed by the IRAM array. Contour of ^{12}CO 2-1 emission at 4.5, 5.75 (systemic) and 7.0 km.s^{-1} (synthesized beam is $1.9'' \times 1.6''$ at P.A. 86°), superposed on the continuum image at 1.3mm in grey scale. The CO contour steps are 250 Jy/Beam. The continuum image (synthesized beam is $0.6'' \times 0.7''$ at P.A. 56°) is shown in the insert; the contours steps are 7.5 mJy/Beam. The cross indicates the position, orientation and aspect ratio of the 1.3mm continuum peak. Both the CO and continuum emission peak at $RA = 4^h55^m10.98^s$ and DEC = $30^o21'59.5''$ (J2000.0). The velocity gradient of the CO emission is along the major axis of the disk as expected for rotation.

al. 1993, Dutrey et al. 1994, etc...). However, such analyses suffer from observational and instrumental biases.

4.2.1. *Confusion with the surrounding medium*

Class II objects are visible stars but most of them are still surrounded by their parent molecular cloud. Observed with a single-dish telescope of resolution ∼10", the CO J=2-1 emission coming from a circumstellar disk is strongly beam diluted and the emission may still be dominated by the surrounding cloud. Using an interferometer is required to filter out the large scales. However this does not solve all problems. If the disk is partially hidden by the cloud, modelling the disk parameters is not easy. For instance, an optically thick CO screen hiding the disk emission around the systemic velocity would artificially increase the disk inclination. For this reason, the best observations of gas disks are obtained when the sources are free of

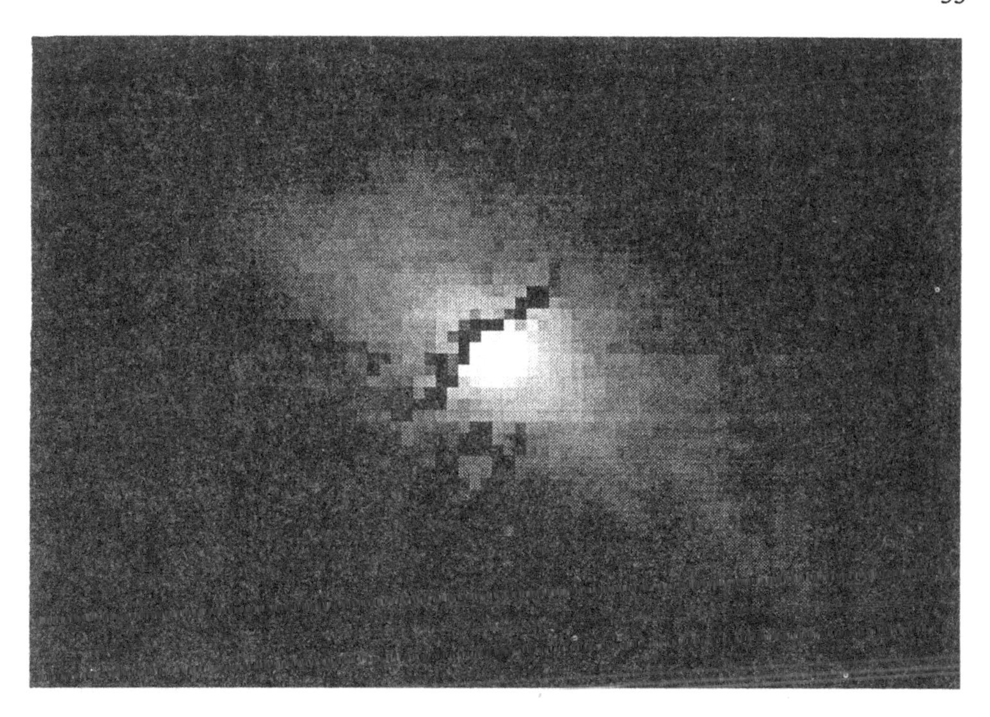

Figure 6. From Stapelfeld et al. 1997, HST image of the GM Aur disk in the V band. The field of view is $3.5 \times 2.4''$. The star has been substracted. The north is up and the east is to the left.

confusion or *located inside holes of the molecular clouds.* This is the case for the best known objects such as GG Tau (see section §5), DM Tau, GM Aur (this section) or MWC480 (Mannings et al. 1998).

As already said (§3.1) for a tilted disk in rotation, the velocity gradient is along the major axis of the disk while it lies along the minor axis for infall. In an intermediate mixed case of rotation and infall, the velocity gradient is tilted and skewed. Unfortunately the use of velocity pattern derived with an elliptical beam would produce a distortion of the velocity field which could easily be mistaken for a combination of infall and rotation. This point is illustrated by Guilloteau & Dutrey 1998 (their fig.3). To properly analyse the velocity field, it is thus essential to use a circular beam or to produce a model which has been convolved by the same beam as the observations.

4.2.2. *Comparing the brightness distribution for dust and gas*

Fig.7 shows the brightness distribution $T_B(r)$ for the dust emission at 1.3mm and the ^{13}CO J=2-1 line (^{13}CO being an optically thinner isotopomer of CO) in a disk having standard properties. $T_B(r)$ must be compared to the sensitivity curves of the interferometer for continuum and line obtained in the same observing time. The sensitivity increases like $\sqrt{\Delta\nu}$

where $\Delta\nu$ is the frequency bandwidth of the system. Increasing the sensitivity in continuum can thus be obtained by increasing the bandwidth of the interferometer. For line observations, the bandwidth is Doppler limited: a significant increase of the sensitivity can only be achieved by adding collecting area.

Fig.7 can be understood as follows. For resolved structures, the brightness sensitivity of the array is constant. For sources of angular size θ smaller than the resolution θ_b, the rms noise in brightness varies as $(\theta_b/\theta)^2$ (beam dilution effect).

In the Rayleigh-Jeans approximation and assuming power law distributions for the kinetic temperature $T_k(r) \propto r^{-q}$ and the surface density $\Sigma(r) \propto r^{-p}$, $T_B(r)$ is $\propto T_k(r) \propto r^{-q}$ in the optically thick regime while it is $\propto T_k(r) \times \Sigma(r) \times \kappa_\nu \propto r^{-q-p}$ in the optically thin case (κ_ν being the absorption coefficient of the dust or the line). In the simulation, the dust becomes optically thin at $R_1 \sim$ 4AU. Outward R_1 the continuum brightness thus falls down like $\sim r^{-p-q}$. Assuming standard CO abundance and dust properties, the absorption coefficient of the ^{13}CO J=2-1 line remains about 2000 times larger than the 1.3 mm dust absorption coefficient (assuming a local linewidth of 1 km.s^{-1}). The ^{13}CO J=2-1 line is optically thick everywhere and its $T_B(r)$ is always $\propto r^{-q}$.

For a source located at 150 pc, one thus concludes that:

1. Due to instrumental limitations, the dust emission and the CO line (or any optically thin or thick line) are not probing the same region of the protoplanetary disk.
2. The partially resolved dust emission probes the disk at the scale of our solar system. Beyond $r > 200$ AU, the dust emission escapes detection because it is too optically thin. This region will be called the *inner disk.*
3. On the contrary, the CO emission arises from the outer disk (this would correspond to a region outside the Kuiper Belt in our solar system). Inside $r \sim 30 - 60$ AU, current CO observations do not provide any information about the gas content (beam dilution effect).
4. The overlap between the dust and CO line emission is too small to allow a proper comparison of the dust and gas distributions. In particular, masses deduced from both tracers are not directly comparable without proper modelling. Note that the case displayed here enlarges the overlap because the disk mass ($\sim 0.05\,\mathrm{M}_\odot$) is relatively high.
5. The observations of optically thin lines are even more difficult. They suffer from beam dilution in the inner disk and are quickly below the threshold of detection in the outer disk.

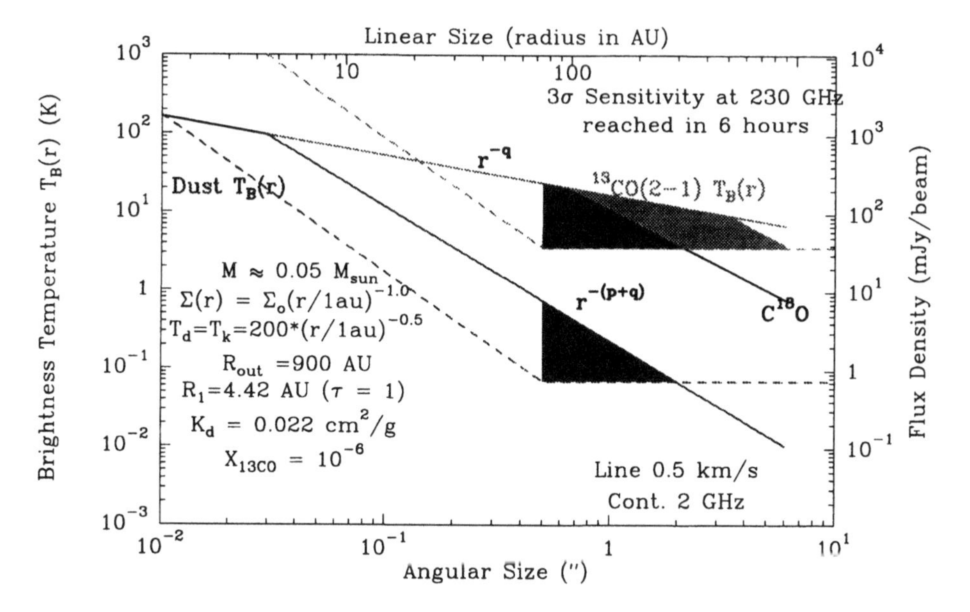

Figure 7. Brightness distribution in Kelvin (Rayleigh-Jeans approximation) of dust at 1.3mm (dark grey), ^{13}CO J=2-1 (grey) and C^{18}O J=2-1 (black) lines in a disk and comparison with the sensitivity curves of the IRAM interferometer at 3σ for the continuum (dashed line in dark grey) and the lines (dashed line in grey). The resolution of the array is 0.5″. The scale is log-log and the angular size (″) is scaled to the Taurus distance (150 pc) with radius in AU. The interferometer is only sensitive to disk emission above the sensitivity curves. Filled areas correspond to the regions of the disk where the emissions are detectable and resolved. Differences between C^{18}O and ^{13}CO emissions come from the fact that C^{18}O is less abundant than ^{13}CO.

4.2.3. *Modelling CO disks*

Even if interferometers are sensitivity limited, they now routinely provide high angular resolution ($\sim 1-2''$) maps of CO lines (MWC480: Mannings et al. 1998, DM Tau: Guilloteau & Dutrey 1998, GM Aur: Koerner et al. 1993, Dutrey et al. 1998). Let us show now on an example (DM Tau) which kind of physical informations can be deduced from CO maps. This however requires 1) high-signal-to-noise maps at high spectral resolution ($\sim 0.2-0.1$ km.s^{-1}), 2) a proper method of analysis and 3) a χ^2 minimisation on the model parameters to evaluate error bars (Guilloteau & Dutrey 1998 for details).

In particular, χ^2 fitting allows to deduce the stellar mass from the dynamical mass of CO because in Keplerian motions, the velocity $v_z(r)$ along the line of sight is given by $v_z(r) = \sqrt{G.M_*/r} \times sin(i)$. If the distance to the source is known and i well constrained, the χ^2 analysis provides an original measurement of the stellar mass, independantly of the star properties.

The DM Tau case: Located inside a hole of the Taurus cloud, DM Tau is a Classical TTauri star of about $\sim 5 \cdot 10^6$yr and spectral type M1, as such it can be considered as a typical TTauri star. Recently, Guilloteau & Dutrey (1998) used the IRAM interferometer to map the gas disk in CO J=1-0. Fig.8 displays the results of their CO analysis, the observed map of CO (top), the best model (middle) and the difference between the data and the model (bottom). They conclude that:

- DM Tau has a large CO disk with $R_{out} \simeq 850$ AU.
- The CO velocity law is in perfect agreement with Keplerian rotation around a central star of $0.50 \pm 0.06 \times (D/150\text{pc})\,M_\odot$.
- The kinematic pattern indicates that the turbulence in the disk is small, of order ~ 0.1 km.s^{-1}.
- The ^{12}CO J=1-0 line being strongly optically thick, the density profile cannot be measured from these data, but the temperature law ($T(r) \propto r^{-0.63}$) is well constrained and consistent with stellar heating in a flared disk.

4.3. SEARCHING FOR OTHER MOLECULAR SPECIES THAN CO

The chemical evolution of the star-forming regions is one of the important challenges of the next decade. Particularly, one would like to understand how the gas and dust evolve and are coupled together in protoplanetary disks. For instance, one expects that in disks associated to low-mass stars, depletion due to the condensation of molecules on grains should occur for many molecules (Aikawa et al. 1996, 1997). Details about the current knowledge of the chemistry in star forming regions can be found in the review by Van Dishoeck & Blake 1998.

So far, there are a few attempts to detect other molecules than CO in circumstellar disks. Around TW Hydra, Kastner et al. 1997 detected ^{13}CO, HCO^+, CS, HCN and CN, and estimated the disk size to be ~50 AU. This result is important because TW Hydra is one of the oldest CTTs ($\sim 10^7$yr) having a CO disk. Using the IRAM 30-m radiotelescope, in the disks surrounding DM Tau and GG Tau, Dutrey et al. 1997 detected HCO^+, CN, CS, HCN, HNC, H_2CO and C_2H, in addition to ^{13}CO and $C^{18}O$. The higher excitation lines detected (CN J=2-1 or HCO^+ J=3-2) provided a minimum value for the H_2 density and the HCN J=3-2 provided an upper limit. Comparing this *direct* measurement of the mean H_2 density with the observed H_2 column density, the authors were able to 1) derive *directly* the gas disk mass (independantly from the dust), and 2) evaluate the molecular depletion (by reference to the Taurus cloud). In DM Tau, CO is depleted by a factor of about ~ 5, the other molecules being depleted up to factors ~ 100 (HCN, H_2CO). The mass of the DM Tau gas disk is about $\sim 4\,10^{-3}\,M_\odot$, still

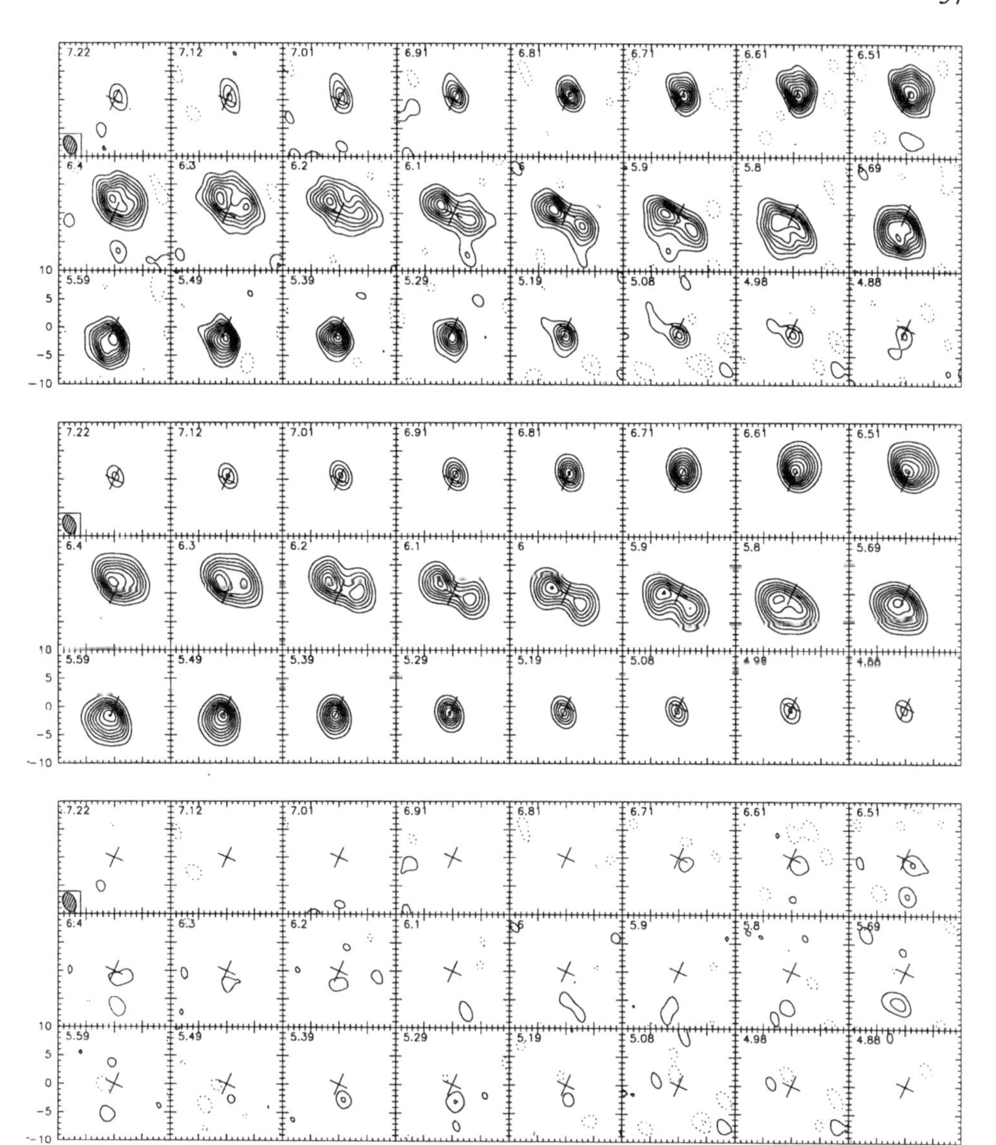

Figure 8. From Guilloteau & Dutrey 1998. Channel map of the ^{12}CO J = 1→0 toward DM Tau. The angular resolution is $3.5 \times 2.4''$ at PA 24°. Coordinates are offsets in $''$ from the continuum position, R.A. 04^h 33^m 48.735^s & Dec. $18^o10'10.2''$ (J2000.0). Contour spacing is 80 mJy/beam, or 0.9 K (2 σ). The LSR velocity is indicated in each panel. The cross indicates the position of the continuum source, and the orientation of the disk axis. Middle: best model, with same contour levels. Bottom: difference between observations and best model.

7 times smaller than the mass derived from the dust emission ($\sim 0.03\,\mathrm{M}_\odot$). However, since the molecular observations are not sensitive to the inner part of the disk, the mass deduced from the molecular analysis is mainly an estimate of the outer disk mass. Therefore, even though the disk of DM Tau is extended, most of the mass seems to be within $\sim$ 200 AU.

4.4. CURRENT PROPERTIES OF GAS AND DUST DISKS

In summary,

- mm arrays are now currently resolving the thermal dust emission of disks around many CTTs. This emission allows measurements of the mass of disks because it is mainly optically thin. Disk masses are typically within the range $10^{-1} - 10^{-3}\,\mathrm{M}_\odot$. The dust emissivity in the mm/submm range correspond to larger grains than those encountered in the interstellar dust and their properties are poorly reproduced by current models and laboratory measurements.
- Optical and NIR observations probe the dust scattering the stellar light at the disk surface. The A_v across the disk, or in front of the star remains difficult to estimate. These observations are only sensitive to smaller grains ($\sim 0.05 - 0.5\mu$m).
- The current CO images only trace the gas in the *outer disk* ($R \geq 30 - 60$ AU). With typical $R_{out} \sim 300 - 800$ AU, CO disks are *large.*
- Contrary to the CO lines, the bulk of the continuum mm emission arises from the *inner disk* and is typically located inside a radius $R \sim 200$ AU. This value is however, a lower limit to the dust disk outer radius and implies at least $R_{out}(\mathrm{mm}) \geq 200$ AU.
- Moreover there are now observational evidences that the CO gas and the dust scattering the stellar light at optical and NIR wavelengths come from the *same disk.*
- Outer CO disks are in Keplerian rotation and the turbulence seems to be small compared to the thermal width. Assuming the distance D is known, the rotation pattern allows an accurate estimate of the stellar mass.
- Temperature laws are consistent with stellar heating in flared disks.
- Surface density laws deduced from mm/submm maps suggest $\Sigma(r) \sim r^{-1-1.2}$, a value of p slightly shallower than expected.
- Detection of other molecular species than CO is possible for a few sources (*e.g.* DM Tau, GG Tau or TW Hydra) but limited to the most abundant molecules. Compared to standard molecular abundance encountered in Taurus clouds, molecules are depleted by factors $\sim$5 for CO to $\sim$100 for HCN and H_2CO.

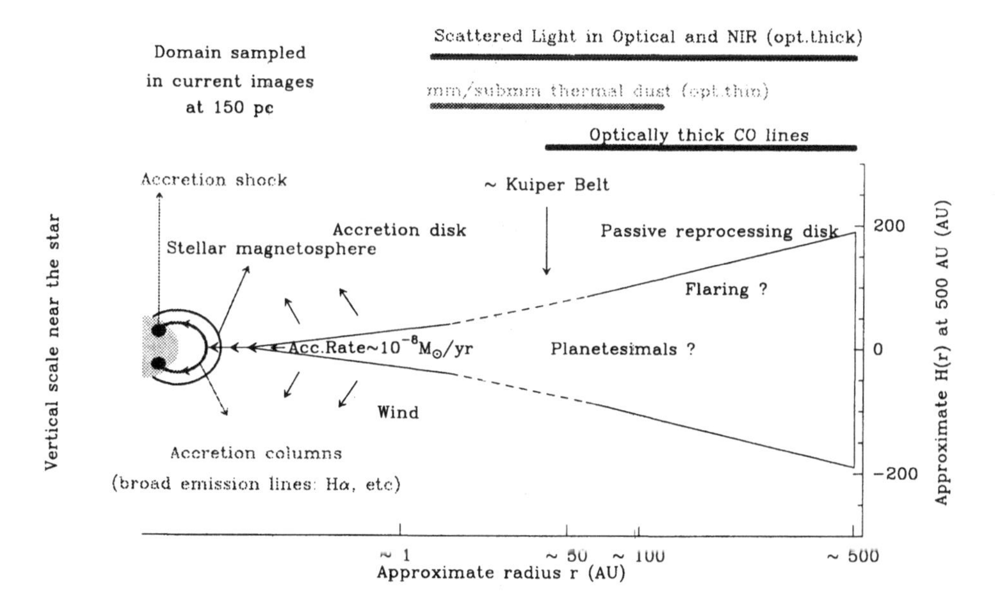

Figure 9. Global picture of a circumstellar disk surrounding a CTTs of $\sim 10^6$ years old (vintage 1998). Partially adapted from Hartmann 1997. Top: domain sampled in images coming from current mm arrays and large optical telescopes, scaled for D$\simeq$ 150 pc. Bottom: The star+disk are seen edge-on. For comparison with our solar system, the approximate location of the Kuiper belt is given.

Fig.9 presents a global picture of what a disk around a Class II CTTs should look like.

5. Environment of multiple stars

In this section, I briefly present our knowledge of the circumstellar environment of binary and multiple PMS low-mass stars. More details about multiplicity and binary properties can be found in the review by Mathieu (1995).

5.1. OVERVIEW

In the solar vicinity the frequency of binaries is about $F \simeq 60\%$ (Duquennoy & Mayor 1991). Surveys of low-mass PMS stars performed in several regions (Ophiuchus, Taurus-Auriga, Chameleon, Corona Australis, Lupus...) and by several groups (Ghez et al. 1993, Leinert et al. 1993, Richichi et al. 1994, Simon et al. 1995) show that more than 60 % of PMS stars form in binary or multiple systems. These surveys suffer from biases such as incompleteness, and involve different technics. However it seems now clear that there is even an excess of PMS binaries within the separation range $\sim 10 - 150$ AU in the Taurus-Auriga clouds. Recent surveys performed on population of more massive objects like Herbig AeBe stars (Corporon et al. 1998) show that the frequency of binary stars is about the same as for TTauri stars.

Multiple systems are very common and seem even to dominate against single star formation.

Since about 60 % of MS stars are indeed binary stars, the problem of planetary formation in binary environment is a real problem which must not be neglected. A first step is to investigate the properties of the material surrounding PMS binary and multiple systems. Let us first discuss the biases introduced when an unresolved binary star is studied as a single star.

Using the example of DF Tau (separation of $0.09''$ in the plane of the sky) Ghez (1995) clearly demonstrates the resulting errors (her fig.1.6). In an HR diagram, the unresolved binary star appears more luminous as a younger, less massive star having a more active accretion disk. Moreover statistically, such errors should also affect the initial mass functions and the estimates of the stellar formation time. Ghez (1995) concludes that *although often overlooked, multiplicity in PMS stars is one of the most important properties to identify.*

5.2. PROPERTIES OF DISKS AROUND MULTIPLE SYSTEMS

Recent mm and submm surveys (Osterloh & Beckwith 1995, Jensen et al. 1996a, Dutrey et al. 1996) show that the flux densities of binary stars is usually significantly weaker than those of single stars. This implies that there is in general a smaller amount of material surrounding binary stars. Moreover this effect depends on the separation (Jensen et al. 1996a). One finds only a few disks for separation $a \leq 100$ AU (or $0.7''$ at 150 pc) while wide binaries ($\sim 500 - 1000$ AU or $3 - 7''$) have often a total flux density which is undistinguishable from what can be found around singles.

Differences in flux density are explained by tidal truncation rather than by different surface density laws. Artimowicz et al. 1991 have shown that star-disk interactions proceed essentially through Linblad resonances with the orbital period of the stars. They performed SPH simulations of the evolution of such disks (Artymowicz et al, 1991, Artymowicz & Lubow 1994). Their simulation started with a filled disk, and continued until it evolved to a quasi-stationnary state (Artymowicz & Lubow 1994, their fig.9 & 10). The main evolution is a change in orbital parameters of the stars, mostly in the eccentricity. The results can be simply understood by considering the tidal effects at apastron: the binary always rotates faster than the disk which tends to delay the binary causing angular momentum and energy losses associated with elongation of its orbits.

The circumstellar material around a binary can be separated in four main regions: two possible circumstellar disks within the Roche lobes and surrounding the individual stars, an "emptied" cavity and a circumbinary ring lying outside the L_2 and L_3 Lagrangian points (see fig.5 from Dutrey et al. 1994); and depending of the eccentricity, the tidal truncation arises at orbital resonances 2:1, 3:1, 4:1, etc...

5.3. OBSERVED DISKS AROUND MULTIPLE SYSTEMS

5.3.1. *Observational biases*

Fig.7 also explains what kind of circumbinary (CB) material an interferometer can detect. For a binary of separation 0.3″, the gap created by tidal effects would be located between 0.1″ and $\sim$ 1″. Hence the interferometer would detect the unresolved continuum emission coming from the circumstellar (CS) inner dust disks (if any) and the CO emission associated to the CB disk. The CO emission cannot be detected on the inner disks due to beam dilution. The dust in the CB disk escapes detection because its emission is too optically thin and is below detection threshold of the array. This simulation also assumes that the disk (CB+CS) is relatively massive (0.05 $M_\odot$) and the CB disk is in this case marginally detected in continuum.

In conclusion, a CB ring can be detectable in CO (optically thick lines) while it escapes the threshold of detection in mm continuum. Observations in the submm range would be preferable. On the contrary, CS inner disks of same mass can be detected in continuum by mm arrays but fail to be detected in CO lines because they are beam diluted.

5.3.2. *Circumstellar medium around multiple systems*

To illustrate the above points, let us now discuss three examples of multiple systems from wide separation to spectroscopic binary. All the sources dicussed above are located at D$\simeq$ 150 pc.

UZ Tau, a quadruple system: UZ Tau includes a wide binary of separation 3.4″ and located East-West. Both components are also binary stars of separation 0.34″ for UZ Tau West, UZ Tau East being a spectroscopic binary with $a.sini \simeq 0.03$ AU (Mathieu et al. 1996). In a recent paper, Jensen et al. 1996b observed UZ Tau with the OVRO interferometer in ^{12}CO 2-1 and in continuum at 1.3mm. These observations (see their fig.1 and fig.2) show that strong continuum and CO emissions arise from the spectroscopic binary UZ Tau East. The CO pattern is consistent with rotation. On the contrary, there is no detection of CB emission neither in continuum nor in CO around the wider binaries UZ Tau West and UZ Tau West/East. On UZ Tau West, the continuum emission is weaker, unresolved and associated to the inner CS disks.

UY Aur: a 0.9″ binary UY Aur is a binary of separation $\sim 0.89''$. Using the IRAM array, Duvert et al. (1998) have observed in ^{13}CO J=1-0 and J=2-1 a CB ring in Keplerian rotation around the stars while the continuum emission at 1.3mm peaks on the binary itself and is associated with the CS disks. Further NIR observations performed by Close et al. (1998, their fig.1) confirm the CB ring. Like in the case of GM Aur, the dust which scatters the stellar light originates from the same location as the CO emission (Close et al. 1998, their fig.3). The ring mass, roughly estimated from the ^{13}CO data is about $\sim 3 - 7 \cdot 10^{-3}\,\mathrm{M}_{\odot}$.

GG Tau: a 0.26″ binary GG Tau is a binary of separation 0.26″. This object is unique because at 1.3mm, its continuum flux is $\sim$ 600 mJy, well above what is usually found around such stars. There is indeed a very massive ($0.15\,\mathrm{M}_{\odot}$) and large CO and dust CB ring which is detected by mm interferometry (Dutrey et al. 1994) and in NIR scattered light (Roddier et al. 1996). The disk is in Keplerian rotation and the radius of the central hole ($R_{in} \simeq 180$ AU) is in perfect agreement with predictions of tidal truncation (see section 5.2 and Artymowicz et al. 1991). Fig.10 is a montage of the 1.4mm and ^{13}CO J=2-1 observations performed at IRAM (from Guilloteau et al. 1998) and the NIR image (Roddier et al. 1996). The central hole, the P.A. and the inclination angle of the disk determined from the mm data or from the NIR image are in perfect agreement. For the first time, this demonstrates that the dust observed 1) in the thermal mm regime and 2) in the scattering regime is mixed, and that 3) the CO emission also originates from *the same disk*.

The UZ Tau case illustrates the complexity of multiple systems and the fact that contrarily to a close separation, a wide separation (here $\sim 3.4''$ or $\sim$ 500 AU) allows to find CS disks (even in CO) around individual components (UZ Tau East). Moreover, UZ Tau East was recently identify as a spectroscopic binary. This does not change its expected mm flux (separation $<$ 1 AU) but it reminds us that many other "robust" single stars could be spectroscopic binaries. With a less massive CB ring ($M \leq 10^{-2}\,\mathrm{M}_{\odot}$) which is not detected at 1.3mm, the gas and dust distribution around UY Aur is more likely representative of disks around binaries than the GG Tau disk which remains exceptionally massive ($0.15\,\mathrm{M}_{\odot}$).

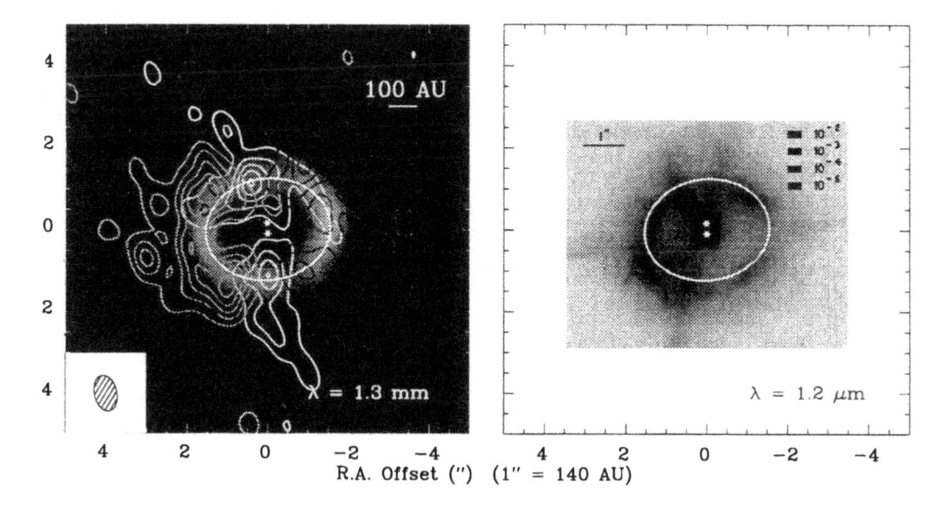

e 10. The GG Tau circumbinary ring seen at 1.4mm (Guilloteau et al. 1998) and in (Roddier et al. 1996). Left: IRAM data, comparison of the ^{13}CO J=2-1 line emission 55, 6.30 (systemic velocity) and 7.05 km.s^{-1}, overlaid on a false colour image of the m continuum emission. The beam size is 0.6″ × 0.9″ at P.A. 13°. The white ellipse ates the ring average radius and the stars give the location of the binary. Right: NIR vations from Roddier et al. 1996 in the J band. The forward scattering dominates mission and the northern part of the disk points towards us, in agreement with the y orbit and the Keplerian rotation.

A FEW COMMENTS ABOUT THE MULTIPLICITY

examples above show how the distribution of material can be "dis-ed" around multiple systems. Multiplicity have also some other influ-s. We have already seen in §5.1 that binarity can strongly affect the ysis of the SEDs by producing strong biases when a binary system is ysed as a single star. In some cases binary stars do not appear co-eval, companion being usually younger (Hartigan et al. 1994). This problem be explained by the accretion which affects the evolutionary status of companion and the evolutionary tracks (Siess et al. 1998). Finally, disk ractions with the binary can also change the orbital parameters, like eccentricity (Artymowicz et al. 1991).

Moreover, in a survey of binary and single TTs performed in the Pleiades, vier et al. (1997b) found that there is no significant differences on the tion speed of single and binary stars. Circumstellar material regulates rotation speed (see section 1.3.2) by the same way in binary than in le stars. In Taurus region, Simon & Prato 1995 found that the timescale inner disks (within a few AU from the stars) dissipation is about the e for single and binary stars; Prato & Simon (1997) showed that for close s (separation ~ 0.3″ − 2.6″), both components are simultaneously CTTs while for wider separations, the number of mixed pairs of CTTs/WTTs is increasing with the separation ($> 3''$).

Since CS disks are small, and assuming standard surface density laws, they would be rapidly exhausted by accretion on stars. The results mentionned above suggest that there might be, at least for close separations, a mechanism which replenishes the CS disks from a circumbinary ring or an envelope. Artymowicz & Lubow (1994) have simulated disks around binary systems and they have shown that this replenishement may be possible (see their fig.8).

This suggests that undetected circumbinary disks exist but have small masses. Jensen et al. 1996 put an upper limit at 3σ on CB rings of $\sim 0.005\,M_\odot$ around binaries with projected separation between 1-100 AU. This value which is slightly smaller than the canonical mass of the protosolar nebula ($\sim 0.01\,M_\odot$), is in agreement with the mass of the UY Aur CB ring and similar to the mass of the outer disk of DM Tau inferred from molecular tracers (see §4.3).

Since 60 % of MS stars are binary or multiple systems, studying the distribution of material and its evolution around PMS multiple stars is of prime importance regarding the problem of planetary formation. Deeper surveys in the mm/submm but also in the NIR (see the UY Aur example) are needed to constrain properly the dust and gas distribution around binary stars not only in the CB disk but also in the CS region where surface density laws (hence the mass) are not yet properly determined.

6. From Class II to Class III: towards the MS phase

On one side, an important conclusion drawn by current mm arrays (§4) is that there are now more and more detected large CO disks around CTTs of a few 10^6 years. On the other side, CO disks have disappeared around stars of about $\sim 10^8$ years. Unfortunately linking both stages is not yet possible. There are, for instance, some detections of cold gas around presumably older TTs (Zuckerman et al. 1995) and a few upper limits obtained on young MS stars in the solar vicinity (*e.g.* Dent et al. 1995). Current radiotelescopes do not have the sensitivity required to allow the detection of the cold gas component (if any) in such objects.

In search for old PMS stars, one needs to mention the existence of the X-rays PMS stars. In the last few years, X-rays satellites have shown the existence of many low-mass stars (Montmerle et al. 1991) in the solar vicinity and in the nearest star forming regions. These stars seem to be old WTTs having a strong magnetic stellar activity. However their X-rays properties do not differ from those of the CTTs. It is important to point out that this activity may significantly affect the chemistry and the physics of the inner part ($\sim 10 - 20$ AU) of the circumstellar disk (Glassgold et al. 1998).

Studying older star forming regions like the Pleiades ($\sim 8 \cdot 10^7 - 10^8$ yr) or the Hyades ($\sim 60 \cdot 10^7$ yr) can help to understand the global evolution of disks. In such clusters, Bouvier et al. (1997b,c) have observed the surface rotation of stars having masses within the range $0.5 - 1.1\,M_{\odot}$. They show that the median lifetime for accretion disks is about $\sim 3 \cdot 10^6$ years. Moreover, at an age of $20 \cdot 10^6$ years, about ~10% of the stars they have observed are still surrounded by disks.

Looking at young MS stars in the solar vicinity, seems also a reasonable approach to understand the gas and dust evolution. In 1983, the discovery by IRAS of dust disks around some Vega-type stars (often called debris disks, because the dust is thought to be formed of the debris of larger size bodies, such as comets or asteroids) has opened a new field of investigation. For example, our current understanding of the β Pictoris disk (see Vidal-Madjar et al. 1998, for a detailed review) suggests that it is a young planetary system in its early phase, of age $\sim$ 100 Myr. According to Vidal-Madjar et al. (1998), the β Pic "dust" disk is understood today as a kind of "cometary" disk, a sort of "post-planetary" disk where there are good evidences of 1) cometary-like objects, 2) kilometer-size or more large bodies while 3) the existence of planets are indirectly inferred to explain many observed phenomena.

It is certainly too early to try a general scheme on the evolution of the gas and dust for sources between a few $10^7 - 10^8$ years. We have seen that current optical telescopes and mm/submm radiotelescopes do not allow to study the circumstellar medium at the scale of our solar system. Understanding the physics of the disks at the scale of the inner Solar System requires the next generation telescopes.

At optical wavelengths, telescopes like the VLT, Keck I and II, Gemini, Subaru or NGST would provide in a near future high angular maps of inner circumstellar disks and even in some cases visual detection of planets. The recent detection with the W.M. Keck telescope of the thermal infrared emission at 20.8 μm of the dust disk surrounding the A0 star HR4796 of age $\sim$ 10Myr (Koerner et al. 1998) is an example. This object opens interesting investigations because it might be less evolved than the debris disks encountered around Vega-type stars.

However, we have also seen that the understanding of planetary formation around low-mass stars requires other wavelength windows because in such disks, most of the material radiates at longer wavelengths, namely in the mm/submm and in Far Infrared. For example, recent bolometer observations performed with SCUBA (JCMT) at 0.8mm provide the first maps of the thermal dust emission around a few Vega-type stars (Waynes et al. 1998) or Solar-type star (ϵEri, Greaves et al. 1998). Furthermore, assuming that the collecting area will be large enough, the next generation of mm/submm arrays would provide images of the dust and gas distribution in disks at the scale of our solar system, bringing key information about the formation of stellar and planetary systems.

Acknowledgement I would like to acknowledge S.Guilloteau, F.Gueth and M.Simon for a careful reading of this manuscript and fruitful collaborations. Ph.André, J.Bouvier, L.Close, G.Duvert, J.Ferreira, E.Forestini, A.Lecavelier des Etangs, F.Malbet, F.Ménard, C.Roddier, F.Roddier, L.Prato, L.Siess, K.Stapelfeld and H.Wiesemeyer are also warmly acknowledged for providing figures and/or papers in advance of publication.

References

Adams F.C., Lada C.J., Shu F.H. (1987) *Astrophys.J*, **Vol. no. 312**, p. 788

Agladze N.I., et al. (1996) *Astrophys.J.*, **Vol. no. 462**, p. 1026

Aikawa J., et al. (1997) *Astrophys.J.*, **Vol. no. 486**, p. L51

Aikawa J., et al. (1996) *Astrophys.J.*, **Vol. no. 476**, p. 684

Akeson R.L., Carlstrom J.E. (1997) *Astrophys.J.*, **Vol. no. 491**, p. 254

André Ph., et al. (1994), in Montmerle Th., Lada Ch.J., Mirabel I.F. & Trân Thanh Vân J. (eds), *The Cold Universe*, Proc. XIIIth Moriond Astrophysics Meetings, p. 179

Artymowicz P., et al. (1991) *Astrophys.J.*, **Vol. no. 370**, p. 35

Artymowicz P., Lubow S. (1994) *Astrophys.J.*, **Vol. no. 421**, p. 651

Bacciotti F. (1997) (1997), in Reipurth B., Bertout C. (eds), *Herbig-Haro Flows and the Birth of Low Mass Stars*, Proc. IAU Symp. 182, p. 73.

Bachiller R. (1996) *Ann.Rev.Astron.Astrophys.*, **Vol. no. 34**, p. 111

Bachiller R., Pérez-Gutiérrez M. (1997) *Astrophys.J.*, **Vol. no. 487**, p. L93

Basri G., Bertout C. (1989) Astrophys.J., **Vol. no. 341**, p. 340

Bastien (1991), in Lada C.J., Kylafis N.D. (eds.), *The physics of star formation and early stellar evolution*, p. 709.

Beckwith S.V.W., et al. (1990) *Astron.J.*, **Vol. no. 99**, p. 924

Beckwith S.V.W., Sargent A.I. (1991) *Astrophys.J.*, **Vol. no. 381**, p. 250

Bertout C., Basri G., Bouvier J. (1988) *Astrophys.J.*, **Vol. no. 330**, p. 350

Blandford R.D., Payne D.G. (1982) *Mon.Not.Roy.Astron.Soc.*, **Vol. no. 199**, p. 883

Bontemps S., et al. (1996) *Astron.Astrophys.*, **Vol. no. 311**, p. 858

Bouvier J., Bertout C., (1989) *Astron.Astrophys.*, **Vol. no. 211**, p. 99

Bouvier J., et al. (1993) *Astron.Astrophys.*, **Vol. no. 272**, p. 176

Bouvier J., et al. (1997a) *Astron.Astrophys.*, **Vol. no. 318**, p. 495

Bouvier J., et al. (1997b) *Astron.Astrophys.*, **Vol. no. 323**, p. 139

Bouvier J., et al. (1997c) *Astron.Astrophys.*, **Vol. no. 326**, p. 1023

Burrows C.J., et al. (1996) *Astrophys.J.*, **Vol. no. 473**, p. 437

Cabrit S., et al. (1996) *Astron.Astrophys.*,**Vol. no. 305**, p. 525

Cabrit S., et al. (1998) Yun J., Liseau R., Eds, *"Star formation with ISO"*, Publisher ASP, **Vol. no. 132**, p. 326

Calvet N. (1997), in Reipurth B., Bertout C. (eds), *Herbig-Haro Flows and the Birth of Low Mass Stars*, Proc. IAU Symp. 182, p. 417.

Cantó J., Raga A.C (1991) *Astrophys.J.*, **Vol. no. 372**, p. 646

Casanova S., et al. (1995), *Astron.Astrophys.*, **Vol. no. 439**, p. 752

Cernicharo J. (1991) (1991), in Lada C.J., Kylafis N.D. (eds.), *The physics of star formation and early stellar evolution*, p. 287.
Chan K.L., Henriksen, R.N. (1980) *Astrophys.J.*, **Vol. no. 241**, p. 534
Close L., et al. (1997a) *Astrophys.J.*, **Vol. no. 478**, p. 768
Close L., et al. (1997b) *Astrophys.J.*, **Vol. no. 489**, p. 210
Close L., et al. (1998) *Astrophys.J.*, **Vol. no. 499**, p. 883
Corporon P., et al. (1998) *Astron.Astrophys.*, in press
Davis C.J., Eislöffel J. (1995) *Astron.Astrophys.*, **Vol. no. 300**, p. 851
Dent C., et al. (1995) *Mon.Not.Roy.Astron.Soc.*, **Vol. no. 277**, p. L25
Duquennoy M., & Mayor M. (1991) *Astron.Astrophys.*, **Vol. no. 248**, p. 485
Duvert G., et al. (1998) *Astron.Astrophys.*, **Vol. no. 332**, p. 867
Dutrey A., Guilloteau S., Simon M. (1994) *Astron.Astrophys.*, **Vol. no. 286**, p. 149
Dutrey A. (1996), Shaver P., Ed. in *ESO proceeding "Science with large mm arrays"* p. 235
Dutrey A., et al. (1996) *Astron.Astrophys.*, **Vol. no. 309**, p. 493
Dutrey A., Guilloteau S., Bachiller R., 1997, *Astron.Astrophys.*, **Vol. no. 325**, p. 475
Dutrey A., et al. (1998) *Astron.Astrophys.*, **Vol. no. 338**, p. L63.
Edwards S., et al. (1993) *Astron.J.*, **Vol. no. 106**, p. 372
Edwards S., et al. (1994) *Astron.J.*, **Vol. no. 108**, p. 1056
Edwards S. (1997), in Reipurth B., Bertout C. (eds), *Herbig-Haro Flows and the Birth of Low Mass Stars*, Proc. IAU Symp. 182, p. 433.
Eislöffel J., Mundt R. (1992) *Astron.Astrophys.*, **Vol. no. 263**, p. 292
Ferreira J., Pelletier G. (1995) *Astron.Astrophys.*, **Vol. no. 295**, p. 807
Ferreira J. (1997) *Astron.Astrophys.*, **Vol. no. 319**, p. 340
Ferreira J. (1998) in Ménard F., Cabrit S., Eds, Actes du GdR 1078 *"La connexion Accrétion-Ejection"*, p. 69
Galli D., Shu F. (1993a) *Astrophys.J.*, **Vol. no. 417**, p. 220
Galli D., Shu F. (1993b) *Astrophys.J.*, **Vol. no. 417**, p. 243
Genzel R. (1990) (1991), in Lada C.J., Kylafis N.D. (eds.), *The physics of star formation and early stellar evolution*, p. 155.
Ghez A., et al. (1993) *Astrophys.J.*, **Vol. no. 106**, p. 2005
Ghez A. (1995) *"Evolutionary Processes in Binary Stars"*, NATO ASI Series, **Vol. no. 477**, p. 1
Glassgold A.E., et al. (1997) *Astrophys.J.*, **Vol. no. 485**, p. 920
Gomez M., et al (1997) *Astron.J*, **Vol. no. 114**, p. 1138
Gueth F., Guilloteau S., Bachiller R. (1995) *Astron.Astrophys.*, **Vol. no. 307**, p. 891
Gueth F., et al. (1997) *Astron.Astrophys.*, **Vol. no. 323**, p. 943
Gueth F., Guilloteau S. (1998), *Astron.Astrophys.*, in press
Guilloteau S. et al. (1992) *Astron.Astrophys.*, **Vol. no. 265**, p. L49
Guilloteau S. et al. (1997), in Reipurth B., Bertout C. (eds), *Herbig-Haro Flows and the Birth of Low Mass Stars*, Proc. IAU Symp. 182, p. 365.
Guilloteau S., Dutrey A. (1998) *Astron.Astrophys.*, **Vol. no. 339**, p. 467
Guilloteau S., Dutrey A., Simon M., (1998), *Astron.Astrophys.*, submitted
Greaves J.S. et al. (1998), *Astrophys.J.Let.*, in press
Hartigan P., Raymond J., Hartmann L. (1987) *Astrophys.J.*, **Vol. no. 316**, p. 323
Hartigan P., et al. (1994) *Astrophys.J.*, **Vol. no. 427**, p. 961
Hartigan P., Edwards S., Ghandour L. (1995) *Astrophys.J.*, **Vol. no. 452**, p. 736
Hartmann L., Kenyon S.J. (1987) *Astrophys.J.*, **Vol. no. 312**, p. 243
Hartmann L., et al. (1994) *Astrophys.J.*, **Vol. no. 430**, p. L49
Hartmann L., Kenyon S.J. (1996) *Ann.Rev.Astron.Astrophys*, **Vol. no. 34**, p. 207
Hartmann L. (1997) (1997), in Reipurth B., Bertout C. (eds), *Herbig-Haro Flows and the Birth of Low Mass Stars*, Proc. IAU Symp. 182, p. 391.
Hartmann L. (1998) *"Accretion Processes in Star Formation"*, Cambridge University Press (Cambridge Astrophysics series; 32)
Henning T., Stognienko R. (1996) *Astron.Astrophys.*, **Vol. no. 311**, p. 291

Herbig G. (1974) *Lick Obs. Bull*, **Vol. no. 658**
Hildebrand (1983) *Royal Astron.Soc.Quart.Jrn.*, **Vol. no. 24**, p. 267
Ho P.T.P., et al. (1977) *Astrophys.J.*, **Vol. no. 215**, p. L29
Jensen E., et al. (1996a) *Astrophys.J.*, **Vol. no. 458**, p. 312
Jensen E., et al. (1996b) *Astron.J.*, **Vol. no. 111**, p. 243
Kastner J., et al. (1997) *Science*, **Vol. no. 277**, p. 61
Koerner D., et al. (1993) *Icarus*, **Vol. no. 106**, p. 2
Koerner D., et al. (1995) *Astrophys.J.*, **Vol. no. 452**, p. L69
Koerner D., et al. (1998) *Astrophys.J.*, Vol. no. 503, p. 83
Kenyon S.J., Hartmann L. (1987) *Astrophys.J* Vol. no. 323, p. 714
Lada E., et al. (1996) *Bull.Astro.Am.Soc.*, **Vol. no. 189**, p. 5301
Langer W., Velusamy T., Xie T. (1996) *Astrophys.J.*, **Vol. no. 468**, p. L41
Lavalley C., et al. (1997) *Astron.Astrophys.*, **Vol. no. 327**, p. 671
Lay O., et al. (1997) *Astrophys.J.*, **Vol. no. 489**, p. 917
Lazareff B., et al. (1990) *Astrophys.J.*, **Vol. no. 358**, p. 170
Leinert C., et al. (1993) *Astron.Astrophys.*, **Vol. no. 278**, p. 129
Looney L., et al. (1997) *Astrophys.J.*, **Vol. no. 484**, p. 157
Lynden-Bell, Pringle J. (1974) *Mon.Not.Roy.Astron.Soc.*, **Vol. no. 168**, p. 603
Mannings V., et al. (1994) *Mon.Not.Roy.Astron.Soc.*, **Vol. no. 267**, p. 361
Mannings V., et al. (1998) *Nature*, **Vol. no. 388**, p. 555
Mathieu R. (1994) *Ann.Rev.Astron.Astrophys.*, **Vol. no. 32**, p. 465
Mathieu R., et al.(1996) *Am.Astron.Soc.*, **Vol. no. 188**, p. 6005
McCaughrean M.J., et al. (1994) *Astrophys.J.*, **Vol. no. 436**, p. L189
Montmerle T. (1991) (1991), in Lada C.J., Kylafis N.D. (eds.), *The physics of star formation and early stellar evolution*, p. 675.
McCaughrean M.J., O'Dell C.R. (1996) *Astron.J.*, **Vol. no. 111**, p. 1977
Mouschovias T.Ch. (1991) (1991), in Lada C.J., Kylafis N.D. (eds.), *The physics of star formation and early stellar evolution*, p. 61.
Motte F., André Ph. & Neri, R. (1998) *Astron.Astrophys.*, **Vol. no. 336**, p. 150
Mundt R., et al. (1983) *Astrophys.J.*, **Vol. no. 265**, p. L71
Mundt R., Fried J. (1983) *Astrophys.J.*, **Vol. no. 274**, p. L83
Mundy L., et al. (1996) *Astrophys.J.*, **Vol. no. 464**, p. 169
Myers P.C., Benson P.J. (1983) *Astrophys.J.*, **Vol. no. 266**, p. 309
Natta A. (1993) *Astrophys.J.*, **Vol. no. 412**, p. 761
Ohashi N., et al.(1991) *Astron.J.*, **Vol. no. 102**, p. 2054
Ohashi N., Hayashi M., Ho P., Momose M. (1997) *Astrophys.J.*, **Vol. no. 475**, p. 211
Osterloh M., Beckwith S.W. (1995) *Astron.Astrophys.*, **Vol. no. 439**, p. 288
Prato L., Simon M., (1997) *Astrophys.J.*, **Vol. no. 450**, p. 455
Pollack J., et al. (1994) *Astrophys.J.*, **Vol. no. 421**, p. 615
Pudritz R.E., Pelletier G., Gomez de Castro A.I., (1991), in Lada C.J., Kylafis N.D. (eds.), *The physics of star formation and early stellar evolution*, p. 539.
Raga A.C., Kofman L. (1992) *Astrophys.J.* **Vol. no. 386**, p. 282
Raga A.C., Biro S. (1993) *Mont.Not.Roy.Astron.Soc.* **Vol. no. 264**, p. 758
Raga A.C., et al. (1993) *Astron.J.*, **Vol. no. 276**, p. 539
Raga A.C., Taylor S., Cabrit S., Biro S. (1995) *Astron.Astrophys.*, **Vol. no. 296**, p. 833
Reipurth B., et al. (1997) *Astron.Astrophys.*, **Vol. no. 114**, p. 757
Richichi A., et al. (1994) *Astron.Astrophys.*, **Vol. no. 287**, p. 145
Roddier C., et al. (1996) *Astrophys.J.*, **Vol. no. 463**, p. 326
Rodriguez L.F. (1997) (1997), in Reipurth B., Bertout C. (eds), *Herbig-Haro Flows and the Birth of Low Mass Stars*, Proc. IAU Symp. 182, p. 83.
Rodriguez L.F., et al. (1998) *Astrophys.J*, in press
Saito et al. (1997) *Astron.Astrophys.*, **Vol. no. 473**, p. 464
Saraceno P., et al. (1998) Yun J., Liseau R., Eds, *"Star formation with ISO"*, Publisher ASP, **Vol. no. 132**, p. 233
Shakura N.I., Sunyaev R.A. (1973) *Astron.Astrophys.*, **Vol. no. 24**, p. 337

Shu F.H. (1991), in Lada C.J., Kylafis N.D. (eds.), *The physics of star formation and early stellar evolution*, p. 365.
Shu F.H., et al. (1994) *Astrophys.J.*, **Vol. no. 429**, p. 781
Siess L. (1996) thèse de doctorat, Université de Grenoble.
Siess L., et al. (1998) *Astron.Astrophys.*, in press
Simon M., et al. (1995) *Astrophys.J.*, **Vol. no. 443**, p. 625
Simon M., Prato L. (1995) *Astrophys.J.*, **Vol. no. 450**, p. 824
Snell R.L., Loren R.B., Plambeck R.L. (1980) *Astrophys.J.*, **Vol. no. 239**, p. L17
Stahler S.W., Walter F.M (1993), Levy E. & Mathews M., Eds. *in Proto-Stars and Planets III*
Stapelfeldt K., et al. (1997), in Reipurth B., Bertout C. (eds), *Herbig-Haro Flows and the Birth of Low Mass Stars*, Proc. IAU Symp. 182, p. 355.
Stocke et al. (1988) *Astrophys.J.Sup.*, **Vol. no. 68**, p. 229
Van Dishoeck E.J. Blake G. A. (1998)*Ann.Rev.Astron.Astrophys.*, **Vol. 36**, in press
Vidal-Madjar A. et al. (1998)*Planet.Space.Sci.*, **Vol. 46**, p. 629
Whitney B. et al. (1992) *Astrophys.J.* **Vol. no. 395**, p. 471
Whitney B. et al. (1997) *Astrophys.J.*, **Vol. no. 485**, p. 703
Wayne S. et al. (1998) *Nature*, in press
Wright E. (1987) *Astrophys.J.*, **Vol. no. 320**, p. 818
Zuckerman B. et al. (1995) *Nature*, **Vol. no. 373**, p. 494

INFALLING MATERIALS ON YOUNG STARS

V.P.GRININ
Astronomical Institute of St. Petersburg University
198904, Petrodvorets St.Petersburg, Russia
Crimean Astrophysical Observatory
Crimea, 334413, Nauchny, Ukraine

Abstract. An accretion of circumstellar (CS) gas onto the young stars is well known phenomenon which accompanied the early stages of stellar evolution and plays an important role in the formation of the biconical outflows and highly collimated jets. In this review I summarise the observational evidence of alternative mechanism of the gas infall onto the stars through the dissipation of the circumstellar dust clouds and planetesimal bodies at their passages in the vicinity of the star. This, so-called "body infall scenario" plays an important role in the later stages of stellar evolution, but can also co-exist with the classical viscous accretion still at the pre-mail-sequence (PMS) phase.

1. Introduction

On the way to the Main Sequence many PMS stars accrete circumstellar matter from the remnants of the protostellar clouds. This process is described in the classical papers by Shakura & Sunjaev (1973) and Lynden-Bell & Pringle (1974) and has numerous observational evidence (Appenzeller & Mundt 1989; Kenyon & Hartmann (1987); Bertout, Basri & Bouvier 1988; Hartigan et al. 1989; Beckwith & Sargent 1993). A central point of this mechanism is the turbulent viscosity which redistributes the angular momentum between different fragments of the CS gas. The gas fragments having lower angular momentum rotate and move to the star. Others, having higher angular momentum move from the star. It is generally assumed that the gas and dust are well mixed and the accreting matter has the normal chemical composition. The gas accretion is frequently accompanied by the anisotropic outflows and highly collimated jets (Mundt and Ray

J.-M. Mariotti and D. Alloin (eds.), Planets Outside the Solar System: Theory and Observations, 51–63.

1994). Theoretical consideration shows that such outflows are the result of interaction of the accreting gas with stellar and CS magnetic fields.

The duration of this phase is of the order of time scale for the circumstellar dust coagulation and the formation of planetary systems. This time scale for the terrestrial planets is about 10^6 yr (Safronov 1969) and for the giant planets like Jupiter is about 10^7 yr (Pollack et al. 1996). According to the estimations by Strom et al. (1993) the life time of the CS disks around PMS stars has the same order. After this period the accretion of CS gas through the disk should become negligible.

However, the observations show that the gas infall onto the stars can continue still at later stages of the life of the star through so-called "body-infall scenario". This mechanism has been first suggested by Lagrange et al. (1988) for the star β Pictoris, a young main-sequence star of spectral type A5.

2. Comet-like activity around β Pictoris

The observational basis for this scenario was the discovery of variable (on time scale of days and hours), red-shifted absorption components in the Ca II resonance lines in the spectrum of this star. Those components were attributed to gas evaporated from comet-like bodies moving in the vicinity of β Pic. Later similar absorption components have been observed in the other metalic lines in the ultraviolet part of its spectrum. The comprehansive reviews of the observations and theoretical considerations of the β Pic phenomenon are given in the proceedings of the 10th IAP Conference "Circumstellar Dust Disks and Planet Formation". Here we note only that the main condition for realization of the "body infall scenario" is the formation of highly eccentric, so-called star-grazing orbits. Moving along such orbits the comet-like bodies can reach the nearest vicinity of the star and dissipate fully or partially due to sublimation, tidal forces etc. The formation of such orbits is a result of the gravitational perturbations produced by planets or other massive bodies in the neighbour- hood of the star.

Recently it was shown that the variable red-shifted absorption components very similar to those observed in the spectrum of β Pic are frequently observed in the UX Ori type stars (see review paper by Perez & Grady 1997 and Figure 1). This and some other observational findings (see below) were the base for suggestion that the β Pic phenomenon can be observed even in the PMS phase of stellar evolution.

Before to go to this topic let us consider briefly what are the UX Ori type stars?

3. UX Ori type stars as precursors of β Pictoris

The UX Ori type stars (or UXORs) belong to a small population of the young stars (predominantly Herbig Ae/Be type) having unusual photometric and polarimetric properties. The main part of the time these stars are usually in the bright state and have sporadic Algol-like minima with amplitudes ΔV up to 2-3 stellar magnitudes and duration from few days to few weeks (Thé 1994; Eaton and Herbst 1995). Simultaneous photometric and polarimetric observations show that linear polarization synchronously increases when the star fades (Grinin et al. 1991; Grinin 1994; Figure 1).

3.1. CORONAGRAPHIC EFFECT OF CS DUST CLOUDS

Such a behaviour of UXORs is explained by the model of variable CS extinction suggested by Wenzel in 1969 and modified by author of this lecture in 1986. In the framework of the modified model the Algol-type minima are the result of the variable screening of the star by optically thick dust clumps intersecting the line of sight from time to time. At the moment of the total (or almost total) eclipse we observe the weak scattered radiation of the protoplanetary disk which can be considered as the analogy of the interplanetary zodiacal light. This explains the observed anti-correlation between brightness and linear polarization and also the color variations in deep minima (Grinin 1986).

¿From this point of view the stars with the Algol-like minima are somewhat similar to β Pic: in both cases, the star should be screened in order one could observe CS disk. In the case of β Pictoris CS disk is spatially resolved and special coronagraphic technique of observations can be used to this purpose (Smith and Terrile 1984). In the case of UXORs we deal with unresolved with telescope CS disks and we have to wait for the moment when the star will be screened by opaque dust cloud.

3.2. EVIDENCE FOR THE EDGE-ON ORIENTATION OF CS DISKS

The high linear polarization (up to 5-7%) systematically observed in deep minima was the first important indication that the CS dust envelopes are flattened and the line of sight is close to their equatorial plane (Grinin et al. 1991). Later the additional arguments in the favour of such orientation of UXORs and their surroundings were found. They are:

a) the edge-on orientation of the CS gaseous envelopes in which the H_α emission line is formed and which are coplanar to the dust disks (Grinin 1992; Grinin & Rostopchina 1996);

b) systematically larger rotation velocities of these stars in the comparison with the photometrically quiet stars of similar spectral types (Grady

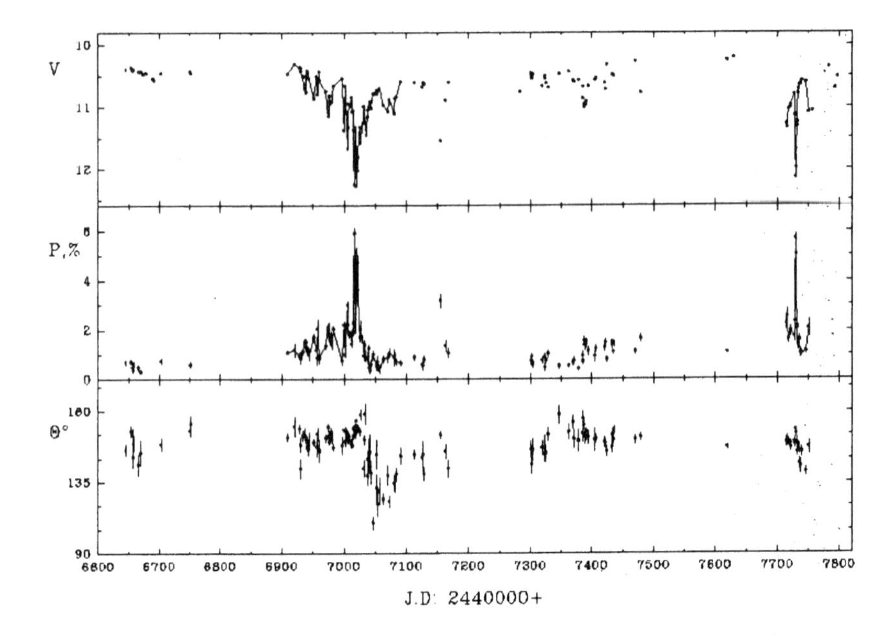

Figure 1. Coronagraphic effect produced by circumstellar dust clouds moving in the neighbourhoods of WW Vul: at the moment of eclipse the highly polarized scattered radiation of the protoplanetary disk dominates. Data are taken from Grinin et al. (1988) and Berdjugin et al. (1992)

et al. 1996);

c) absence of any dependence of the CS dust mass (estimated from mm-observations) on the amplitudes of the optical variability of Herbig Ae stars (Natta et al. 1997). The last result gives the most important evidence that UXORs do not differ principally from other PMS stars and their activity is coused by their "optimal" orientation in space. The high frequency of UXORs (about 25-30% of all Herbig Ae stars) argues for a moderately flattened distribution of the obscuring dust matter.

4. Comet-like activity in the neighbourhood of UXORs

Similarity of the orientations of CS disks around UXORs and β Pic relatively to an observer stimulated searches of the spectroscopic signatures of the gas infall onto UXORs and such signatures were found in the different lines of metals both in the visual part of the spectrum (Graham 1992; Grinin et al. 1994; de Winter 1996) and in UV (Pérez et al. 1993; Grady et al. 1996) (see review by Pérez & Grady 1997 for more details). Some examples of the red-shifted absorption components in the D Na I resonance

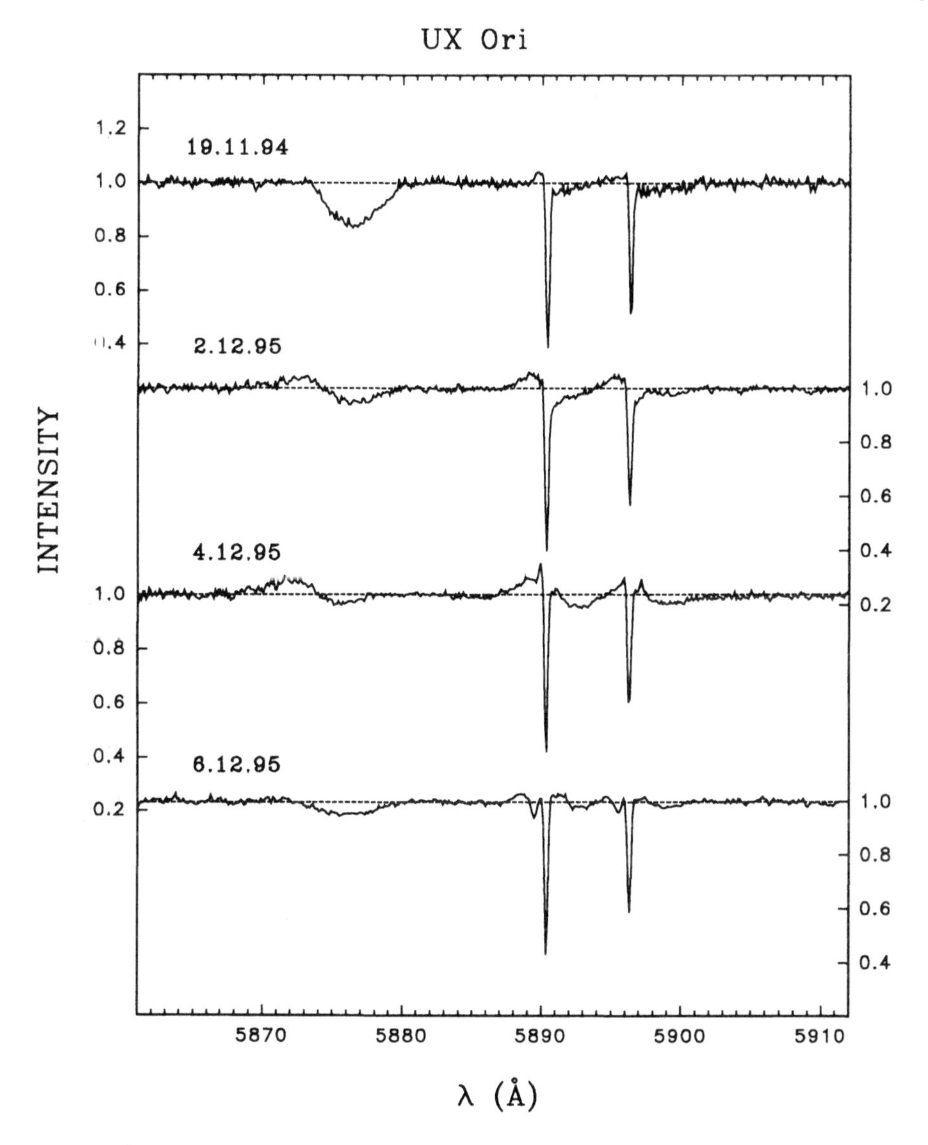

Figure 2. The examples of the variable red-shifted absorption components in the sodium D Na I resonance lines in the spectrum of UX Ori observed with the Nordic Optical Telescope (Grinin et al. 1998). The blue-shifted absorption componets are also observed sometimes.

lines are given in Figure 2.

We see that these components are highly variable and sometime disappear almost fully. Their maximal radial velocities are about 300 km/s that coincides practically with the maximal infall velocities in the case of the β Pic. However, in the case of UX Ori type star the CS disks are much younger than in β Pic and their gas component did not still dissipate. The presence of the H_α emission in the their spectra (Figure 3) indicates clearly that the classical mechanism of the disk accretion can still work. The numerical modeling shows (Sorelli et al. 1996) that the red-shifted absorption componets in the metalic lines can be also explained in the framework of the magnetospheric accretion model. This means that existence of the red-shifted absorption components in the spectra of UXORs cannot be considered as a simple evidence for the comet-like activity. The main indicators of such activity follow from photometric, polarimetric and infrared observations.

4.1. THE DUST CLOUDS IN THE VICINITY OF HAEBE STARS

An important indirect argument in the favour of the comet-like activity is the observational evidence of the nearest to the star passages of the dust clouds which produce the Algol-like variability and related phenomena (Grinin 1992,1994; Sitko et al. 1994).

The observations show (Grinin et al. 1988) that the large-amplitude brightness variations of UXORs are sometimes observed on the time scale of one day that corresponds to the tangential velocity of the dust cloud intersecting the line of sight about 30-40 km/s. In the case of the circular Keplerian orbit it gives the distanse of the order 1.5 - 2 AU from the star with 3 $M_\odot$. In the case of the highly eccentric orbit this distance should be even smaller. The appearance of such clouds in the vicinity of fairly hot and luminous star is the non-trivial event.

The point is that the characteristic sizes of the dust particles inside CS clouds should be quite small (of order 0.1 μm) in order to explain the reddening of UXORs observed at the initial phase of Algol-like minima. Such particles should be accelerated and sweped out by the radiative force. Therefore in order to reach the stellar vicinity *the dust clouds have to include the dense, optically thick central condensations or nuclei, or to be the part of the gaseous stream moving to the star.* In the last case the gas drag can compensate the radiative force on the dust clouds since the gas component of the CS matter is much more massive then the dust one (the "standart" ratio $M_{gas}/M_{dust} = 100$).

Appearance of the dust clouds near to the star can explain several important observational facts which it is difficult to explain by the other way.

1). One of them is the strong fluctuations of the Stokes parameters of scattering radiation of UXORs on the time scale of about 1-2 weeks which cannot be explain by the simple coronagraphic effect (Grinin 1994; Rostopchina et al. 1998). The dissipation of the dust clouds produced by the comet-like objects in the neighbourhood of the UX Ori type stars due to the termal stress, sublimation and tidal forces seems to be the most favorable explanation of this observational fact.

2). Similar effect explains the variability of the near IR fluxes (up to 10 μm) which has been observed in some UXORs independently by Sitko et al. (1994) and Hutchinson et al. (1994). An important detail of the IR variability was that it has been observed in the anti-correlation with the optical brightness changes: when the star faded in visual region its IR luminosity increased. According to Sitko et al. such variations can be explain by disruption of the large bodies, such as comets, in front of the star. If this idea is correct, the shortest time scale of such variability (about one week: Hutchinson et al. 1994) indicates on the existence of star-grazing comets or planetesimal bodies. Dissipating in the vicinity of the star they screen it from an observer. At the same time the nearest to the star and therefore hottest dust associated with such a body can be strong IR emitter.

3). An interesting argument in the favour of the comet-like activity around young stars was suggested by Baade & Stahl (1989). Investigating the photometric activity of the UX Ori type star HR 5999 they have found the small-amplitude brightness variations with the periods P – 48 days and found that the periodic dissipation of the giant comet-like body near the periastron of its orbit is the most plausible explanation among others. Similar idea was used later by Shevchenko et al. (1993) for explanation of the wave-like variability of several other HAEBE stars.

4). An important evidence for similarity of the CS dust and the dust coma of the Hale-Bopp comet were obtained with ISO by Malfait et al. (1998) for the star HD 100546. According to Grady et al. (1997) this star has also the spectroscopic signatures of the gas infall and seen edge-on.

5. What are the "protocomets" and how do they reach the inner parts of the protoplanetary disks?

According to the classical scenario of the planet formation (Safronov 1969) the planetesimal bodies are formed due to coagulation of the dust particles and their initial keplerian orbits are almost circular. In order to reach the nearest neibourhood of the star the dust clumps or planetesimals have to move along highly eccentric orbits. Formation of such orbits is possible only if the gravitational perturbations produced by protoplanets or secondary components of the young binary system exist.

The analysis of the long-term activity of UXORs (see below) shows that many of them have probably companions or protoplanets.

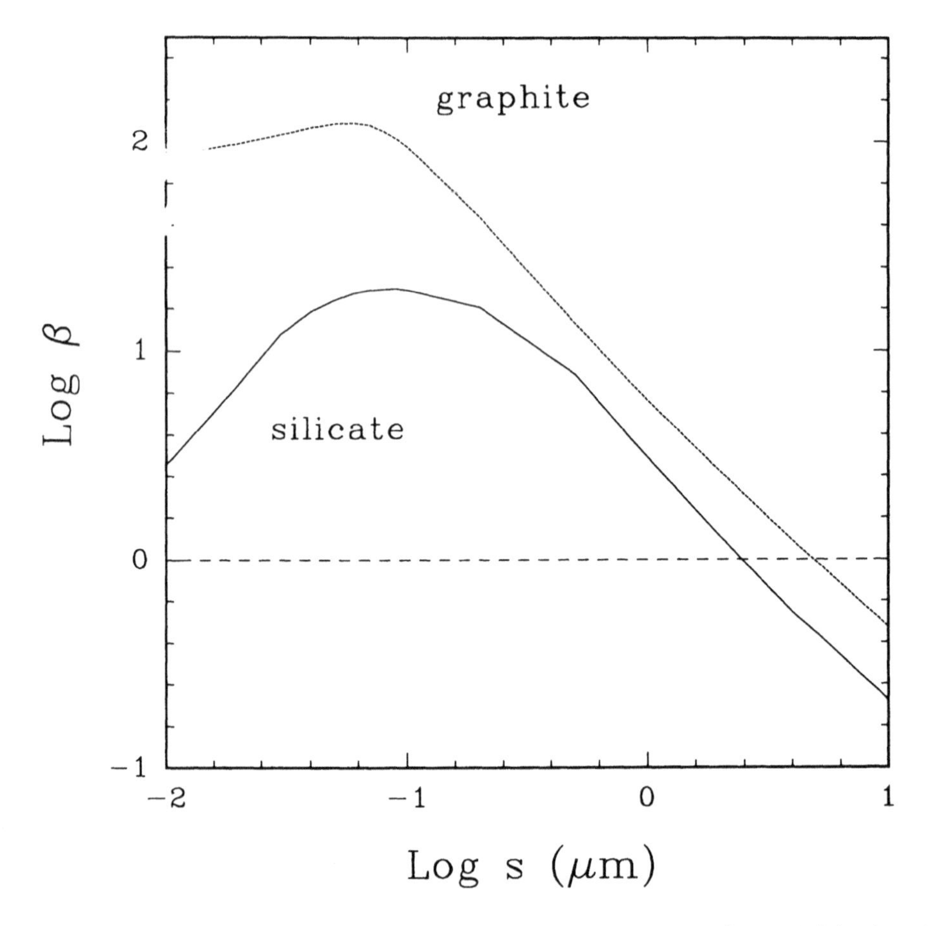

Figure 3. The ratio of radiative force to the gravity for silicate (the solid line) and graphite (dots) for the different grain sizes.

5.1. LONG-TERM CYCLIC VARIABILITY OF UXORS AS INDICATOR OF THEIR BINARITY

As we noted above the UX Ori type stars are known among variables as the stars with non-periodic Algol-type minima. Many attempts have been made in order to find a periodicity in the minima, but none have been confirmed by observations. At the same time many observers noted that the probability to observe the minima can change from year to year and in some cases the hints on the periodicity were found (Shevchenko et al. 1993; Friedemann et al. 1993).

In the recent papers by Grinin et al. (1998), Rostopchina et al. (1998) and Shakhovskoy et al. (1998) we have analysed our long-term photoelectric observations of several UXORs and the published results of the other authors. We have not confirmed many periods estimated earlier. At the same

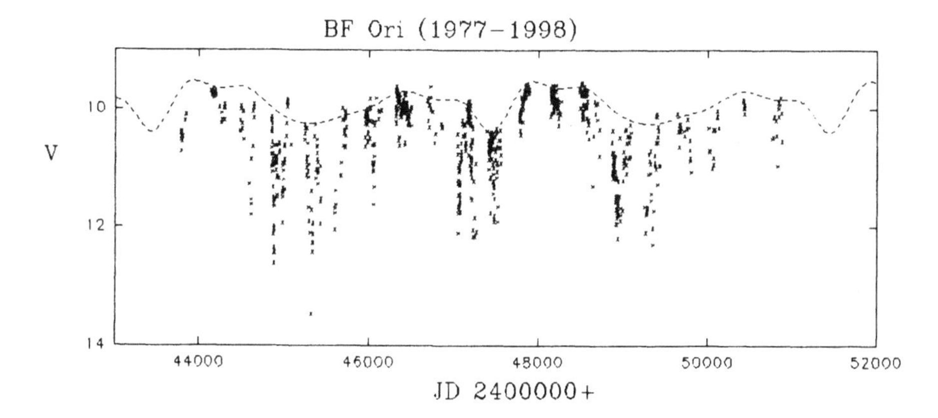

Figure 4. An example of the cyclic variability of CO Ori from the paper by Grinin et al. (1998)

time we have found that the long-term cyclic activity of UXORs really exists. In two cases (BF Ori and CO Ori) there are indications of the two-component structure of cycles (Figures 5 and 6). Their duration is about 11 yrs for BF Ori and 11.2 yrs for CO Ori indicating the presence of the companions at the distances of 6.6 AU and 7.2 AU from the stars correspondingly. In the case of UX Ori itself the duration of the cycle is about 6.6 yrs (Rostopchina et al. 1998). It has the one-component structure and there is indication that the secondary component can be the protoplanet.

5.2. ORIGIN OF CYCLES

It is well known that existence of a companion in the neighbourhood of the young star can strongly influence the disk evolution. The tidal forces create gaps in the inner part of the circumbinary (CB) disks (Lin & Papaloizou 1979) and can disrupt disks whose sizes are comparable with the binary separation. The recent numerical simulations by Artimowicz & Lubow (1996) show that the streams of CS matter from CB disk have to be formed. In general case two unequal streams exist (depending on the mass ratio of the components). They penetrate the disk gap and supply an accretion activity around the components of the system.

The presence of dust clumps in such streams can naturally explain not only the long-term cyclic activity of UXORs, but also the important property of such cycles as existence of the two-component and one-component cycles. According to Artimowicz & Lubow one stream of CS matter from CB disk exists in the limiting case of the low-mass component, perhaps the protoplanet.

Moving within the gas stream the dust clumps evaporate when approa-

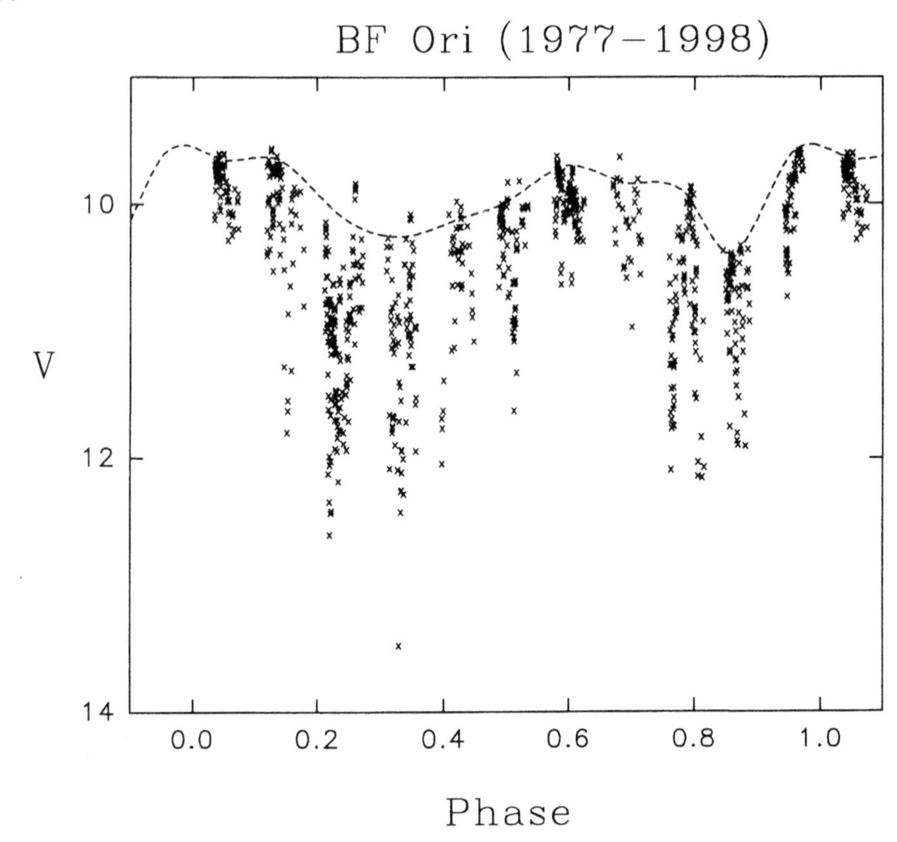

Figure 5. The phase curve of CO Ori for the period 11.2 yrs from the paper by Grinin et al. (1998).

ching the star. As a result such a steam will be locally enriched by heavy elements. The fluctuations of the metal abundances on the line of sight can explain the observed variability of the red-shifted absorption components in the spectra of UX Ori stars. In this case the chemical composition of the accreting gas will depend on the ratio of the gas to dust in the streams and can deviate from the normal one.

Another possibility to explain the "β Pictoris phenomenon" among young stars is the capture of the planetesimal bodies from CB disk by secondary component or a protoplanet. Existence of such bodies on the "inner" orbits can explain the appearance of the short living low aplitude waves in the light curves of the UXORs similar to those found by Baade & Stahl (1989).

6. Conclusion

We have summarized above the different observational evidence for the the comet-like activity in the neighbourhoods of PMS stars. They are not too numerous, but they exist. The most plausible mechanism of the "capture" of the dust clumps and their transport to the nearest vicinity of the star is the formation of the streams of gas and dust from circumbinary disk due to the tidal perturbations in the young binary (or planetary) system. This mechanism can be considered as the precursor of the well known mechanism: the gravitational "capture" of the comets from the Oort cloud by giant planets in our Solar System.

Aknowledgement. I am thankful to Antonella Natta and Malcolm Walmsley for very usefull discussion and comments.

References

1. Appenzeller I, and Mundt, R. (1989), T Tauri Stars, *Astron. Astrophys. Rev.*, **1**, 291-334
2. Artimovicz, P. and Lubow, S.H. (1996), Mass flow through gaps in circumbinary disks, *Astrophys. J.*, **467**, L77-L80
3. Baade, D. & Stahl, O. (1989), New Aspects of the Variability of the Probable Pre-Main-Sequence Star HR 5999, *Astron. Astrophys.*, **209**, 255-268
4. Beckwith S., & Sargent, A.I. (1993), The Occurrence and Properties of Disks around Young Stars, in *Protostars and Planets III*, eds. E.H.Levy and J.I.Lunine, Tucson, Univ. of Arizona, pp. 521-541
5. Berdjugin, A.V., Grinin, V.P., and Minikhulov, N.Kh. (1992), Results of Synchronous Photometric and Polarimetric Observations of WW Vul, *Bull. Crimean Astrophys. Obs.*, **86**, 69-82
6. Bertout, C., Basri, G., and Bouvier, J. (1988), Accretion Disks around T Tauri Stars, *Astrophys. J.*, **330**, 350-373
7. de Winter, D. (1996), *Observational Aspects of Herbig Ae/Be stars and Candidate Young A and B Stars*, Ph. D. Thesis, University of Amsterdam,
8. Friedemann, C., Gurtler, J. Reimann, H.-G., and Toth, V. (1993), The Cloudy Circumstellar Dust Shell of WW Vul Revisited, *Astron. Astrophys.*, **277**, 184-191
9. Grady C., Sitko M.L., Bjorkman, K.S. Pérez M.R., et al. (1997), The Star-Grazing Comets in the HD 100546, *Astrophys. J.*, **483**, pp. 449-456
10. Graham, J.A., (1992), Clumpy Accretion onto Pre-Main-Sequence Stars, *PASP*, **104**, 479-488
11. Grady, C., Pérez, M.R. Talavera, A. et al. (1996), The β Phenomenon among Young Stars. UV and optical high dispersion spectra, it Astron. Astrophys. Suppl. S., **120**, 157-177
12. Grady, C.A., Sitko, M.L., Bjorkman K.S., Pérez, M.R. (1997), The Star-Grazing Extrasolar Comets in the HD 100546 System, *Astrophys. J.*, **483**, 449-456
13. Grinin, V.P. (1986), On the Nature of Blue Emission Visible during Deep Minima of Young Irregular Variables, Unpublished poster paper presented at IAU Symp. 122 "*Circumstellar Matter*", see *Soviet Astron. Lett.*, (1988), **14**, pp. 27-28
14. Grinin, V.P. (1992), Young Stars with Non-Periodic Algol-Type Minima, *Astron. Astrophys. Trans.*, **3**, 17-32
15. Grinin, V.P., Kiselev, N.N., Minikhulov N.Kh., and Chernova G.P. (1988), Observations of Linear Polarization in Deep Minima of WW Vulpeculae, *Sov. Astron. Lett.*, **14**, 219-223

16. Grinin, V.P., Kiselev, N.N., Minikhulov N.Kh., Chernova G.P. and Voshchinnikov N.V. (1991), The Investigations of "Zodiacal Light" of Isolated Ae-Herbig Stars with Non-Periodic Algol-type Minima, *Astrophys. Space Sci.*, **186**, 283-298
17. Grinin, V.P. (1994), Polarimetric Activity of Herbig Ae/Be Stars, in *The Nature and Evolutionary Status of Herbig Ae/Be Stars*, eds. by P.S.Thé, M.R.Pérez, and P.J. van den Heuvel, PASPC 62, 63-70
18. Grinin, V.P., Thé, P.S., de Winter D., et al. (1994), The β Pictoris Phenomenon among Young Stars. I. The Case of Ae Herbig Star UX Ori, *Astron. Astrophys.*, **292**, 165-174
19. Grinin, V.P. and Rostopchina A.N. (1996), Orientation of Circumstellar Disks and the Statistics of H_α Line Profiles of Ae/Be Herbig Stars, *Astron. Rep.* **40**, 171-178
20. Grinin, V.P., Shakhovskoy, D.N., and Rostopchina, A.N. (1998), On the Nature of Cyclic Variability of UX Ori type Stars, *Letters to Astron. J.*, in press
21. Grinin, V.P., Natta, A., Kozlova, O.V., Il'in I., Tuominen, I., Thé, P.S., Rostopchina, A.N., Shakhovskoy, D.N. (1998) Non-Stationary Gas Accretion onto UX Ori, in preparation
22. Eaton, N.L. and Herbst, W. (1995), An Ultraviolet and Optical Study of Accreting Pre-Main-Sequence Stars: UXORs, *Astron. J*, **110**, 2369-2377
23. Hutchinson, M.G., Albinson, J.S., Barrett, P. Davies, J.K., Evans, A., Goldsmith, M.J., and Maddison, R.C. (1994), Photometry and Polarimetry of Pre-Main-Sequence Stars, *Astron. Astrophys.*, **285**, 883-896
24. Kenyon, S.J. and Hartmann, L. (1987), Spectral Energy Distributions of T Tauri Stars: Disk Flaring and Limits on Accretion, *Astrophys. J*, **323**, 714-733
25. Lagrange-Henry, A.-M., Vidal-Madjar, A., and Ferlet, R. (1988), The β Pictoris Circumstellar Disk. VI. Evidence for material falling onto the star. *Astron. Astrophys.*, **190**,275-282
26. Lin, D.C., and Papaloizou, J.C.B. (1979), Tidal Torques on Accretion Disks in Binary Systems with Extreme Mass Ratio, *Mon. Not. Roy. Astron. Soc*, **186**, 799-812
27. Lynden-Bell D., and Pringle, J.E. (1974), The Evolution of Viscous Disks and the Origin of the Nebulae Variables, *MNRAS*, **168**, 603-637
28. Hartigan, P., Hartmann, L., Kenyon, S.J., et al. (1989), How to unveil a T Tauri Star? *Astrophys. J. Suppl.*, **70** 899-914
29. Mundt R., and Ray, T. (1994), Optical Outflows from Herbig Ae/Be Stars and Other High Luminosity Young Stellar Objects, in *The Nature and Evolutionary Status of Herbig Ae/Be Stars*, eds. by P.S.Thé, M.R.Pérez, and P.J. van den Heuvel, PASPC 62, 237-252
30. Malfait, K., Waelkens C., Waters, L.B.F.M., Vandenbussche, B. Huygen, E. and de Graauw, M.S. (1998), The Spectrum of the Young Star HD 100546, Observed with the Infrared Space Observatory, **Astron. Astrophys. 1998, L25-L28.**
31. Natta, A., Grinin, V.P., Mannings, V. and Ungerechts H. (1997), On the Evolutionary Status of UX Ori type Stars, *Astrophys. J*, **491**, 885-894
32. Pérez, M.R., Grady, C.A., and Thé, P.S. (1993), UV Spectral Variability in Herbig Ae Star HR 5999 XI. The Accretion Interpretation, *Astron. Astrophys.*, **274**, 381-390
33. Pérez, M.R., and Grady, C. (1997), Observational Overview of Young Intermediate-Mass Objects: Herbig Ae/Be Stars, *Space Science Rev.*, **82**, 407-450
34. Pollack J.B., Hubickyj, O., Bodenheimer, P., Lissauer, J.J., Podolak, M., and Greenzweig, Y. (1996), Formation of the Giant Planets by Concurent Accretion of Solids and Gas, *Icarus*, **124**, 62-85
35. Rostopchina, A.N. (1998), The Position of UX Ori type star on the H-R diagram, *Astron. Rep.*, in press.
36. Rostopchina, A.N., Grinin, V.P., and Shakhovskoy, D.N. (1998) Cyclic Variability of UX Ori Type Stars SV Cep, WW Vul and UX Ori, *Letters in Astron. J.*, in press
37. Safronov, V.S. 1969, *Evolution of the Protoplanetary Cloud and Formation of the Earth and Planets*, Moscow, Nauka

38. Shakhovskoy, D.N., Rostopchina A.N. and Grinin V.P. (1998), Cyclic variability of UX Ori type stars VX Cas, RZ Psc and CQ Tau, in preparation
39. Shakura, N.I., and Sunjaev, R.A. (1973), Black holes in binary systems: Observational appearance. *Astron. Astrophys*, **24**, 337-355
40. Shevchenko V.S., Grankin K.N., Ibragimov, M.A. et al. (1993), Periodic Phenomena in Ae/Be Herbig Stars Light Curves, *Astrophys. Space Sci.*, **202**, 121-154
41. Sitko, M.L., Halbedel, E.M., Lawrence G.F., Smith, J.A., and Yanow, K. (1994), Variable Extinction in HD 45677 and the Evolution of Dust Grains in Pre-Main-Sequence Disks, *Astrophys. J.*, 432, 753-762
42. Sorelli, C., Grinin, V.P. and Natta, A. (1996), Infall in Herbig Ae/Be stars: what Na D lines tell us, *Astron. Astrophys.* **309**, 155-162
43. Smith, B.A. and Terrile, R.J. (1984), A Circumstellar Disk around β Pictoris: Recent Optical Observations, *Science*, **226**, 1421-1424
44. Strom, S.E., Edwards, S., and Skrutskie, M.F. (1993), Evolutionary Time Scale for Circumstellar Disks Associated with Intermediate and Solar-type Stars, in *Protostars and Planets III*, eds. E.H.Levy and J.I.Lunine, Univ. of Arizona Press, Tucson, 837-866
45. Thé, P.S. (1994), Photometric Behaviour of Herbig Ae/Be Stars and its Interpretation, in *The Nature and Evolutionary Status of Herbig Ae/Be Stars*, eds. by P.S.Thé, M.R.Pérez, and P.J. van den Heuvel, PASPC 62, 37-29
46. Wenzel, W. (1969), in *Non-Periodic Phenomena in Variable Stars*, IAU Coll., Ed. by L. Detre, Budapest, 61-70

PROTOSTELLAR DISCS AND PLANET FORMATION

J. C. B. PAPALOIZOU
Astronomy Unit, Queen Mary and Westfield College
Mile End Road, London E14NS, UK

C. TERQUEM
Lick Observatory, University of California
Santa Cruz, CA 95064, USA

AND

R. P. NELSON
Astronomy Unit, Queen Mary and Westfield College
Mile End Road, London E14NS, UK

1. Introduction

The planets in the solar system are for the most part in near circular orbits which approximately lie in the same plane. The hypothesis that they were formed in a flattened differentially rotating gaseous disc was originally proposed by Laplace (1796) to account for the origin of these dynamical properties and it has been the subject of much theoretical development in recent times (e.g. Lin and Papaloizou, 1985, 1993 and references therein).

2. Rotating Discs

When discussing the large scale physical properties of such a disc, it is convenient to make the idealization that it is infinitesimally thin or completely flat. A gaseous element at radius r rotates in circular orbit with angular velocity Ω. When, as in the present solar system, most of the mass lies in a central star of mass M_*, the local gravitational acceleration $g = GM_*/r^2$. The angular velocity is , to a good approximation, obtained from centrifugal balance by equating $g = r\Omega^2$. This gives Kepler's law

$$\Omega^2 = \frac{GM_*}{r^3}. \tag{1}$$

J.-M. Mariotti and D. Alloin (eds.), Planets Outside the Solar System: Theory and Observations, 65–85.

To a first approximation forces other than that due to the gravity of the central star may be neglected. Then material rotates in circular orbit conserving its angular momentum, just as in the case of an isolated planet. The angular momentum per unit mass, or the specific angular momentum, $j = r^2\Omega$, changes only slowly under viscous or other perturbative forces.

2.1. OBSERVATIONS OF PROTOSTELLAR DISCS

A well observed protostellar gaseous disc is that around HL Tau. The ^{13}CO map of HL Tau shows an elongated image extending to 2,000 AU (Sargent and Beckwith, 1987, 1991). The velocity field on a scale of $\sim 10^3$ AU is compatible with that of a slowly rotating and mainly collapsing flattened structure in which there is a mass inflow rate of $\dot{M} \sim 5 \times 10^{-6}$ $M_\odot$ yr^{-1} (Hayashi *et al.*, 1993). This indicates a young object with an age of about 10^5 yr, being the time required for a typical stellar mass to flow to the centre. The specific angular momentum carried by the infalling gas is adequate to cause the formation of a centrifugally supported disc with a size comparable to that expected for the primordial solar nebula. The inner disc at $\sim$ 50–100 AU from the centre is resolved in submillimeter and millimeter interferometer maps (Lay *et al.*, 1994; Koerner and Sargent, 1995; Mundy *et al.*, 1996).

Between 25 and 75 percent of young stellar objects in the Orion nebula appear to have discs (McCaughrean and Stauffer, 1994). Masses of $\sim 10^{-2\pm1}$ $M_\odot$, and dimensions $\sim 40 \pm 20$ AU have been estimated (Beckwith and Sargent, 1996).

The presence of discs on the scale of astronomical units has been inferred from the infrared excesses observed in many, $\sim$ 50% of T Tauri stars. The colors and magnitudes of these stars are best fitted by those expected for young pre–main sequence stars with ages $\sim 10^6$ yr (Strom *et al.*, 1993). The infrared emission may be produced by the liberation of the gravitational potential energy of matter flowing inwards at a rate $\dot{M} \sim 10^{-8\pm1}$ $M_\odot$ yr^{-1}. The non observation of discs around older T Tauri stars together with these values of $\dot{M}$ suggests a disc lifetime of $\sim 10^7$ yr.

2.2. DISC FORMATION

It is believed that protostars and protoplanetary discs originate from interstellar matter contained in molecular clouds. Observations (Goodman *et al.*, 1993) indicate that typical star–forming dense cores in dark molecular clouds have specific angular momentum $j > 6 \times 10^{20}$ cm^2 s^{-1}. When these clouds undergo gravitational collapse, because the collapse is fast, j is initially approximately conserved. Gas in the outer parts is thus unable to fall directly to the centre ($r = 0$) if j is not zero.

To obtain an estimate of the initial size of the disc, we consider the idealized situation when the pre–collapse cloud is a cold rotating sphere of mass M and radius R, with a single axis of rotation. Matter located on the rotation axis has no angular momentum so it can fall directly to the centre. An estimate of the time required, t_{ff}, is given by the time required to free fall from rest through a distance R under the initial inward gravitational acceleration at the surface, $g = GM/R^2$. This gives $t_{ff} = \sqrt{2R/g} = \sqrt{2R^3/(GM)}$. This may be expressed in terms of the initial mean density $\overline{\rho}$ of the cloud as $t_{ff} = \sqrt{3/(2\pi G\overline{\rho})}$. For molecular cloud cores with $M = 1\ M_\odot$ and $R = 0.1$ pc, this gives typically $t_{ff} \sim 6 \times 10^5$ yr.
It is clear that, while matter located on the rotation axis can move unimpeded to the centre, matter in the equatorial plane at the surface of the sphere will be at the outermost radius R_d of a disc after collapse. We can find R_d from the conservation of specific angular momentum. Assuming the total mass, M, to be concentrated at the centre, the angular velocity at the outer edge of the disc will be given by Kepler's law such that

$$\Omega^2 = \frac{j^2}{R_d^4} = \frac{GM}{R_d^3}. \tag{2}$$

Thus R_d is given in terms of the conserved quantities j and M by $R_d = j^2/(GM)$. Adopting $j = 10^{21}$ cm^2 s^{-1} and $M = 1\ M_\odot$, we find that $R_d \sim$ 500 AU. The characteristic dimension of such a disc is about an order of magnitude larger than our present solar system but is similar to those of protostellar discs now being observed by direct imaging.

3. Evolution of protostellar discs

The formation of a protostellar disc through the collapse of a molecular cloud core takes a time of 10^5–10^6 yr. During the early stages when the disc is still embedded (class 0/1 object) and has a significant mass compared to the central star, there may exist strong disc winds and bipolar outflows (e.g. Reipurth *et al.*, 1997) these being associated with magnetic fields. During this stage a hydromagnetic disc wind may be an important means of angular momentum removal for the system (see Papaloizou and Lin, 1995, and references therein).

When the mass of the disc is significant compared to that of the star, there may be a short period ($\sim 10^5$ yr) of non axisymmetric global gravitational instability with associated outward angular momentum transport (Papaloizou and Savonije, 1991; Heemskerk *et al.*, 1992; Laughlin and Bodenheimer, 1994; Pickett *et al.*, 1998) that enables additional mass growth of the central star. This redistribution may occur on the dynamical timescale (a few orbits) of the outer part of the disc and so may be quite rapid, on the order of 10^5 yr for $R = 500$ AU. The parameter governing the importance of disc self–gravity is the Toomre parameter, $Q = M_* H/(M_d r)$, with M_d being the disc mass contained within radius r and H being the disc semi–thickness. In this paper we shall take H to be the distance between the disc mid–plane and surface. This is usually a factor of two or so larger than the height reached by a gas particle moving vertically through the disc mid–plane with the local sound speed c_s. Thus $H \sim (2c_s)/\Omega$. Typically $H/r \sim 0.1$ (Stapelfeldt *et al.*, 1998) such that the condition for the importance of self–gravity, $Q \sim 1$, gives $M_d \sim 0.1 M_*$.

The characteristic scale associated with growing density perturbations in a disc undergoing gravitational instability with $Q \sim 1$ is $\sim H$, and the corresponding mass scale is $M_d(H/r)^2 \sim M_*(H/r)^3$, which is ~ 1 M_J for $H/r \sim 0.1$ and $M_* = 1$ $M_\odot$, with M_J being Jupiter's mass. Gravitational instability does not necessarily lead to fragmentation (e.g. Pickett *et al.*, 1998), nonetheless it has been proposed as a mechanism for directly forming giant planets by Cameron (1978) and recently by Boss (1998).

During the period of global gravitational instability, it is reasonable to suppose that the disc mass is quickly redistributed and reduced such that global gravitational stability is restored ($Q > 1$), after which further disc evolution occurs on a longer timescale governed by viscosity with effects due to self–gravity being small.

3.1. VISCOUS EVOLUTION

During this phase, the disc reaches a condition proposed for the minimum mass primordial solar nebula, $M_d \sim 10^{-2}$–10^{-1} $M_\odot$. Planets have been proposed to form out of such a disc by a process of growth through planetesimal accumulation followed, in the giant planet case, by gas accretion (Safronov, 1969; Wetherill and Stewart, 1989).

At this stage, it is reasonable to regard the disc as an axisymmetric configuration in which, to a first approximation, material orbits in circles, conserving specific angular momentum, with centrifugal acceleration balanced by gravitational attraction towards the centre. However, other weaker forces due to internal pressure, viscosity, or magnetic fields may also operate in the disc. These forces can redistribute angular momentum on a longer timescale. In order to flow inwards, material has to transport its angular momentum outwards to matter at larger radii. The angular momentum transport process determines the timescale on which mass accretion can occur and hence the evolutionary timescale of the disc.

Historically, the first angular momentum transport mechanism to be considered was through the action of viscosity (von Weizsäcker, 1948). This acts through the friction of neighbouring sections of the disc upon each other.

The inner regions rotate faster than the outer regions and thus viscous friction tends to communicate angular momentum from the inner parts of the disc to the outer parts. In order to result in evolution on astronomically interesting timescales, it is necessary to suppose that an anomalously large viscosity is produced through the action of some sort of turbulence. The magnitude of the viscosity is usually parameterized through the Shakura and Sunyaev (1973) α prescription.
In this we suppose the kinematic viscosity coefficient $\nu = \alpha c_s^2/\Omega \sim \alpha H^2 \Omega$, where α is a dimensionless constant which must be less than unity. Currently the most likely mechanism for producing turbulence is through hydromagnetic instabilities (Balbus and Hawley, 1991) which might produce $\alpha \sim 0.01$, provided the disc has adequate ionization.

Gammie (1996) has proposed a model in which viscosity only operates in the surface layers where external sources of ionization such as cosmic rays can penetrate. Such layered models may have an interior dead zone of material for $r > 0.1$ AU and, although they can be considered as models with a variable α, they will behave somewhat differently from the standard models with α taken to be constant.

4. Time dependent diffusion equation

The evolution of a viscous disc is controlled by angular momentum conservation. The equation governing the specific angular momentum may be written in standard conservation law form (Papaloizou and Lin, 1995)

$$\rho \frac{Dj}{Dt} = -\nabla \cdot \mathbf{F} = -\frac{1}{r}\frac{\partial}{\partial r}(rF_r) - \frac{\partial}{\partial z}(F_z), \tag{3}$$

where $\mathbf{F} = (F_r, F_z)$ is the angular momentum flux, with the vertical coordinate being denoted by z, and ρ is the mass density.
The process of vertical averaging applied to equation (3) in the case of a Keplerian disc yields

$$\Sigma \langle v_r \rangle \frac{dj}{dr} = -\frac{1}{r}\frac{\partial}{\partial r}(r\langle F_r \rangle), \tag{4}$$

where v_r is the radial velocity in the disc. The vertical average for a quantity Q is defined by

$$\langle Q \rangle = \frac{\int_{-H}^{H} \rho Q dz}{\int_{-H}^{H} \rho dz}, \tag{5}$$

where $\Sigma = \int_{-H}^{H} \rho dz$ is the surface mass density.

We assume $\mathbf{F}$ vanishes on the disc boundaries ensuring that the total angular momentum of the disc is conserved. For differential rotation the radial angular momentum flux arising from viscosity is given by

$$F_r = -r^2 \rho \nu \frac{d\Omega}{dr}, \tag{6}$$

and thus

$$\langle F_r \rangle = -r^2 \langle \nu \rangle \Sigma \frac{d\Omega}{dr}, \tag{7}$$

We remark that in a Keplerian disc the angular velocity decreases outwards so that the angular momentum flux is directed outwards as required in order that mass accretion may occur.
Using equation (7) in equation (4) and solving for $\langle v_r \rangle$ gives

$$\langle v_r \rangle = \frac{1}{r\Sigma} \left(\frac{dj}{dr} \right)^{-1} \frac{\partial}{\partial r} \left(r^3 \langle \nu \rangle \Sigma \frac{d\Omega}{dr} \right).$$

For a Keplerian disc in which $\Omega = \sqrt{GM_*/r^3}$ and $j = \sqrt{GM_* r}$, we get

$$\langle v_r \rangle = -\frac{3}{r^{1/2}\Sigma} \frac{\partial}{\partial r} \left(r^{1/2} \langle \nu \rangle \Sigma \right). \tag{8}$$

This gives an expression for the radial velocity in the disc in terms of the kinematic viscosity. For the simple example of a steady state disc where $\langle \nu \rangle \Sigma$ is constant, we have

$$\langle v_r \rangle = -\frac{3\langle \nu \rangle}{2r}.$$

This velocity is negative corresponding to mass flowing inwards to be accreted by the central object.
To obtain general governing equation for the disc, the velocity (8) is substituted into the vertically averaged continuity equation

$$\frac{\partial \Sigma}{\partial t} + \frac{1}{r} \frac{\partial}{\partial r} \left(\Sigma r \langle v_r \rangle \right) = 0.$$

After doing this we see that the global evolution of the disc is governed by a single diffusion equation for the surface density which takes the form (Lynden–Bell and Pringle, 1974)

$$\frac{\partial \Sigma}{\partial t} = \frac{3}{r} \frac{\partial}{\partial r} \left(r^{1/2} \frac{\partial}{\partial r} \left(\Sigma \langle \nu \rangle r^{1/2} \right) \right). \tag{9}$$

According to this the characteristic evolution timescale for the disc will be the global diffusion timescale. The diffusion coefficient is $3\langle\nu\rangle$ and, adopting a characteristic radius r, the corresponding global diffusion timescale is

$$t_e = \frac{r^2}{3\langle\nu\rangle}.$$

For viscosity coefficient ν which is parameterized through the α prescription in which $\nu = \alpha c_s^2/\Omega \sim \alpha H^2 \Omega$, we obtain for the disc evolution time

$$t_e\Omega = (1/3)(r/H)^2\alpha^{-1}$$

For a protostellar disc of size 200 AU with $H/r \sim 0.1$, and a central solar mass, we find $t_e = 1.5 \times 10^6(0.01/\alpha)$ yr. This gives lifetimes comparable to those estimated for discs circulating around T Tauri stars if $\alpha \sim 10^{-3}$–10^{-2}.

Equation (9) enables the evolution of the disc to be calculated if $\langle\nu\rangle$ is specified as a function of Σ. This can be done by calculating the vertical structure of the disc.

5. Vertical structure

5.1. BASIC EQUATIONS

We solve the equation of vertical hydrostatic equilibrium:

$$\frac{1}{\rho(z)}\frac{\partial P}{\partial z} = -\Omega^2 z, \tag{10}$$

and the equation of vertical thermal equilibrium, which states that the rate of energy dissipation by radiation is locally balanced by the rate of energy production by viscous dissipation:

$$\frac{\partial \mathcal{F}}{\partial z} = \frac{9}{4}\rho(z)\nu(z)\Omega^2, \tag{11}$$

where $\mathcal{F}$ is the radiative flux of energy through a surface of constant z and is given by:

$$\mathcal{F}(z) = \frac{-16\sigma T^3(z)}{3\kappa(\rho, T)\rho(z)}\frac{\partial T}{\partial z}. \tag{12}$$

Here P is the pressure, T is the temperature, κ is the opacity, which in general depends on both ρ and T, and σ is the Stefan–Boltzmann constant. For the sake of brevity, we have indicated only the z–dependence of the variables, but of course they also depend on r.

To close the system of equations, we relate P, ρ and T through the equation of state of an ideal gas:

$$P(z) = \frac{\rho(z)kT(z)}{\mu m}, \tag{13}$$

where k is the Boltzmann constant, μ is the mean molecular weight and m is the mass of the hydrogen atom. Since the main component of protostellar disks at the temperatures we consider is molecular hydrogen, we take $\mu = 2$. We denote the isothermal sound speed by c_s ($c_s^2 = P/\rho$).

As above, we adopt the α–parametrization of Shakura and Sunyaev (1973), so that the kinematic viscosity is written $\nu(z) = \alpha c_s^2(z)/\Omega$. In general, α is a function of both r and z. However, we shall limit our analysis below to a uniform α. With this formalism, equation (11) becomes:

$$\frac{\partial \mathcal{F}}{\partial z} = \frac{9}{4}\alpha\Omega P(z). \tag{14}$$

5.1.1. *Boundary conditions*

Since we have to solve three first order ordinary differential equations for the three variables $\mathcal{F}$, P (or equivalently ρ), and T as a function of z at a given radius r, we need three boundary conditions at each r. We shall denote with a subscript s values at the disk surface.

• One boundary condition is obtained by integrating equation (11) over z between $-H$ and H. Since by symmetry $\mathcal{F}(z=0) = 0$, this gives:

$$\mathcal{F}_s = \frac{3}{8\pi}\dot{M}_{st}\Omega^2, \tag{15}$$

where we have defined $\dot{M}_{st} = 3\pi\langle\nu\rangle\Sigma$. If the disk were in a steady state, $\dot{M}_{st}$ would not vary with r and would be the constant accretion rate through the disk. In general however, this quantity does depend on r.

• To get another boundary condition, we use the fact that very close to the surface of the disc, since the optical depth τ_{ab} above the disc is small, we have:

$$P_s = g_s \tau_{ab}/\kappa_s.$$

This condition is often used in stellar structure (e.g. Schwarzschild, 1958). Using $g_s = \Omega^2 H$, we thus obtain

$$P_s = \frac{\Omega^2 H \tau_{ab}}{\kappa_s}. \tag{16}$$

Provided $\tau_{ab} \ll 1$, the results do not depend on the value of τ_{ab} we choose (see below).

• The third boundary condition we use is

$$2\sigma\left(T_s^4 - T_b^4\right) - \frac{9\alpha k T_s \Omega}{8\mu m \kappa_s} - \frac{3}{8\pi}\dot{M}_{st}\Omega^2 = 0. \tag{17}$$

Here we assume the disc to be in a background medium with temperature T_b. The surface opacity κ_s in general depends on T_s and ρ_s and we have used $c_s^2 = kT/(\mu m)$.

The boundary condition (17) is the same as that used by Levermore and Pomraning (1981) in the Eddington approximation (their eq. [56] with $\gamma = 1/2$). In the simple case when $T_b = 0$, and the surface dissipation term involving α is set to zero, it simply relates the disc surface temperature to the emergent radiation flux.

5.1.2. *Vertical structure models*

At a given radius r and for a given values of the parameters $\dot{M}_{st}$ and α, we solve equations (10), (12) and (14) with the boundary conditions (15), (16) and (17) to get the vertical structure. The opacity we use is given in the Appendix of Bell and Lin (1994). Contributions due to dust grains, molecules, atoms and ions are included. It is given in the form $\kappa = \kappa_i \rho^a T^b$ where κ_i, a and b are constants which depend on the range of temperature considered.

The equations are integrated using fifth–order Runge–Kutta method with adaptive step length (Press *et al.*, 1992). Given $\dot{M}_{st}$, we guess a value of H, the vertical height of the disc surface. It is then possible, after satisfying the surface boundary conditions, to integrate the equations down to the mid–plane $z = 0$. We then check whether $\mathcal{F} = 0$ at $z = 0$. If not, an iteration procedure (e.g. the Newton–Raphson method) is used to generate a new value of H and the integration is repeated. The whole procedure is iterated until the solution has $\mathcal{F} = 0$ at $z = 0$ to a specified accuracy.

An important point to note is that an output of the vertical structure model is the surface density Σ for a given $\dot{M}_{st} = 3\pi\langle\nu\rangle\Sigma$. Thus a $\langle\nu\rangle, \Sigma$ relation is derived.

In the calculations presented here, we have taken the optical depth of the atmosphere above the disk surface $\tau_{ab} = 10^{-2}$ and a background temperature $T_b = 10$ K. Our calculations are limited to temperatures lower than about 4,000 K, so that, at the densities we consider, hydrogen is not dissociated and the mean molecular weight $\mu = 2$.

In the optically thick regions of the disk, the final value of H does not depend on the value of τ_{ab} we choose. However, this is not the case in optically thin regions where we find that, as expected, the smaller τ_{ab}, the larger H. However, this dependence of H on τ_{ab} has no physical significance, since in all cases, the mass is concentrated towards the disk mid–plane in a layer with thickness independent of τ_{ab}.

For instance, if we fix $\dot{M}_{st} = 10^{-8}$ M$_\odot$ yr^{-1} and $\alpha = 10^{-2}$, we find, at $r =$ 100 AU, that $H/r = 0.08$ and 0.24 for $\tau_{ab} = 10^{-2}$ and 10^{-5}, respectively. However, in both cases, the surface density, the optical thickness and the mid–plane temperature are the same. Only the mid–plane pressure varies slightly (by about 30%) from one case to the other.

In Figures 1a–c, we plot H/r, Σ and the mid–plane temperature T_m versus r for $\dot{M}_{st}$ between 10^{-9} and 10^{-6} M$_\odot$/year (assuming this quantity is the same at all radii, i.e. the disk is in a steady state) and for illustrative purposes we have adopted $\alpha = 10^{-2}$.

Figure 1a indicates that the outer parts of the disk are shielded from the radiation of the central star by the inner parts. This is in agreement with the results of Lin and Papaloizou (1980) and Bell *et al.* (1997). For $\alpha = 10^{-2}$, the radius beyond which the disk is not illuminated by the central star varies from 0.1 AU to about 2 AU when $\dot{M}_{st}$ goes from 10^{-9} to 10^{-6} M$_\odot$ yr^{-1}. These values of the radius move to 0.2 and 3 AU when $\alpha = 10^{-3}$. Since reprocessing of the stellar radiation by the disk is not an important heating factor below these radii, this process will in general not be important in protostellar disks. We note that this result is independent of the value of τ_{ab} we have taken. Indeed, as we pointed out above, only the thickness of the optically thin parts of the disk (which do not reprocess any radiation) gets larger when τ_{ab} is decreased.

The values of H/r, Σ and T_m we get are in qualitative agreement with those obtained by Lin and Papaloizou (1980), who adopted a simple prescription for viscosity based on convection, and Bell *et al.* (1997). Our values of H/r are somewhat larger though, since H is measured from the disk mid–plane to the surface such that τ_{ab} is small, and not 2/3 as usually assumed. However, as we commented above, this has no effect on the other physical quantities. We also recall that H, as defined here, is about 2–3 times larger than c_s/Ω, with c_s being the mid–plane sound speed, a quantity often used to define the disc semi–thickness.

5.1.3. *Time dependent evolution of the radial structure*

To compute the radial structure of a non–steady α–disk, we have to solve the diffusion equation (9). To do that, we need to know the relation between $\dot{M}_{st} = 3\pi\langle\nu\rangle\Sigma$ and Σ at each radius. Interpolation of or piece–wise power law fits to numerical data may be used to represent this relation and more details of these will be published elsewhere. We note that they can be used either to compute Σ from $\dot{M}_{st}$ or $\dot{M}_{st}$ from Σ.

In Figures 2a–b we plot both the curves $\dot{M}_{st}\,(\Sigma)$ that we get from the vertical structure integrations as described above and those obtained from piece–wise power law fits. Figures 2a and 2b are for $\alpha = 10^{-2}$ and 10^{-4}, respectively. In each case the radius varies between 0.01 and 100 AU. How-

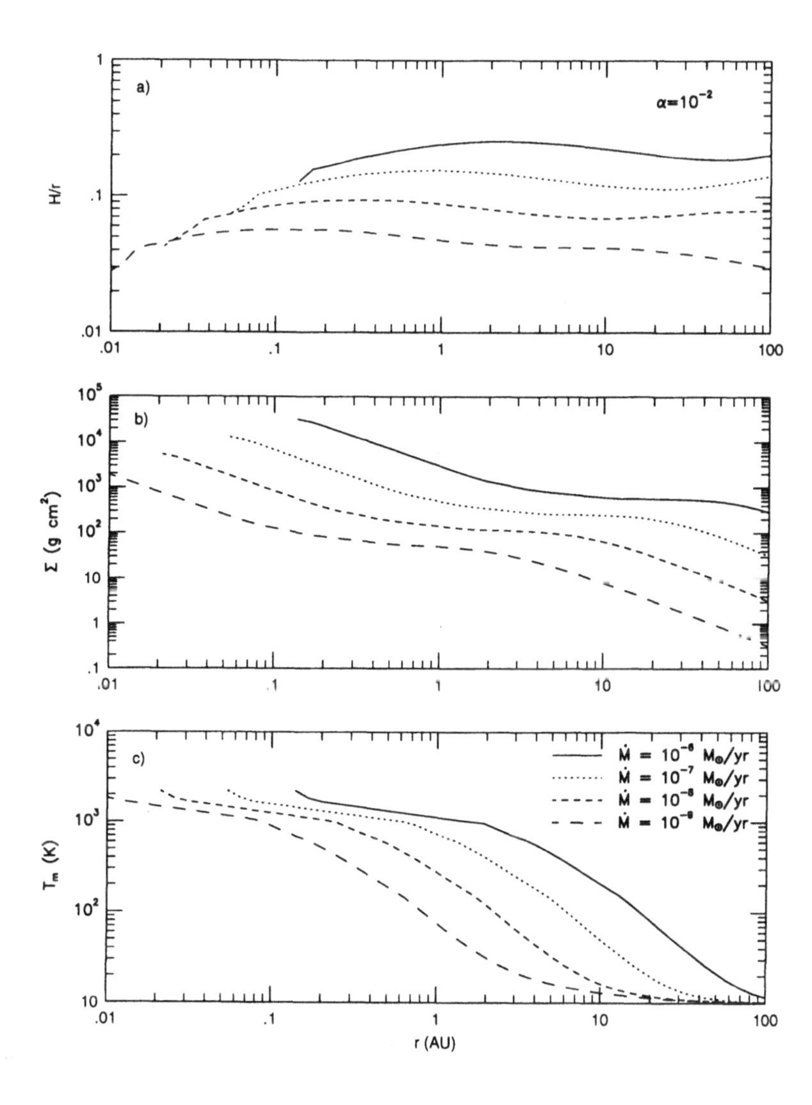

Figure 1. Shown is H/r (*plot a*), Σ in units g cm^{-2} (*plot b*) and T_m in K (*plot c*) vs. r/AU using a logarithmic scale for $\dot{M}_{st}$ (in units $M_\odot$ yr^{-1}) = 10^{-6} (*solid line*), 10^{-7} (*dotted line*), 10^{-8} (*short–dashed line*) and 10^{-9} (*long–dashed line*) and for $\alpha = 10^{-2}$.

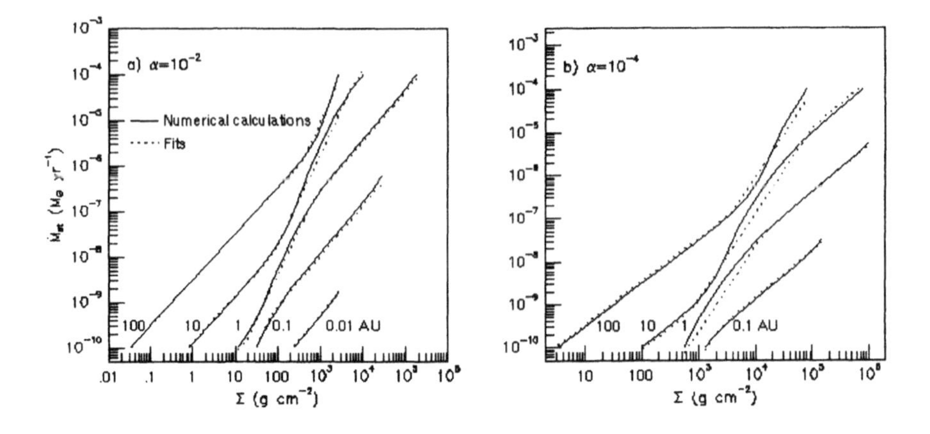

Figure 2. $\dot{M}_{st}$ in units $M_\odot$ yr^{-1} vs. Σ in units g cm^{-2} using a logarithmic scale for $\alpha = 10^{-2}$ (*plot a*) and 10^{-4} (*plot b*). Both the curves corresponding to the numerical calculations (*solid line*) and the fits (*dashed line*) are shown. The label on the curves represents the radius, which varies between 0.01 and 100 AU.

ever, for $\alpha = 10^{-4}$, the temperature in the disk at $r = 0.01$ AU gets larger than 4,000 K for values of $\dot{M}_{st}$ larger than 10^{-10} $M_\odot$ yr^{-1}. Since we have limited our calculations to temperatures smaller than 4,000 K, there is no curve corresponding to $r = 0.01$ for this value of α. If we calculate $\dot{M}_{st}$ using the fits with Σ as an input parameter, the average error is 36, 22, 18, 13 and 16% whereas the maximum error is 107, 55, 48, 42 and 103% for $\alpha = 10^{-5}, 10^{-4}, 10^{-3}, 10^{-2}$, and 10^{-1}, respectively. We see that, apart from the cases $\alpha = 10^{-5}$ and 10^{-1}, the fits give a good approximation.

Using the $\left(\dot{M}_{st}, \Sigma\right)$ relation derived from the vertical integrations, we have solved equation (9) using standard explicit finite difference techniques. We considered the situation of a disc with initially 0.1 $M_\odot$ for which $\Sigma \propto r^{-1}$, extending to 100 AU. The central star had $M_* = 1$ $M_\odot$ and for illustrative purposes we took $\alpha = 10^{-3}$.

In Figure 3 we show the evolution of Σ as a function of time. After a time $\sim 10^6$ yr, the Σ profile resembles that of a steady disc in the inner parts justifying the assumption of a steady state disc model. After a time 5×10^6 yr, the model is similar to that assumed for the primordial solar nebula with $\Sigma \sim 300$ g cm^{-2} at 5 AU.

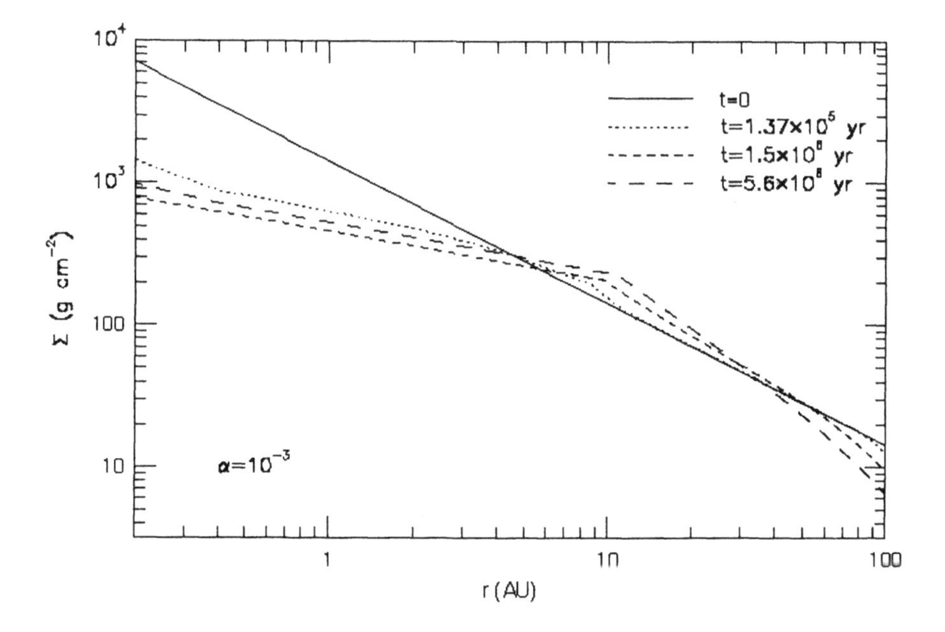

Figure 3. Solution of the diffusion equation. Shown is Σ in units g cm^{-2} vs. r in AU using a logarithmic scale plotted at times $t = 0$ (*solid line*), $t = 1.37 \times 10^5$ yr (*dotted line*), $t = 1.5 \times 10^6$ yr (*short-dashed line*) and $t = 5.6 \times 10^6$ yr (*long-dashed line*). This run has $\alpha = 10^{-3}$. The total disc mass decreases because of accretion onto the central object

6. Planets and Planetesimals

6.1. FORMATION

It is generally believed that planetesimals are built up from μm sized particles through processes of collision, sticking and accumulation all occurring in a gaseous medium with some degree of turbulence (see the review by Weidenschilling and Cuzzi, 1993). The idea is that particles with a size distribution ranging up to $\sim$ a few km can be produced on timescales on the order of 10^4 yr at 1AU. The dynamics of these processes are very uncertain depending as they do on sticking probabilities and the degree of particle settling in a turbulent medium etc. Their efficiency may depend on the location in the nebula, being more effective beyond a few AU, where ice has condensed. Here we assume that planetesimals with mass $m_p \sim 10^{18}$ g may form on a sufficiently rapid timescale anywhere in the nebula.

6.2. DYNAMICAL EVOLUTION OF A PLANETESIMAL SWARM

Suppose material with surface density Σ_p is in the form of planetesimals with number distribution $n(m) \propto m^{-q}$, for some index q. Here the number of planetesimals in the mass range $(m, m + dm)$ is $n(m)dm$. Then most of the mass is distributed in the most/least massive objects according as $q < 2$ or $q > 2$.

Suppose that the masses lie between 10^{18} g and 10^{24} g. The characteristic number density, $n(m_p)$, for characteristic mass m_p is $\Sigma_p/(2m_pH_p)$. Here H_p is the scale height of the distribution ($\sim \sqrt{2\pi/3}v/\Omega$), where v is the root mean square velocity dispersion.

For $\Sigma_p = 1$ g cm^{-2}, $m_p = 10^{18}$ g, and $H_p = 10^{12}$ cm, the mean distance between planetesimals is $\sim 10^{10}$ cm. This suggest use of a local box model (e.g. Stewart and Wetherill, 1988). In this box with centre which orbits with the local circular velocity, planetesimals are assumed to move with constant velocity between encounters. Thus the effect of the central mass is ignored. and the planetesimals are treated using the methods of kinetic theory. The idea (Safronov, 1966) is that the velocity dispersion of the planetesimals is increased by gravitational scattering, enhancing direct collisions between them through which they accumulate and grow.

The local box model will fail when the largest planetesimals go into isolation. That happens when they have accumulated to the stage where there are so few left that they are in non–overlapping orbits such that they cannot perturb similar mass objects in neighbouring circular orbits into collision. At this stage the effects of the central mass cannot be neglected. To get an approximate idea of when this happens at radius r, equate

$$\Sigma_p = \frac{m_p}{8\pi(m_p/M_*)^{1/3}r^2}.$$

This means there is one object in an annulus of width equal to four times its Roche lobe size. Then

$$m_p = (8\pi\Sigma_p r^2)^{3/2}(M_*)^{-1/2}.$$

This gives $m_p = 10^{25}$ g at 1 AU for $\Sigma_p = 1$ g cm^{-2} with obvious scalings to other surface densities and radii (a few earth masses may be reached at 5 AU). This argument assumes circular orbits which is probably reasonable for the largest objects which are circularized through dynamical friction. After the isolation stage, planetesimal evolution is probably best followed by global N–body methods (e.g. Aarseth *et al.*, 1993). In the gas free case the ultimate result is expected to be the formation of terrestrial planets.

6.3. GRAVITATIONAL SCATTERING AND BUILD UP OF VELOCITY DISPERSION

For a mass interacting gravitationally with a swarm of other bodies of equal mass, the cross section for elastic scattering is larger than that for a direct impact provided the velocity dispersion is less than the escape velocity. Thus if the velocity dispersion is initially small it will be built up by the effect of encounters producing elastic scattering.

The interaction radius for two masses m_p to give significant scattering is $r_x = 2Gm_p/v^2$. The time between encounters is then $t_c = 1/(n(m_p)\pi r_x^2 v \log(\Lambda))$. Here the $\log(\Lambda) \sim 10$ term accounts for more distant collisions (Binney and Tremaine, 1987). Using the above, one gets

$$t_c = \frac{v^3}{4\pi n(m_p) G^2 m_p^2 \log(\Lambda)} = \sqrt{\frac{3}{2\pi}} \frac{3}{40\pi^2} \left(\frac{H_p}{r}\right)^4 \frac{M_*^2}{m_p \Omega \Sigma_p r^2}.$$

For $m_p = 10^{20}$ g, one finds at 1 AU that $t_c = 2 \times 10^{17} (H_p/r)^4$ yr.

Note that H_p is small because the planetesimal dispersion velocities are expected to reach the escape velocity for the characteristic mass. At this point inelastic geometric collisions become as important as scattering and damp the random motions. Then for $m_p = 10^{20}$ g, $v = 1.7 \times 10^3$ cm s^{-1}, $H_p/r = 8 \times 10^{-4}$ and $t_c \sim 9 \times 10^4$ yr at 1 AU.

Thus early planetesimal build up occurs on a rapid timescale.

6.4. RUNAWAY ACCRETION

From the above arguments, the collision time of m_p with $m_{p'}$ is inversely proportional to $n(m_{p'})(m_p + m_{p'})^2$. If this does not decrease with $m_{p'}$, then collisions with larger masses are dominant and we expect velocity dispersions to build up to the escape velocity of the largest body (Safronov, 1966). This occurs for $q < 2$.

For $q > 2$, the largest bodies collide with predominantly smaller ones and are circularized by dynamical friction. They can then accrete efficiently from the smaller ones which move with a velocity dispersion small compared to the escape velocity from them.

The accretion rate from planetesimals with mass $m_{p'}$ and velocity dispersion v_p' is enhanced by gravitational focusing. Thus the growth of the largest mass m_p say, with radius R_p is given by

$$\frac{dm_p}{dt} = n(m_{p'}) m_{p'} \pi R_p^2 v_{p'} \left(1 + \frac{2Gm_p}{R_p v_{p'}^2}\right).$$

Runaway is caused both by the increase of R_p with m_p and gravitational focusing (the term in brackets). The timescale for growth to isolation mass is comparable to the encounter time, t_c, indicated above. But note that growth may be slowed down by encounters between neighbouring runaways producing an increase in the velocity dispersion of all components in the system (Ida and Makino, 1993)

7. Formation of giant planets

As the core increases in mass, a critical value of around a few earth masses is attained beyond which the surrounding gaseous atmosphere can no longer grow quasi–statically in mass. At this stage a process of collapse ensues possibly leading to dynamical accretion and mass growth to values characteristic of giant planets. We review the theory of this below. The critical core mass model is supported by the indication from models of Jupiter that it has a solid core of $\sim$ 5–15 $M_{\oplus}$ (Podolak *et al.*, 1993).

7.1. BASIC EQUATIONS GOVERNING A PLANETARY CORE AND ATMOSPHERE

We define the variable ϖ to be the spherical polar radius in a frame with origin at the planet's core center. We neglect the rotation of the planet around both its own spin axis and the disk spin axis. We assume that the envelope is in hydrostatic equilibrium and spherically symmetric, so that:

$$\frac{dP}{d\varpi} = -g(\varpi)\rho(\varpi), \tag{18}$$

where $g = GM(\varpi)/\varpi^2$ is the acceleration due to gravity, $M(\varpi)$ being the mass contained in the sphere of radius ϖ (this includes the core mass if ϖ is larger than the core radius). Mass conservation gives:

$$\frac{dM}{d\varpi} = 4\pi\varpi^2\rho(\varpi). \tag{19}$$

The thermodynamic variables in the planet atmosphere are generally such that the equation of state for an ideal gas does not apply there. Here we then adopt the state–of–the–art Chabrier *et al.* (1992) equation of state for a hydrogen and helium mixture. We fix the abundances of hydrogen and helium to be 0.7 and 0.28 respectively.

The equation of radiative transport is:

$$\frac{dT}{d\varpi} = \frac{-3\kappa(\rho, T)\rho(\varpi)}{16\sigma T^3(\varpi)} \frac{L(\varpi)}{4\pi\varpi^2}, \tag{20}$$

where L is the luminosity carried by radiation. We denote the radiative and adiabatic temperature gradients by ∇_{rad} and ∇_{ad} respectively. We have:

$$\nabla_{rad} = \left(\frac{\partial \ln T}{\partial \ln P}\right)_{rad} = \frac{3\kappa(\rho, T)L_{core}P(\varpi)}{64\pi\sigma GM(\varpi)T^4(\varpi)}, \tag{21}$$

and

$$\nabla_{ad} = \left(\frac{\partial \ln T}{\partial \ln P}\right)_s, \tag{22}$$

with the subscript s meaning that the derivative has to be evaluated at constant entropy.

We assume that only energy source comes from the core which outputs the core luminosity L_{core}, given by:

$$L_{core} = \dot{M}_{core}\frac{GM_{core}}{r_{core}}, \tag{23}$$

where M_{core} and r_{core} are respectively the mass and the radius of the core, and $\dot{M}_{core}$ is the rate of accretion of planetesimals onto the core. We note that it is customary to take, instead of L_{core}, the luminosity supplied by the gravitational energy which the planetesimals entering the planet atmosphere release near the surface of the core (see, e.g., Mizuno, 1980; Bodenheimer and Pollack, 1986). However, when the mass of the atmosphere is small compared to that of the core, these two luminosities are comparable, providing we take for $\dot{M}_{core}$ the rate of accretion of planetesimals onto the core during the phase of accretion of the atmosphere and not during the phase of the core formation. These two rates may be different if the core has migrated in the disk before accreting the atmosphere.

If $\nabla_{rad} < \nabla_{ad}$, the medium is convectively stable and the energy is transported only by radiation. In that case we simply have $L = L_{core}$.

When $\nabla_{rad} > \nabla_{ad}$, energy is also transported by convection. In that case, $L_{core} = L + L_{conv}$, where L_{conv} is the convective luminosity. We use the expression of L_{conv} given by the mixing length theory (Cox and Giuli, 1968):

$$L_{conv} = \pi\varpi^2 C_p \Lambda_{ml}^2 \left[\left(\frac{\partial T}{\partial \varpi}\right)_s - \left(\frac{\partial T}{\partial \varpi}\right)\right]^{3/2} \sqrt{\frac{1}{2}\rho g \left|\left(\frac{\partial \rho}{\partial T}\right)_P\right|}, \tag{24}$$

where $\Lambda_{ml} = |\alpha_{ml} P/(dP/d\varpi)|$ is the mixing length, α_{ml} being a constant of order unity, $(\partial T/\partial \varpi)_s = \nabla_{ad} T\,(d\ln P/d\varpi)$, and the subscript P means that the derivative has to be evaluated for a constant pressure. The quantities $(\partial \rho/\partial T)_P$ and ∇_{ad} are given by Chabrier *et al.* (1992).

7.2. BOUNDARY CONDITIONS

We suppose that the planet core has a uniform mass density ρ_{core}, here taken to be $3.2 gm cm^{-3}$. The core radius, which is the inner boundary of the atmosphere, is then given by:

$$r_{core} = \left(\frac{3M_{core}}{4\pi \rho_{core}} \right)^{1/3} . \tag{25}$$

The outer boundary of the atmosphere is at the Roche lobe radius r_L of the planet:

$$r_L = \frac{2}{3} \left(\frac{M_{pl}}{3M_*} \right)^{1/3} r, \tag{26}$$

where $M_{pl} = M_{core} + M_{atm}$ is the planet mass, M_{atm} being the mass of the atmosphere, and r is the location of the planet in the disk (i.e. the separation between the planet and the central star).

To avoid confusion, we will denote the disk mid–plane temperature, pressure and mass density at the distance r from the central star by T_{mid}, P_{mid} and ρ_{mid}, respectively.
At $\varpi = r_L$, the mass is equal to M_{pl}, the pressure is equal to P_{mid} and the temperature is given by:

$$T = \left(T_{mid}^4 + \frac{\tau_L L_{core}}{4\pi\sigma r_L^2} \right)^{1/4} , \tag{27}$$

where

$$\tau_L = \frac{3}{4} \kappa \left(\rho_{mid}, T_{mid} \right) \rho_{mid} r_L .$$

The condition at $\varpi = r_{core}$ is that the mass is equal to M_{core} there.

7.3. NUMERICAL RESULTS

At a given disk radius r and for a given core mass M_{core}, we solve the equations (18), (19) and (20) with the boundary conditions described above to get the structure of the envelope. The opacity laws we use are the same as those used for the disk models. We note that when the mass density gets large, the interior of the envelope becomes convective so that the choice of opacity does not matter there.
The equations are integrated using the fifth–order Runge–Kutta method with adaptive step–size control (Press *et al.*, 1992). We guess a starting value of M_{atm} and integrate the equations from $\varpi = r_L$ down to the core

surface $\varpi = r_{core}$. We then iterate the integration, adjusting M_{atm} at each step, until the solution gives $M = M_{core}$ at $\varpi = r_{core}$ with some accuracy.

At each radius r, for a fixed $\dot{M}_{core}$, there is a critical core mass M_{crit} (which increases as $\dot{M}_{core}$ increases) above which no solution can be found, i.e. there can be no atmosphere in hydrostatic and thermal equilibrium confined between the radii r_{core} and r_L around cores with mass larger than M_{crit}. This is because when the core mass is too large, the atmosphere has to collapse onto the core in order to supply adequate luminosity to support itself. For masses below M_{crit}, there are (at least) two solutions, corresponding, respectively, to a low–mass and a high–mass atmosphere.

In Figure 4 we plot M_{pl} versus M_{core} for different $\dot{M}_{core}$ (between 10^{-6} and 10^{-11} $M_\oplus$ yr^{-1}) at a radius of 5 AU and for T_{mid} = 140.05 K and P_{mid} = 0.13 dyn cm^{-2}. These values of the temperature and pressure are obtained from the vertical structure integrations described above when the parameters $\alpha = 10^{-2}$ and $\dot{M}_{st} = 10^{-7}$ $M_\odot$ yr^{-1} are used at r = 5 AU.

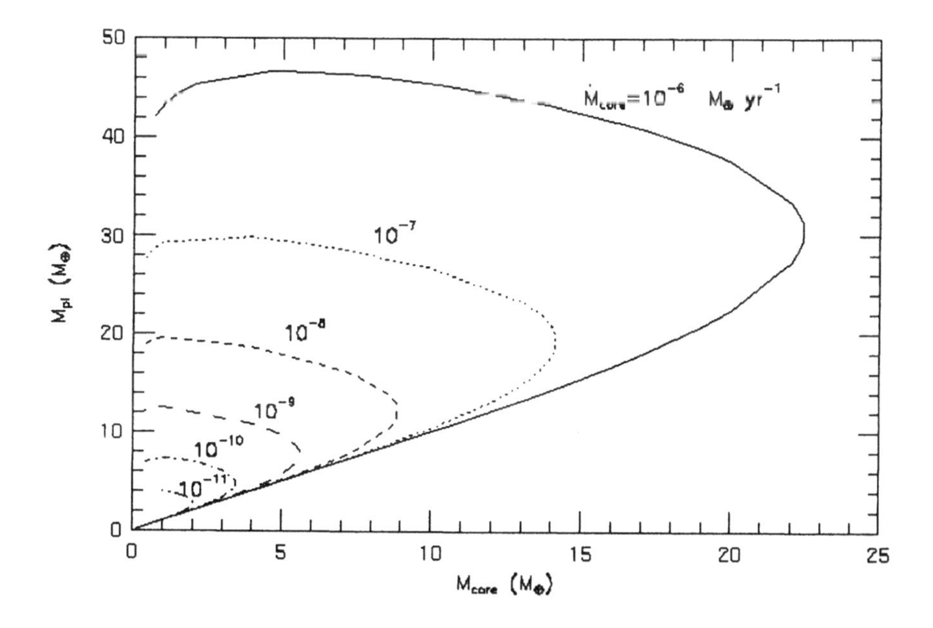

Figure 4. M_{pl} in $M_\oplus$ versus M_{core} in $M_\oplus$ for $\dot{M}_{core}$ between 10^{-6} and 10^{-11} $M_\oplus$ yr^{-1} at a radius r = 5 AU and for T_{mid} = 140.05 K and P_{mid} = 0.13 dyn cm^{-2}. From one curve to another, starting from the right, $\dot{M}_{core}$ decreases by a factor 10. The critical core mass increases with $\dot{M}_{core}$, varying between ~ 2 and 22.5 $M_\oplus$.

The critical core mass, which decreases when $\dot{M}_{core}$ decreases, is found to be 22.5 $M_\oplus$ for $\dot{M}_{core}$ = 10^{-6} $M_\oplus$ yr^{-1} and 2.5 $M_\oplus$ for $\dot{M}_{core}$ = 10^{-11} $M_\oplus$ yr^{-1}. These values are slightly larger than those found by Bodenheimer and Pollack (1986). The difference may be accounted for by the fact that we do not calculate the luminosity in exactly the same way as they do. Also we use a slightly different boundary condition for the temperature at the surface of the planet.

8. Final stages

Once the planetary mass has attained values of around an earth mass or higher, dynamical interactions with the surrounding disc matter become important, leading to phenomena such as inward orbital migration and tidal truncation (Lin and Papaloizou, 1993; Ward, 1997; Lin *et al.*, 1998). These effects, together with tidal interaction with the central star (Terquem *et al.*, 1998) can lead to massive planets in circular orbits, with periods of a few days, as observed (Butler *et al.*, 1997). The reader is referred to the article by Lin *et al.* (1998) for an account of the dynamical phenomena which can play a role in determining the final orbital configuration of planetary systems.

References

1. Aarseth, S. J., Lin, D. N. C., and Palmer, P. L. 1993, ApJ, 403, 351
2. Balbus, S.A. and Hawley, J. F. 1991, ApJ, 376, 214
3. Beckwith, S. V. W., and Sargent, A. I. 1996, Nature, 383, 139, 144
4. Bell, K. R., Cassen, P., Klahr, H. H., Henning, T. 1997, ApJ, 486, 372
5. Bell, K. R., and Lin, D. N. C. 1994, ApJ, 427, 987
6. Binney, J., and Tremaine, S. 1987, *Galactic Dynamics*, 3d. ed., Princeton: Princeton Univ. Press
7. Bodenheimer, P., and Pollack, J. B. 1986, Icarus, 67, 391
8. Boss, A. P. 1998, ApJ, 503, 923
9. Butler, R. P., Marcy, G. W., Williams, E., Hauser, H., and Shirts, P. 1997. ApJ, 474, L115
10. Cameron, A. G. W. 1978, Moon Planets, 18, 5
11. Chabrier, G., Saumon, D., Hubbard, W. B., and Lunine, J. I. 1992, ApJ, 391, 817
12. Cox, J. P., and Giuli, R. T. 1968, *Principles of Stellar Structure: Physical Principles*, New York: Gordon and Breach
13. Gammie, C. F. 1996, ApJ, 457, 355
14. Goodman, A. A., Benson, P. J., Fuller, G. A., and Myers, P. C. 1993, ApJ, 406, 528
15. Hayashi, M., Ohashi, N., Miyama, S. M. 1993, ApJ, 418, L71
16. Heemskerk, M. H. M., Papaloizou, J. C. B., and Savonije, G. J. 1992, A&A, 260, 161
17. Ida, S., and Makino, J. 1993, Icarus, 106, 210
18. Koerner, D. W., Sargent, A. I. 1995, ApJ, 109, 2138
19. Laplace, P. S. 1796, *Exposition du Système du Monde*, Paris
20. Laughlin, G., and Bodenheimer, P. 1994, ApJ, 436, 335
21. Lay, O., Carlstrom, J. E., Hills, R. E. and Phillips, T. G. 1994 ApJ, 434, L75
22. Levermore, C. D., and Pomraning, G. C. 1981, ApJ, 262, 768
23. Lin, D. N. C., and Papaloizou, J. 1980, MNRAS, 191, 37
24. Lin, D. N. C., and Papaloizou, J. 1985, in *Protostars and planets II*, p. 981–1072, University of Arizona Press

25. Lin, D. N. C., and Papaloizou, J. C. 1993, in *Protostars and planets III*, p. 749, University of Arizona press
26. Lin, D. N. C., Papaloizou, J. C. B., Bryden, G., Ida, S., and Terquem, C. 1998, in Protostars and planets IV, University of Arizona Press, *in press*
27. Lynden–Bell, D., and Pringle, J. E. 1974, MNRAS, 168, 60
28. McCaughrean, M. J., and Stauffer, J. R. 1994, AJ, 108, 1382
29. Mizuno, H. 1980, Prog. Theor. Phys., 64, 544
30. Mundy, L. G., Looney, L. W., Erickson, W., Grossman, A., Welch, W. J., Forster, J. R., Wright, M. C. H., Plambeck, R. L., Lugten, J., and Thornton, D. D. 1996, ApJ, 464, L169
31. Papaloizou, J. C. B. , and Lin, D. N. C. 1995, ARA&A, 33, 50
32. Papaloizou, J. C., and Savonije, G. J. 1991, MNRAS, 248, 35
33. Pickett, B. K., Cassen, P., Durisen, R. H., and Link, R. 1998, ApJ, 504, 468
34. Podolak, M., Hubbard, W. B., and Pollack, J. B. 1993, in *Protostars and planets III*, p. 1109, University of Arizona press
35. Press, W. H., Flannery, B. P., Teukolsky, S. A., and Vetterling, W. T. 1992, *Numerical Recipes: The Art of Scientific Computing*, 3d ed., Cambridge: Cambridge Univ. Press
36. Reipurth, B., Baly, J. and Devine, D. 1997, AJ, 114, 2708
37. Safronov, V. S. 1966, Soviet Astron. J. 9, 987
38. Safronov, V. S. 1969, in *Evoliutsiia doplanetnogo oblaka*, Moscow
39. Sargent, A. I. and Beckwith, S. V. W. 1987 ApJ, 323, 294
40. Sargent, A. I. and Beckwith, S. V. W. 1991 ApJ, 382, 31
41. Schwarzschild, M. 1958, *Structure and Evolution of the Stars*, Princeton: Princeton University Press
42. Shakura, N. I., and Sunyaev, R. A. 1973, A&A, 24, 337
43. Stapelfeldt, K. R., Krist, J. E., Menard, F., Bouvier, J., Padgett, D. L., and Burrows, C. J. 1998, ApJ, 502, 65
44. Stewart, G. R. and Wetherill, G. W. 1988, Icarus, 74, 553
45. Strom, S. E., Edwards, S., and Skrutskie, M. F. 1993, in *Protostars and Planets III*, p. 837, University of Arizona Press
46. Terquem, C. , Papaloizou, J. C. B. , Nelson, R. P., and Lin, D. N. C. 1998, ApJ, 502 ,788
47. von Weizsäcker, C. F. 1948, Z. Naturforsch, 3a, 524
48. Ward, W. 1997, Icarus, 126, 261
49. Weidenschilling, S. J., Cuzzi, J. N. 1993, in *Protostars and Planets III*, p. 1031, University of Arizona Press
50. Wetherill, G. W., and Stewart, G. R. 1989, Icarus, 77, 330

ZODIACAL DUST IN THE EARTH SCIENCES

BERTIL OLSSON
Penn State University
Dep. of Geosciences
436 Deike Building
University Park, PA 16802

Abstract. Zodiacal dust will be a problem for infrared interferometric planetary searches, and much work is therefore in progress to improve our understanding of the dynamics and structure of circumstellar dust clouds in mature planetary systems. A serendipitous result from this work is the emergence of a new area in the Earth sciences, dedicated to studies of variations in terrestrial dust accretion and possible induced environmental effects. Some problems from this area are described, and illustrated with examples from current research.

1. Introduction

The solar zodiacal cloud has been observed to contain density-enhanced bands correlating with known asteroid families [1-3]. This shows that the asteroid belt produces a significant amount of dust, and that the production rate is determined by the spatial density of orbits. The continuous depletion of the asteroid belt [4] thus implies a high dust density in the early solar system, and it has been suggested that IR-interferometric planetary searches should be concentrated to old systems, in order to avoid the thermal zodiacal background [5-7]. Unfortunately, the age-criterion cannot be set with confidence, and even mature systems could contain dense zodiacal clouds if most dust is produced in a few collisions [8].

Many uncertainties would be reduced by a better known history of the asteroidal dust in the solar system, and we are therefore fortunate in living on Earth, which has collected dust since the beginning of geologic time. The current terrestrial dust accretion of about 40,000 tons per year [9] is probably geologically insignificant, but asteroidal dust could be abun-

J.-M. Mariotti and D. Alloin (eds.), Planets Outside the Solar System: Theory and Observations, 87–94.

dant in 3.8 billion years old sediments [10], since the asteroidal zodiacal cloud may originally have been a million times denser than today [5]. In the simplest approximation, this would give a terrestrial accretion of about 80 g m^{-2} yr^{-1}, which is actually *larger* than typical deep-sea sedimentation rates of ≤ 25 g m^{-2} yr^{-1} [11]. Still, few attempts have been made to determine the abundance of asteroidal dust in very old sediments, mainly because no one knows exactly what to look for! The composition of stratospheric micrometeors is too diverse for an attribution of grains to specific sources [12], and it is obviously impossible to identify particles of uncertain composition and abundance mixed up in metamorphosed rocks of often uncertain origin. Ideally, one would like to find dust of known asteroidal origin in younger sediments, and use its properties as identification criteria. The density-enhanced band structures in the zodiacal cloud make this a promising idea, since they suggest an orbitally modulated terrestrial accretion rate [13]. From a geological point of view, the problems are then to determine how variations in the terrestrial dust accretion rate can occur, and how such variations can be detected in sedimentary rocks of different ages. These highly non-trivial problems include areas like celestial mechanics, the optics of small particles, stratigraphy, and geochemistry.

2. Dust dynamics

The evolution of a zodiacal cloud depends ultimately on the forces acting upon individual particles, which for small grains include a significant component from radiative interactions with sunlight. When a dust particle is ejected from an asteroid it immediately responds to radiation pressure, which gives a radial force according to

$$F_{rp} = \left(\frac{S_0}{r^2 c}\right) AQ \tag{1}$$

where A is the cross-sectional area of the dust particle, S_0 the solar constant (≈ 1367 W/m^2), r the heliocentric distance in astronomical units, and Q a radiation pressure coefficient that depends on the absorption and scattering efficiencies for the particle [14]. The effect of radiation pressure can be seen as a decrease of the effective solar mass "felt" by a particle of mass m

$$F_g' = GmM_{Sun}(1-\beta) \quad \rightarrow \quad M_{sun}' \equiv M_{Sun}(1-\beta) \tag{2}$$

where β is the ratio of the radiation pressure force to the solar gravitational force [15]. The effective solar mass becomes negative for $\beta > 1$, but smaller β can still lead to ejection of grains from the solar system if their initial velocity is high. Dust particles levitated from comet Halley at perihelion will, for example, achieve instantaneous solar system escape velocity for $\beta > 0.016$ [15].

Radiation pressure instantaneously changes the heliocentric distances of particles, but their long-term dynamical evolution is dominated by the Poynting-Robertson and solar wind drags, which continuously remove energy and angular momentum. This shrinks and circularizes the dust orbits, and most 10-100 μm grains eventually enter the inner solar system. The Poynting-Robertson drag can be separated into an angular and a radial component [15]

$$F_{pr,\theta} = -\left(\frac{S_0}{r^2c^2}\right) AQ\dot{\theta}, \quad F_{pr,r} = -2\left(\frac{S_0}{r^2c^2}\right) AQ\dot{r} \tag{3}$$

where the terms are the same as in equation 1. The solar wind drag is analogous to the Poynting-Robertson drag, but only about a third as strong [14, 16]. A useful estimate for the maximum time it could take for a particle to move from an initial circular orbit of radius a_i to a smaller orbit of radius a_f is

$$T_{pr} = K(a_i{}^2 - a_f{}^2)\frac{m}{A} \tag{4}$$

where $K \approx 7.4 \cdot 10^{-10}$ s/kg, and the last term includes the mass and cross-sectional area of the particle.

In an unperturbed system the dust would spiral inward along a surface, since all radiative forces act in the plane of motion, but in reality the planets gravitationally perturb the particles in three dimensions. The combination of forces leads to an extremely complicated dynamical system that must be analyzed numerically [16-18].

3. The Lilliput dust-integrator

Lilliput is an orbital dust-model specifically designed for field geologists with moderate knowledge of celestial mechanics. The code consists of a gravitational integrator based on a mixed-variable symplectic algorithm taken from the package SWIFT [19], and a separate routine in which positions and velocities are corrected for radiative perturbations. Lilliput integrates 1000 dust particles from the asteroid belt to Earth in 4 hours on a Macintosh-G3 laptop computer, with 8 planets included and a step-size of 20 days.

The initial planetary and asteroidal coordinates are calculated from osculating elements collected for a single date, in order to ensure consistency [20]. The dust grains are created in a virtual reality environment, where the shadows of rotating particles are measured to give their average cross-sectional area, and their optical properties are approximated with values calculated for astronomical silicate spheres [21] of equal cross-sectional area, typically giving $Q \approx 1.2$. The small mass/area-ratio of irregular particles gives a β that is about twice as large as for spheres of equal mass.

The Lilliput output can be presented in diagrams or as a virtual-reality model of the resulting dust cloud, and is here exemplified with a simulation of dust produced by the Eos asteroid family [22]. An important observation is that the inclination relative the Jovian osculating orbital plane is conserved for most particles during the transfer through the solar system.

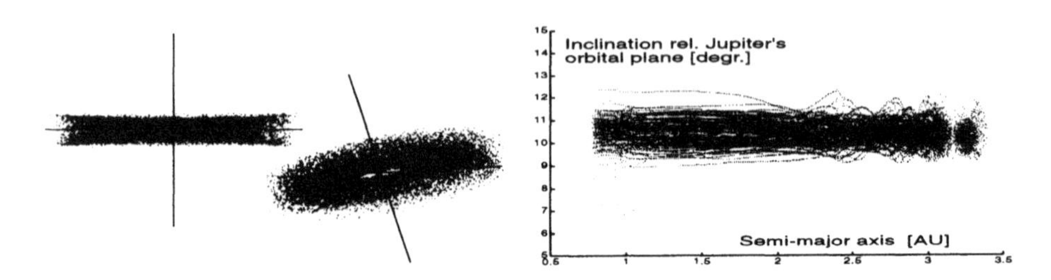

Figure 1. Lilliput-output: Two views of a dust cloud formed by the Eos asteroidal family, and the inclinations relative Jupiter's osculating orbital plane for all cloud particles.

Lilliput can also simulate dust clouds as seen from Earth (fig. 2a). The direction and the parallactic curvature of the modeled dust bands correlate well with the observations (fig. 2b).

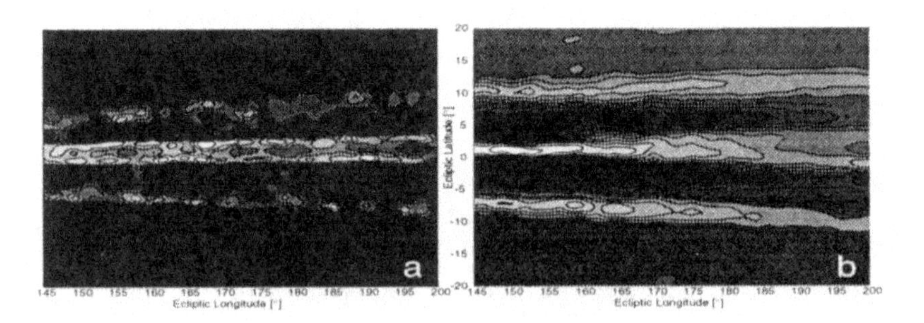

Figure 2. The Eos/Koronis dust bands in the sky ($\lambda = 25\mu m$) in January 1990, simulated by Lilliput (a) and as seen in a Fourier-filtered DIRBE-observation (b). (Reduction method taken from Reach, Franz & Weiland (1997) [23].)

4. Terrestrial accretion of asteroidal dust

A dust particle in a near-Earth orbit does not have to intercept the physical planet in order to be captured, since gravitational focusing increases the effective cross section of Earth according to

$$\sigma_{\oplus} = A_{\oplus} \left[1 + \left(\frac{v_{\oplus}}{v_p} \right)^2 \right] \tag{5}$$

where $A_{\oplus}$ is the physical cross-sectional area of Earth, $v_{\oplus}$ the escape velocity for Earth (≈11 km/s) and v_p the geocentric velocity of the dust particle before acceleration in Earth's gravitational field [24]. This defines a closeness-factor C as the fraction of a dust orbit that lies within a distance $d = \sqrt{\sigma_{\oplus}/\pi}$ from a point on the orbit of Earth.

The gravitational focusing and the closeness-factor depend on the inclination and eccentricity of the orbit of Earth, with some modulation due to the ascending node and argument of perihelion. It is trivial that the relative velocity depends on the direction in which the dust particle approaches Earth, and the dependence of C can be seen in a simplified picture where the gravitational focusing is held constant (figure 3).

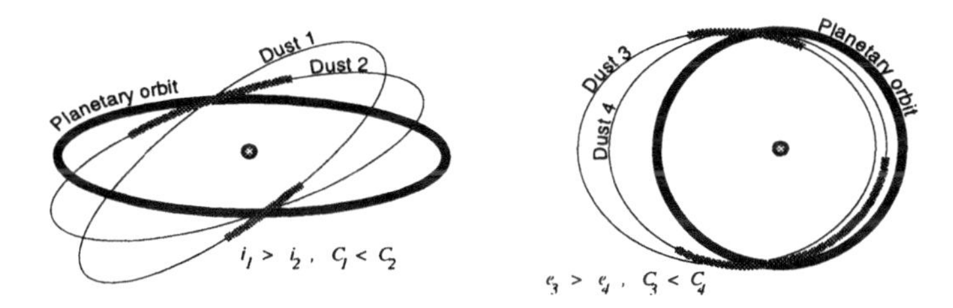

Figure 3. Schematic picture showing the relation between inclination, eccentricity and the fraction C of the dust orbits in which planetary capture can take place (shaded, for clarity only one region shown for each particle).

The orbital dependence ensures that the long-term terrestrial accretion is dominated by asteroidal dust emitted at low inclinations [24], but also implies a predictable temporal variation of the accretion rate, since Earth's orbital elements change with time due to planetary perturbations [13]. The variation of Earth's orbital eccentricity is ~ 6% with fundamental periods of 95, 125, and 400 kyr, and has been modeled to give an anticorrelated accretion rate variation of a factor 2-3 in a toroidal near-Earth dust environment [8]. The relative inclination of Earth's and Jupiter's orbital planes varies between about 0 and 3 degrees with a period of about 100,000 years, which implies an inclination-forced accretion rate variation in the form of recurrent peaks [18].

5. Detection of accretion increases

Small particles suffer little melting during atmospheric entry, and hence retain much Helium-3 collected from solar wind [25]. The extraterrestrial fraction of He-3 in marine sediments has been found to correlate with Earth's orbital elements [26], but the relative importance of the eccentricity and the inclination is still an open question. The periodicity of the He-3 variation correlates well with the eccentricity, but the phases are reversed; when the models predict an increased accretion the measured He-3 abundance is decreasing [8]. The inclination-forced accretion events are difficult to confirm chemically, since samples have to be relatively large to yield detectable abundances, and the events will therefore be confirmed (or refuted) through direct observations of dust in Antarctic ice.

5.1. DUST-INDUCED ENVIRONMENTAL EFFECTS?

An important part of stratigraphical work is the deduction of sedimentological responses to environmental forcings. The possibility for very high dust accretion rates makes it necessary to consider effects with no analogies on the modern Earth.

Inclination-driven accretion has been suggested as a driving force for the 100,000 year periodicity of Earth's ice-ages [27] through periodic dust-induced cooling, and a similar eccentricity-driven dust cooling has been proposed as an explanation for the demise of dinosaurs preceding the K/T-boundary extinction [8]. Aerosol climatology is, however, not well understood, and the inclusion of mineral dust in climate models of the last glacial maximum led to a 2.4°C *warming* of high latitudes, even if the global effect was a small cooling [28]. Moreover, the albedo of snow decreases noticeably for dust mass concentrations of only 1 ppm, assuming 5 μm particles [29], so even a small dust injection could lead to a surface melting on an ice sheet initially in steady state. This increases the surface dust concentration further, since particles larger than a few microns are immobile during the melt [30]. The Greenland snow accumulation during the last glacial maximum was ≈ 70 kg m^{-2} yr^{-1} [31], and a positive melt-dust feedback loop could hence be initiated with an interplanetary dust flux of about 1000 times the present, even without the atmospheric warming mentioned above. The dust-induced surface warming is purely hypothetical, but shows that much work remains to be done before the climatological relevance of interplanetary dust can be evaluated.

A high accretion rate also creates a possibility for more direct dust-induced effects on Earth's biology. This is a problem related to new findings in biogeochemistry, and can be exemplified with iron-fertilization. Plankton in waters far from continental mineral sources are often growth-limited by the amount of available iron [32]. In the modern equatorial Pacific Ocean most iron is supplied through atmospheric fall-out, which amounts to about 10^{-3} g yr^{-1} m^2 [33]. This is about 65 times larger than the current interplanetary iron input, assuming a dust iron abundance of $\sim 20\%$ [34], so it is reasonable to expect that large accretion rate would affect many marine organisms by providing otherwise sparse nutrients.

6. A note about the future

The area of "zodiacal geology" is too young to have produced any far-reaching conclusions, but much work is in progress. During the next decade we will no doubt see an enormous progress in our understanding of dust-planet interactions.

Regardless of results from specific investigations, it is already clear that planets cannot be viewed as closed systems *a priori*. We can only speculate about life on a planet where any wind fills the atmosphere with clouds of interplanetary dust, where a frozen ocean turns black in weeks, and where asteroidal debris provides important nutrients for organisms. Nonetheless, this scenario may be common in many planetary systems. A better understanding of such environments will require a great deal of interdisciplinary collaboration, and probably a highly creative methodology since we have only a vague idea about what to expect. Indeed, many relevant questions may not even have been asked yet.

References

1. Low, F.J., Young, E., *et al.* (1984) Infrared Cirrus: New Components of the Extended Infrared Emission, *Astrophys. J.*, **278**, pp. L19-L22
2. Spiesman, W.J., Hauser, M.G., . (1995) Near and Far Infrared Observations of Interplanetary Dust Bands from the COBE Diffuse Infrared Background Explorer, *Astrophys. J.*, **442**, pp. 662-667
3. Dermott, S.F., Nicholson, P.D., Burns, J.A., and Houck, J.R. (1984) Origin of Solar System Dust Bands Discovered by IRAS, *Nature*, **312**, pp. 505-509
4. Wisdom, J. (1982) The Origin of the Kirkwood Gaps: A Mapping for Asteroidal Motion Near the 3/1 Commensurability, *Astron. J.*, **87**, pp. 577-593
5. Backman, D., Beichman, C., *et al.* (1998) Extrasolar Zodiacal Emission - NASA Panel Report, in D.E. Backman, L.J. Caroff, S.A. Sanford, and D.H. Wooden (eds.), *Exozodiacal Dust Workshop*, NASA, Moffett Field, CA, pp. 297-328
6. Legér, A., Mariotti, J.M., *et al.* (1996) Could We Search for Primitive Life on Extra-solar Planets in the Near Future?, *Icarus*, **123**, pp. 249-255
7. Woolf, N.J. (1997) Bringing Planet Finder Closer: 1AU Missions for Terrestrial Planet Finder, Precursor and Follow-on, *Bull. American Astron. Soc.*, **191**, #64.04
8. Kortenkamp, S.J. and Dermott, S.F. (1998) A 100,000-Year Periodicity in the Accretion Rate of Interplanetary Dust, *Science*, **280**, pp. 874-876
9. Love, S.G. and Brownlee, D.E. (1993) A Direct Measurement of the Terrestrial Mass Accretion Rate of Cosmic Dust, *Science*, **262**, pp. 550-553
10. Mojzsis, S.J., Arrhenius, G., *et al.* (1996) Evidence for Life on Earth Before 3,800 Years Ago, *Nature*, **384**, pp. 55-61

11. Kennett, J. (1982) *Marine Geology*, Prentice-Hall, Inc., Englewood Cliffs, N.J.
12. Brownlee, D.E. (1996) The Elemental Composition of Interplanetary Dust, in B.A.S. Gustafsson and M.S. Henner (eds.), *Physics, Chemistry, and Dynamics of Interplanetary Dust*, ASP, San Fransisco, CA, pp. 261-273
13. Kortenkamp, S.J. and Dermott, S.F. (1996) Naturally Occurring Selection Effects of the Terrestrial Accretion of Interplanetary Dust Particles, in B.A.S. Gustafsson and M.S. Hanner (eds.), *Physics, Chemistry and Dynamics of Interplanetary Dust*, ASP, Gainsville, Fl, pp. 167-170
14. Gustafson, B..S. (1994) Physics of Zodiacal Dust, *Annu. Rev. Of Earth Planetary Sci.*, **22**, pp. 553-595
15. Burns, J.A., Lamy, P.L., and Soter, S. (1979) Radiation Forces on Small Particles in the Solar System, *Icarus*, **40**, pp. 1-48
16. Jackson, A.A. and Zook, H.A. (1992) Orbital Evolution of Dust Particles from Comets and Asteroids, *Icarus*, **97**, pp. 70-84
17. Dermott, S.F., Nicholson, P.D., Gomes, R.S., and Malhotra, R. (1990) Modeling the IRAS Solar System Dust Bands, *Adv. Space Res.*, **10**, pp. 171-180
18. Olsson, B. (1998) Direct Measurements of Upper Limits for Transient Density Fluctuations in the Zodiacal Cloud, *Bull. American Astron. Soc.*, **191** , #69.04
19. Levison, H.F. and Duncan, M.J. (1994) The Long-Term Dynamical Behavior of Short-Period Comets, *Icarus*, **108**, pp. 18-36
20. Laskar, J. and Simon, J.L. (1988) Fitting a Line to a Sine, *Cel. Mech.*, **43**, pp. 37-45
21. Laor, A. and Draine, B.T. (1993) Spectroscopic Constraints on the Properties of Dust in Active Galactic Nuclei, *Astrophys. J.*, **401**, pp. 441-468
22. Zappala, V., Cellino, A., Farinella, P., and Milani, A. (1994) Asteroid Families: Extension to Unnumbered Multiopposition Asteroids, *Astron. J.*, **107**, pp. 772-801
23. Reach, W.T., Franz, B.A., and Weiland, J.L. (1997) The Three-Dimensional Structure of the Zodiacal Dust Bands, *Icarus*, **127**, pp. 461-484
24. Flynn, G.J. (1990) The Near-Earth enhancement of Asteroidal over Cometary dust, in *Proceedings of the 20th Lunar and Planetary Science Conference*, Lunar and Planetary Institute, Houston, pp. 363-371
25. Farley, K.A., Love, S.G., and Patterson, D.B. (1997) Atmospheric Entry Heating and Helium Retentivity of Interplanetary Dust Particles, *Geochim. Cosmochim. Acta*, **61**, pp. 2309-2316
26. Farley, K.A. and Patterson, D.B. (1995) A 100-kyr Periodicity in the Flux of Extraterrestrial 3He on the Sea Floor, *Nature*, **378**, pp. 600-603
27. Muller, R.A. and MacDonald, G.J. (1995) Glacial Cycles and Orbital Inclination, *Nature*, **377**, pp. 107-108
28. Overpeck, J., Rind, D., Lacis, A., and Healy, R. (1996) Possible Role of Dust-Induced Warming in Abrupt Climate Change During the Last Glacial Period, *Nature*, **384**, pp. 447-449
29. Warren, S.G. and Wiscombe, W.J. (1980) A Model for the Spectral Albedo of Snow. II: Snow Containing Atmospheric Aerosols, *J. Atmos. Sci.*, **37**, pp. 2734-2745
30. Conway, H., Gades, A., and Raymond, C.F. (1996) Albedo of Dirty Snow During Conditions of Melt, *Water Resources Research*, **32**, pp. 1713-1718
31. Cuffey, K.M. and Clow, G.D. (1997) Temperature, Accumulation, and Ice Sheet Elevation in Central Greenland Through the Last Deglacial Transition., *J. Geophys. Res.*, **102**, pp. 26383-26396
32. Martin, J.H. and Gordon, R.M. (1988) Northeast Pacific Iron Distributions in Relation to Phytoplankton Productivity, *Seep-Sea Res.*, **35**, pp. 177-196
33. Duce, R.A. and Tindale, N.W. (1991) Atmospheric Transport of Iron and its Deposition in the Ocean, *Limn. Oceanogr.*, **36**, pp. 1715-1726
34. Schramm, L.S., Brownlee, D.E., and Wheelock, M.M. (1989) Major Element Composition of Stratospheric Micrometeorites, *Meteoritics*, **24**, pp. 99-112

CIRCUMSTELLAR DISKS AND OUTER PLANET FORMATION

A. LECAVELIER DES ETANGS
Institut d'Astrophysique de Paris
98 Bld Arago, F-75014 Paris, France

Abstract.

The dust disk around β Pictoris must be produced by collision or by evaporation of orbiting Kuiper belt-like objects. Here we present the Orbiting-Evaporating-Bodies (OEB) scenario in which the disk is a gigantic multi-cometary tail supplied by slowly evaporating bodies like Chiron. If dust is produced by evaporation, a planet with an eccentric orbit can explain the observed asymmetry of the disk, because the periastron distribution of the parent bodies are then expected to be non-axisymmetric. We investigate the consequence for the Kuiper belt-like objects of the formation and the migration of an outer planet like Neptune in Fernández's scheme (1982). We find that bodies trapped in resonance with a migrating planet can significantly evaporate, producing a β Pictoris-like disk with similar characteristics like opening angle and asymmetry.

We thus show that the β Pictoris disk can be a transient phenomenon. The circumstellar disks around main sequence stars can be the signature of the present formation and migration of outer planets.

1. Introduction

The infrared excess Vega-like stars and their circumstellar dusty environment have been discovered by IRAS in the 80's (Aumann et al. 1984). Among these infrared excess stars, β Pictoris has a very peculiar status because images have shown that the dust shell is in fact a disk seen edge-on from the Earth (Smith & Terrile 1984) and have given unique information on the dust distribution. The disk morphology and the inferred spatial distribution of the dust have been carried out in great details (Artymowicz et al. 1989, Kalas & Jewitt 1995). The morphological properties can be summarized as follows (see Lecavelier des Etangs et al., 1996, hereafter LVF):

J.-M. Mariotti and D. Alloin (eds.), Planets Outside the Solar System: Theory and Observations, 95–103.

First, the gradient of the scattered light follows a relatively well-known power law. But the slope of this power law changes at about 120 AU from the star (Golimowski et al. 1993). Second, the disk has an inner hole with a central part relatively clear of dust (Lagage & Pantin 1994). In the third dimension, the disk is a "wedge" disk: the thickness increases with radius (Backman & Paresce 1993). More surprisingly, the disk is not symmetric with one branch brighter than the other (see details in Kalas & Jewitt, 1995). Finally, the inner part of the disk ($\sim$ 40 AU) seems to be warped. This warp has been well-explained by Mouillet et al. (1997) as due to an inclined planet inside the disk.

As the dust particle life-time is shorter than the age of the system, one must consider that the observed dust is continuously resupplied (Backman & Paresce 1993). In order to explain the origin of the dust in the β Pictoris disk and these well-known morphological properties, we have proposed the *Orbiting-Evaporating-Bodies* model (hereafter OEB) (see LVF).

After a brief summary of the OEB scenario (Sect. 2), we present its consequences on the explanation for the presence of CO (Sect. 3) and the asymmetry (Sect. 4). Then, we will see that the βPictoris disk can be a natural consequence of the formation of Neptune-like outer planets (Sect. 5).

2. Summary of the Orbiting-Evaporating-Bodies Scenario

The observed dust is continuously resupplied. Two mechanisms can produce dust in this low density disk: collision or evaporation of kilometer-sized parent bodies. In both cases, because of the radiation pressure, the particles ejected from the parent bodies follow very eccentric orbits whose eccentricity is related to the grain size (Burns et al. 1979). If we assume a zero-order model of a narrow ring of bodies producing dust, the particles are then distributed on a disk-like structure presenting three morphological similarities with the β Pictoris disk. First, the central region of the disk is empty of dust, its limit corresponds to the inner radius of the parent bodies' orbits. Second, this zero-order model disk is open because the distribution of the particles inclinations are the same as that of the parent bodies. Last, the dust density is decreasing with the distance to the star, moreover this density distribution follows a power law. Consequently, if seen edge-on from the Earth, the radial brightness profile along the mid-plane of this disk follows also a power law: $F(r) \propto r^{-\alpha}$, with $\alpha \sim 5$ (LVF).

We can conclude that a ring of parent bodies on circular orbits can naturally produce a disk with an inner hole, which is open, and if seen edge-on, the scattered light distribution follows a power law.

But the slope of this zero order model is steeper than the observed slope of the power law in the β Pictoris disk ($\alpha \sim 4$, Kalas & Jewitt 1995).

To explain this distribution, it is possible that the dust is produced by collisions of Kuiper belt-like objects spread in a wide range of distances. But, an alternate solution is also possible in keeping the assumption that the parent bodies remain in a narrow ring close to the star. In that case, a large quantity of small particles is needed, because these particles have larger apoastron and can explain the less steep slope in the power law. These small particles can typically be produced by the evaporation of parent bodies of size $\gtrsim$ 10km and located at large distances. Indeed, if the evaporation rate is small enough, there is a cut-off on the maximal size of the particles which can be ejected from the bodies gravitational field by the evaporating gas. This slow evaporation and peculiar particle size distribution is observed in the Solar System around Chiron (Elliot et al. 1995, Meech et al. 1997).

Several arguments are in favor of this scenario in the case of the βPictoris disk. First, it is obviously easy to explain any asymmetry even at large distances, because a planet in the inner disk (on eccentric orbit) can have influence on the distribution of nearby parent bodies, and this non-axisymmetric distribution is projected outward by the particle on very eccentric orbits. Of course the CO/dust ratio is one of the arguments which is in favour of the OEB scenario (see Sect. 3). Finally, the connection between the inner radius of the disk and evaporation limit is a direct consequence of this scenario because the periastron distances of the particles are similar to the periastron of the parent bodies. Any hypothetical planet at this limit is no more needed to explain the presence of the inner void in the disk.

3. The Carbon Bearing Gas: a Clue to Evaporation

3.1. A NEEDED SOURCE OF CO

An important characteristic of the β Pictoris gaseous disk is the presence of cold CO and C I (Vidal-Madjar et al. 1994). CO is cold with a typical temperature of less than 30 K which corresponds to the temperature of CO-evaporation; for instance, with an albedo of 0.5 this temperature corresponds to an evaporating body located between 100 and 200 AU. CO and C I are destroyed by UV interstellar photons and have lifetime shorter than the star age ($t_{CO} \sim t_{CI} \sim$200 years). A permanent replenishment mechanism must exist. To estimate the supplying rate of CO, one must assume a cloud geometry which gives the connection between the observed column density and the total CO mass. Assuming a disk geometry with an opening angle similar to the dust disk ($\theta = 7$ degrees), and a characteristic distance given by the CO temperature ($r_0 = 150$ AU), we get a mass of CO: $M_{CO} \approx 4\pi\theta\mu_{CO}N_{CO}r_0^2 \approx 7 \times 10^{20}$kg, where μ_{CO} is the molecular weight and $N_{CO} = 2 \times 10^{15}$ cm^{-2} is the column density of CO. Then, the known photodissociation rate of CO, $\tau_{CO} = 2 \cdot 10^{-10}$s^{-1} (Van Dishoeck & Black,

1988) gives a relation between the total CO mass and the corresponding supplying rate. We obtain $\dot{M_{CO}} = M_{CO}\tau_{CO} \approx 10^{11}$kg s^{-1}.

We can also estimate the needed supplying rate of dust $\dot{M_d} = M_d/t_d \approx 10^{11}$kg s^{-1}, where M_d is the mass of the dust disk ($M_d \sim 10^{23}$kg), t_d is the dust life-time ($t_d \approx 10^4$yr). It is very interesting to see that the dust/CO supplying rate is consequently $\dot{M_d}/\dot{M_{CO}} \approx 1$. This very similar to the ratio in the material supplied by evaporation in the solar system. This provide an *independent* evidence that the β Pictoris dust disk can be supplied by *Orbiting-Evaporating-Bodies*.

We can now estimate the number of bodies producing this CO. If we take an evaporation rate of CO per body $Z_{\text{body}} \sim 5 \times 10^{28}$body^{-1}s^{-1}, N_{CO} the number of bodies now evaporating CO around β Pictoris must be $N_{CO} = (M_{CO}\tau_{CO})/(Z_{\text{body}}\mu_{CO}) \approx 6 \cdot 10^7$ bodies. This number is extremely large but unavoidable because CO is *observed*. These $\sim 10^7$–10^8 objects must be compared to the 10^8–10^9 objects believed to be present between 30 and 100 AU from the Sun as the source of the Jupiter Family Comets. Anyway, the mass of parent bodies required by the evaporation process (about one Earth mass, provided that some process is able to start its evaporation) is well below the mass needed to supply the β Pictoris disk only by collision (30 Earth, Backman et al. 1995).

3.2. β PICTORIS A TRANSIENT PHENOMENON ?

It could be difficult to imagine that $\sim 10^8$ bodies have always been active for $\sim 10^8$ years. $M_{CO} \times 10^8$years $\times\, \tau_{CO} = 20 M_{\text{Earth}}$ of CO should have been evaporated ! It seems unlikely that this large number of bodies have been active since the birth of the system. This gives evidence that either that β Pictoris is very young or that it is a transient phenomenon. There is in fact no reason to believe that the β Pictoris system was always as dusty as observed today. Of course, the idea that this disk is not transient is a consequence of any model of collisional erosion from asteroid to dust. But with other scenarios, we can easily imagine that a particular phenomenon occurred recently, and that the density of the β Pictoris disk must be significantly smaller during the quiescent phase of simple collisional erosion during which the density can be similar to the characteristic density of the more common Vega-like stars.

4. The Asymmetry Problem

In contrast to a production by a set of collisional bodies at very large distances where planets have no influence, if the dust is produced by a narrow ring of orbiting evaporating bodies, these bodies must be close to the star where the planetary perturbations can be important. In this case,

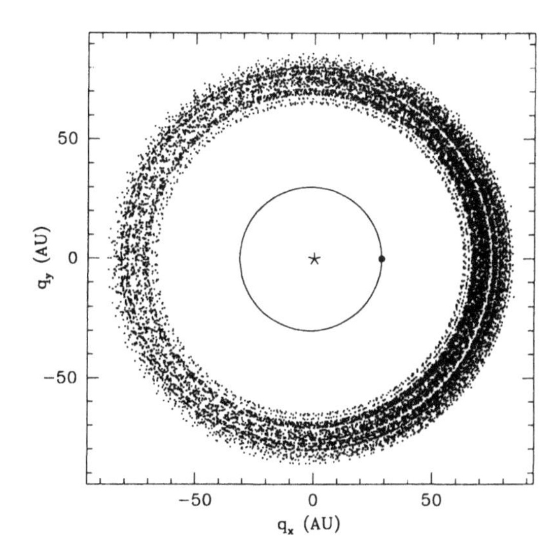

Figure 1. Plot of the spatial distribution of the periastron of a set of bodies located between 70 and 90 AU and perturbed by a planet on eccentric orbit ($M_p = 3 \cdot 10^{-4} M_{\odot}$, $e_p = 0.05$, $a_p = 30$ AU). We see that the density of bodies with periastron in the direction of the planet periastron (black dot) is very large. Moreover the periastron distances are also smaller. For these two reasons, the dust production by evaporation must be larger in this direction. If these bodies evaporate, they produce a dust disk which must be asymmetric.

the asymmetry can simply be due to an eccentric orbit of the perturbing planet. For instance, one major planet on an eccentric orbit can cause a modulation of the precessing rate of the periastron of the OEBs. It is thus well-known that the distribution of the perihelion of the asteroids in the Solar System is not axisymmetric, and is closely related to the Jupiter longitude of the perihelion (Kiang 1966). The density of asteroids with the same longitude of perihelion as Jupiter is thus ~ 2.5 times larger than that with periastron in opposite direction. This is simply because when the periastron of an asteroid is located at 180 degrees from the periastron of Jupiter, the precessing rate is quicker and the density is smaller.

Such an effect would obviously cause an asymmetry in a disk *if it is produced by evaporation* of bodies with a distribution of periastron perturbed in this way. As the dust is mainly produced at the periastron of the parent bodies and principally observed during the apoastron, the part of the disk at 180 degrees from the perturbing planet periastron could be more dense (an example of such a situation is given in Fig 1).

5. Resonances with a Migrating Planet

Possible origin of these OEBs, or more exactly the perturbations necessary to explain their evaporation, have to be explored. Indeed, evaporation takes place only when a body is formed beyond a vaporization limit of a volatile and its periastron distance then decreases below this limit.

5.1. THE FORMATION OF URANUS AND NEPTUNE

To solve the problem of time scales for the formation of the outer planet of the Solar System, Fernández (1982) suggested that the accumulation and scattering of a large number of planetesimals is the origin of the migration of the outer planets during their formation. This migration is essentially due to the exchange of angular momentum between Jupiter and the proto-Uranus and proto-Neptune, via the accretion and gravitational scattering of planetesimals, the orbit of Jupiter loses angular momentum and shifts slightly inward, while those of Saturn, Uranus and Neptune move outwards by several AU. This model successfully explains the formation of the two outer planets of the Solar System, in short time scale ($2 \cdot 10^8$ to $3 \cdot 10^8$ years), their mass and their actual position (Fernández & Ip 1996).

The consequences of this scenario on the structure of the outer Solar System has been investigated by Malhotra (1993, 1995) who showed that this also explain the particular orbit of Pluto with its large eccentricity and inclination, and its resonance with Neptune. In short, Pluto was trapped in the orbital commensurability moving outward during the expansion phase of Neptune's orbit. The outward migration of Neptune can also explain the fact that numerous Kuiper belt objects are observed in Pluto-like orbit in 2:3 resonance with Neptune (Malhotra 1995).

5.2. PLANET MIGRATION AND PERTURBATION ON PARENT BODIES.

With this in mind, it is interesting to evaluate the possible link between the migration of outer planets and the β Pictoris-like circumstellar disks for which we know that the age is similar to the time scale of formation of these planets (10^8 years is about the age of βPictoris and α PsA). Following Malhotra, we numerically investigate the consequence of the migration of the planets in the Fernández's scheme on the dynamical evolution of the planetesimals, and their possible trapping in resonant orbits which allow evaporation of frozen volatiles. For simplicity we consider only one outer massive planet supposed to suffer an exchange of orbital angular momentum as a back-reaction on the planet itself of the planetesimal scattering. Of course, at least a second inner planet must be there. Here, we consider only the principal outer perturbing planet which is supposed to migrate because

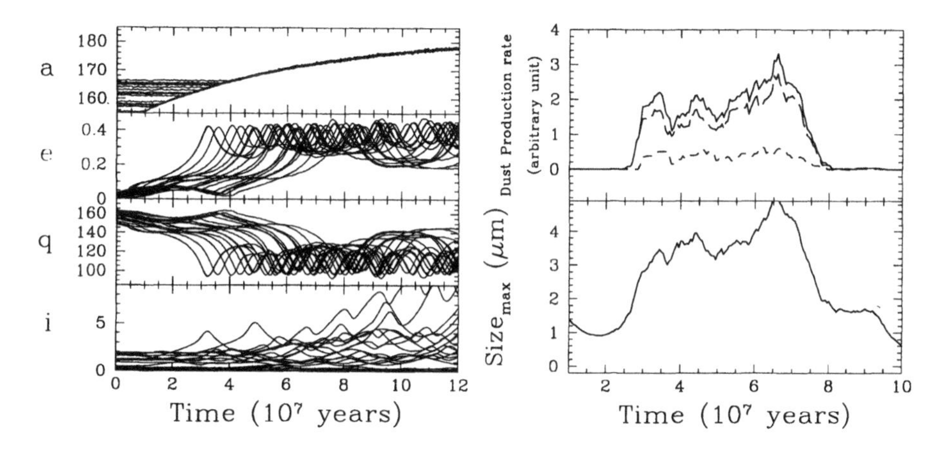

Figure 2. Left panel. Plot of the orbital parameters semi-major axis (a), eccentricity (e), periastron (q) and inclination (i) of 19 bodies trapped in the 1:4 resonances with a migrating planet. We see that although the semi-major axis of the bodies increase, their periastron decrease. These bodies can start to produce dust by evaporation.
Right panel. Plot of the dust production as a function of time for evaporating bodies trapped in 1:4 resonance with a migrating planet. This is the dust production for grains larger than $2\mu m$, because smaller grains are supposed to be quickly expelled by the radiation pressure. The production of dust starts when the periastron distance of the parent bodies is small enough for the CO production rate to allow ejection of grains larger than $2\mu m$. Then, it stops when the parent bodies are exhausted and have no more volatile. Consequently, the dust production is transient and is large only when the bodies trapped in the resonances are entering in the evaporation limit. The production rate is also not axisymmetric. As in Sect. 4, we see that the production is larger in the direction of the periastron of the perturbing planet (long dashed) than in the opposite direction (short dashed).
The right bottom panel gives the corresponding maximal size of grains ejected from the bodies by the evaporating gas. Because the periastrons are still larger than 100 AU, the evaporation produces only small particles. In this simulation, the maximal particle size is around $4\mu m$ as expected to explain the slope $\alpha \sim 4$ observed in the β Pictoris case.

of a force equivalent to a drag force decreasing with time: $F_D \propto e^{-t/\tau}$, where τ is the characteristic time of the migration.

In fact, if the migrating planet is moving inward, the planetesimals are not trapped in the resonances. Their semi-major axis remain unchanged and their eccentricities are only slightly increased. Consequently, the decrease of the periastron distance is too small to allow the volatiles to evaporate.

On the contrary, if the outer planet is moving outward, as found in the models of Fernández & Ip (1996), a fraction of bodies can be trapped in resonances. Their semi-major axis and eccentricities increase significantly and the net result is a decrease of their periastron. This can start the evaporation of the trapped bodies.

We have tested several configurations of outward migration and have evaluated the effect on planetesimals in the zones swept by first order resonances. The 1:2 and 1:3 resonances does not allow to explain the observed characteristics of the β Pictoris disk. The 1:4 resonance give the most interesting results (Fig. 2). The trapping has been found to be efficient if the mass of the planet is $M_p \gtrsim 0.5 M_J$, where M_J is the mass of Jupiter, and if the migration rate is $\tau \gtrsim 5 \cdot 10^7$ years. With these conditions, the periastron of trapped bodies significantly decreased. A significant increase of the inclination has also been observed after few τ as well as a large asymmetry in the distribution of the periastrons longitude. The 1:5 resonance is efficient in trapping only if the parameters of the migrating planet are extreme with $M_p \gtrsim M_J$, $e_p \gtrsim 0.1$ and $\tau \gtrsim 5 \cdot 10^7$ years.

5.3. ASYMMETRY AND OPENING ANGLE

If the bodies are trapped in a resonance with a planet on eccentric orbit, there can be an asymmetry in the distribution of the periastron as already seen in Sect. 4. For example, the Fig. 2 gives the dust production rate by the bodies trapped in the 1:4 resonance with a planet on an eccentric orbit ($e_p = 0.05$). The production rate is larger in the direction of the periastron of the planet than in the opposite direction. The disk thus produced must be asymmetric with a larger density in the direction of the apoastron of the migrating planet.

From Fig. 2, we also conclude that the production of dust can take place with the inclination of the parent bodies larger than the initial inclination, up to several degrees. Moreover, with several giant planets, the precession of the ascending nodes can produce an additional increase of the parent bodies inclination. In all cases, this migrating and resonance trapping process gives a large increase in the inclinations and consequently a large opening angle of the associated dust disk.

6. Conclusion

Collisions and evaporation are the two main processes believed to be able to supply disks like the β Pictoris one. These two processes are not exclusive. However, the β Pictoris disk is more likely produced by the evaporation process. The CO and C I gas detected with HST definitely shows that evaporation takes place around β Pictoris, even if its consequence on dust replenishment in comparison to the collisional production is still a matter of debate. The dust spatial distribution with the slope of the power law, and the central hole can be explained by the characteristic distances of evaporation. Finally, the asymmetry at large distances can easily be explained in evaporating scenarios because the parent bodies are maintained close to the star where planets' influences are important. The asymmetry is then simply a consequence of the non-axisymmetry of the perturbation by planet(s) on eccentric orbits.

We have shown the possibility that bodies trapped in resonances with a migrating planet can evaporate. The large number of CO evaporating bodies is explained by a transient evaporation during a short period.

From another point of view, if the migration of the outer planets took place in the Solar System, why not around other stars ? This is in fact a simple consequence of the presence of a forming planet inside a disk of residual planetesimals. Here we have explored a new consequence of this migration of a forming planet. Some planetesimals can be trapped in resonances, enter inside evaporation zone and finally become parent bodies of β Pictoris-like disks. In short, as a direct consequence of the formation of outer planets in the Fernández's scheme, evaporation of Kuiper belt-like objects around bright stars can be expected to be common. This allows us to look at the circumstellar disks around main sequence stars as a possible signature of outer planet formation.

Acknowledgements

I am particularly grateful to J.M. Mariotti and D. Alloin for organizing this very interesting and fruitful meeting.

References

1. Artymowicz P., Burrows C., Paresce F., 1989, ApJ 337, 494
2. Aumann H.H., Gillett F.C., Beichman C.A., et al., 1984, ApJ 278, L23
3. Backman D.E., Paresce F., 1993, in *Protostars and Planets III*, Eds. E.H. Levy, J.I. Lunine & M.S. Matthews (Tucson: Univ. Arizona Press), pp 1253
4. Backman D.E., Dasgupta A., Stencel R.E., 1995, ApJ 450, L35
5. Burns J., Lamy P., Soter S., 1979, Icarus 40, 1
6. Elliot J.L., Olkin C.B., Dunham E.W., et al., 1995, Nature 373, 46
7. Fernández J.A., 1982, A.J. 87, 1318
8. Fernández J.A., Ip W.H., 1996, Planet. Space Sci. 44, 431
9. Golimowski D.A., Durrance S.T., Clampin M., 1993, ApJ 411, L41
10. Kalas P., Jewitt D., 1995, AJ 110, 794
11. Kiang T., 1966, Icarus 5, 437
12. Lagage P.O., Pantin E., 1994, Nature 369, 628
13. Lecavelier des Etangs A., Vidal-Madjar A., Ferlet R., 1996, A&A 307, 542, (LVF)
14. Lecavelier Des Etangs A., Vidal-Madjar A., Burki G., et al., 1997, A&A 328, 311
15. Malhotra R., 1993, Nature 365, 819
16. Malhotra R., 1995, AJ 110, 420
17. Meech K.J., Buie M.W., Samarasinha N.H., et al., 1997, AJ 113, 844
18. Mouillet D., Larwood J.D., Papaloizou J.C.B., Lagrange A.M., 1997, MNRAS 292, 896
19. Smith B.A., Terrile R.J., 1984, Sci 226, 1421
20. Sylvester R.J., Skinner C.J., Barlow M.J., Mannings V., 1996, MNRAS 279, 915
21. Van Dishoeck E.F., Black J.H., 1988, ApJ 334, 771
22. Vidal-Madjar A., Lagrange-Henri A.M., Feldman P.D., et al., 1994, A&A, 290, 245

DYNAMICAL INTERACTION OF PLANETS IN THE CIRCUMSTELLAR DISK

P. ARTYMOWICZ
Stockholm Observatory
13336 Saltsjobaden, Sweden

Manuscript not provided.

J.-M. Mariotti and D. Alloin (eds.), Planets Outside the Solar System: Theory and Observations, 105.

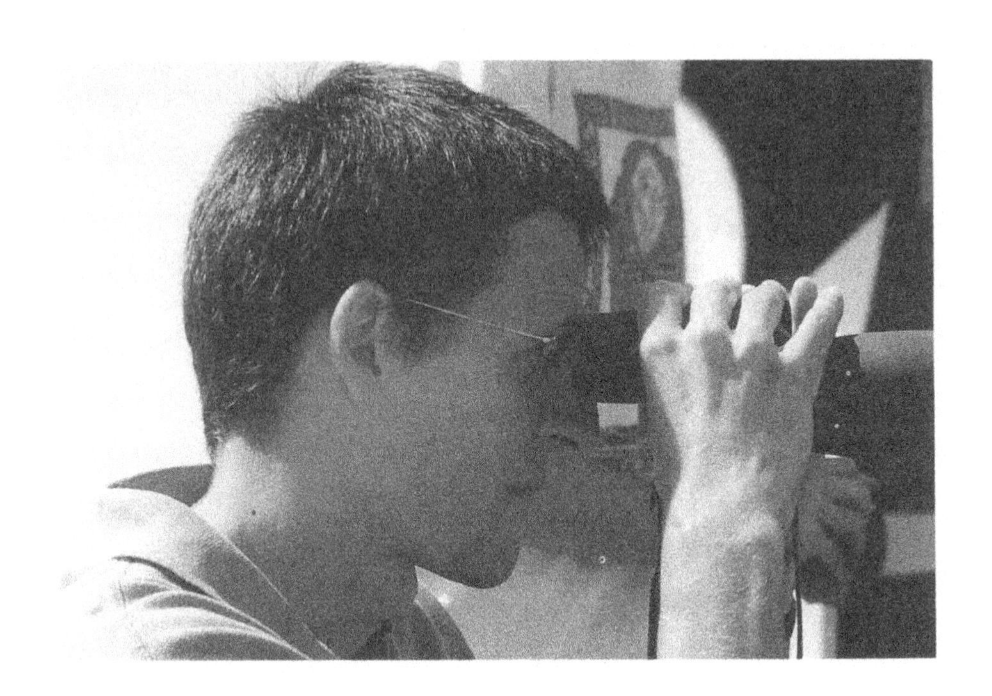

THE NEW PLANETARY SYSTEMS

D. QUELOZ
Observatoire de Genève
51 ch des Maillettes
1290 Sauverny
Switzerland †

Abstract. A summary on recent results about planet detections are presented. Some results about brown-dwarfs are also discussed in the context of the maximum mass for a giant planet and the realm of planetary formation scenarios. The comparison between companion orbital parameter distributions and the ones for stellar and substellar mass objects are made. This places interesting constraints on recent planetary formation models. A tentative binary mass function for substellar companions and a first estimate of the occurence of 51Peg like planets are also derived.

1. Introduction

The recent growth of interest in the detection and observation of extra-solar planets has been strongly driven by the recent successes in the detection of planets orbiting solar-type stars and the detection of proto-planetary disks. Within the last 5 years starting from the "...murky depths of the main sequence." (Henry *et al.*, 1994) to now, where almost 10 planetary candidates are known, huge progress – at least from the psychological point of view – has been achieved. Some of the recent detections were unexpected, mostly because they could not be predicted by the canonical theory of planetary formation. Now we know that the picture of planetary formation we had from the study of our own solar system is far from complete. The detection of giant planets very close to their stars such as 51 Peg (Mayor & Queloz, 1995), the detection of an eccentric giant planet orbiting the star 16 Cyg B (Cochran *et al.*, 1997), the detection of super Jupiter objets such

†Now Visiting Scientist at JPL, 4800 Oak Grove Drive, Pasadena, CA 91109, USA

J.-M. Mariotti and D. Alloin (eds.), Planets Outside the Solar System: Theory and Observations, 107–119.

as 70 Vir (Marcy & Butler, 1996) and the detection of a planetary system orbiting the pulsar PSR1257+12 (Wolszczan, 1994) suggest a wide variety of phenomena in the planetary formation and evolution processes of such systems which had not all been envisioned. In this chapter is reviewed the status of the search for extrasolar planets including the recent observations of brown-dwarfs. From recent results some general conclusions will be drawn about the nature and the formation of these systems.

2. Isolated (free floating) brown dwarfs

Direct detections of young brown dwarfs have been announced in the Pleiades cluster (Rebolo *et al.*, 1995) as well as in the field (Delfosse *et al.*, 1997) (Ruiz *et al.*, 1997) (Tinney *et al.*, 1997). The substellar nature of each of these objects relies upon the mass-luminosity relationship stemming from the atmosphere computation and the modeling of the internal structure of these objects as well as on the"lithium test" (see below).

The lithium (Li) is a very good tracer of the internal temperature of stellar objects because it burns very quickly at rather low temperature (25 million degree). Therefore, the presence of Li at the surface of a convective object can be used to set a lower limit to its internal temperature and therefore to its mass (Rebolo *et al.*, 1994).

The decrease in luminosity of stellar objects at the very end of the main sequence is a very steep function of their mass and of the object metallicity (Baraffe & Chabrier, 1998) (Burrows *et al.*, 1998). The age is also of prime importance in the sub-stellar regime since, for these objects, the Hydrogen burning does not provide enough support against self-gravitation. The luminosity decreases with time.

The presence of clouds (or dusts) at the surface of low temperature objects highly complicates the modeling of the mass-luminosity relationship. The presence of dust adds uncertainties to the expected critical T_e and L_bol between stars and brown-dwarfs. Moreover, for surface temperatures less than 1600K, the Li may also condense into clouds and, for old brown-dwarfs, the "Li-test" may be useless (see Burrows in this issue).

Since the first detection of a brown dwarf in the Pleiades (PPL15, which later turned out to be a binary system of two close brown-dwarfs (Basri & Martín, 1998)) a lot more substellar objects have been detected in this cluster (Zapatero Osorio *et al.*, 1997) (Bouvier *et al.*, 1998). Current surveys are now aimed at the detection and follow-up of the faintest candidates in order to measure the mass function in the sub-stellar regime. They are all trying to go fainter so as to reach the bottom of the brown dwarf sequence.

3. Binary systems

One of the main avantages of binary systems is that they provide – if the whole orbit can be resolved – a dynamic measurement of the mass of the companion. From the inferred mass and the observed luminosity of the companion, the mass-luminosity relationship can be measured and compared to models.

So far no substellar companion, with a small enough orbital separation to measure its dynamical mass, has been observed close enough to us to exhibit a large angular separation with respect its parent star. A very dim companion at 45 AU to the star Gl229 has however been detected (Nakajima *et al.*, 1995). From spectral analysis and constraints from luminosity, it turns out that this companion has a mass small enough (between 30 and 55 M_J) to be a brown-dwarf (Marley *et al.*, 1996). This important discovery proves to be the sole detection of a substellar companion amongst a sample of 120 nearby stars (< 8pc sample)(Oppenheimer *et al.*, 1999). Spectral analysis of this object has revealed water Methane and dust features. Such spectral features had never been detected on star-like objects before. It offers an unique oportunity to better understand the structure of brown-dwarfs and the internal structure of giant planets (see Burrows for details in this issue).

When the companion is too close to its bright primary star or too faint to be directly seen, we must rely on indirect techniques by using the primary star only to detect it. By looking at the reflex motion of the star due to the gravity interaction with its low mass companion, the radial velocity techniques (Doppler effect measurement) and the astrometric techniques can detect planetary companions invisible to direct detection. The best accuracies reached so far on Doppler measurements, by the few teams achieving large surveys to search for planets, are already good enough to detect giant planets (see Queloz this issue). This is not yet the case for astrometry. No astrometric instrument is achieving sub-milliarcsecond accuracy –required for planet detection– . (See in Colavita this issue for a description of performance improvements expected in the next few years by interferometry techniques.)

The detection of a low mass companion by radial velocity techniques provides us with a number of interesting parameters[1]: a, the distance to its star; e, the eccentricity of the orbit and $m_2 \sin i$, the minimum mass of the companion. Interestingly, the mass and the orbital parameters of each system might tell us something about their nature and their history (formation and evolution). The study of binary stellar systems has already

[1] if the period is short enough to be measured

provided some interesting results which are helpful for the interpretation of the orbital behaviour of substellar companions.

From the study of nearby G dwarfs (Duquennoy & Mayor, 1992), we know that at least 50% of nearby solar type stars are in fact the primary component of a multiple system. The binary mass ratio distribution (or q function, $q = f(m_2/m_1)$) is almost flat and seems to rise below $q = 0.2$ (Mazeh *et al.*, 1992), (Henry & McCarthy , 1992). The period distribution of nearby binary G dwarfs exhibits a log-normal shape centered at $\mu \approx 30$AU. This means that the chance to observe a binary star and to measure its complete orbital motion is not so high, since only 30% of binary systems have periods less than 10 yr. Most binary systems have separations too large to permit measurements of their dynamical mass by resolving their whole orbit either from radial velocity technique or from astrometry.

4. Substellar companions detected by radial velocity surveys

In the past two years, two different sets of surveys have been yielding all recent sub-stellar companion detections. A large number of stars (over 1000) have been monitored at a precision of 300-500 ms^{-1} (Carney & Latham, 1987) (Mayor *et al.*, 1992) and more restricted samples have been measured with 5-20 ms^{-1} accuracies ((Walker *et al.*, 1995) (Mayor & Queloz, 1996), (Queloz *et al.*, 1998),(Marcy & Butler, 1992), (Butler & Marcy, 1998), (Noyes *et al.*, 1998), (Cochran & Hatzes, 1994)). The big surveys, carried out with moderate accuracies were directed at objects lying at the bottom end of the Main Sequence and within the brown dwarf realm. The smaller but much more accurate surveys were directed towards planet detections. As results of these long term efforts we now have an handful of companions (see Table 1) with masses ranging from Jupiter to stellar boundary masses.

From all these first results concerning extra-solar planets, one gets a rough minimum estimate of 2-3% for the occurence of a massive planet ($< 5M_J$) orbiting a solar-type star. Suprisingly, some of these planets, lie much too close to their star to be explained by the"classical" Jupiter formation by agglomeration of ice cores (see Papaloizou, in this issue). New formation mechanisms including orbital migration have been proposed (Artymowicz, 1993), (Lin *et al.*, 1996), (Ward, 1997), (Trilling *et al.*, 1998), (Murray *et al.*, 1998). Moreover, the discovery of objects sitting in the confused borders between planets and brown-dwarfs have also brought up the issue of building a super planet by accreting matter through the disk gap cleared by the planet (Artymowicz & Lubow, 1996) (see also Artymowicz, in this issue).

TABLE 1. Table of planetary and brown-dwarf candidate companions detected by radial velocity measurements.

star	Sp type	$m_2 \sin i$ (M_J)	Period (days)	eccentricity	references
51 Peg	G2 V	0.45	4.23	0	(Mayor & Queloz, 1995)
υ And[†]	F8 V	0.6	4.61	0.11:	(Butler *et al.*, 1997)
ρ Cr B	G0 V	1.1	39.6	0	(Noyes *et al.*, 1997)
16 Cyg B	G2 V	1.6	804	0.66	(Cochran *et al.*, 1997)
Gl 876	M4 V	2.0	60.8	0.3:	(Marcy *et al.*, 1999), (Delfosses *et al.*, 1998)
55 ρ Cnc[*]	G8 V	2.3	14.7	0	(Butler *et al.*, 1997)
47 UMa	G0 V	2.4	1090	0	(Butler & Marcy, 1996)
τ Boo[‡]	F7 V	4.0	3.31	0	(Butler *et al.*, 1997)
70Vir	G4 V	7.4	117	0.37	(Marcy & Butler, 1996)
HD 114762	F9 V	9.4	84	0.35	(Latham *et al.*, 1989)
HD 110833	K3 V	17	270	0.69	(Mayor *et al.*, 1997)
HD 112758	K0 V	35	103	0.16	(Mayor *et al.*, 1997)
HD 127506	K3 V	38	2500	0.63	(Mayor *et al.*, 1998)
HD 18445	K2 V	39	553	0.54	(Mayor *et al.*, 1997)
HD 29587	G2 V	40	1472	0.37	(Mayor *et al.*, 1997)
BD -4 782	K5 V	45	710	0.07	(Mayor *et al.*, 1998)
HD 140913	G2 V	46	148	0.61	(Mayor *et al.*, 1997)
HD 89707	G1 V	54	298	0.95	(Mayor *et al.*, 1997)
HD 217580	K4 V	60	455	0.52	(Tokovinin *et al.*, 1994)

([†]) The eccentricity value is very uncertain. An extra scattering is observed on the orbital solution for this star. This indicates that an other signal is superimposed to the star. The 4d rotation period may well produce some radial velocity scattering but the light curve doesn't show any variation (G. Henry priv. comm.). Moreover the star is a late F like τBoo which doesn't show an unusual scattering in its orbital motion fit to the radial velocity data. The superimposed signal to the υ And orbital motion may be due to the presence of a third body (Marcy *et al.*, 1999).
([*]) A long term radial velocity variation of 120 ms^{-1} has also been detected. A companion with a period longer than 12 yr and a minimum mass larger than 5 M_J may also be orbiting that star.
([‡]) The rotation period matches well the orbital period. However since the orbital motion is very stable on a very long term (Mayor *et al.*, 1997) and no photometric variability or line bissector changes has been detected (Brown, 1998), (Hatzes & Cochran, 1998), the planet is the most reasonable interpretation for the observed radial velocity changes. The agreement between the two rotation values may well be explained by the fact that the planet is close enough and massive enough to tidally lock the system.

4.1. THE ECCENTRICITY DISTRIBUTION

The eccentricity of orbits has to be somehow related to the formation and evolution of systems. We can expect that objets formed inside a disk by agglomeration of particules may have different eccentricities than object formed by condensation during the gas collapse of a protostellar cloud.

The tidal interaction between the star and its companion, if they are close enough, can easily transform an originally very eccentric orbit into a circular one (circularization process) and can also lock the rotation of each object with their orbital motion (synchronization). The typical circularization time τ_e can be defined as: $1/\tau_e = \dot{e}/e$. The tidal effect from the star, scales as $1/\tau_e \sim (m_2/m_1)(R/a)^8$ with R being the star radius and a being the semi-major axis (Zahn, 1989) . Therefore we can define a critical distance $a_{\rm crit}$ for a given age t and convert it into a period: $P_{\rm crit} \sim (t\, m_2/m_1)^{3/16}$ ([2]) . In the case of 51 Peg we have $P_{\rm crit}(\text{planet}) \approx P_{\rm crit}(\text{star})/3.6$. Since the critical period of a star like the Sun is about 10 days, this leads us to a critical period of about 2-3 days for a Jupiter mass object. Therefore, following the result from this calculation, the system is separated enough not to be tidally circularized during its main-sequence life. However the tidal effect of the planet is not taken into account (see (Marcy *et al.*, 1997) and (Rasio *et al.*, 1997) for an estimation of the Q parameter of the planet) and the pre-main-sequence history of the system, when the star was bigger, is neglegted. Therefore, the use of observed circular orbits of any 51 Peg-like systems, as a tracer of their *formation* processes, has to be handled with caution.

The presence of a remote third body may also secularly change the eccentricity of the planet orbit. This phenomenon might well be the reason of the highly eccentric orbit of 61 Cyg B . This star has a companion (61 Cyg A) at 800 AU. The three-body gravity interaction between the planet and the two stars 61 Cyg A and B may easily increase the eccentricity of the orbital motion, if the orbital plane of the 2 stars is well inclined compared to the plane of the planet motion (Mazeh *et al.*, 1997) (Holman *et al.*, 1997). In studies of triple systems, such configurations (orbital pairs on 2 different planes) are always found for systems with one wide binary (Hale, 1994).

The n-body interaction between giant planets in a planetary system is also a way to build eccentric orbits. A set of giant planets in a close configuration is highly unstable (the bodies can easily interact between each other). Some planets can even be ejected from the system. The net result of such strong interactions are usually eccentric systems (Rasio & Ford, 1996), (Lin *et al.*, 1997).

On Fig. 1 the eccentricity distribution of substellar objects and solar system giant planets is compared to the eccentricity distribution of stellar companions from G and K dwarfs in solar neighborhood. The stellar distribution shows a large scattering with a mean eccentricity of about 0.4 with almost no low eccentricity objects ($e < 0.1$) except the ones tidally circu-

[2]The power index may differ a little between authors but they all agree on a value close to 1/4 (Terquem *et al.*, 1998)

Figure 1. The eccentricity distribution of planets, substellar objects and companions to solar-type stars is displayed. The dashed lines indicated the limit between the stellar and substellar objects (at $80M_J$) and the dotted line the believed location of the brown dwarf bottom-end at about $10M_J$. The empty symbols are short period orbits believed to be circularized by tidal effect. The uncertainties from the $\sin i$ is displayed with cuneiform symbols. The area of each indicates the probability to have m_2 larger than m_2 min. The location of Jupiter and Gl229 B (triangle, lower limit) is also indicated

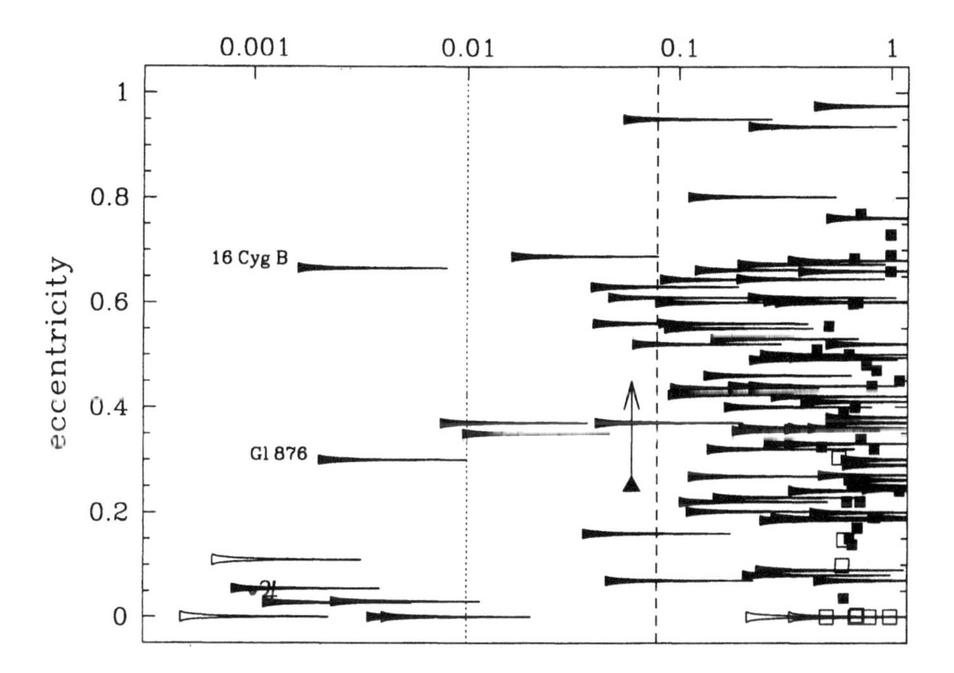

larized. For the very low mass companions in the planet range, excepted 16 Cyg B whose eccentricity may be explained by the presence of a third body and the very recent detection of an eccentric planet orbiting the star Gl 876, all objects with $m \sin i < 5M_J$ seem to have almost circular orbits ($e \approx 0$). The fact that the two object at the bottom end of the brown-dwarf sequence 70 Vir and HD114762, with respectively $m \sin i$=7.4 and 9.4 both have eccentric orbits has to be somehow related to their formation process. Since the distinction between a low mass brown-dwarf and a massive planet is still obscure, the exact nature and formation history of each of these two companions remain a puzzle.

4.2. THE MASS DISTRIBUTION OF LOW MASS COMPANIONS

The interpretation of the observed binary mass function stemming from radial velocity surveys is limited by the unknown orbital projection ($\sin i$). In particular, the number of detected sub-stellar mass objects is too small to allow us to statistically remove the $\sin i$ projection. Moreover, given the lack of knowledge of the binary mass function close to brown dwarf-stellar boundaries, it is unwise to draw early conclusions about the nature of an objet, even at the very bottom of the brown dwarf sequence. So far, the only "safe" brown-dwarf companion known to be the companion of an old star is Gl229 B. However already from the current sample surveyed in radial velocity, the very few number of brown dwarfs compared to the number of $\mathrm{m} \sin i < 5\mathrm{M_J}$ object is striking (see Fig. 2). It seems, in other words, that with a survey limited to $10\,\mathrm{ms}^{-1}$ precision it is easier to detect a planet orbiting a star than a brown-dwarf, because a planetary mass companion is observed to be more frequent than a brown-dwarf companion. This apparent rarity of brown dwarfs orbiting solar type stars helps in a sense to interpret $\mathrm{m} \sin i < 5\mathrm{M_J}$ objects as real planetary-like objects.

For an object like HD114762, it is not unrealistic to imagine that a massive brown-dwarf or even a stellar companion, rather than a planet, is orbiting that star. Some arguments, like the absence of an observed rotational broadening, suggest a small $\sin i$ value (Hale, 1995). For the companion to 70 Vir, it is more difficult to imagine (from $\sin i$ statistics) a stellar companion but maybe not a brown-dwarf. Direct imaging attempts and astrometric measurements may allow us to determine the real mass of that object in the near future. Such a result would provide us some insights into the maximum mass of a giant planet (or the minimum mass of a brown dwarf).

4.3. SEPARATION DISTRIBUTION

The separation distribution best emphasises the strong constraints steemming from the observational limits given by the Doppler technique. The amplitude of radial velocity changes (K) is proportional to the mass but decreases with the distance to the star ($K \sim \mathrm{m_2} \sin i\, \mathrm{P}^{-1/3}$). Given a detection threshold K_{lim} the $(\mathrm{m_2} \sin i)_{\mathrm{lim}}$ is thus tied to the period of the system. In other words, for a given mass, the closest planets to their stars are the easiest to detect.

On Fig. 3 one sees that a large fraction of detected planets lie closer than 0.3 AU. We also note that all these discoveries seem to lie along an invisible line not far from the $30\,\mathrm{ms}^{-1}$ detection threshold. Therefore, given our limits, it is premature to study the planet distribution as a function of distances to their stars. With the accuracy improvement of current surveys

Figure 2. Binary mass function for nearby G and K dwarfs at the low-end of the main sequence and in the substellar domain. The low and high precision surveys have been both respectively scaled to be displayed on the same scale of a 1000 stars sample. Per mass-bin the planetary-mass objects ($m_2 \sin i < 10M_J$) are cleary more numerous than sub-stellar companions within the brown-dwarf domain. The vertical dashed lines indicated the limit between the stellar and substelar objects (at $80M_J$).

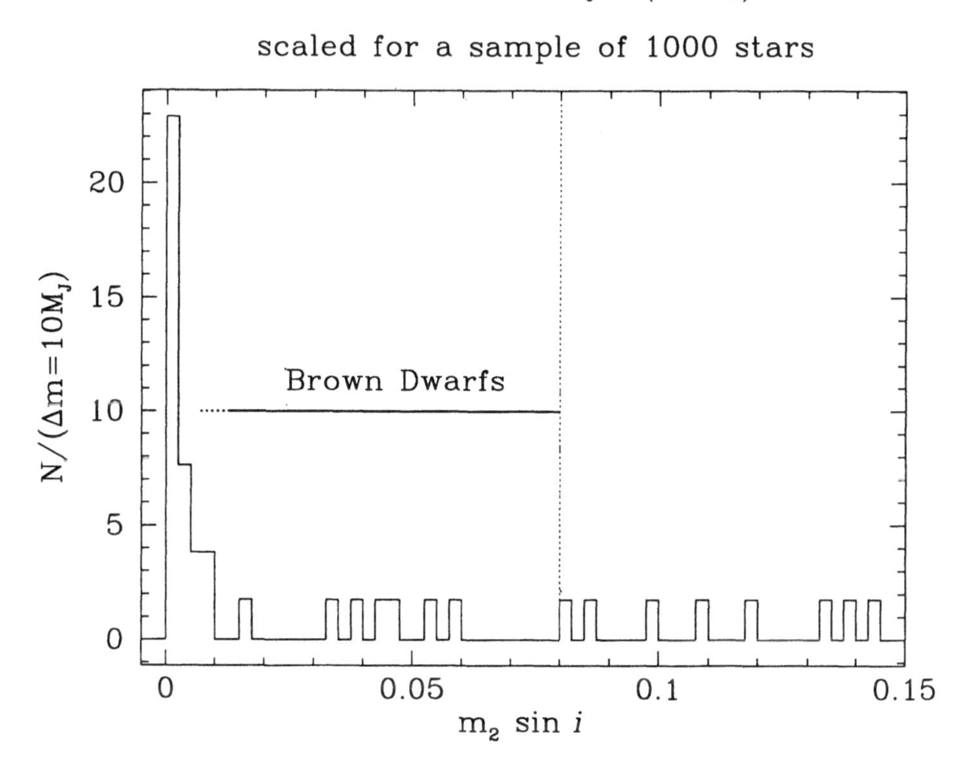

we may soon expect some progress. We should be able to figure out if their is a real stacking of giant planets between 0.05 and 0.2 AU and if the separation range from 0.1 to few AU is sparcely populated. Such results are crucial to challenge the migration theory. In the next years we should have a much better understanding of the formation of giant planets and maybe some cluses about the likelihood of finding tellurique planets like the earth.

Figure 3. Separation distribution of a sample of stellar companions to nearby G and K dwarfs observed with CORAVEL and all substellar companions to nearby stars so far detected. The uncertainties from the $\sin i$ is displayed with cuneiform symbols. The two diagonals indicates the locations of $30\,\mathrm{ms}^{-1}$ and $300\,\mathrm{ms}^{-1}$ radial velocity amplitude ($\sin i = 1$) coresponding to the detection thresholds of respectively low and high accuracy current surveys. Note that a Jupiter-like planetary companion could not have been detected.

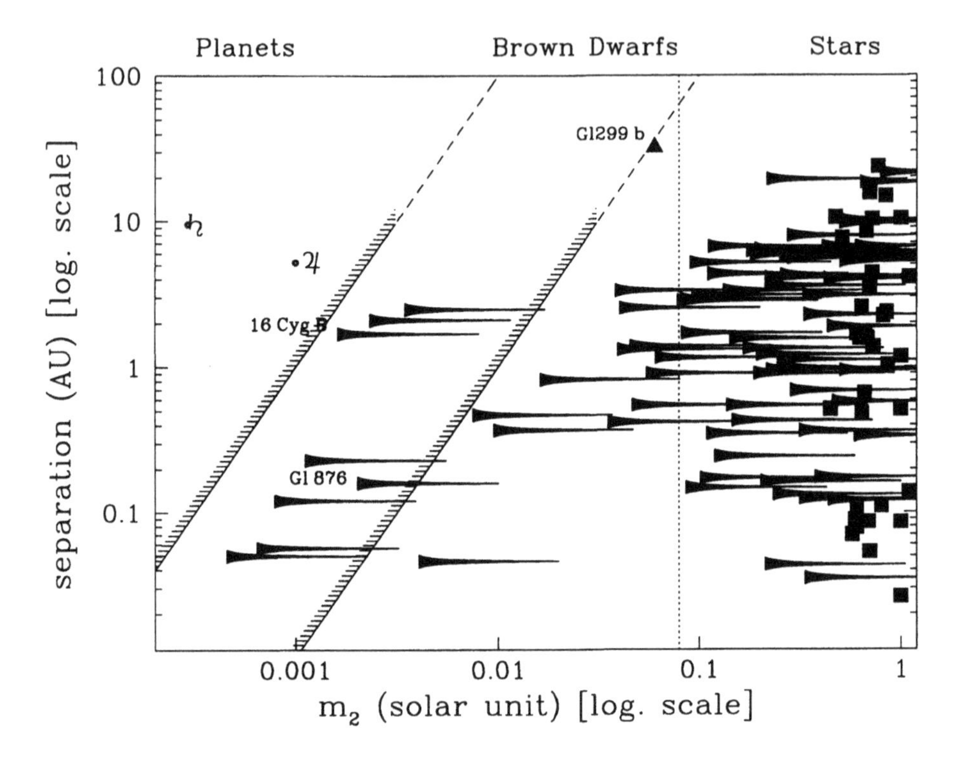

References

Artymowicz, P., 1993, "Disk-Satellite Interaction via Density Waves and the Eccentricity Evolution of Bodies Embedded in Disks", ApJ, vol. 419, 166.

Artymowicz, P., Lubow, S., 1996, "Mass flow through gaps in circumbinary disks", ApJ, vol. 467, L77.

Baraffe, I. & Chabrier, G., 1998, "Structure and Evolution of low-mass stars and brown dwarfs", Proceeding of the workshop: "Brown dwarfs and extrasolar planets", ASP Conf. Ser., vol 134, Eds: R. Rebolo, E.L. Martin and M.R. Zapatero Osorio, 345.

Basri, G., Martín, E. L., 1998, "PPL15: the first binary brown dwarf system?", Proceeding of the workshop: "Brown dwarfs and extrasolar planets", ASP Conf. Ser., vol 134, Eds: R. Rebolo, E.L. Martin and M.R. Zapatero Osorio, 284.

Bouvier, J., Stauffer, J. R., Martin, E. L., et al., 1998, "Brown dwarfs and very low-mass stars in the Pleiades cluster: a deep wide-field imaging survey", A&A, vol. 336, 490

Boss, A. P., 1988, "Protostellar formation in rotating interstellar clouds. VII -Opacity and fragmentation", ApJ, vol 331, 370.

Brown, T. M., Kotak, R., Horner, S. et al., 1998, "A search for line shape and depth variations in 51 Peagsi and τ Bootis", ApJ, 494, L85.

Burrows, A., Sudarsky, D., Sharp, C., et al., 1998, "Structure and Evolution oof low-mass stars and brown dwarfs", Proceeding of the workshop: "Brown dwarfs and extrasolar planets", ASP Conf. Ser., vol 134, Eds: R. Rebolo, E.L. Martin and M.R. Zapatero Osorio, 354.

Butler, R. P., Marcy, G. W., 1996, "A planet orbiting 47 Ursae Majoris", ApJ, vol. 464, L153.

Butler, R. P., Marcy G. W., Williams, E., Hauser, HH., Shirts, P., 1997, "Three new 51 Pegasi-type planets", ApJ, vol. 474, L115.

Butler R. P., Marcy, G. W., 1998, "The near term futur of extrasolar planet searches", in "Brown Dwarfs and Extrasolar Planets", ASP Conf. Ser., vol. 134, R. Rebolo, E. L. Martín and R. Zapatero Osorio (eds.), 162.

Carney, B. W., Latham, D. W. 1986, "A survey of proper-motion stars. I - UBV photometry and radial velocities", AJ, vol. 93, 116.

Cochran, W. D., Hatzes, A. P., 1994, "A high-precision radial-velocity survey for other planetary systems", Astrophysics and Space Science, vol. 212, 281.

Cochran, W. D., Hatzes, A. P., Butler, R.P., Marcy, G. W., 1997, "The discovery of a planetary companion to 16 Cygni B", ApJ, 483, 457.

Delfosse, X., Tinney, C.G., Forveille, et al., 1997, "Field brown dwarfs found by DENIS", A&A, vol. 328, L25.

Delfosse, X., Forveille, T., Mayor, M., et al. , 1998, "The closest extrasolar planet. A giant planet around the M4 dwarf Gl 876", A&A, vol. 338, L67

Duquennoy, A., Mayor, M., 1992, "Multiplicity among solar-type stars in the solar neighbourhood", A&A, vol. 248, 485.

Hale, A., 1994, "Orbital coplanarity in solar-type binary systems: Implications for planetary system formation and detection", AJ, vol. 107, 306.

Hale, A., 1995, "On the nature of the companion to HD114762", PASP, vol. 107, 22.

Hatzes A. P., Cochran W. D., "A search for variability in the spectral line shapes of Tau Bootis: Does this star really have a planet?", ApJ 502, 944.

Henry, T.J., McCarthy, D.W., 1992,"Complementary Approaches to Double and Multiple Star Research". IAU Coll. 135, ASP Conf. Ser., vol 32, eds. H. McAlister and W. Hartkopf, 10.

Henry, T.J, 1994,"The solar Neighbors in the Murky Depths of the Main Sequence", Proceeding of the ESO Workshop: The bottom of the main sequence and beyond, held in Garching Germany, Ed. C.G. Tinney, 79.

Holman, M., Touma, J., Tremaine, S., 1997, "Chaotic variations in the eccentricity of the planet orbiting 16 Cyg B.", Nature, vol. 386, 254.

Latham D. W., Stefanik, R. P., Mazeh, T., Mayor, M., Burki, G., 1989, "The unseen

companion of HD114762 - A probable brown dwarf", Nature, vol. 339, 38.
Lin, D. N. C., Bodenheimer, P., Richardson, D. C., 1996, "Orbital migration of the planetary companion of 51 Pegasi to its present location", Nature, vol. 380, 606.
Lin, D. N. C., Ida, S., 1997, "On the Origin of Massive Eccentric Planets", ApJ, vol. 477, 781.
Marcy, G. W., Butler, R. P., 1992, "Precision radial velocities with an iodine absorption cell", PASP, vol. 104, 270.
Marcy, G.W., Butler, R.P., 1996, "A planetary companion to 70 Virginis", ApJ, vol. 464, L147.
Marcy, G. W., Butler R. P., Williams, E., et al., 1997, "The planet around 51 Pegasi", ApJ, 481, 926.
Marcy, G. W.; Butler, P., Vogt, S., 1999, "Searches for planetary companions with Lick, Keck and AAT", in IAU Coll. 170, "Precise stellar radial velocities", J.B. Hearnshaw and C.D.Scarfe (eds.), ASP Conference Series, in press.
Marley, M.S., Saumon, D., Guillot, et al., 1996, "Athmospheric, Evolutionary, and Spectral Models of the Brown Dwarf Gliese 229 B", Science, vol. 272, 1919.
Mayor, M., Duquennoy, A., Halbwachs, J.-L., Mermilliod, J.-C., 1992,"Complementary Approaches to Double and Multiple Star Research". IAU Coll. 135, ASP Conf. Ser., vol. 32, ed. H. McAlister and W. Hartkopf, 32.
Mayor, M., Queloz, D., 1995, "A Jupiter-mass companion to a solar-type star", Nature, vol. 378, 355.
Mayor, M., Queloz, D., 1996, "A search for Substellar Companions to Solar-type Stars via precise Doppler Measurements: A first Jupiter Mass Comppanion Detected", in Cool Star, Stellar Systems, and the Sun, 9th Cambridge Workshop, ASP Conf. Ser. vol. 109, R. Pallavicini and K. Dupree (eds.), 35.
Mayor, M., Queloz, S., Halbwachs, J.-L., 1997, "The mass function below the substellar limit", in IAU Coll 161 on "Astronomical and Biochemical Origins and the search for life in the Universe", Editrice Compositori Bologna, C.B. Cosmovici, S. Bowyer and D. Werthimer, 313.
Mayor M., 1998, private comm. from unpublished CORAVEL data.
Mazeh, T., Goldberg, D., Duquennoy, A., Mayor, M.,1992, "On the mass-ratio distribution of spectroscopic binaries with solar-type primaries", ApJ, vol. 401, 265.
Mazeh, T., Krymolowski, Y., Rosenfeld, G., 1997, "The high eccentricity of the planet orbiting 16 Cygni B", ApJ, 477, L103.
Murray, N., Hansen, B., Holman, M., Tremaine, S., 1998, "Migrating Planets", Science, vol. 279, 69.
Nakajima, T., Oppenheimer, B. R., Kulkarni, S. R., et al. 1995, "Discovery of a cool brown dwarf", Nature, vol. 378, 463.
Noyes, R. W., Saurabh, J., Korzennik, S. G., et al., 1997, "A planet orbiting the star rho Coronae Borealis", ApJ, vol. 483, L111.
Noyes et al., 1997, "The AFOE Program of Extra-Solar Planet Research", in "Planets Beyond the Solar System and the Next Generation of Space Missions", Space Telescope Science Institute Workshop, PASP, vol. 119, D. Soderblom (ed.), 119.
Oppenheimer, B. R., Golimowski, D. A., Kulkarni, S. R. and Matthews, K., 1999, "An Optical Coronagraphic and Near IR Imaging Survey of Stars for Faint Companions", AJ, submitted.
Queloz, D., Mayor, M., Sivan, J. P., et al. 1998, "The Observatoire de Haute-Provence Search for Extrasolar Planets with ELODIE", in "Brown Dwarfs and Extrasolar Planets", ASP Conf. Ser., vol. 134, R. Rebolo, E. L. Martín and R. Zapatero Osorio (eds.), 324.
Rasio, F. A., Ford, E. B., 1996, "Dynamical instabilities and the formation of extrasolar planetary systems", Science, vol. 274, 954.
Rasio, F. A., Tout, C. A., Lubow, S. H., Livio, M., 1996, "Tidal decay of close planetary orbits", ApJ, vol. 470, 1187.
Rebolo, R., Magazzù, A., Martìn, E.L, 1994, "Lithium and the nature of brown dwarf

candidates", Proceeding of the ESO Workshop: The bottom of the main sequence and beyond, Ed. C.G. Tinney, 159.

Rebolo, R., Zapatero Osorio M. R., Martìn, E. L., 1995, "Discovery of a brown dwarf in the Pleiades star clustter", Nature, vol 377, 129.

Ruiz, M. T., Leggett, S.K., Allard, F., 1997, "KELU-1: A free-floating brown dwarf in the solar neighborhood", ApJ vol. 491, L107.

Terquem, C., Papaloizou, J. C. B., Nelson, R. P., Lim, D. N. C, 1998, "On the Tidal Interaction of a Solar-Type Star with an Orbiting Companion: Excitation of g-Mode Oscillation and Orbital Evolution", ApJ, vol. 502, 788.

Tinney, C. G., Delfosse, X., Forveille, T., 1997, "DENIS-P J1228.2-1547 a new benchmark brown dwarf", ApJ, vol 490, L95.

Tokovinin, A. A., Duquennoy, A., Halbwachs, J.-L., Mayor, M., 1994, "Duplicity in the solar neighborhood. VII: Spectroscopic orbits of three K-dwarf stars", A&A, vol. 282, 831.

Trilling, D.E., Benz, W., Guillot, T., Lunine, J. I., Hubbard, W. B., Burrows, A., "Orbital Evolution and Migration of Giant Planets: Modeling Extrasolar Planets", ApJ, vol. 500, 428.

Ward, W. R., 1997, "Survival of Planetary Systems", ApJ, vol. 482, L211.

Walker, G. A. H., Walker, A. R., Irwin, A. W., 1995, "A search for Jupiter-mass companions to nearby stars", Icarus, 116, 359

Wolszczan, A., 1994, "Confirmation of earth-mass planets orbiting the millisecond Pulsar PSR B 1257+12", Science, vol. 264, 538.

Zahn, J.-P., 1989, "Tidal evolution of close binary stars", A&A 220, 112.

Zapatero Osorio, M. R., Rebolo, R., Martin, E. L., et al., 1997, "New Brown Dwarfs in the Pleiades Cluster", ApJ, vol. 491, L81.

EXTRASOLAR GIANT PLANET AND BROWN DWARF THEORY

ADAM BURROWS
The University of Arizona
Tucson, Arizona, 85721, U.S.A.

1. Introduction

With little doubt, we are in the midst of a renaissance in the study of extrasolar planets and brown dwarfs. High–resolution Doppler spectroscopy of nearby stars has netted $\sim$14 giant planets/brown dwarfs within 2.5 A.U. of their primaries (Table 1), seven within 0.25 A.U.[1, 2, 3, 4, 5, 6, 7]. Surveys in stellar clusters and in the field are discovering scores of objects near the stellar/substellar boundary, with exotic spectra such as have never before been catalogued or studied[8, 9, 10, 11]. There are numerous searches now underway to directly detect brown dwarfs, one succeeding spectacularly with the discovery of Gliese 229B[12, 13, 14, 15, 16]. The spectrum of Gl 229B is similar to that of Jupiter, in that it has methane bands. In fact, it has features in common with both stars and giant planets. As a missing link, Gl 229B is heralding a new era in the theoretical study of stellar/planetary atmospheres, brown dwarfs, and planets.

With all this activity, there is a need for a reflective look at the general physics and chemistry of substellar mass objects. In these few pages, I will attempt to summarize their salient characteristics. This effort is in part in response to the growing interest in this emerging field among students and practitioners alike. However, here I can touch on only the highlights and will of necessity emphasize the work with which I am most familiar and directly involved. With this caveat, I have assembled discussions, some from recent papers of immediate relevance [17, 18, 19], that serve to inform the reader of the major progress and of the many puzzles that enrich the current theory of extrasolar giant planets (EGPs) and brown dwarfs. In §2, I discuss simple analytic models and the equation of state. In §3, I review the history of the theory of EGPs and some of the results from recent evolutionary and spectral calculations. In §4, I discuss the chemistry of substellar

J.-M. Mariotti and D. Alloin (eds.), Planets Outside the Solar System: Theory and Observations, 121–140.

atmospheres and in §5 I summarize some of the general conclusions reached to date. Since objects with the same mass and composition, whatever their origin or mode of formation, should evolve and radiate in very much the same way, I make no substantive distinction in what follows between EGPs and brown dwarfs. Nevertheless, the two classes of objects may have characteristics (mass spectrum, composition, rotation, eccentricity distribution, etc.) that distinguish them. Ultimately, researchers may be able to separate substellar objects with the same mass, but different modes of formation, by their secondary properties or spectroscopically and this possibility should be borne in mind in what follows.

TABLE 1. The "Planetary" Bestiary

Object	Star	[Fe/H]	M (M_J)	a (AU)	P (days)	e
τ Boo b	F7V	+0.34	$\geq$3.44	0.046	3.313	0.0162
51 Peg b	G2.5V	+0.21	$\geq$0.45	0.05	4.23	0.0
υ And b	F7V	+0.17	$\geq$0.68	0.058	4.61	0.109
55 Cnc b	G8V	+0.29	$\geq$0.84	0.11	14.76	0.051
Gl 876 b	M4V	?	$\geq$1.9	~0.2	61.1	0.35
ρ CrB b	G0V	-0.19	$\geq$1.13	0.23	39.65	0.028
HR7875 b	F8V	-0.46	$\geq$0.69	~0.25	42.5	0.429
HD114762	F9V	-0.60	$\geq$10	0.38	84	0.25
70 Vir b	G4V	-0.03	$\geq$6.9	0.45	116.7	0.40
HR5568 b	K4V	~0.0	$\geq$0.75	~1.0	400	?
HR810 b	G0V	?	$\geq$2.0	~1.2	599.4	0.492
16 Cyg Bb	G2.5V	0.11	$\geq$1.66	1.7	2.19 yrs	0.68
47 UMa b	G0V	+0.01	$\geq$2.4	2.1	2.98 yrs	0.03
Gl 614 b	K0V	+0.31	$\geq$3.3	2.5	4.4 yrs	0.36
55 Cnc c	G8V	+0.29	$\geq$5	3.8	>8 yrs	?
Jupiter	G2V	0.0	1.00	5.2	11.86 yrs	0.048
Saturn	G2V	0.0	0.3	9.54	29.46 yrs	0.056
Gl 229 B	M1V	+0.20	30-55	$\geq$44.0	$\geq$400	?

2. Basic Brown Dwarf Theory

Brown dwarfs are compact "stars," predominantly of hydrogen, that are not massive enough ($M \leq 0.075\,M_\odot$, for solar metallicity) eventually to ignite hydrogen stably on the hydrogen main sequence. This is not to say that brown dwarfs do not burn hydrogen during some phase of their lives. In fact, the most massive ($M \gtrsim 0.07\,M_\odot$) can burn hydrogen for billions of years and objects more massive than ~13 M_J (~0.013 $M_\odot$) burn their stores of "primordial" deuterium. However, in a brown dwarf the surface energy losses are never completely compensated by thermonuclear burning in the core and the light–hydrogen nuclear fires are eventually extinguished as it cools. The surface and central temperatures are never stabilized.

2.1. BROWN DWARFS AS POLYTROPES

Composed predominantly of metallic hydrogen and helium and with mass densities in the vicinity of 10–1000 g cm^{-3}, brown dwarfs are supported in hydrostatic equilibrium by electron degeneracy pressure. It is only for a brief time ($\leq 10^8$ years), early in their cooling-regulated contraction, when the central densities are sufficiently low and the entropies are sufficiently high, that thermal ideal gas pressure is important. But no matter the character of the pressure support, a brown dwarf contracts along a Hayashi track and is convective. This implies that the entropy is the same throughout the star (save for in the very thin outer radiative skin). The P–ρ relation is approximately a power law:

$$P = K\rho^{1+\frac{1}{n}} \quad (= K\rho^{5/3}, \quad if \quad n = 1.5), \tag{1}$$

where P is the pressure, ρ is the mass density, n is the polytropic index, and K is a constant. In eq. (1), K depends only on the composition and the specific entropy. Stars which satisfy eq. (1) are called polytropes of index n. (Note that it is the non-relativistic nature of the pressure support that sets $n = 1.5$.) They form a sequence of objects with self-similar structure and radii (R), central densities (ρ_c), and central pressures (P_c) that depend on one another and on K and M (the total mass) via simple power laws. The general equations of hydrostatic equilibrium

$$\frac{dP}{dr} = -\frac{GM(r)\rho}{r^2} \quad \text{and} \tag{2}$$

$$\frac{dM(r)}{dr} = 4\pi r^2 \rho, \tag{3}$$

where G is Newton's universal constant of gravitation ($G = 6.6726 \times 10^{-8}$ cgs), can be combined for polytropes to yield the Lane-Emden equation [20],

$$\frac{1}{\xi^2}\frac{d}{d\xi}\left(\xi^2 \frac{d\theta}{d\xi}\right) = -\theta^n \; ; \qquad \theta(0) = 1, \quad \left.\frac{d\theta}{d\xi}\right|_{\xi=0} = 0 \tag{4}$$

where

$$\rho = \rho_c \theta^n \; ; \qquad P = P_c \theta^{n+1} \tag{5}$$

and

$$\xi = \frac{r}{a} \; ; \qquad a = \left[\frac{(n+1)K}{4\pi G}\rho_c^{\left[\frac{1-n}{n}\right]}\right]^{1/2}. \tag{6}$$

The solution of eq. (4) for $\theta(\xi)$ subject to the zero-pressure outer boundary condition solves the problem, given ρ_c, K, and n or M, K, and n. Useful results are

$$\rho_c = \delta_n \langle\rho\rangle = \delta_n \left[\frac{3M}{4\pi R^3}\right] \propto M^{\frac{2n}{3-n}}, \tag{7}$$

$$P_c = W_n \frac{GM^2}{R^4} \propto M^{\frac{2(1+n)}{3-n}}, \tag{8}$$

and

$$R = \gamma_n \left(\frac{K}{G}\right)^{\frac{n}{3-n}} M^{\frac{1-n}{3-n}}, \tag{9}$$

where δ_n, W_n, and γ_n are functions of n alone. For $n = 1.5$, $\delta_n = 5.991$, $W_n = 0.770$, and $\gamma_n = 2.357$ and, importantly, $\rho_c \propto M^2$, $P_c \propto M^{10/3}$, and $R \propto M^{-\frac{1}{3}}$. For $n = 1.0$, $\delta_n = \frac{\pi^2}{3} = 3.290$, $W_n = 0.393$, and $\gamma_n = \sqrt{\frac{\pi}{2}} = 1.253$, while $\rho_c \propto M$, $P_c \propto M^2$, and R is independent of M.

At the low-mass end of the brown-dwarf branch, the Coulomb corrections to the P-ρ law compete with the degeneracy component and shift the effective polytropic index from 1.5 for the more massive brown dwarfs to 1.0. As stated above, when $n = 1.0$, the radius of the object is independent of the mass. Even above $M \cong 5\,\mathrm{M_J}$ when $n \sim 1.5$ and $R \propto M^{-1/3}$, the dependence of the brown dwarf's radius on mass is weak. Zapolsky and Salpeter [21] have provided a useful analytic fit to the radius-mass relation that includes both planets and cold brown dwarfs:

$$R = 2.2 \times 10^9 \left(\frac{M_\odot}{M}\right)^{1/3} \Big/ \left(1 + \left(\frac{M}{0.0032\,M_\odot}\right)^{-1/2}\right)^{4/3} \mathrm{cm}\,. \tag{10}$$

In fact, as the mass ranges from 1 $\mathrm{M_J}$ to $\sim$80 $\mathrm{M_J}$ along the full cold brown dwarf and EGP branches, the radius changes by no more than 50% and is always near $R_J \left(= 8.26 \times 10^9\ \mathrm{cm}\ \sim 0.12\,\mathrm{R_\odot}\right)$. Furthermore, in a structural, but not an evolutionary sense, cold brown dwarfs are at the hydrogen-rich (high Z/A), low-mass end of the white dwarf family.

Equations (7) through (9) allow us to derive most of the salient features of old brown dwarfs. Their R's are near $0.1\,\mathrm{R_\odot}$, ρ_c ranges from $\sim$10 g cm^{-3} for the light to 1000 g cm^{-3} for the heavy, and P_c can be found above, near, and below 10^5 Mbars (1 Mbar $= 10^{12}$ dynes cm^{-2}). However, the polytropic analysis alone does not illuminate the thermal structure and evolution of a brown dwarf. The time-dependence of brown dwarf characteristics is a crucial dimension in brown-dwarf theory. The effective photospheric temperature, luminosity, and central temperature (T_c) at a given time require a knowledge of the specific heat, atmospheric opacities, phases, etc. that can not be extracted from eqs. (2) through (9). Supplemented with this ancillary knowledge, the polytropic approximation can indeed provide an analytic model of brown dwarf theory (versus M and t). Burrows & Liebert [17] and Stevenson [22] have developed such models. However, an accurate theory must incorporate the thermodynamic and atmospheric details. An analytic model reveals the systematics, but only the full simulations should be used for numbers.

2.2. THE EQUATION-OF-STATE AND THERMODYNAMICS OF HYDROGEN

More than fifty years ago, Wigner and Huntington [23] postulated that under high pressures molecular hydrogen would undergo a phase transition to the metallic state. This metal-insulator transition has received sporadic attention ever since. Pressure ionization and metallization of hydrogen and hydrogen/helium mixtures is unavoidable given suitable pressure ($P \gtrsim 3$ Mbars) and such an alloy must comprise the bulk of Jupiter (~85%), Saturn (~50%), and brown dwarfs (>99.9%). Theoretical treatments of the thermodynamics and energetics of such strongly-coupled Coulomb plasmas in astrophysical contexts have been attempted by many, but for this synopsis we have been guided by the work of Stevenson [22], Magni and Mazzitelli [24], and Saumon and Chabrier [25, 26, 27].

2.2.1. *The Low-Density Regime*

At low densities ($10^{-5}\,\mathrm{g\,cm^{-3}} < \rho < 0.1\,\mathrm{g\,cm^{-3}}$) and low temperatures ($T \leq 4000$ K) hydrogen is predominantly in the molecular state and is a fluid. Between ~0.1 g cm^{-3} and ~1.0 g cm^{-3} and far below the critical point (according to Saumon and Chabrier at $P_c = 0.614$ Mbar, $T_c = 1.53 \times 10^4$ K, and $\rho_c = 0.35\,\mathrm{g\,cm^{-3}}$), hydrogen is a solid. As the temperature increases above $\sim 10^4$ K, molecular hydrogen first dissociates and then ionizes. Temperature ionization does not occur via a phase transition. At low densities and high temperatures, hydrogen is a very weakly-coupled gas. The pressure law throughout the low-density regime is given approximately by the ideal gas law:

$$P = \frac{\rho N_A k T}{\mu}, \tag{11}$$

where N_A is Avogadro's number, k is Boltzmann's constant, and μ is the mean "molecular" weight. When hydrogen and helium are completely ionized (ignoring for the moment "metals"),

$$\frac{1}{\mu} = 2X + \frac{3}{4} Y_\alpha\,, \tag{12}$$

but when they are in the molecular state

$$\frac{1}{\mu} = \frac{X}{2} + \frac{Y_\alpha}{4}, \tag{13}$$

where X and Y_α are the hydrogen and helium mass fractions, respectively. For $X = 0.75$ and $Y_\alpha = 0.25$, $\mu = 0.593$ and 2.29 in the two regimes, but it is always of order unity.

The photosphere of a brown dwarf is located approximately at the $\tau = 2/3$ surface, where τ is the optical depth given by

$$\tau = \int_r^\infty \kappa_R \rho dr. \tag{14}$$

κ_R is the Rosseland mean opacity. Using the equation of hydrostatic equilibrium (eq. 2) and eq. (14), we obtain

$$P_{Ph} \cong \frac{2}{3}\frac{g}{\kappa_R} \sim 10^7\,\text{dynes}\,\text{cm}^{-2} \equiv 10\,\text{bars}, \tag{15}$$

where g is the surface gravity and has been set equal to $10^5\,\text{cm}\,\text{s}^{-2}$ and κ_R has been set equal to $10^{-2}\,\text{cm}^2\text{g}^{-1}$.

In a region such as the photosphere, where it can be assumed that g is constant, eqs. (2) and (3) can be rewritten as

$$\Delta M = \frac{4\pi R^2}{g} P, \tag{16}$$

where ΔM is the total mass that overlies the shell at pressure P and R is the total brown dwarf radius. Setting $P = P_{Ph}$, we derive that $\Delta M_{Ph} \sim 10^{-11}\,M_\odot$. In addition, eq. (8) can be used to give us

$$\frac{\Delta M}{M} = \left[4\pi W_n\right]\frac{P}{P_c}, \tag{17}$$

where P_c is the central pressure and M is the total stellar mass. Eq. (17) gives us $\frac{\Delta M_{Ph}}{M} \sim 10^{-10}$. The radiative region of a brown dwarf is indeed a prodigiously small fraction of the object.

At a photospheric temperature (T_{eff}) of ~1500 K, eq. (11) implies that $\rho_{Ph} \sim 10^{-4}\,\text{g}\,\text{cm}^{-3}$ and puts the photosphere squarely in the gaseous molecular hydrogen region. In and around this low-density, low-temperature region, the temperatures are sufficient to excite the rotational levels of the H_2 molecule into equipartition, but not its vibrational levels. This fact implies that the adiabatic indices (Γ_1, Γ_2, Γ_3) are all near 7/5 (=1.4), not 5/3. Therefore, $P \propto \rho^{1.4}$ and $T \propto \rho^{0.4}$ along an adiabat in the low-density molecular region.

2.2.2. *The High–Density Regime*

Most of a brown dwarf can be found in the high-density, liquid metallic region. The characteristics of a strongly-coupled Coulomb plasma are determined by the dimensionless Coulomb parameter, the ratio of the Coulomb energy per ion to kT:

$$\Gamma = \frac{Z^2 e^2}{r_s kT} = 0.227 \left(\frac{\rho}{\mu_e} \right)^{1/3} Z^{5/3} \Big/ T_6 \,, \tag{18}$$

where Ze is the ionic charge ($Z = 1$ for H), $T_6 = T/10^6$ K, ρ is in cgs, and μ_e is the number of baryons per electron given by

$$\frac{1}{\mu_e} = X + \frac{1}{2} Y_\alpha = \frac{1+X}{2}. \tag{19}$$

For $Y_\alpha = 0.25$, $\mu_e = 1.143$. r_s is the radius of a sphere that contains one nucleus on average and along with r_e, the electron spacing parameter, is the relevant length scale. r_s and r_e are defined as,

$$r_s = \left(\frac{3Z}{4\pi n_e} \right)^{1/3} = \left(\frac{3\mu_e Z}{4\pi N_A \rho} \right)^{1/3} \tag{20}$$

and

$$r_e = \left(\frac{3}{4\pi n_e} \right)^{1/3} = \left(\frac{3\mu_e}{4\pi N_A \rho} \right)^{1/3} = \frac{r_s}{Z^{1/3}}, \tag{21}$$

where n_e is the electron density ($n_e = \rho N_A / \mu_e$). The weakly-coupled regime is characterized by $\Gamma \ll 1$ and is called the Debye-Hückel limit. The transition to the strongly-coupled liquid state occurs at $\Gamma \sim 1$. Brown–dwarf matter lies between $\Gamma \sim 1$ and 10 in the liquid metal regime. Note that Γ in a brown dwarf achieves its *minimum* at its center and gradually grows to its highest value near the metallic transition.

Not only does one need to include the simple Coulomb part of the ion-ion interaction, but also the electron exchange and correlation terms. Furthermore, electron screening and density fluctuation corrections are required. Some of the various approaches that have been used in the past include the Wigner-Seitz approximation, the Thomas-Fermi-Dirac theory, and dielectric function theory. The latter in particular, which focuses on the dielectric response of the plasma to calculate the effect of fluctuations in n_e on the Coulomb binding, has been quite fruitful [28].

The atomic unit of pressure ($P_o = e^2/a_o^4$, where a_o is the Bohr radius) is 294 Mbars. For pressures significantly above P_o, the degeneracy pressure of free electrons dominates and $P \sim P_D$ where

$$P_D \simeq 10^{13} \left(\frac{\rho}{\mu_e}\right)^{5/3} \text{dynes cm}^{-2} = 10\,\text{Mbars} \left(\frac{\rho}{\mu_e}\right)^{5/3}. \tag{22}$$

This echoes the polytropic form of eq. (1) with $n = 1.5$. The temperature corrections to P are generally small along brown dwarf tracks (the electron component is especially small) and can be approximated by the Grüneisen [29] theory:

$$P_T = \gamma E_T = \frac{3\gamma N_A kT}{V}, \tag{23}$$

where γ is the Grüneisen parameter (0.6 $\rightarrow$ 0.5 for H from 1 to 10^6 Mbar), E_T is the thermal energy of the ions in the liquid, and V is the molar volume. Core brown dwarf temperatures exceed the Debye temperature and the Dulong and Petit formula for the molar specific heat, $C_V = 3R$, is a satisfactory approximation [30]. With the pressure in degenerate electrons and the specific heat, entropy, and thermal energy in the ions, the equation-of-state at high pressures and densities is conceptually clean. However, at pressures at and below P_o, the Coulomb interaction is crucial. A characteristic length, the Bohr radius, appears along with a characteristic density ($\sim$ the density of "terrestrial" matter). Since the central pressure of Jupiter (P_{CJ}) is $\sim$100 Mbar and $P_{CJ} < P_o$, Coulomb effects make a qualitative difference for objects near Jupiter's mass. For a pure hydrogen plasma, the density-dependent Coulomb correction to the pressure is a factor of $(1 - 0.85/\rho^{1/3})$, which below $\rho \sim 50\,\text{g cm}^{-3}$ is significant. It is at the onset of Coulomb dominance, with its characteristic density and length scale, that the effective polytropic index slides from $n = 1.5$ through $n = 1.0$ to $n \sim 0.0$ ($\sim$ incompressible).

2.3. SIMPLE POWER LAW MODEL OF BROWN DWARF COOLING

Because brown dwarfs are convective and, hence, isentropic and their equation of state is polytropic both above and below the metallic transition, all the essentials of brown dwarf cooling theory can be analytically obtained. Burrows & Liebert [17] derive $T_{\text{eff}}(t, M)$ and $L(t, M)$ using only the scaling laws of a polytrope, the entropy matching condition at the transition between metallic liquid hydrogen at high densities and molecular hydrogen at low densities, and a crude model of the photosphere. They obtain:

$$T_{\text{eff}} = 1551K \left(\frac{10^9\,\text{yrs}}{t}\right)^{0.324} \left(\frac{M}{0.05\,M_\odot}\right)^{0.827} \left(\frac{\kappa_R}{10^{-2}}\right)^{0.088} \tag{24}$$

and

$$L = 3.82 \times 10^{-5}\,\mathrm{L}_\odot \left(\frac{10^9\,\mathrm{yrs}}{t}\right)^{1.297} \left(\frac{M}{0.05\,M_\odot}\right)^{2.641} \left(\frac{\kappa_R}{10^{-2}}\right)^{0.35} \tag{25}$$

Equations (24) and (25) encapsulate the analytic theory of brown dwarf cooling and are surprisingly accurate, given the approximations they embody. The powers and trends in eqs. (24) and (25) are realistic and the numbers obtained are good to better than 50%, when compared to, for example, numerical model G of Burrows, Hubbard, and Lunine [31], for all but the earliest epochs. The weak dependence on κ_R, (especially for $T_{\rm eff}$), encourages us that this most problematic piece of brown dwarf theory won't, when completely understood, force qualitative changes in our theories. The stiff dependence of L on M reflects the stiff dependence of the binding energy ($\sim \frac{GM^2}{R} \sim M^{7/3}$) on mass, but is only slightly steeper than the corresponding relation above the transition region on the M dwarf branch ($L \propto M^{2.2-2.6}$).

3. Calculations of the Evolution and Emissions of EGPs

3.1. EARLY GRAY MODELS

Burrows *et al.* [32] and Saumon *et al.* [33] calculated a suite of models of the evolution and emissions of EGPs, under the problematic black body assumption. They derived fluxes and dimensions as a function of age, composition, and mass, both as a guide for giant planet searches and as a tool for interpreting the results of any positive detections. Extrasolar giant planets (EGPs) can arbitrarily be said to span the mass range from 0.25 $\mathrm{M_J}$ to 10–20 $\mathrm{M_J}$, but are merely low–mass extensions of the brown dwarf family. It has long been recognized that essentially the same physics governs the structure and evolution of the suite of electron-degenerate and hydrogen-rich objects ranging from brown dwarfs (at the high–mass end) to Jupiters and Saturns (at the low–mass end). Surprisingly, no one had accurately mapped out the properties of objects between the mass of giant planets in our solar system and the traditional brown dwarfs. This is precisely the mass range for the newly–discovered planets listed in Table 1.

Earlier work generally consisted of evolutionary models of planets of 1 $\mathrm{M_J}$ and below, beginning with Graboske *et al.* [34]. Graboske *et al.* calculated the evolution of the low-mass objects Jupiter and Saturn from an age of 10^7 years to the present ($\sim$4.5 Gyr). Working down from higher masses, Grossman and Graboske [35] extended their calculations of brown dwarf evolution to as low as 12 $\mathrm{M_J}$, but had to limit their study to ages less than about 0.1 Gyr. Black [36] used the results of Grossman & Graboske [35] and Graboske *et al.* [34] to infer simple power-law relations for the variation of luminosity, L, and radius, R, as a function of mass, M_p, and time, t. Black's relations are roughly valid for objects close in mass to 1 M_J and close in age to 4.5 Gyr. However, Black's formulae become inaccurate at earlier ages and at larger masses.

EGPs radiate in the optical by reflection and in the infrared by the thermal emission of both absorbed stellar light and the planet's own internal energy. To calculate EGP cooling curves from 0.3 M_J (the mass of Saturn) through 15 M_J, Burrows *et al.* [32] used the Henyey code previously constructed to study brown dwarfs and M dwarfs [31, 37]. Below effective temperatures ($T_{\rm eff}$) of 600 K, they employed the atmospheres of Graboske *et al.* [34], who included opacities due to water, methane, ammonia, and collision-induced absorption by H_2 and He. The gravity dependence of the EGP atmospheres was handled as in Hubbard [38]. Above $T_{\rm eff} = 600\,{\rm K}$, they used the X model of Burrows *et al.* [37]. The two prescriptions were interpolated in the overlap region. They employed the hydrogen/helium equation of state of Saumon & Chabrier [26, 27] and ignored rotation and the possible presence of an ice/rock core [39, 40]. Furthermore, they assumed that EGPs are fully convective at all times and included the effects of "insolation" by a central star of mass M_*. Whether a 15 M_J object is a planet or a brown dwarf is largely a semantic issue, though one might distinguish gas giants and brown dwarfs by their mode of formation (*e.g.*, in a disk or "directly"). Physically, compact hydrogen–rich objects with masses from 0.00025 $M_\odot$ through 0.25 $M_\odot$ form a continuum.

3.2. NON–GRAY MODELS

However, to credibly estimate the infrared band fluxes and improve upon the black body assumption made in Burrows *et al.* [32] and Saumon *et al.* [33], Burrows *et al.* [18] have recently performed non–gray simulations at solar–metallicity of the evolution, spectra, and colors of isolated EGP/brown dwarfs down to $T_{\rm eff}$s of 100 K. Figure 1 portrays the luminosity versus time for objects from Saturn's mass (0.3 M_J) to 0.2 $M_\odot$ for this model suite. The early plateaux between 10^6 years and 10^8 years are due to deuterium burning, where the initial deuterium mass fraction was taken to be 2×10^{-5}. Deuterium burning occurs earlier, is quicker, and is at higher luminosity for the more massive models, but can take as long as 10^8 years for a 15 M_J object. The mass below which less than 50% of the "primordial" deuterium is burnt is $\sim$13 M_J. On this figure, we have arbitrarily classed as "planets" those objects that do not burn more than 50% of their deuterium and as "brown dwarfs" those that do burn deuterium, but not light hydrogen. While this distinction is physically motivated, we do not advocate abandoning the definition based on origin. Nevertheless, the separation into M dwarfs, "brown dwarfs", and giant "planets" is useful for parsing by eye the information in the figure.

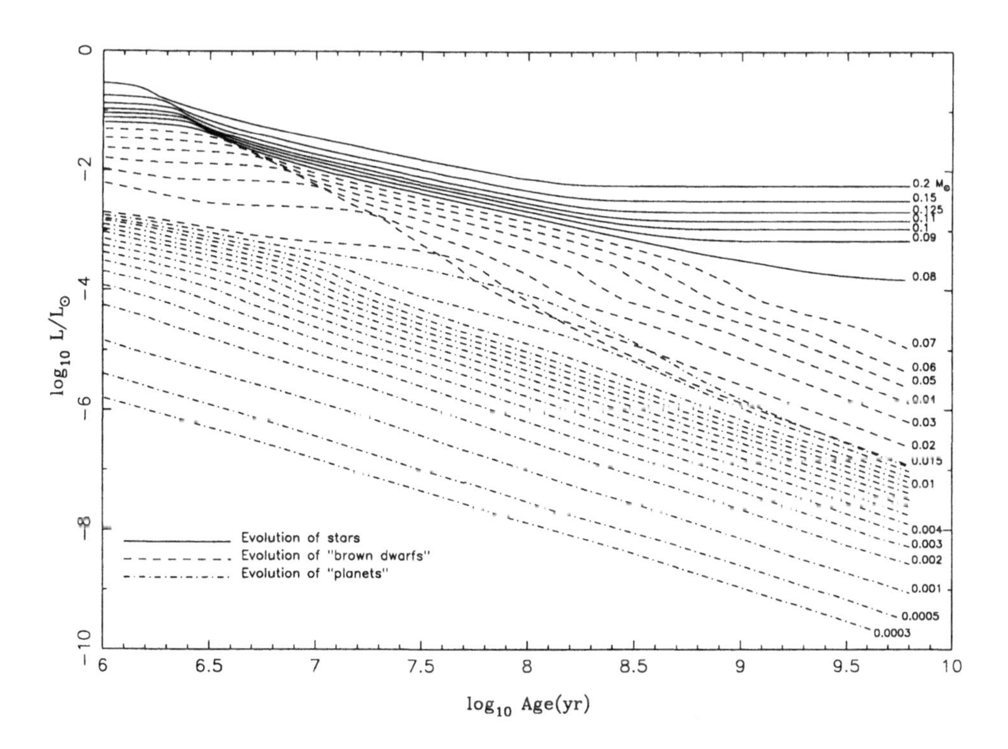

Figure 1. Evolution of the luminosity (in L$_\odot$) of solar–metallicity M dwarfs and substellar objects versus time (in years) after formation. The stars, "brown dwarfs" and "planets" are shown as solid, dashed, and dot–dashed curves, respectively. In this figure, we arbitrarily designate as "brown dwarfs" those objects that burn deuterium, while we designate those that do not as "planets." The masses in M$_\odot$ label most of the curves, with the lowest three corresponding to the mass of Saturn, half the mass of Jupiter, and the mass of Jupiter (from Burrows *et al.* 1997).

In Figure 1, the bumps between 10^{-4} $L_\odot$ and 10^{-3} $L_\odot$ and between 10^8 and 10^9 years, seen on the cooling curves of objects from 0.03 $M_\odot$ to 0.08 $M_\odot$, are due to silicate and iron grain formation. These effects, first pointed out by Lunine *et al.* [41], occur for $T_{\rm eff}$s between 2500 K and 1300 K. The presence of grains affects the precise mass and luminosity at the edge of the main sequence. Since grain and cloud models are problematic, there still remains much to learn concerning their role and how to model them [19, 41, 42].

To constrain the properties of the brown dwarf Gl 229B [13, 14, 12, 15, 16], Marley *et al.* [43] constructed a grid of atmospheres with $T_{\rm eff}$ ranging from 600 K to 1200 K and 1.0×10^4 cm s^{-2} < gravity < 3.2×10^5 cm s^{-2}. For each case, they computed a self-consistent radiative-convective equilibrium temperature profile and the emergent radiative flux. By comparing their theoretical spectra with the UKIRT [15] and Keck [16] data, they derived an effective temperature of 960 ± 70 K and a gravity between 0.8×10^5 and 2.2×10^5 cm s^{-2}. These results translate into masses and ages of 20–55 M_J and 0.5–5 Gyr, respectively. Gravity maps almost directly into mass, and ambiguity in the former results in uncertainty in the latter. The Marley *et al.* [43] study was nicely complemented in the literature by the Gl 229B calculations of Allard *et al.* [44] and Tsuji *et al.* [45].

The studies of Burrows *et al.* [18] and Marley *et al.* [43] revealed major new aspects of EGP/brown dwarf atmospheres that bear listing and that uniquely characterize them. Below $T_{\rm eff}$s of 1300 K, the dominant equilibrium carbon molecule is CH_4, not CO, and below 600 K the dominant nitrogen molecule is NH_3, not N_2 [46]. The major opacity sources are H_2, H_2O, CH_4, and NH_3. For $T_{\rm eff}$s below ~400 K, water clouds form at or above the photosphere and for $T_{\rm eff}$s below 200 K, ammonia clouds form (*viz.*, Jupiter). Collision–induced absorption of H_2 partially suppresses emissions longward of ~10 μm. The holes in the opacity spectrum of H_2O that define the classic telluric IR bands also regulate much of the emission from EGP/brown dwarfs in the near infrared. Importantly, the windows in H_2O and the suppression by H_2 conspire to force flux to the blue for a given $T_{\rm eff}$. The upshot is an exotic spectrum enhanced relative to the black body value in the J and H bands (~1.2 μm and ~1.6 μm, respectively) by as much as *two* to *ten* orders of magnitude, depending upon $T_{\rm eff}$. Figure 2 depicts spectra between 1 μm and 10 μm at a detector 10 parsecs away from solar–metallicity objects with age 1 Gyr and masses from 1 M_J through 40 M_J. Superposed are putative sensitivities for the three NICMOS cameras [47], ISO [48], SIRTF [49], and Gemini/SOFIA [50, 51]. Figure 2 demonstrates how unlike a black body an EGP spectrum is. For example, the enhancement at 5 μm for a 1 Gyr old, 1 M_J extrasolar planet is by four orders of magnitude. As $T_{\rm eff}$ decreases below ~1000 K, the flux in the M band (~5 μm) is progressively enhanced relative to the black body value. While at 1000 K there is no enhancement, at 200 K it is near 10^5. Hence, the J, H, and M bands are the premier bands in which to search for cold substellar objects. The Z band (~1.05 μm) flux is also above the black–body value over this $T_{\rm eff}$ range. Eventhough K band (~2.2 μm) fluxes are generally higher than black body values, H_2 and CH_4 absorption features in the K band decrease its importance *relative* to J and H. As a consequence of the increase of atmospheric pressure with decreasing $T_{\rm eff}$, the anomalously blue

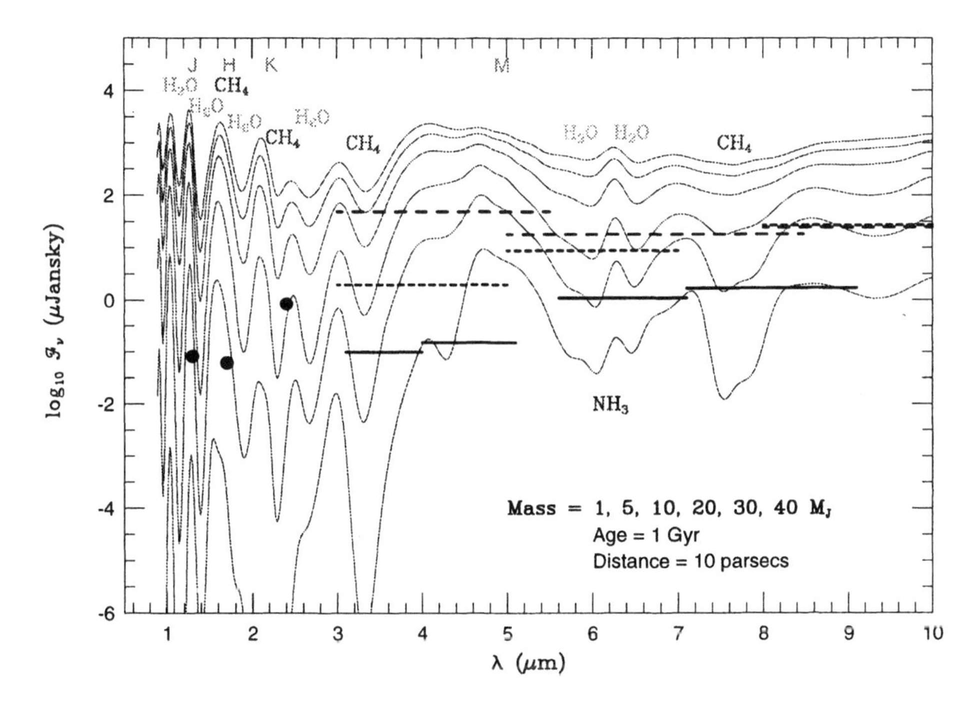

Figure 2. The flux (in μJanskys) at 10 parsecs versus wavelength (in microns) from 1 μm to 10 μm for 1, 5, 10, 20, 30, and 40 M_J solar-metallicity models at 1 Gyr. Shown are the positions of the J, H, K, and M bands and various molecular absorption features. Superposed for comparison are the putative sensitivities of the three NICMOS cameras, ISO, Gemini/SOFIA, and SIRTF. NICMOS is denote with large black dots, ISO with long dashes, Gemini/SOFIA with short dashes, and SIRTF with solid lines. At all wavelengths, SIRTF's projected sensitivity is greater than ISO's. SOFIA's sensitivity overlaps with that of ISO around 10μm. For other wavelength intervals, the order of sensitivity is SIRTF > Gemini/SOFIA > ISO, where > means "is more sensitive than" (from Burrows *et al.* 1997).

$J - K$ and $H - K$ colors get *bluer*, not redder.

To illustrate this, Figure 3 shows a representative color–magnitude diagram (J versus $J - K$) for solar–metallicity objects with masses from 3 M_J to 40 M_J, for ages of 0.5, 1.0, and 5.0 Gyr. Included are the corresponding black body curves, hot, young brown dwarf or extremely late M dwarf candidates such as LHS2924, GD 165B, Calar 3, and Teide 1 [52, 53, 11], and a sample of M dwarfs from Leggett [54]. This figure illustrates the unique color realms occupied by extrasolar giant planets and brown dwarfs. Figure 3 portrays the fact that the K and J versus $J - K$ infrared H–R diagrams loop back to the blue below the edge of the main sequence and are not continuations of the M dwarf sequence into the red. The difference between the black body curves and the model curves is between 3 and 10 magnitudes for J versus $J-K$, more for K versus $J-K$. Gl 229B fits nicely among these theoretical isochrones. The suppression of K by H_2 and CH_4 features is largely responsible for this anomalous blueward trend with decreasing mass and $T_{\rm eff}$.

Ignoring for the moment the question of angular resolution, one can compare the theoretical solar–metallicity spectrum and color predictions of Burrows *et al.* [18] with putative detector sensitivities to derive encouraging detection ranges. For example, at 5 μm, SIRTF might see a 1 Gyr old, 1 M_J object in isolation out to nearly 100 parsecs. The range of NICMOS in H for a 1 Gyr old, 5 M_J object is approximately 300 parsecs, while for a coeval 40 M_J object it is near 1000 parsecs. Furthermore, SIRTF will be able to see at 5 μm a 5 Gyr old, 20 M_J object in isolation out to $\sim$400 parsecs and NICMOS will be able to see at J or H a 0.1 Gyr old object with the same mass out to $\sim$2000 parsecs. These are dramatic numbers that serve to illustrate both the promise of the new detectors and the enhancements theoretically predicted.

4. Chemical Abundances in Substellar Atmospheres

The molecular compositions of low–ionization, substellar atmospheres can serve as diagnostics of temperature, mass, and elemental abundance and can help define a spectral sequence, just as the presence or absence of spectral features associated with various ionization states of dominant, or spectroscopically active, atoms and simple molecules does for M through O stars. However, the multiplicity of molecules that appear in their atmospheres lends an additional complexity to the study of substellar mass objects that is both helpfully diagnostic and confusing. Nowhere is the latter more apparent than in the appearance at low temperatures of refractory grains and clouds. These condensed species can contribute significant opacity and can alter an atmosphere's temperature/pressure profile and its albedo. Grain and cloud droplet opacities depend upon the particle size and shape distribution and these are intertwined with the meteorology (convection) in complex ways. Furthermore, condensed species can rain out and deplete the upper atmosphere of heavy elements, thereby changing the composition and the observed spectrum. Burrows & Sharp [19] have estimated the consequences of rainout and conclude that the "abundant" low–temperature refractories are predominantly the chlorides and the sulfides. In particular, KCl, NaCl, NaF, and SiS_2 stand out, with SiS_2 dominating for some assumptions. Without depletion, SiS_2 would not form in abundance, but progressive depletion starting at the highest temperatures is needed for it to

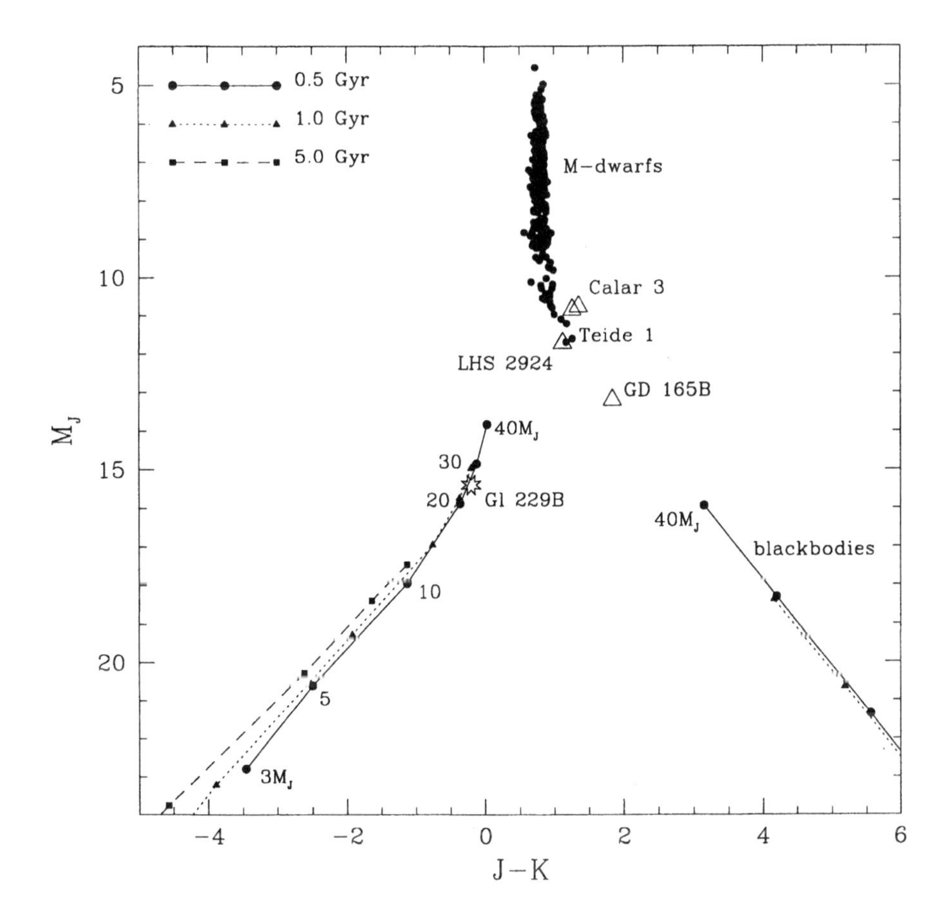

Figure 3. Absolute J vs. $J - K$ color–magnitude diagram, from Burrows *et al.* 1997. Theoretical isochrones are shown for $t = 0.5$, 1, and 5 Gyr, along with their blackbody counterparts. The difference between blackbody colors and model colors is striking. The brown dwarf, Gliese 229B (Oppenheimer *et al.* 1995), the young brown dwarf candidates Calar 3 and Teide 1 (Zapatero-Osorio, Rebolo, & Martin 1997), and late M dwarfs LHS 2924 and GD165B (Kirkpatrick, Henry, & Simons 1994,1995)) are plotted for comparison. The lower main sequence is defined by a selection of M–dwarf stars from Leggett (1992).

become important. If SiS_2 forms, it does so around 743 K for solar metallicity, whereas NaF forms near 920 K, NaCl forms near 800 K, and KCl forms near 740 K. Be that as it may, the sodium and potassium salts emerge as aerosol candidates in Gl 229B's upper atmosphere, with grammages that may range from $\sim 10^{-4}$ to $\sim 4 \times 10^{-6}$ g cm^{-2}. These calculations suggest that thin clouds of non–silicate, low–temperature refractories can and do exist in the upper atmosphere of Gl 229B. Furthermore, they may also reside in the atmospheres of brown dwarfs with $T_{\rm eff}$s below about 1400 K. Hence, it is expected that many cloud/aerosol layers with different compositions are formed both above and below the photospheres of brown dwarfs and EGPs. At times, from the lowest effective temperatures up, clouds of either NH_3 ($T_{\rm eff} \leq 250$ K), H_2O ($T_{\rm eff} \leq 400$ K), chlorides, sulfides, iron, or silicates should be found in these exotic substellar and late M dwarf atmospheres.

The formation of refractory silicate grains below 2500 K was already shown by Lunine *et al.* [41] and Burrows *et al.* [18] to influence the evolution of late M dwarfs and young brown dwarfs through their "Mie" opacity. The blanketing effect they provide lowers the effective temperature and luminosity (L) of the main sequence edge mass from about 2000 K and 10^{-4} $L_\odot$ to about 1750 K and 6×10^{-5} $L_\odot$, an effect recently verified by Chabrier *et al.* [55]. In addition, grain opacity slightly delays the cooling of older brown dwarfs, imprinting a slight bump on their luminosity/age trajectories (see Figure 1). The presence of grains in late M dwarf spectra was invoked to explain the weakening of the TiO bands and the shallowing of their H_2O troughs in the near infrared [45, 56]. Tsuji and collaborators concluded that titanium was being depleted into refractories.

Gl 229B is a Rosetta stone for brown dwarf research. With a $T_{\rm eff}$ of $\sim$950 K, a luminosity below 10^{-5} $L_\odot$, and spectra or photometry from the R Band through 5 μm, Gl 229B hints at or exemplifies all of the unique characteristics of the family [43, 44]: metal (Fe, Ti, V, Ca, Mg, Al, Si) depletions, the dominance of H_2O vapor, the appearance of CH_4 and alkali metals, and the signatures of clouds. Clouds of low–temperature condensible species above the photosphere are the most natural explanation for the steep drop below 1 μm in the Keck spectra between 0.83 μm and 1 μm [14]. These clouds may not be made up of the classic silicate refractories formed at much higher temperatures, since these species have probably rained out. From simple Mie theory, their mean particle size must be small (~ 0.2 μm) in order to influence the "optical" without much perturbing the near infrared. In addition, such a cloud of small droplets can help explain why Gl 229B's near–infrared troughs at 1.8 μm and 3.0 μm are not as deep as theory would otherwise have predicted. Just as Tsuji and collaborators have shown that silicate grains at higher temperatures can shallow out the H_2O troughs, so too can species that condense at lower temperatures (≤ 1000 K

?) explain the shallower–than–predicted Gl 229B H_2O troughs. What those species might be can be illuminated by further chemical abundance studies. Note that a cloud grammage in these small–radius low–temperature refractories of only $\sim 10^{-5}\ \mathrm{g\,cm^{-2}}$ would be adequate to explain the anomalies.

As $T_{\rm eff}$ decreases (either as a given mass cools or, for a given age, as we study objects with lower masses), the major atmospheric constituents of brown dwarfs and EGPs change. This change is reflected in which spectral features are most prominent and in the albedos of substellar objects near their primaries. Hence, specific mixes of atoms, molecules, and clouds can serve as approximate $T_{\rm eff}$ and temperature indicators and a composition scale can be established. In order to do this definitively, self–consistent synthetic spectra are required. However, the composition trends Burrows & Sharp [19] have identified are suitably dramatic that reasonable molecular indicators of spectral type can be suggested. A workable sequence might be: TiO disappears (at 2000–2300 K), refractory silicates and Fe(c) appear (at 2300–2000 K), Mg_2SiO_4 appears (at 1900 K), VO(g) disappears (at 1700–1900 K), $MgSiO_3$ appears (at 1700 K), silicates rainout (at ~1400 K (?)), CrH disappears (~1400 K), Li $\rightarrow$ LiCl ($\lesssim$1400 K), CO $\rightarrow$ CH_4 (1200–1500 K), (Rb,Cs,K) $\rightarrow$ chlorides ($\leq$ 1200 K), $PH_3 \rightarrow (P_4O_6, Mg_3P_2O_8)$ ($\leq$ 1000 K), formation of NaF, NaCl, and KCl clouds and various sulfide clouds (~700–1100 K), $N_2 \rightarrow NH_3$(g) (~700 K), H_2O(g) $\rightarrow$ H_2O(c) (~350 K), and $NH_3 \rightarrow$ (NH_3(c), NH_4SH(c)) (~200 K).

Disequilibrium chemistry and convection and differences in the spectroscopic strengths of the various indicators will no doubt partially alter the $T_{\rm eff}$ order of this spectral scale. Note that objects with $T_{\rm eff}$s higher than the temperatures quoted above may nevertheless, because of the lower temperatures that can be achieved at lower optical depths, manifest lower–temperature compounds and/or transitions in their atmospheres and spectra. Gl 229B with a $T_{\rm eff}$ of ~950 K is a case in point, if lower–temperature chloride and sulfide refractories are to be invoked to explain its spectrum. How much higher the appropriate $T_{\rm eff}$ should be for a given chemical transition to be fully manifest will depend upon consistent atmosphere models. Clearly, the above temperatures for the appearance or disappearance of species should be used with great caution when estimating $T_{\rm eff}$. Very crudely, the "L" spectral type suggested by Kirkpatrick *et al.* [10] would correspond to $T_{\rm eff}$s between about ~1500 K and ~2200 K. None but the youngest and most massive brown dwarfs and only the very youngest EGPs could have this proposed spectral designation. Most brown dwarfs and EGPs will be of an even later spectral type, yet to be coined, a spectral type that would include Gl 229B.

5. Conclusions

There now exists a general non–gray theory of objects from 0.3 M_J to 70 M_J below ~1300 K using the best input physics and some of the best numerical tools available, but much remains to be done. In particular, the opacity of CH_4 and a proper treatment of silicate/iron, H_2O, and NH_3 clouds are future challenges that must be met before the theory can be considered mature. Furthermore, the effects of stellar insolation, addressed only approximately in Saumon *et al.* [33] and Guillot *et al.* [57], must be incorporated consistently. Since the near IR signature of proximate sub-stellar companions will be significantly altered by a reflected component, a theory of albedos in the optical and in the near IR must be developed [58]. It will be useful to predict the signatures of specific systems with known orbital characteristics, primaries, and ages, such as τ Boo, 51 Peg, υ And, 55 Cnc, ρ CrB, 70 Vir, 16 Cyg, and 47 UMa.

Nevertheless, the recent theoretical calculations lead to certain general conclusions:

1. H_2O, H_2, and CH_4 dominate the spectrum below T_{eff}~1200 K. For such T_{eff}s, most or all true metals are sequestered below the photosphere.
2. Though EGP colors and low–resolution spectra depend upon gravity, this dependence is weak. However, high–resolution spectra may provide useful gravity diagnostics.
3. The primary bands in which to search are Z, J, H, K, M and N. K is not as good as J or H.
4. Enhancements and suppressions of the emergent flux relative to black body values can be by many orders of magnitude.
5. Objects that were considered from their low T_{eff}s ($\leq$ 600 K) to be undetectable in the near IR may not be.
6. The infrared colors of EGPs and brown dwarfs are much bluer than the colors previously derived using either the blackbody assumption or primitive non–gray models.
7. In some IR colors (*e.g.*, $J-K$), an object gets bluer, not redder, with age and for a given age, lower–mass substellar objects are bluer than higher–mass substellar objects.
8. For a given composition, only two observables are necessary to constrain a substellar object's parameters. For instance, given only T_{eff} and gravity, one can derive mass, age, and radius.
9. Clouds of H_2O and NH_3 are formed for T_{eff}s below ~400 K and ~200 K, respectively, and clouds of sulfides and chlorides form between 700 K and 1100 K. Their formation will affect the colors, spectra, and albedos of EGPs and brown dwarfs in ways not yet fully characterized.

Acknowledgements

I would like to thank Bill Hubbard, Jonathan Lunine, David Sudarsky, Didier Saumon, Jim Liebert, Chris Sharp, Gilles Chabrier, Richard Freedman, Mark Marley, and Tristan Guillot for helpful comments and conversations. This work was supported under NASA grants NAG5-7499, NAG5-7073, and NAG5-2817.

References

1. Mayor, M. & Queloz, D. (1995), *Nature*, **378**, 355
2. Marcy, G. W. & Butler, R. P. (1996), *Ap. J.*, **464**, L147
3. Butler, R. P. & Marcy, G. W. (1996), *Ap. J.*, **464**, L153
4. Butler, R. P., Marcy, G. W., Williams, E., Hauser, H., & Shirts, P. (1997), *Ap. J.*, **474**, L115
5. Cochran, W.D., Hatzes, A.P., Butler, R.P., & Marcy, G. (1997), *Ap. J.*, **483**, 457
6. Noyes, R.W. *et al.* (1997), *Ap. J.*, **487**, 195
7. Latham, D. W., Mazeh, T., Stefanik, R.P., Mayor, M., & Burki, G. (1989), *Nature*, **339**, 38
8. Delfosse, X., Tinney, C.G., Forveille, T., Epchtein, N., Bertin, E., Borsenberger, J., Copet, E., De Batz, B., Fouqué, P., Kimeswenger, S., Le Bertre, T., Lacombe, F., Rouan, D., & Tiphène, D. (1997), *Astr. Ap.*, **327**, L25
9. Tinney, C.G., Delfosse, X., & Forveille, T. (1997), *Ap. J.*, **490**, L95
10. Kirkpatrick, J.D., Reid, I.N., Liebert, J., Cutri, R.M., Nelson, B., Beichman, C.A., Dahn, C.C., Monet, D.G., Skrutskie, M.F., & Gizis, J. (1998), submitted to *Ap. J.*
11. Zapatero-Osorio, M.R., Rebolo, R., & Martin, E.L. (1997), *Astr. Ap.*, **317**, 164
12. Nakajima, T., Oppenheimer, B.R., Kulkarni, S.R., Golimowski, D.A., Matthews, K. & Durrance, S.T. (1995), *Nature*, **378**, 463
13. Oppenheimer, B.R., Kulkarni, S.R., Matthews, K., & Nakajima, T. (1995), *Science*, **270**, 1478
14. Oppenheimer, B.R., Kulkarni, S.R., Matthews, K., & van Kerkwijk, M.H. (1998), submitted to *Ap. J.*
15. Geballe, T. R., Kulkarni, S. R., Woodward, C. E., & Sloan, G. C. (1996), *Ap. J.*, **467**, L101
16. Matthews, K., Nakajima, T., Kulkarni, S.R., & Oppenheimer, B.R. (1996), *A. J.*, **112**, 1678
17. Burrows, A.& Liebert, J. (1993), *Rev. Mod. Phys.*, **65**, 301
18. Burrows, A., Marley M., Hubbard, W.B. Lunine, J.I., Guillot, T., Saumon, D. Freedman, R., Sudarsky, D. & Sharp, C. (1997), *Ap. J.*, **491**, 856
19. Burrows, A. & Sharp, C., (1998), submitted to *Ap. J.*
20. Chandrasekhar, S., 1939, *An Introduction to the Study of Stellar Structure* (Dover Publications, New York).
21. Zapolsky, H. S. & Salpeter, E.E. (1969), *Ap. J.*, **158**, 809
22. Stevenson, D. J. (1991), *Ann. Rev. Astr. Ap.*, **29**, 163
23. Wigner, E. & Huntington, H.B. (1935), *J. Chem. Phys.*, **3**, 764
24. Magni, G. & Mazzitelli, I. (1979), *Astr. Ap.*, **72**, 134
25. Saumon, D. & Chabrier, G. (1989), *Phys. Rev. Letters*, **62**, 2397
26. Saumon, D. & Chabrier, G. (1991), *Phys. Rev.*, **A44**, 5122
27. Saumon, D. & Chabrier, G. (1992), *Phys. Rev.*, **A46**, 2084
28. Hubbard, W. B. & DeWitt, H.E. (1985), *Ap. J.*, **290**, 388
29. Grüneisen, E. (1926), in *Handbuch der Physik*, edited by H. Greiger and K. Scheel (Springer, Berlin), **10**, pp. 1–59
30. Debye, P. (1912), *Ann. Physik*, **39**, 789

31. Burrows, A., Hubbard, W.B., & Lunine, J.I. (1989), *Ap. J.*, **345**, 939
32. Burrows, A., Saumon, D., Guillot, T., Hubbard, W.B., & Lunine, J.I. (1995), *Nature*, **375**, 299
33. Saumon, D., Hubbard, W.B., Burrows, A., Guillot, T., Lunine, J.I., & Chabrier, G. (1996), *Ap. J.*, **460**, 993
34. Graboske, H. C., Pollack, J. B., Grossman, A. S., & Olness, R. J. (1975), *Ap. J.*, **199**, 265
35. Grossman, A.S. & Graboske, H.C. Jr. (1973), *Ap. J.*, **180**, 195
36. Black, D. C. (1980), *Icarus*, **43**, 293
37. Burrows, A., Hubbard, W.B., Saumon, D., & Lunine, J.I. (1993), *Ap. J.*, **406**, 158
38. Hubbard, W. B. (1977), *Icarus*, **30**, 305
39. Pollack, J.B., (1984), *Ann. Rev. Astr. Ap.*, **22**, 389
40. Bodenheimer, P. & Pollack, J. B. (1986), *Icarus*, **67**, 391
41. Lunine, J.I., Hubbard, W.B., Burrows, A., Wang, Y.P., & Garlow, K. (1989), *Ap. J.*, **338**, 314
42. Allard, F., Hauschildt, P.H., Alexander, D.R. & Starrfield, S. (1997), *Ann. Rev. Astr. Ap.*, **35**, 137
43. Marley, M., Saumon, D., Guillot, T., Freedman, R.S., Hubbard, W.B., Burrows, A., & Lunine, J.I. (1996), *Science*, **272**, 1919
44. Allard, F., Hauschildt, P.H., Baraffe, I., & Chabrier, G. (1996), *Ap. J.*, **465**, L123
45. Tsuji, T., Ohnaka, W., Aoki, W., & Nakajima, T. (1996), *Astr. Ap.*, **308**, L29
46. Fegley, B. & Lodders, K. (1996), *Ap. J.*, **472**, L37
47. Thompson, R. (1992), *Space Science Reviews*, **61**, 69
48. Benvenuti, P. *et al.* (1994), in *ESA's Report to the 30th COSPAR Meeting, ESA SP-1169, Paris*, p. 75
49. Erickson, E. F. & Werner, M.W. (1992), *Space Science Reviews*, **61**, 95
50. Mountain, M., R. Kurz, R., & Oschmann, J. (1994), in *The Gemini 8-m Telescope Projects, S.P.I.E. Proceedings on Advanced Technology Optical Telescopes V*, **2199**, p. 41
51. Erickson, E. F. (1992), *Space Science Reviews*, **61**, 61
52. Kirkpatrick, J.D., Henry, T.J., & Simons, D.A. (1994), *A. J.*, **108**, 1437
53. Kirkpatrick, J.D., Henry, T.J., & Simons, D.A. (1995), *A. J.*, **109**, 797
54. Leggett, S.K. (1992), *Ap. J. Suppl.*, **82**, 351
55. Chabrier, G. *et al.* (1998), to be published in the proceedings of the first Euroconference on *Stellar Clusters and Associations*, held in Los Cancajos, La Palma, Spain, May 11–15, ed. R. Rebolo, V. Sanchez–Bejar, and M.R. Zapatero-Osorio
56. Jones, H.R.A. & Tsuji, T. (1997), *Ap. J.*, **480**, L39
57. Guillot, T., Burrows, A., Hubbard, W.B., Lunine, J.I., & Saumon, D. (1996), *Ap. J.*, **459**, L35
58. Marley, M.S., Gelino, C., Stephens, D., Lunine, J.I., Freedman, R., (1998), submitted to *Ap. J.*

FROM THE INTERSTELLAR MEDIUM TO PLANETARY ATMOSPHERES VIA COMETS

T. OWEN
University of Hawaii, Institute for Astronomy
2680 Woodlawn Drive, Honolulu,
Hawaii 96822 USA

AND

A. BAR-NUN
Tel-Aviv University
Tel-Aviv 69978, ISRAEL

Abstract. Laboratory experiments on the trapping of gases by ice forming at low temperatures implicate comets as major carriers of the heavy noble gases to the inner planets. Recent work on deuterium in Comet Hale-Bopp provides good evidence that comets contain some unmodified interstellar material. However, if the sample of three comets analyzed so far is typical, the Earth's oceans cannot have been produced by comets alone. The highly fractionated neon in the Earth's atmosphere also indicates the importance of non-icy carriers of volatiles, as do the noble gas abundances in meteorites from Mars.

1. Introduction

Ever since the pioneering suggestion by Oro (1961) that comets could have been an important source of organic material on the early Earth, numerous investigations have suggested that comets could have brought in a variety of volatile elements and compounds. Recent work on this question has focused on models of the dissipation of icy planetesimals from the Uranus-Neptune region as these planets finished forming, or attempts to calculate the extent of the terrestrial cometary influx through analyses of the cratering record of the moon. All of these studies suffer from the absence of evidence for a uniquely identifiable contribution in the Earth's volatile inventory. Until

J.-M. Mariotti and D. Alloin (eds.), Planets Outside the Solar System: Theory and Observations, 141–153.

recently, there was little incentive for discovering such evidence since models invoking meteoritic sources for terrestrial volatiles appeared perfectly adequate, requiring no additional contributions.

The difficulty in identifying the source of the Earth's volatiles is compounded by the 4.5 billion year history of the planet, during which chemical reactions with the crust, escape of gases from the upper atmosphere and the origin and evolution of life have completely changed the composition of the atmosphere and hydrosphere. The central problem of life's origin is intimately dependent on the early composition of the atmosphere, making our inability to define that composition all the more frustrating.

It is this dilemma that the heavy noble gases have the potential to resolve, since they are chemically inert, do not easily escape from the atmosphere, and are not involved in the activities of living organisms. The firmly entrenched idea that the atmospheric noble gas abundances are the result of delivery by meteorites stems from the recognition of a so-called "planetary pattern" in the noble gases found in these objects many years ago. This idea gained widespread support, despite the fact that it has never been possible to explain why the abundances of krypton and xenon in the meteorites are about the same, while xenon is much lower than krypton in the earth's atmosphere (Figure 1). Attempts to find the "missing " xenon buried in shells, clathrates, or ice have failed. It was this discrepancy coupled with the arguments for early cometary bombardment that encouraged us to pursue the idea that icy, rather than rocky planetesimals may have delivered these gases.

2. A COMETARY MODEL FOR DELIVERY OF VOLATILES

Accordingly, we set out to determine whether or not the abundance patterns of the heavy noble gases in the atmospheres of the inner planets could be accounted for by comets. The first step was to apply the same laboratory techniques for trapping gases in amorphous ice deposited at low temperatures that had been used with N_2, CO, H_2, etc. to a mixture of heavy noble gases: argon, krypton and xenon. These experiments were designed to imitate the formation of comets in the outer solar nebula. The idea is that interstellar ice grains probably sublimated as they fell toward the mid-plane of the nebula and recondensed on cold refractory cores. In this recondensation process, they could trap ambient gas according to the local temperature. The laboratory results showed that temperature-dependent fractionation of the gas mixtures occurs, which suggested that trapping in ice might indeed be responsible for the patterns of noble gas abundances found in the atmospheres of Mars, Earth and Venus. This initial analysis was not conclusive, however, since it did not seem possible to account for the variety of patterns observed on the three planets with a single model of cometary combardment.

The SNC meteorites provided another clue. It is now widely accepted that these meteorites come from Mars. Yet a three-isotope plot(Figure 2) of $^{36}Ar/^{132}Xe$ vs $^{84}Kr/^{132}Xe$ shows that the abundances of these gases are very different in the different meteorites, forming a straight line on a log-log plot. This line passes through points corresponding to the noble gases found in the atmospheres of the Earth and Mars, but it is widely separated from the field of abundances in chondritic meteorites or the values found in the solar wind.

We have interpreted this distribution of noble gas abundances in the SNC meteorites as an effect of the mixing of two different reservoirs: one that represents gases trapped in the rocks that form the planets, the second a contribution from impacting icy planetesimals. This interpretation is based on our data for the fractionation of noble gases by trapping in amorphous ice. At a temperature of $\sim$ 50 K, ice formed by condensation of water vapor traps argon, krypton and xenon in relative abundances that fall on an extrapolation of a mixing line drawn through abundances from the atmospheres of Earth and Mars on the same three-isotope plot (Figure 2). The analysis is complicated by the fact that the linear relation on the log-log plot illustrated by Ott and Begemann, is very close to the mixing line. It may well be that more than one process is at work here. However, we have pointed out that the increase in $^{129}Xe/^{132}Xe$ that occurs along this same mixing line demonstrates that solubility in a melt (for example) cannot explain the data by itself.

We have further tested the applicability of the laboratory data to natural phenomena by examining the abundances of CO and N_2 in comets. We pointed out that the apparently mysterious depletion of nitrogen in comets probably results from the inability of ice to trap N_2 efficiently when the ice forms at T$\sim$35 K. Our assumption is that most of the nitrogen that was present in the outer solar nebula was in the form of N_2, just as it is in the interstellar medium. Hence for comets to acquire a solar ratio of N/O, the condensing ice that formed the comets would have to trap N_2 in the same fashion as the noble gases. We also showed that the relative abundances of CO^+/N_2^+ derived from observations of comet tails are consistent with the formation of the icy nuclei at temperatures near 50 K. Further work on the trapping of CO and N_2 in ice has substantiated our conclusions.

These consistencies between the laboratory results and the observed abundances in comets lent support to our effort to extrapolate laboratory work on ice trapping of noble gases to noble gas abundances in planetary atmospheres. Accordingly, we developed an "icy impact" model for the contribution of comets to inner planet atmospheres. An immediate consequence of this model was the realization that if comets contributed the terrestrial heavy noble gases, they would not supply enough water to fill the oceans. Impact erosion might have solved this problem by removing noble gases (plus carbon monoxide, carbon dioxide and nitrogen), but we also suggested that some water might have been contributed from another source, such as the rocks making up most of the mass of the planet.

Turning to Mars, we used the present atmospheric abundance of ^{84}Kr to predict that only the equivalent 75 mb of CO_2 could be accounted for. This amount of CO_2 would give a ratio of C/^{84}Kr in the atmosphere equal to the value found in the Earth's volatile reservoir and in the present atmosphere of Venus. While 75 mb of CO_2 is 10 times the present value in the Martian atmosphere, it corresponds to a layer of water only 12 m thick over the planet's surface, an insufficient amount to account for the observed fluvial erosion. Accordingly, we invoked impact erosion as the responsible agent for diminishing the atmosphere to the 75 mb level. We pointed out that impact erosion appears to be the only way to explain the relatively high abundance of ^{129}Xe/^{132}Xe in the Martian atmosphere which is trapped in the meteorites EETA 79001 and Zagami. Production of massive amounts of carbonates would remove CO_2 but it would not account for excess ^{129}Xe and it would leave behind large amounts of noble gases and nitrogen, which are not observed.

We thus suggested a scenario in which the early atmosphere of Mars could have passed through several episodes of growth and decay, depending on the planet's bombardment history. We suggested that it is possible to see the effects of cometary bombardment in the gases trapped in the SNC meteorites, although other interpretations are certainly possible.

3. DEUTERIUM IN COMETS: A TIE TO THE INTERSTELLAR MEDIUM

Clearly one of the assumptions in this model is that comets have retained a good memory of interstellar chemical conditions. One way to test this assumption is to investigate the abundances of stable isotopes of common elements, as isotope ratios should not change during the comet-formation process we have described.

The recent appearance of two bright comets has allowed us to add to the data collected by in situ measurements of Comet Halley using ground-based techniques. Observations at radio wavelengths have been especially helpful.

The value of D/H is particularly sensitive to physical processes, since the mass ratio of these two isotopes is the largest in the periodic table. In Halley's comet, two independent investigations led to the result that

$D/H=3.2\pm0.3\times10^{-4}$ in the comet's H_2O, twice the value found in standard mean ocean water (SMOW) on Earth. These studies were carried out in situ, with mass spectrometers on the Giotto spacecraft. Observations of Comet Hyakutake at radio and infrared wavelengths from Earth allowed the detection of lines of HDO and H_2O that led to a value of $D/H=3.3\pm1.5\times10^{-4}$.

In the case of Comet Hale-Bopp, it was possible to use these same remote sensing techniques to determine D/H in both H_2O and HCN. The results were quite different: (D/H) $H_2O=3.2\pm1.2\times10^{-4}$, (D/H) HCN= $2.3\pm0.4\times10^{-3}$. This is exactly the kind of difference that shows up in interstellar molecular clouds, providing a strong case for the preservation of interstellar chemistry in comet ices. If comets had formed from solar nebula gas that was warmed and homogenized by radial mixing, the values of D/H in both molecules would be much lower and identical, as the HCN and H_2O would exchange D with the huge reservoir of H_2 in the solar nebula, in which $D/H=2\text{-}4\times10^{-5}$.

4. THE ORIGIN OF WATER

We might also expect to see signs of cometary bombardment in the water that is present on the inner planets. Note that all three of the comets that have been studied for this exhibit a value of D/H in ice that is twice the value found in ocean water on Earth. All of these comets came from the Oort Cloud, a spherical shell of cometary nuclei whose average radius is 50,000 times the Earth's distance from the sun. Three is certainly a tiny number compared to the 10^{11} comets that may be present in the Oort Cloud, so one might speculate that other comets may have different values of D/H. In particular, we have not yet sampled any comets from the Kuiper Belt, a disk of comet nuclei extending outward from Pluto. The Kuiper Belt is currently thought to be the likely source of short period comets, which are unfortunately too faint to yield values of D/H to present ground-based techniques.

It is difficult to imagine a process that would selectively change the isotope ratios in comets forming in different regions of the solar nebula. Nevertheless, we should continue to measure D/H in all comets that are bright enough and we should take advantage of proposed spacecraft visits to short period comets to extend this analysis to that family as well.

Meanwhile, we shall assume that the three comets are indeed representative and see whether we can find traces of cometary water on the inner planets. Mars is the best case, because the atmosphere is so thin. The total amount of water that exchanges seasonally through the Martian atmosphere from pole to pole is just 2.9×10^{15} gm, equivalent to a single comet nucleus (density r = 0.5 g/cm3) with a radius of only 1 km. Hence the impact of a relatively small comet could have a significant effect on the surface water supply. Indications of such an effect can be found in studies of D/H in hydrous phases of the SNC meteorites. Watson et al. reported a range of values of D/H in kaersutite, biotite and apatite in Chassigny, Shergotty and Zagami, with the highest values, 4 to 5.5 times the terrestrial ocean standard, occurring in Zagami apatite. Infrared spectroscopy from Earth has determined D/H in Martian atmospheric water vapor to be 5.5 $\pm$ 2 times terrestrial. The overlap between the SNC and Mars atmosphere values indicates that some mixing between atmospheric water vapor and the crustal rocks must occur. It is therefore arresting to note that the lowest values of D/H measured in the SNC minerals do not cluster around 1 time terrestrial, but rather around 2 times terrestrial. This is also true in the most recently discovered shergottite QUE 94201. The significance of this lies in the fact that D/H in the water in our three comets is also 2 times terrestrial. It thus appears possible that most of the near-surface water on Mars was contributed primarily by cometary impact, rather than by magma from the planet or by meteoritic bombardment.

This interpretation is consistent with the geochemical analysis by Carr and Waenke who concluded that Mars is much drier than the Earth, with roughly 35 ppm water in mantle rocks as opposed to 150 ppm for Earth. These authors suggested that one possible explanation for this difference is the lack of plate tectonics on Mars, which would prevent a volatile-rich veneer from mixing with mantle rocks. This is exactly what the D/H values in the SNC minerals appear to signify. In whole rock samples, D/H in the Shergottites is systematically higher than in Nakhla and Chassigny. This could be a result of the larger fraction of atmospheric H_2O incorporated in the Shergottites by shock, a process Nakhlites evidently avoided.

Furthermore, higher values of $\Delta^{17}O$ are found in these samples of Nakhla and Chassigny compared with the Shergottites, reinforcing the idea that the Shergottites sampled a different source of water. Finally, the oxygen isotope ratios in water from whole rock samples of the SNCs are also systematically different from the ratios found in silicates in these rocks again suggesting a hydrosphere that is not strongly coupled to the lithosphere.

On Earth, the oxygen isotopes in sea water match those in the silicates, indicating thorough mixing for at least the last 3.5 BY. The Earth has also lost relatively little hydrogen into space after the postulated hydrodynamic escape, so the value of D/H we measure in sea water today must be close to the original value. If Halley and Hyakutake and Hale-Bopp are truly representative, of all comets, then we can't make the oceans out of melted comets alone. This is a very different situation from Mars. It suggests that water from the inner reservoir, the rocks making up the bulk of the Earth, must have mixed with incoming cometary water to produce our planet's oceans.

Explanations for the relatively high value of O/C (12 ±6) in the terrestrial volatile inventory also suggest such mixing. The solar value of O/C = 2.4, which was also the value found in Halley's comet. Impact erosion on the Earth could remove CO and CO_2 while having little effect on water in the oceans or polar caps thereby raising the value of O/C. This process would not affect the value of D/H, however. The average value of D/H in chondritic meteorites is close to that in sea water, so mixing meteoritic and cometary water would not lead to the right result. We need a contribution from a reservoir with D/H less than 1.6×10^{-4}. Lecluse and Robert have shown that water vapor in the solar nebula at 1 AU from the sun would have developed a value of D/H $\sim 0.8 \times 10^{-4}$ to 1.0×10^{-4}, depending on the lifetime of the nebula (2×10^5 to 2×10^6 years). An ocean made of roughly 35% cometary water and 65% water from the local solar nebula (trapped in planetary rocks) would satisfy the D/H constraint and would also be consistent with the observed value of O/C = 12 ±6.

To accept this idea, we should be able to demonstrate that water vapor from the solar nebula was adsorbed on grains that became the rocks that formed the planets. Our best hope for finding some of that original inner-nebula water appears to be on Mars, where mixing between the surface and the mantle has been so poor. The test is thus to look for water incorporated in SNC meteorites that appear to have trapped mantle gases, to see if D/H $< 1.6\times10^{-4}$.

The best case for a such a test among the rocks we have is Chassigny, which exhibits no enrichment of ^{129}Xe and thus appears not to have trapped any atmospheric gas. However, there is no evidence of water with low D/H in this rock. It may be that contamination by terrestrial water has masked the Martian mantle component in Chassigny. This is a good project for a sample returned from Mars by spacecraft, where such contamination can be avoided.

5. THE IMPORTANCE OF NEON

We have concentrated our analysis on water and the heavy noble gases: argon krypton and xenon. Any model for the origin of the atmosphere must also account for neon. This gas has about the same cosmic abundance as nitrogen relative to hydrogen viz., 1.2×10^{-4} vs 1.1×10^{-4}. Hence we expect any atmosphere that consists of a captured remnant of the solar nebula to exhibit a ratio of $Ne/N_2 \simeq 2$. On Earth, Mars and Venus, $Ne \ll N_2$. The neon is not only deficient in these atmospheres, the isotopes have been severely fractionated. The solar ratio is of $^{20}Ne/^{22}Ne$ = 13.7, on Earth $^{20}Ne/^{22}Ne$

= 9.8, on Venus 11.8 ± 0.7 and on Mars 10.1± 0.7. Concentrating on the Earth, we must ask how neon can be so severely fractionated, whereas nitrogen is not.

The answer may again lie with the comets, albeit in a paradoxical manner. The laboratory work shows that neon is not trapped in ice that forms at temperatures above 20 K. As the overwhelming majority of the icy planetesimals that formed in the solar system condensed at temperatures higher than this, we have assumed that comets carry no neon. This assumption is supported by the apparent absence of neon in the atmospheres of Titan, Triton, and Pluto, where the upper limits on Ne/N_2 are typically about 0.01. These three objects in the outer solar system, especially Triton and Pluto, represent giant icy planetesimals, which can be thought of as the largest members of the Kuiper Belt. Hence the absence of neon in their atmospheres may be taken as a good indication that ice condensing in the outer solar nebula did not trap this gas, and thus we do not expect to find it in comets. This prediction is consistent with new observations of Comet Hale-Bopp carried out with the Extreme Ultraviolet Explorer satellite by Krasnopolsky et al.. These authors established an upper limit of Ne/O <1/200 solar.

If the comets don't carry neon, how did this gas reach the inner planets? Once again the meteorites don't help. Even if they brought in all the xenon, the neon they could deliver would be less than 10% of what we observe. Instead it seems likely that neon was brought in by the rocks. In fact, we have evidence that this was the case, because we can still find neon whose isotope abundances approach the solar ratio in rocks derived from the mantle. Unlike the other noble gases, neon cannot be subducted into the Earth's interior. Thus it is not possible to dilute the original trapped gas with highly fractionated atmospheric neon. The record of original emplacement is preserved.

If neon, which diffuses so easily through solids, was retained by the Earth's rocks from the time of the planet's accretion, we can reasonably assume that some water was also kept in the interior, to emerge after the catastrophic formation of the moon, mixing with incoming water from comets to form the oceans we find today. This perspective supports the idea that we may yet find evidence of this original water in mantle-derived rocks on Mars.

Returning to the Earth, it appears that the atmospheric neon bears a record of an early fractionating process that sharply reduced the ratio of $^{20}Ne/^{22}Ne$ from the solar value. This process must have affected all of the other species in the atmosphere at the time. It may have been the massive, hydrodynamic escape of hydrogen produced by the reduction of water by contemporary crustal iron. The fact that we do not see evidence of any such fractionation in atmospheric nitrogen in the atmosphere today suggests that the nitrogen reached the Earth after this process had ended. Cometary delivery of nitrogen (and other volatiles) but not neon offers an easy means of achieving this condition. In this case, neon is a kind of atmospheric fossil, a remnant of conditions that existed on the Earth before the volatiles that produced the bulk of the present atmosphere were in place.

6. CONCLUSIONS

How unique is the Earth? This is a perennial question in efforts to estimate the possibilities for abundant life in the universe. We have argued here that the source of our planet's atmosphere can be found in a combination of volatiles trapped in the rocks that made the planet and a late-accreting veneer of volatile-rich material delivered by icy planetesimals. The volatiles composing the atmosphere include the carbon, nitrogen, and water essential to life. The close similarities between the elemental and isotopic abundances found in comets with those in the interstellar medium imply that icy planetesimals that form in any planetary system originating from an interstellar cloud will carry these same biogenic materials. Hence this model for the origin of our planet's atmosphere suggests that there is nothing unique about the inventory of volatiles that was delivered to the early Earth. Current studies of the Martian atmosphere, aided by study of the SNC meteorites, reinforce this idea by indicating a similar inventory on that planet.

Nevertheless, we are still in the stage of finding "similarities" and "indications." We do not yet have a rigorous proof of the validity of the icy impact model. We have stressed the constraints provided by the abundances and isotope ratios of the noble gases and determinations of cometary D/H. It is clear that meteoritic delivery of volatiles cannot satisfy the constraints set by our present knowledge of noble gas abundances and isotope ratios. However, the cometary alternative that we have emphasized will remain conjectural until noble gases are actually measured in a comet. Although laboratory studies strongly suggest that comets can deliver the correct elemental abundances, trapping of gas in ice does not affect isotopic ratios. We are therefore forced to assume that comets carry xenon whose isotopes resemble the distribution found in the terrestrial and Martian atmospheres rather than that found in the solar wind. It is not at all obvious why this should be the case. If the cometary xenon in fact resembles solar wind xenon, it will be necessary to invoke a fractionating process that acted to produce identical results for xenon on Mars and Earth followed by subsequent selective replacement of other volatiles.

How can we move forward from this unsatisfactory situation?There are a number of possible sources of new data on the horizon:

6.1. MARS

The Japanese Planet B Mission to be launched in 1998 will carry instruments that will teach us much more about possible non-thermal escape processes on Mars. This knowledge will allow a more confident reconstruction of the early mass and composition of Martian the atmosphere. These parameters can then be used (again!) to test our understanding of the origin of our own atmosphere.

If present plans mature, the step forward achieved from Planet B will soon be overshadowed by information obtained from Mars Sample Return Missions, scheduled to begin in 2005. These missions will bring back samples of Martian rocks and atmosphere for analysis on Earth, enabling far more accurate measurements of isotopic ratios than we can expect from missions to the planet. These measurements will include not only the noble gases, but also isotopes of carbon, nitrogen, and oxygen, the last in both H_2O and CO_2. With the kind of precision obtainable in laboratories on Earth, great progress should be achieved in unraveling the history of the Martian atmosphere from these isotope measurements, including estimates of the size and location of contemporary reservoirs of H_2O and CO_2. Another goal of this research would be a search for low values of D/H in water from mantle-derived rocks, which should also contain neon with $^{20}Ne/^{22}Ne$ approaching the solar value of 13.7.

6.2. COMETS

Both NASA and ESA are planning missions that will rendezvous with comets and deploy landers to explore their nuclei. These missions will have the capability to detect and measure the abundances of the heavy noble gases and their isotopes. This will be the most definitive test of the icy impact model. It is especially important to have this information from several comets, as we already know that the composition of comets can vary, both from the laboratory work on the trapping of gas in ice (Figure 2) and from observations of variations in carbon compounds in comets.

The ESA mission to Comet Wirtanen is called Rosetta, it should arrive at its destination in 2012. The NASA missions have not yet been approved, but are planned for approximately the same time period. There is even hope for a new atmospheric probe to Venus in this time frame, which would tell us more about the apparently anomalous noble gas abundances on this planet. We await all these new results with great interest.

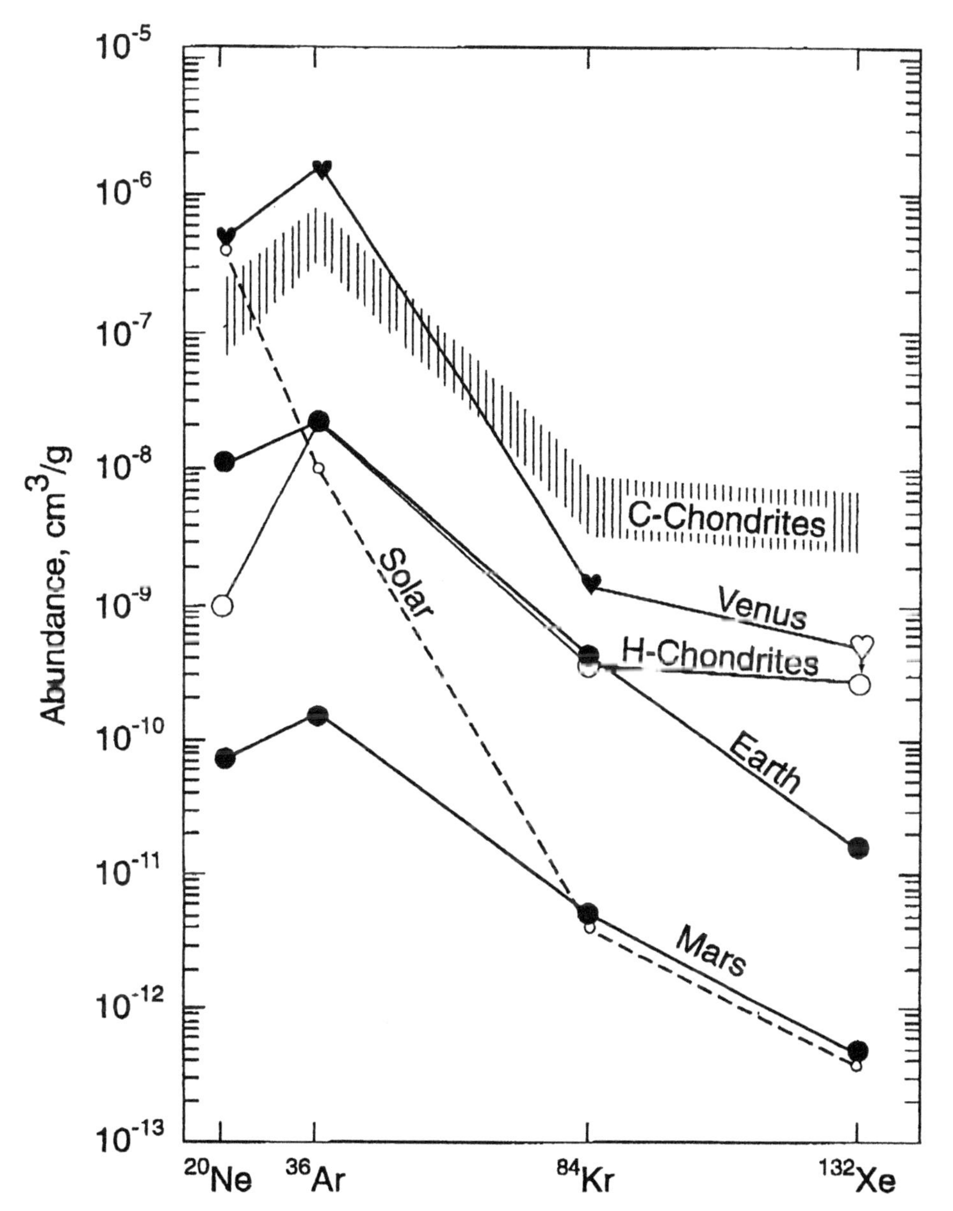

Figure 1. Chondritic meteorites contain about as much xenon as krypton. The meteoritic noble gas abundances therefore do not match the abundance patterns found in inner planet atmospheres, despite the apparent agreement for Ne, Ar, and Kr. (Solar values are normalized for ^{84}Kr on Mars.)

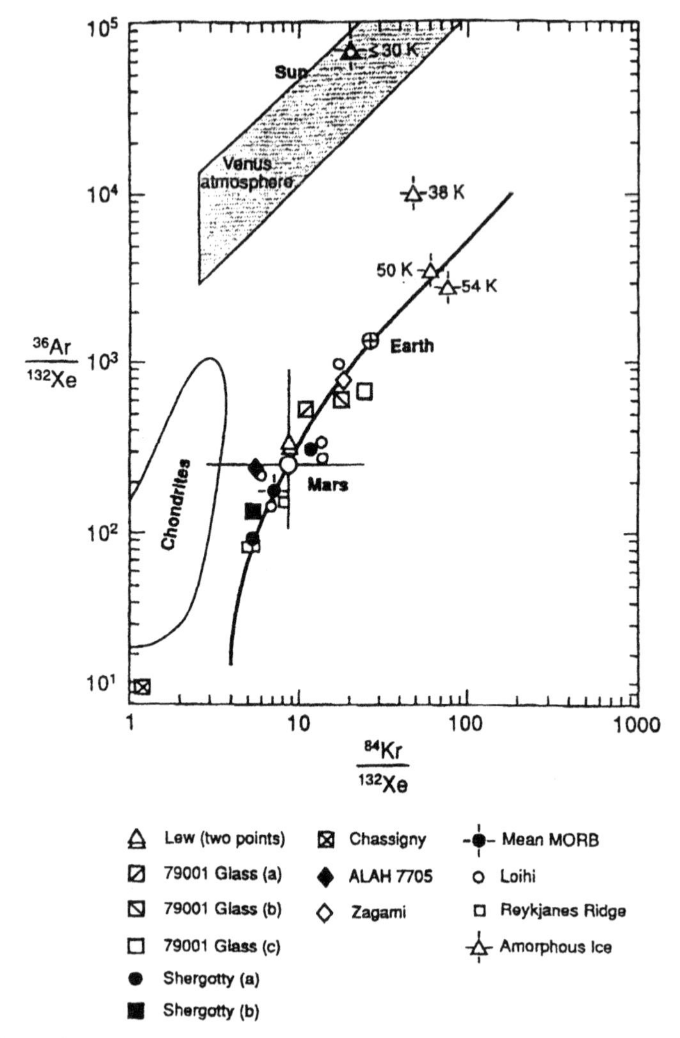

Figure 2. In a three isotope plot, noble gas abundances in the atmospheres of Mars and Venus can be used to define a mixing line between internal and external volatile reservoirs. The internal reservoir lies below Mars on this plot, and consists of the rocks that formed the planet. We suggest icy planetesimals as the external reservoirs is lying above the Earth at the opposite end of the line. The external reservoir is represented here by the noble gases trapped in amorphous ice in laboratory experiments (the open triangles). Noble gas abundances in Shergottites (meteorites from Mars) fall along this line, as do gases from mantle-derived rocks on Earth (Mean MORB, Loihi, Reykjanes Ridge). The gases on Venus could have been delivered by comets from the Kuiper Belt that formed at temperatures of 30-35 K. The abundance of Xe on Venus is not yet known, hence the stippled trapezoid.

References:
J. Oro, 1961, Nature 190, 389
U. Ott, F. Begemann, 1985, Nature 317, 509
L.L.Watson, I.D. Hutcheon, E.M. Stopler, 1994, Science 265, 86
M.H. Carr, H. Waenke, 1992, Icarus 98, 61
C. Lecluse, F. Robert, 1994, Geochim et Cosmochim, Acta.58, 2927

Reprinted with permission from Faraday Discussions No.109, (1998) published by The Royal Society of Chemistry, Cambridge, UK.

GIANT PLANET FORMATION

Formation and Growth of Massive Envelopes

G. WUCHTERL
Institut für Astronomie der Universität Wien
Türkenschanzsrtaße 17, A-1180 Wien, Austria
wuchterl@amok.astro.univie.ac.at

1. Introduction

Our four giant planets contain 99.5% of the angular momentum of the Solar System, but only 0.13% of its mass[1]. On the other hand, more than 99.5% of the mass of the planetary system is in those four largest bodies. The angular momentum distribution can be understood on the basis of the 'nebula hypothesis' (Kant 1755). The nebula hypothesis assumes concurrent formation of a planetary system and a star from a centrifugally supported flattened disk of gas and dust with a pressure supported, central condensation (Laplace 1796, Safronov 1969, Lissauer 1993). Theoretical models of the collapse of slowly rotating molecular cloud cores have demonstrated that such preplanetary nebulae are the consequence of the observed cloud core conditions and the hydrodynamics of radiating flows, provided there is a macroscopic angular momentum transfer process (Cassen and Moosman 1981, Morfill et al. 1985, Laughlin and Bodenheimer 1994). Assuming turbulent viscosity to be that process, dynamical models have shown how mass and angular momentum separate by accretion through a viscous disk onto a growing central protostar (Tscharnuter 1987, Tscharnuter and Boss 1993). Those calculations, however, do not yet reach to the evolutionary state of the nebula where planet formation is expected. Interestingly the most far reaching calculations point to nebulae, that are initially below the so called minimum reconstituted mass (Morfill et al. 1985). Observational evidence for solar system sized disks of dust and gas around young stars has accumulated in an impressive way in the last years. Observationally inferred disk sizes and masses are overlapping theoretical expectations and fortify

[1]This work has been supported by the Österreichischer Fonds zur Förderung der wissenschaftlichen Forschung (FWF) under project numbers S-7305–AST, S-7307–AST.

J.-M. Mariotti and D. Alloin (eds.), Planets Outside the Solar System: Theory and Observations, 155–172.

the nebula hypothesis. High resolution observations at mm-wavelengths are now sensitive to disk conditions at orbital distances > 50AU (Dutrey, this volume, Dutrey et al. 1999, Guilloteau and Dutrey 1999).

However, observations thus far provide little information about the physical conditions in the respective nebulae on scales of 1 to 40 AU, where planet formation is expected to occur. Midplane-densities and temperatures in a preplanetary nebula at a few AU from the star have not yet been measured. Planet formation studies therefore obtain plausible values for disk conditions from nebulae that are reconstructed from the present planetary system and disk physics. The so obtained 'minimum reconstituted nebula masses' *defined as the total mass of solar composition material needed to provide the observed planetary/satellite masses and compositions by condensation and accumulation*, are a few percent of the central body for the solar nebula *and* the circumplanetary protosatellite nebulae (Kusaka et al. 1970 Hayashi 1980, Stevenson 1982a). The total angular momenta of the satellite systems, however are only about 1% of the spin angular momentum of the respective giant planet (Podolak et al. 1993), in strong contrast with the planetary system/Sun ratio. Assembling planets from a nebula disk and advecting the angular momentum due to keplerian shear until the present giant planet masses are reached results in total angular momenta overestimating the present spin angular momenta of the giant planets only by small factors 2.4,1.5,2.9 and 4.5 for J,S,U,N, resp., cf. (Götz 1993). Even if giant planets would have kept this angular momentum they still would not rotate critically! Giant planets, unlike stars therefore do not have an angular momentum problem. That has important consequences for the formation process: while planetary systems and protostellar disks highlight the importance of angular momentum redistribution in protostellar evolution, satellite systems do not provide similar evidence for the giant planet formation process. This may justify why most studies of proto-giant planets neglect rotation or treat it as a small perturbation.

Giant planet formation is discussed here mostly in the context of the *nucleated-instability hypothesis*, i.e. as the consequence of the formation and growth of the solid building blocks of rocky solar system objects as terrestrial planets, moons and asteroids. According to the 'planetesimal hypothesis' these objects grow within circumstellar disks via pairwise accretion of km-sized solid bodies, the so-called planetesimals. Sufficiently massive planetesimals embedded in a gravitationally stable preplanetary nebula can capture large amounts of gas and become the cores of giant planets (e.g. Wuchterl 1995a, Tajima and Nakagawa 1997). I will focus here on how the envelopes grow for given cores and simple core-growth rates. For a discussion of solid body accretion and quasi-hydrostatic giant planet formation models with detailed treatment of the planetesimal accretion rate see Lissauer (1993), Pollack et al. (1996), Wuchterl et al. (1999), Bodenheimer et al. (1999) and Papaloizou, this volume. For the discussion of the late phases of giant planet growth (masses $> 300\,\mathrm{M}_\oplus$), where the proto giant planets perturbe the ambient nebula, I refer to the chapter by Artymowicz.

The 'nucleated instability hypothesis' is the only model for the formation of Uranus and Neptune at the moment, while other models also exist for Jupiter and Saturn. Presently advocated models of giant planet are briefly scetched in the next section.

2. Current Hypothesis about Giant Planet Formation

The key problem in giant planet formation is that preplanetary disks are only weakly self-gravitating equilibrium structures supported by gas pressure and the shearing keplerian motion (see Papaloizou, this volume). Any isolated, orbiting object below the Roche density is pulled apart by the stellar tides. Typical nebula densities are more than two orders of magnitude below the Roche density, so compression is needed to confine a condensation of mass M inside its tidal or Hill-radius[2] at orbital distance a:

$$R_T = a\left(\frac{M}{3\mathrm{M}_\odot}\right)^{1/3} . \tag{1}$$

A local enhancement of self-gravity is needed to overcome the counteracting gas pressure. Giant planet formation theories may be classified by how they provide this enhancement: (i) the *nucleated instability* model relies on the extra gravity field of a sufficiently large solid core (condensed material represents a gain of ten orders of magnitude in density and therefore self-gravity compared to the nebula gas) (Mizuno 1980, Bodenheimer and Pollack 1986, Wuchterl 1991a,b, Götz 1993, Wuchterl 1993,1995a, Pollack et al. 1996), Tajima and Nakagawa 1997, Wuchterl 1997, 1998) (ii) a *disk instability* may operate on length scales between short scale pressure support and long scale tidal support (DeCampli and Cameron 1979, Boss 1997, 1998), or (iii) an *external perturber* could compress an otherwise stable disk on its local dynamical time scales, e.g., by accretion of a clump onto the disk or rendezvous with a stellar companion. (e.g. Brandl and Sterzig 1998). If the gravity enhancement is provided by a dynamical process as in the latter two cases, the resulting nebula perturbation, (say of a Jupiter mass of material) is compressionally heated because it is optically thick under nebula conditions. Giant planet formation would then involve a transient

[2]The tidal radius and the Hill-radius are conceptually different, but their values are, to the first order in $(M/3M_\odot)^{1/3}$, equal to each other and to the distance to the first Lagrangian point, L_1.

phase of tenuous giant gaseous protoplanets, that would be essentially fully convective and contract on a timescale of typically $\sim 10^6$ yr (Bodenheimer 1986). Another mechanism of forming stellar companions (iv) *fragmentation during collapse*, is plausible for binary stars and possibly brown dwarfs, but it is unlikely to form objects of planetary masses because opacity limits the process to masses above $\sim 10\,M_J$ for the reason just given.

2.1. NEBULA STABILITY

Preplanetary nebulae with minimum reconstituted mass are stable. Substantially more massive disks resulting from the collapse of cloud cores are self-stabilizing due to transfer of disk mass to the stabilizing central protostar (Papaloizou, this volume, Papaloizou and Savonije 1991, Laughlin and Bodenheimer 1994, Bodenheimer et al. 1993). Nevertheless a moderate mass nebula disk might be found that can develope a disk instability leading to a strong density perturbation, especially when forced with a finite external perturbation. Giant gaseous protoplanets (GGPPs) might form when the instability has developed into a clump (DeCampli and Cameron 1979, see Bodenheimer 1985, and Bodenheimer 1986 for a discussion). Boss (1997, 1998) has constructed such an unstable disk with $0.13\,M_{\odot}$ within 10 AU and obtained maximum density enhancements (by a factor ~ 20) with ~ 10 Jupiter masses above the background for a few orbital periods. (The density enhancement at the surface of a $1\,M_{\oplus}$ core is between 10^5 to 10^7, for comparison). These clumps, provided they are stable on a few cooling times are candidates to become proto-giant planets via an intermediate state as tenuous giant gaseous protoplanet (GGPP). A key issue, as in any theory involving an instability of the disk gas, is then the a posteriori formation of a core. Only metals that are present initially would rain out to form a core, while material added later by impacts of small bodies after the GGPPs had formed would be soluble in the envelope (Stevenson 1982a). Boss (1998) outlined how a core corresponding to the solar composition high-Z material ($6\,M_{\oplus}$ and $2\,M_{\oplus}$ for Jupiter- and Saturn-mass respectively) might form if the density enhancements he found are long lived, need no more pressure confinement and evolve into GGPPs that are *non-turbulent.* It should be noted here (see Guillot 1998, Wuchterl et al. 1999) that although interior models of Saturn do not rule out the possibility that the planet has no core (or, equivalently a $2\,M_{\oplus}$ core), this is not the favored solution. Also GGPP models would probably predict that Jupiter should have a bigger core than Saturn which is only marginally consistent with present interior models. Finally, Saturn and certainly Jupiter contain a lot of heavy elements that are substantially enriched relative to the composition of the preplanetary nebula. To account for these bulk heavy element compositions planetesimal accretion has to occur anyway after the GGPPs have formed their cores.

If GGPPs need pressure confinement they also require the presence of an (undepleted) nebula and pose a lifetime constraint for the nebula, namely that nebula dispersal can only begin after a cooling time, that is $\sim 10^6$ yr (Bodenheimer 1986). To determine whether GGPPs are convectively stable so that the non-turbulent core growth scenario can be applied, a detailed calculation of their thermal structure during contraction is necessary De-Campli and Cameron (1979) performed such a study and found largely convective GGPPs for their, Jeans-unstable initial conditions in a massive preplanetary nebula.

2.2. NUCLEATED INSTABILITY

Planetesimals in the solar nebula are small bodies surrounded by gas. A rarified equilibrium atmosphere forms around such objects. Early work in the nucleated instability hypothesis that assumes that such solid 'cores' trigger giant planet formation was motivated by the idea that at a certain critical core-mass the atmosphere could not be sustained and isothermal, shock-free accretion (Bondi 1952) would set in. Determinations of this critical mass were made for increasingly detailed description of the envelopes: adiabatic (Perri and Cameron 1974), isothermal (Sasaki 1989), isothermal-adiabatic (Harris 1978, Mizuno et al. 1978), radiative and convective energy transfer (Mizuno 1980). By then modelling the formation and evolution of a proto-giant planet had become essentially a miniature stellar structure calculation with energy dissipation of impacting planetesimals replacing the nuclear reactions as the energy source. Present results on the critical mass are reviewed in the next section. Already Safronov and Ruskol (1982) pointed out that *the rate of gas accretion following instability [at the critical mass] is determined not by the rate of delivery of mass to the planet [as in Bondi-accretion] but by the energy losses from the contracting envelope.*[3] Consequently the energy budget of the envelope has been modelled more carefully taking into account the heat generated by gravitational contraction (quasi-hydrostatic models by Bodenheimer an Pollack 1986).

Present quasi-hydrostatic studies of giant planet formation use planetesimal dynamics to calculate the core accretion-rate and model the capture, dissolution and sinking that determines how much and where in the envelope the planetesimal kinetic energy is liberated (Pollack et al. 1996). That made possible the first study of the coupling between gas accretion and solid

[3]Safronov and Ruskol (1982) also give a careful discussion of the effects governing the embedding of the protoplanet in the nebula, that determines which material can be attributed to the protoplanet. The resulting definitions of radii of proto giant planets have been used in the subsequent detailed modelling (tidal radius, accretion radius, corrections for disk scale-height).

accretion. On the other hand hydrodynamic studies treat the mechanics of contraction more carfully and determine the flow velocity of the gas by solving an equation of motion for the envelope gas in the framework of convective radiation-fluid-dynamics (Wuchterl 1991a,b, 1993, 1995a, 1997, 1998). That allows the study of collapse of the envelope, accretion with finite Mach-number and an access to the study of linear adiabatic (Tajima and Nakagawa 1995) and nonlinear, nonadiabatic pulsational stability and pulsations of the envelope. Furthermore, the treatment of convective energy transfer has been improved by calculations using a time dependent mixing length theory of convection (Wuchterl 1995b, 1996, 1997) in hydrodynamics. The first hydrodynamic calculations with rotation in the quasispherical approximation have been undertaken by (Götz 1993).

3. Early Growth of Proto Giant Planets

Most aspects of early envelope growth, up to $\sim 10\,\mathrm{M}_\oplus$ can be understood on the basis of a simplified analytical model given by Stevenson (1982a) for a protoplanet with constant opacity, κ_0, core-mass accretion-rate $\dot{M}_{\mathrm{core}}$, core-density, ρ_{core} inside the tidal radius r_T. The key properties of Stevenson's model come from the 'radiative zero solution' for spherical protoplanets with static, fully radiative envelopes, i.e. in hydrostatic and thermal equilibrium. The solution, that describes the structure of an envelope in the gravitational potential of a constant mass and for zero external temperature and pressure is given here (cf. Wuchterl et al. 1999) for generalized opacity laws of the form $\kappa = \kappa_0 P^a T^b$ (Kippenhahn and Weigert 1990). The critical mass, defined as the largest mass a core can grow to with the envelope kept static is then given by

$$M_{\mathrm{core}}^{\mathrm{crit}} = \left[\frac{3^3}{4^4} \left(\frac{\mathcal{R}}{\mu} \right)^4 \frac{1}{4\pi G} \frac{4-b}{1+a} \frac{3\kappa_0}{\pi\sigma} \left(\frac{4\pi}{3} \rho_{\mathrm{core}} \right)^{\frac{1}{3}} \frac{\dot{M}_{\mathrm{core}}}{\ln \frac{r_T}{r_{\mathrm{core}}}} \right]^{\frac{3}{7}}, \qquad (2)$$

and $M_{\mathrm{core}}^{\mathrm{crit}}/M_{\mathrm{tot}}^{\mathrm{crit}} = 3/4$; $\mathcal{R}$, G, σ denote the gas-, the gravitational-, and the Stefan-Boltzmann constant respectively. The critical mass does not depend on the midplane density, ϱ_{neb} nor temperature, T_{Neb}, of the nebula in which the core is embedded. The outer radius, r_T, enters only logarithmically. The strong dependence of the analytic solution on molecular weight, μ, led Stevenson (1984) to propose 'superganymedean puffballs' with atmospheres assumed to be enriched in heavy elements and a resulting low critical mass as a way to form giant planets rapidly (see also Lissauer et al. 1995).

The solution permits also a glimpse on the effect of the run of opacity via the power law exponents a and b. Except for the weak dependences discussed above, a proto-giant planet essentially has the same global properties for a given core wherever it is embedded in a nebula. Even the dependence on $\dot{M}_{\rm core}$ is relatively weak: Detailed radiative/convective envelope models show that a variation of a factor of 100 in $\dot{M}_{\rm core}$ leads only to a 2.6 variation in the critical core mass[4].

This similarity in the static structure of proto-giant envelopes yields similar dynamical behaviours characterized by pulsation driven mass loss for solar composition nebula opacities. That originally lead to an apparent problem to form Jupiter mass giant planets (Wuchterl 1991b). Wuchterl (1993), in search for a way to form Jupiter, without the ad-hoc assumption that hydrodynamic effects have to be neglected, looked for a new class of protoplanets that would escape from this constancy of the critical mass and of proto giant planet envelope structure. By deducing conditions that invalidate the radiative zero solution Wuchterl (1993) found protoplanets with *convective* outer envelopes, for conditions with somewhat higher mid-plane densities than in minimum mass nebulae. These largely convective proto-giant planets have larger envelopes for a given core and a reduced critical core mass. Their properties can be illustrated by a simplified analytical solution for fully convective, adiabatic envelopes with constant first adiabatic exponent, Γ_1:

$$M_{\rm core}^{\rm crit} = \frac{1}{\sqrt{4\pi}} \frac{\sqrt{\Gamma_1 - \frac{4}{3}}}{(\Gamma_1 - 1)^2} \left(\frac{\Gamma_1}{G} \frac{\mathcal{R}}{\mu} \right)^{\frac{3}{2}} T_{\rm neb}^{\frac{3}{2}} \rho_{\rm Neb}^{-\frac{1}{2}}, \qquad (3)$$

and $M_{\rm core}^{\rm crit}/M_{\rm tot}^{\rm crit} = 2/3$. In this case, the critical mass depends on the nebula gas properties and therefore the location in the nebula, but it is independent of the core accretion rate. Of course both the radiative zero and fully convective solutions are approximate because they only roughly estimate envelope gravity and all detailed calculations show radiative *and* convective regions in proto-giant planets. In Figure 1 the transition from 'radiative' to 'convective' protoplanets is shown by results from detailed static radiative/convective calculations for $\dot{M}_{\rm core} = 10^{-6}\,{\rm M}_\oplus\,{\rm yr}^{-1}$ (Wuchterl 1993). Nebula conditions are varied from low densities resulting in radiative outer envelopes to enhanced densities that result in largely convective proto-giant planets. The critical mass can be as low as $1\,{\rm M}_\oplus$, and subcritical static envelopes can grow to $48\,{\rm M}_\oplus$.

[4]According to the analytical model the variation in $\dot{M}_{\rm core}$ could account for a factor ~ 15 variation in the critical core mass between 5 and 30 AU, because $\dot{M}_{\rm core}$ should scale as $r^{-7/2}$, for same core mass, isolation mass and velocity dispersion; a contribution of r^{-2} from a plausible radial variation of the planetesimal surface-density, σ and $r^{-3/2}$ from the variation in orbital frequency, Ω.

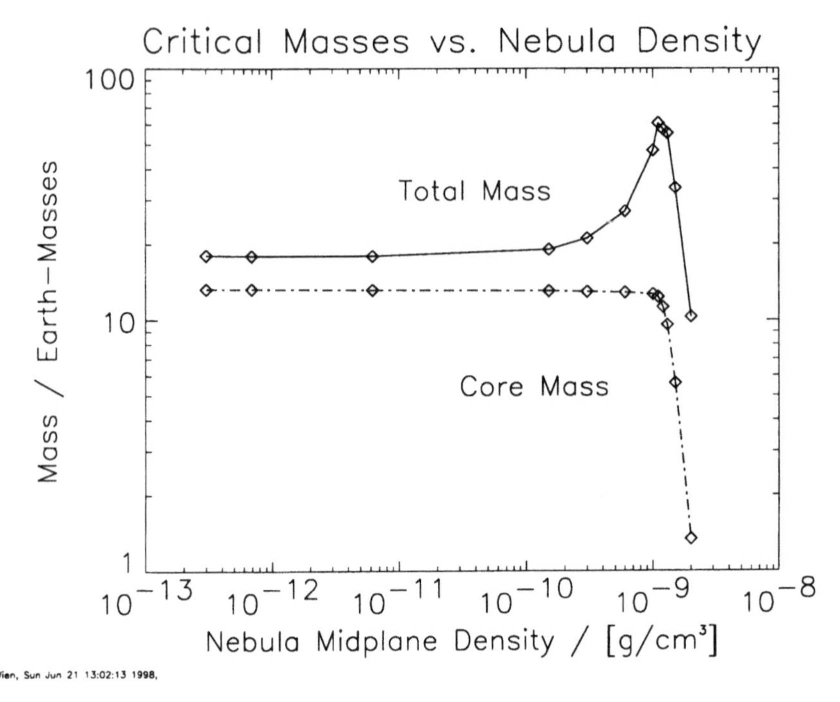

Figure 1. Masses of static protoplanetary envelopes as a function of nebula midplane density. Observe the increased envelope masses, and decreased core masses for the convective outer envelopes occurring at larger nebula densities. Results from Wuchterl (1993). The lowest three densites are calculated for Mizuno's Neptune, Uranus and Saturn positions and conditions respectively an illustrate the constancy of the critical mass for the case of radiative outer envelopes.

3.1. THE CRITICAL MASS AND ENERGY TRANSFER

In this section I present envelope-mass calculations that are based on a new, time-dependet convection model (Wuchterl and Feuchtinger 1999) that is derived from the model of Kuhfuß (1987). It reduces to standard mixing length theory in a local, static limit and has been succesfully tested in calculations of the Sun and RR-Lyrae pulsations. To determine the critical mass, the protoplanetary envelopes are assumed to be in hydrostatic and thermal equilibrium. The static structure equations are solved with a protoplanetary version of the VIP-code (Dorfi and Feuchtinger 1995) where convection is added as outlined by Wuchterl (1995b). The extended Kuhfuß model is used with the standard parameters given by Wuchterl and Feuchtinger (1999). For the formulation of equations and boundary conditions see Wuchterl (1991a). Calculations are made for Mizuno's (1980), minimum mass Jupiter conditions and constant core accretion rate: $10^{-6}\, M_{\oplus}/yr$. Those assumptions are identical to the 'Jupiter' cases of Wuchterl (1993, 1995a). The new

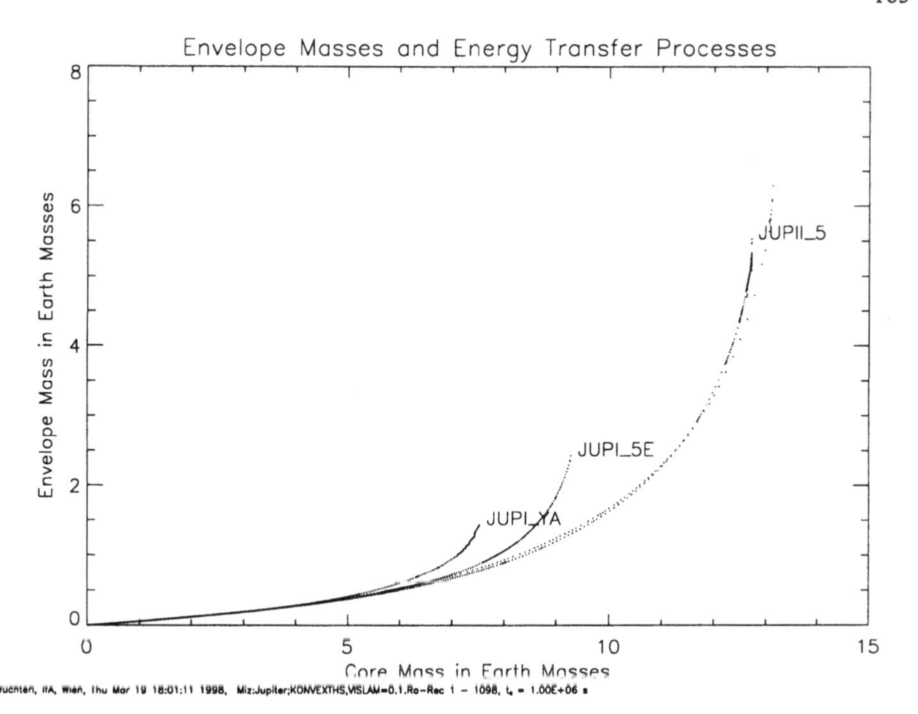

Figure 2. Envelope masses of proto giant planets and energy transfer processes. The masses of static protoplanetary envelopes are plotted as a function of the core mass for a standard 'Jupiter' case. The two curves labelled JUPII_5 are a result from Wuchterl (1993) and a recalculation with the planetary VIP-code, respectively, both for zero entropy convection. JUPI_5E is the same, but for mixing length convection, and JUPI_YA for mixing length theory and updated molecular opacities.

calculations, however are made for three cases with different treatment of energy transfer: JUPII_5: zero entropy gradient convection, JUPI_5E: mixing length convection, JUPI_YA: mixing length convection and 'Alexander' molecular opacities (Alexander and Ferguson 1994), in a compilation by Götz (1993). Results are shown in Fig. 2. Mixing length convection reduces the critical mass relative to the value for zero entropy convection. The updated (higher) molecular opacities result in a further reduction. Altogether the value of the critical core mass is reduced by roughly a factor of two relative to the value obtained for zero entropy gradient convection. Together the improvements of the energy transfer result in: (1) larger envelopes at given core mass, (2) smaller critical envelopes and (3) a reduction of the *critical corc mass* from about 13 to 7.2 Earth-masses for proto giant planets with radiative other envelopes. Earlier calculations that include rotational effects in the quasi-spherical approximation (Götz 1993) point in the same direction. The new lower values for the core masses are in better agreement with new interior models (cf. Guillot 1998, 1999, Wuchterl et al. 1999).

4. Rapid Envelope Evolution

The early phases of giant planet formation discussed above are dominated by the growth of the core. The envelopes adjust much faster to the changing size and gravity of the core than the core grows. As a result the envelopes of proto-giant planets remain very close to static and in equilibrium below the critical mass (Mizuno 1980, Wuchterl 1993). This has to change when the envelopes become more massive and cannot reequilibrate as fast as the cores grow. The onset of more rapid envelope-evolution is characterized by the so-called critical and crossover masses. Typical values for the critical mass (largest static envelope for growing core) are in the range of 7 to 25 earth masses, for standard (minimum mass) assumptions about the preplanetary nebula. Pollack et. al. (1996) obtain crossover masses ($M_{\text{high Z}} = M_{\text{low Z}}$) in the range of 10 to 30 $M_\oplus$ for various assumptions about planetesimal accretion.

At the critical mass the nucleated instability was assumed to set in, originally as a hydrodynamic instability analogously to the Jeans instability. With the recognition that energy losses from the proto-giant planet envelopes control the further accretion of gas, it followed that quasi-hydrostatic contraction of the envelopes would play a key role. Consequently quasi-hydrostatic studies (Bodenheimer and Pollack 1986, Pollack et al. 1996, Tajima and Nakagawa 1997, Bodenheimer et al. 1999) determined the envelope growth by including heat gains and losses due to slow (very subsonic) contraction and expansion of the envelopes. An important result of these studies is that envelopes grow to large ($> 100\,\mathrm{M}_\oplus$) masses, whenever the cores reach the crossover mass. Uranus/Neptune-type giant planets, that are mostly core are formed when the crossover mass is not reached until the protoplanetary nebula is dispersed. The detailed modelling of planetesimal accretion in quasi-hydrostatic models allows a detailed discussion of the time required to form a giant planet. Assumptions made, mostly about solid surface-densities and planetesimal properties are related to the time-scale constraint for giant planet formation that is set by observationally inferred nebula lifetimes of a few Myr (see Pollack et al. (1996) and Papaloizou, this volume).

5. Hydrodynamic Evolution Beyond the Critical Mass

Static and quasi-hydrostatic models discussed so far rely on the assumption that gas accretion from the nebula onto the core is very subsonic, and the inertia of the gas and dynamical effects as dissipation of kinetic energy do not play a role. To check whether hydrostatic equilibrium is achieved and whether it holds, especially beyond the critical mass, hydrodynamical investigations are necessary. Two types of hydrodynamical investigations have been undertaken: (1) linear adiabatic dynamical stability analysis of envelopes evolving quasi-hydrostatically (Tajima and Nakagawa 1997) that found the hydrostatic equilibrium to be stable in the case they investigated and (2) nonlinear, convective radiation hydrodynamical calculations of core-envelope proto-giant planets (Wuchterl 1991a,b, 1993, 1995a, 1996, 1997, 1998) that follow the evolution of a proto-giant planet without a priori assuming hydrostatic equilibrium and which *determine* whether envelopes are hydrostatic do pulsate or collapse and at which rates mass flows onto the planet. They solve the flow-equations for the envelope gas essentially only assuming that spherical symmetry holds. They determine the net gain and loss of mass from the solution of the equations of motion for the gas in spherical symmetry while quasi-hydrostatic calculations add mass according to some prescription and then calculate the new equilibrium structure for the updated mass, imposing that there is a new equilibrium. While the other assumptions made in the hydrodynamic calculations agree with those of the quasi-hydrostatic models in the previous section there is a second important difference: the core accretion-rate is, for simplicity assumed to be either constant or calculated according to the particle in a box approximation (see e.g. Lissauer 1993). The first hydrodynamical calculation of the nucleated instability (Wuchterl 1991a,b) started at the static critical mass and brought a surprise: Instead of collapsing, the proto-giant planet envelope started to pulsate after a very short contraction phase (see Wuchterl 1990 for a simple discussion of the driving κ-mechanism). The pulsations of the inner protoplanetary envelope expanded the outer envelope and the outward travelling waves caused by the pulsations resulted in a mass loss from the envelope into the nebula. The process can be described as a pulsation-driven wind. After a large fraction of the envelope mass had been pushed back into the nebula, the dynamical activity faded and a new quasi-equilibrium state was found that resembled Uranus and Neptune in core and envelope mass. As a typical example the hydrodynamical calculation for Mizuno's (1980) 'Neptune' case is shown in Fig. 3. It turned out that the mass loss process occurred in a very similar way for nebula conditions at Jupiter to Neptune positions (Wuchterl 1995a) and core mass accretion rates from 10^{-7} to $10^{-5}\,M_\oplus\,yr^{-1}$. Starting the hydrodynamics at low core mass rather than at the critical mass does not change this result with respect to the mass loss (Wuchterl 1995a).

Pulsations and mass loss do not occur when 'no dust' zero metallicity opacities are used (Wuchterl 1995a); the lack of dust makes conditions as favorable as possible for energy loss from the envelope and therefore for accretion. It is interesting to note that even for zero metallicity opacities, the static critical core mass is between 1.5 and $3\,M_\oplus$ for $\dot{M}_{\rm core} = 10^{-8}$ to $10^{-6}\,M_\oplus\,yr^{-1}$, respectively. Envelope accretion becomes independent of

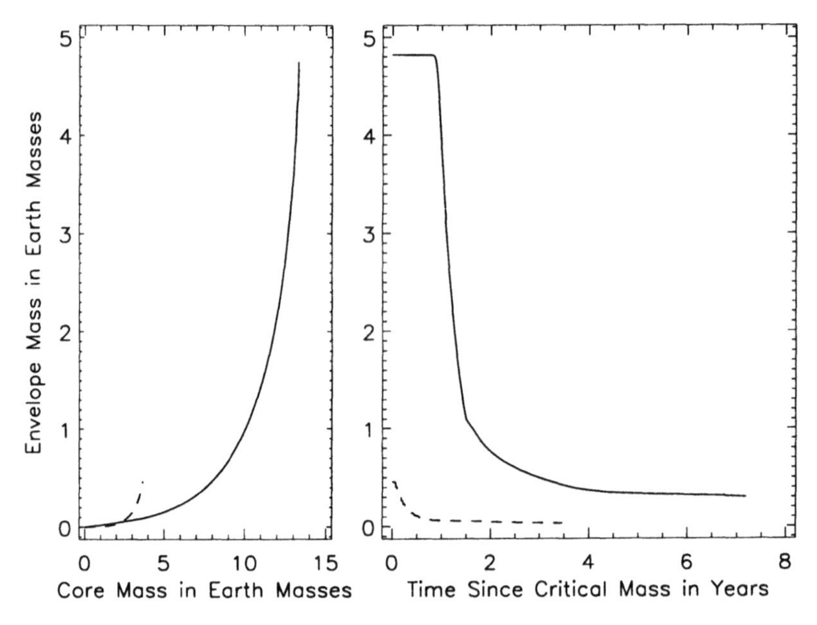

Figure 3. Hydrodynamical envelope dilution due to pulsation driven mass-loss. The evolution is shown for Mizuno's (1980) Neptune conditions and $\dot{M}_{\rm core} = 10^{-6}\,{\rm M}_\oplus\,{\rm yr}^{-1}$ (full line). The dashed line is for time-dependent MLT-convection, updated molecular opacities, as discussed in the text and a particle-in-a-box core-accretion-rate (Wuchterl 1997) with a planetesimal surface density of 1 g cm^{-3} and a Safronov-number of 1000.

the core accretion at about 15 $\rm M_\oplus$, the quasi hydrostatic assumptions holds until inflow velocities reach a Mach-number of 0.01 at about 50 $\rm M_\oplus$. At a total mass of about 100 $\rm M_\oplus$ the nebula gas influx approaches the Bondi accretion rate and at 300 $\rm M_\oplus$ the envelope collapses overall (cf. Wuchterl 1995a). This result shows that there must be an opacity dependent transition from pulsational driven winds to efficient gas accretion at the critical mass.

The main question concerning the hydrodynamics was then to ask for conditions that allow gas accretion, i.e., damp envelope pulsations for 'realistic' solar composition opacities that include dust. Wuchterl (1993) derived conditions for the breakdown of the radiative zero solution by determining nebulae conditions that would make the outer envelope of a 'radiative' critical mass proto-giant planet convectively unstable. The resulting criterion gives a minimum nebula density that is necessary for a convective outer envelope:

$$\varrho_{\rm neb}^{\rm crit} = [2.2\,10^{-10}\,{\rm g\,cm^{-3}}]\frac{\left(\frac{T}{[100\,{\rm K}]}\right)^3 \frac{\nabla_s}{[2/7]}\frac{\mu}{[2.2]}\left(\frac{M_{\rm core}}{[10\,{\rm M}_\oplus]}\right)^{\frac{1}{3}}}{\frac{\kappa}{[{\rm cm^2\,g^{-1}}]}\frac{\dot{M}_{\rm core}}{[10^{-6}\,{\rm M}_\oplus\,{\rm yr}^{-1}]}\left(\frac{\varrho_{\rm core}}{[5.5]}\right)^{\frac{1}{3}}}, \tag{4}$$

where μ is the mean molecular weight, T, the nebula midplane temperature, ∇_s, the isentropic temperature gradient with respect to pressure, κ, the Rosseland-mean of the mass absorption coefficient, and $M_{\rm core}$, $\dot{M}_{\rm core}$, $\varrho_{\rm core}$ the core mass, mass accretion-rate, and density, respectively. Interestingly the critical density is close to the Kusaka et al. minimum mass 'Jupiter' conditions. Protoplanets that grow under nebula conditions above that density have larger envelopes for a given core and a reduced critical mass as described in Section 3, cf. Fig. 1.

Since convection is of great importance in damping stellar pulsations of RR-Lyrae and δ-Cepheid stars at the cool, so-called 'red end' of the stellar instability strip, a similar behaviour may be expected in proto-giant planet envelopes. Wuchterl (1995a) calculated the growth of giant planets from low core masses hydrodynamically for a set of nebula conditions reaching from below the critical density to somewhat above. As the density was increased, the envelopes became increasingly more convective at the critical mass, but still showed the mass loss. At a nebula density of $10^{-9}\,{\rm g\,cm^{-3}}$, i.e., increased by a factor 6.7 relative to Mizuno's (1980) minimum reconstituted mass nebula value, the dynamical behaviour was different: the pulsations were damped and rapid accretion of gas set in and proceeded to $300\,{\rm M}_\oplus$. Apparently the spreading of convection in the outer envelope had damped the pulsations, thereby inhibiting the onset of a wind and leading to accretion. The critical core masses required for the formation of this class of proto-giant planets are significantly smaller than for the Uranus/Neptune type (see Wuchterl 1993, 1995).

6. Hydrodynamics with Time Dependent Convection

Most giant planet formation studies use zero entropy gradient convection, i.e., set the temperature gradient to the adiabatic value in convectively unstable layers of the envelope. That is done for simplicity but can be inaccurate (cf. Sect. 3.1), especially when the evolution is rapid and hydrodynamical waves are present (see Wuchterl 1991b). It was therefore important to develope a time-dependent theory of convection that can be solved together with the equations of radiation hydrodynamics. Such a time-dependent convection model (Kuhfuß 1987) has been reformulated for self-adaptive grid radiation hydrodynamics (Wuchterl 1995b) and applied to giant planet formation (Götz 1993, Wuchterl 1996, 1997). In a

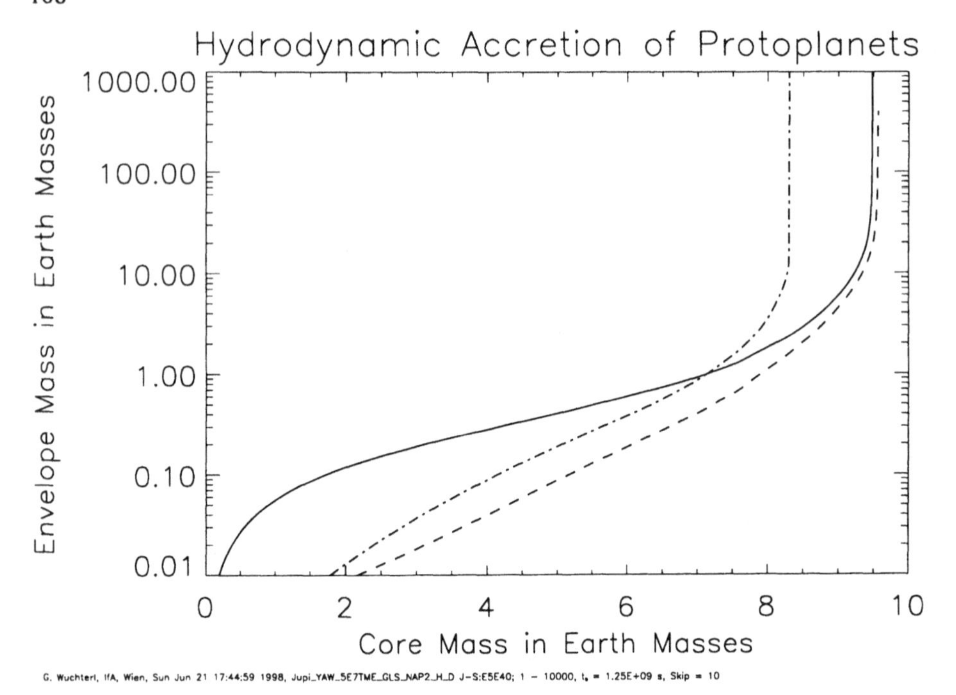

Figure 4. Evolution of proto-giant planets with time-dependent mixing lenght convection. Envelope masses as obtained from hydrodynamic accretion calculations are plotted as functions of core mass for for locations at 0.05 AU in the Hayashi et al. (1985) nebula (dash-dot) and for Mizuno's (1980) 'Jupiter' (full) and 'Neptune' (dashed) cases, see text for details.

reformulation by Wuchterl and Feuchtinger (1999) it reduces to standard mixing length theory in a static local limit and accurately describes the solar convection zone and RR-Lyrae lightcurves (Wuchterl and Feuchtinger 1999, Feuchtinger 1999). In addition updated molecular opacities (Alexander and Ferguson 1994) are used in a compilation by Götz (1993) to improve the accuracy of radiative transfer in the proto-giant planet envelopes. An interesting question is then to ask for the core mass that is needed to initiate gas accretion to a few hundred earth masses at various orbital radii, with the improvments made in energy transfer. The evolution of the envelope mass of proto giant planets located at 0.05, 5.2 and 17.2 AU is shown in Fig. 4. Envelope accretion has been calculated with the same set of hydrodynamic equations as in the last section, but now with the fully time-dependent mixing-length description of convective energy transfer. Orbital radii and nebula temperatures and densities for the three cases are 'Vulcan' (0.052 AU,1252 K, $5.3\,10^{-6}\,\mathrm{g\,cm^{-3}}$), 'Jupiter' (5.2 AU, 97 K, $1.5\,10^{-10}\,\mathrm{g\,cm^{-3}}$), and 'Neptune'(17.2 AU, 45 K, $3.0\,10^{-13}\,\mathrm{g\,cm^{-3}}$), respectively. The core accretion rate is $10^{-6}\,\mathrm{M_{\oplus}/yr}$, and standard value that is consistent with detailed studies of core accretion after the initial runaway phase (cf. Pollack et. al. 1996). For reference: the critical mass (last static envelope) for the 'Jupiter' case is $M_{\rm core} = 7.54 \rm M_\oplus$; the crossover mass ($M_{\rm core} = M_{\rm env}$), at $M_{\rm core} = 9.21 \rm M_\oplus$, is reached 2.17 Myr after $M_{\rm crit}$. A total mass of $100 \rm M_\oplus$ has been accumulated $\sim$ 5200 yr later. The proto giant planet arrives at 300 $\rm M_\oplus$ about 670 yr thereafter. The typical limit for mass influx into the Hill-Sphere limited by gap opening etc. is 0.01 $\rm M_\oplus \, yr^{-1}$ and is exceeded around $100 \rm M_\oplus$. So the elapsed time value from 100 to 300 $\rm M_\oplus$ is probably too small and should be roughly 20000 yr. The overall effect of the improvements in energy transfer is that the core masses needed to initiate gas accretion to a few hundred Earth masses at various orbital radii reduces to 8.30, 9.48 and 9.56 $\rm M_\oplus$ at 0.052, 5.2 and 17.2 AU, respectively, cf. Fig. 4.

7. Hydrodynamical Models of Giant Planet Formation Near Stars

A major result of the hydrodynamical studies is that proto-giant planets may pulsate and develop pulsation driven mass loss. Only if the pulsations are damped can gas-accretion produce Jupiter-mass envelopes. Since all extrasolar planets discoverd so far have minimum masses $>\sim 0.5 \, M_J$ they probably require efficient gas accretion and should satisfy the convective outer envelope criterion (Wuchterl 1993). A glance at Wuchterl's (1993), Fig. 2. shows that proto-giant planets, somewhat inside of Mercurys orbit in the Hayashi et al. 1985 minimum mass nebula fullfill that condition. Convective radiation hydrodynamical calculations of core-envelope growth at 0.05 AU, for particle-in-a-box core mass accretion, at nebula temperatures of 1250 and 600 K, show gas accretion beyond 300 $\rm M_\oplus$ at core masses of 13.5 $\rm M_\oplus$ and 7.5 $\rm M_\oplus$, respectively, (Wuchterl 1996, 1997). It is interesting to apply the arguments based on the convection-controlled bifurcation in hydrodynamic accretional behaviour to an ensemble of preplanetary nebula models, to simulate a variety of initial conditions for planet formation that might be present around other stars. Wuchterl (1993) has shown that almost all nebula conditions, from a literature collection of nebula models, result in radiative outer envelopes at the critical mass and as a consequence should produce Uranus/Neptune type giant planets and Jupiter-mass planets should be the exception. The first calculations with time-dependent mixing lenght convection, discussed in the preceeding section do now show gas accretion to beyond a Jupiter mass for a much wider range of nebula conditions. Apparently the improved description of convection (and the updated opacities) have shifted the instability-strip for pulsations and mass loss at the critical mass. Further calculations and a reanalysis of the conditions for efficient gas accretion for mixing lenght convection have to be undertaken before an updated expectation concerning the mass-distribution of extrasolar planets can be given. An important requirement for that is an extensive study of plausible preplanetary nebulae, both theoretical and observational.

8. Conclusion

The application of time-dependent mixing length theory, that correctly describes the solar convection zone and RR Lyrae pulsations, to giant planet formation, results in lower critical core masses than obtained previously with simplified zero entropy gradient convection. The core masses that are necessary to initiate hydrodynamic gas accretion of Jupiter-mass envelopes are reduced to 8.30, 9.48 and 9.56 earth masses at 0.052, 5.2 and 17.2 AU respectively.

References

Alexander, D.R. and J.W. Ferguson (1994) Low-temperature Rosseland opacities, *Astrophys. Journ.* **437**, 789–891.

Bodenheimer, P. (1985) Evolution of the giant planets, in D.C.Black and M.S.Matthews (eds.), *Protostars and Planets II* Univ. of Arizona Press, Tucson, pp. 873–894.

Bodenheimer, P. and Pollack J.B. (1986) Calculations of the accretion and evolution of giant planets: The effects of solid cores, *Icarus* **67**, 391–408.

Bodenheimer, P., Hubickyj, O. and Lissauer, J.J. (1999) Models of the In Situ Formation of Observed Extrasolar Giant Planets, *Icarus*, submitted.

Bondi, H. (1952) On spherical symmetric accretion. *Mon. Not. Roy. Astron. Soc.* **112**, 195–204.

Boss, A.P. (1997) Formation of giant gaseous protoplanets by gravitational instability. *Science* **276**, 1836–1839.

Boss, A.P. (1998) Evolution of the Solar Nebula IV – Giant Gaseous Protoplanet Formation. *Ap.J.* **503**, 923.

Brandl, A. and Sterzig, M.F. (1998) The Occurence and Ejection of Substellar Companions: TMR-1C, in *Astronomische Gesellschaft Meeting Abstracts, Abstracts of Contributed Talks and Posters presented at the Annual Scientific Meeting of the Astronomische Gesellschaft at Heidelberg, September 14–19, 1998*, talk #A04.

Cassen, P.M. and Moosman, A. (1981) On the formation of protostellar disks, *Icarus* **48**, 353–376.

DeCampli, W.M., and Cameron, A.G.W. (1979) Structure and evolution of isolated giant gaseous protoplanets, *Icarus* **48**, 353–376.

Dorfi, E.A. and Feuchtinger, M.U. (1995) Adaptive Radiation Hydrodynamics of Pulsating Stars, *Comp. Phys. Comm.* **89**, 69–90.

Dutrey, A., Guilloteau, S., Prato, L., Simon, M., Duvert, G., Schuster, K., and Mnard, F., (1999) CO Study of the GM Aurigae Keplerian Disk. *Astron. Astrophys.*, in press.

Feuchtinger, M.U. (1999) *Astron. Astrophys*, in prep.

Götz, M. (1993): *Die Entwicklung von Proto-Gasplaneten mit Drehimpuls — Strahlungshydrodynamische Rechnungen*, Dissertation, Univ. Heidelberg.

Guillot, T. (1998) A comparison of the interiors of Jupiter and Saturn. In preparation for *Planet. Space Sci.*.

Guilloteau, S. and Dutrey, A. (1999) Physical parameters of the Keplerian protoplanetary disk of DM Tau. *Astron. Astrophys.*, in press.

Hayashi, C. (1980) Structure of the solar nebula, growth and decay of magnetic fields and effects of turbulent and magnetic viscosity on the nebula. *Prog. Theor. Phys. Suppl.*

70, 35–53.
Hayashi, C., K. Nakazawa and Y. Nakagawa (1985): Formation of the Solar System, in D.C.Black and M.S.Matthews (eds.), *Protostars and Planets II* Univ. of Arizona Press, Tucson.
Harris, A.W. (1978) The formation of the outer planets, *Lunar Planet. Sci.* **IX**, 459-461, (abstract).
Kant, I. (1755) *Allgemeine Naturgeschichte und Theorie des Himmels*, Johann Friedrich Petersen, Königsberg und Leipzig.
Kusaka T., Nakano, T. and Hayashi, C. (1970) Growth of solid particles in the primordial solar nebula. *Prog. Theor. Phys.* **44**, 1580–1596.
Kuhfuß, R. (1987) *Ein Modell für zeitabhängige, nichtlokale Konvektion in Sternen.* Ph. D. thesis, Tech. Univ. Munich.
Kippenhahn, R., and Weigert, A. (1990) *Stellar Structure and Evolution*, Springer-Verlag, Berlin.
Laplace, P. S. (1796) *Exposition du Systéme du Monde* Circle-Sociale, Paris.
Laughlin, G., and P. Bodenheimer (1994) Nonaxisymmetric evolution in protostellar disks, *Astrophys. J.* **436**, 335–354.
Lissauer, J.J., Pollck, J.B., Wetherill, G.W. and Stevenson, D.J. (1995) Formation of the Neptune system, in *Neptune and Triton*, D.P. Cruikshank (ed.), Univ. Arizona Press, Tucson, pp. 37 108.
Lissauer, J.J. (1993) Planet formation, *Ann. Rev. Astron. Astrophys.* **31**, 129–174.
Mizuno, H. (1980) Formation of the Giant Planets, *Prog. Theor. Phys.* **64**, 544–557.
Mizuno, H., Nakazawa, K. and Hayashi, C. (1978) Instability of a gaseous envelope surrounding a planetary core and formation of giant planets, *Prog. Theor. Phys.* **60**, 699–710.
Morfill, G., Tscharnuter, W.M. and Völk, H. (1985), in D.C. Black and M.S. Matthews (eds.) *Protstars and Planets II*, Univ. Arizona Press, Tucson, pp. 493–533.
Perri, F. and Cameron, A.G.W. (1974) Hydrodynamic instability of the solar nebula in the presence of a planetary core, *Icarus* **22**, 416–425.
Papaloizou, J. C., Savonije, G. (1991) Instabilities in self-gravitating gaseous discs, *Month. Not. Roy. Astron. Soc.* **248**, 353–369.
Podolak, M., Hubbard, W. B., and Pollack, J. B. (1993) Gaseous accretion and the formation of giant planets, in E. H. Levy and J. I. Lunine (eds.) *Protostars & Planets III*, Univ. of Arizona Press, Tucson, pp. 1109–1147.
Pollack, J.B., O. Hubickyj, P. Bodenheimer, J.J. Lissauer, M. Podolak and Y. Greenzweig (1996) Formation of the Giant Planets by Concurrent Accretion of Solids and Gas. *Icarus*, **124**, 62–85.
Sasaki,S. (1989) Minimum planetary size for forming outer Jovian-type planets - Stability of an isothermal atmosphere surrounding a protoplanet, *Astron. Astrophys.* **215**, 177-180.
Safronov, V. S. (1969) *Evolution of the protoplanetary cloud and formation of the earth and planets.* Nauka Press, Moscow (also NASA–TT–F–677, 1972).
Safronov, V.S. and Ruskol, E.L. (1982) On the origin and initial temperature of Jupiter and Saturn, *Icarus* **49**, 284–296.
Stevenson, D.J. (1982a) Formation of the Giant Planets, *Planet. Space Sci.* **30**, 755–764.
Stevenson, D.J. (1984) On forming giant planets quickly ('superganymedian puffballs'), *Lunar Planet. Sci.* **XV**, 822–823 (abstract).
Tajima, N. and Y. Nakagawa (1997), Evolution and Dynamical Stability of the Proto-giant-planet envelope, *Icarus* **126**, 282–292.
Tscharnuter,W.M. (1987) A Collapse Model of the Turbulent Presolar Nebula, *Astron. Astrophys.* **188**, 55–73.
Tscharnuter, W.M. and Boss, A.P. (1993) Formation of the protosolar nebula, in E.H. Levy and J.I. Lunine (eds.) *Protostars & Planets III*, Univ. of Arizona Press, Tucson, pp. 921–938.
Wuchterl, G. (1990) Hydrodynamics of Giant Planet Formation I. I - Overviewing the

kappa-mechanism, *Astron. Astrophys.*, **238**, 83–94.
Wuchterl, G. (1991a) Hydrodynamics of Giant Planet Formation II: Model Equations and Critical Mass, *Icarus*, **91**, 39–52.
Wuchterl, G. (1991b) Hydrodynamics of Giant Planet Formation III: Jupiter's Nucleated Instability, *Icarus* **91**, 53–64.
Wuchterl, G. (1993) The Critical Mass for Protoplanets Revisited: Massive Envelopes Through Convection, *Icarus* **106**, 323–334.
Wuchterl, G. (1995a) Giant Planet Formation. A comparative View of Gas-Accretion, in M.T. Chahine, M.F. A'Hearn, J.H. Rahe, (Eds.), *Comparative Planetology*, Kluwer Acad. Publ., Dordrecht, and *Earth, Moon, and Planets* **67**, 51–65.
Wuchterl, G. (1995b) Time dependent convection on self adaptive grids, *Comp. Phys. Comm.* **89**, 119–126.
Wuchterl, G. (1996) Formation of Giant Planets Close to Stars. *Bull. Amer. Astron. Soc.* **28**, 1108.
Wuchterl, G. (1997) Giant Planet Formation and the Masses of Extrasolar Planets, in F. Paresce (ed.) *Science with the VLT Intereferometer* Springer, Berlin, pp. 71–74.
Wuchterl, G. (1998) Convection and Giant Planet Formation. in, *Euroconference on Extrasolar Planats*, Lisbon, in press.
Wuchterl, G. and Feuchtinger M.U. (1999) *Astron. Astrophys.*, **341**, in press.
Wuchterl, G., Guillot, T. and Lissauer, J.J. (1999) Giant Planet Formation, in A.P. Boss, S. Russel (eds.), *Protostars and Planets IV*, Univ. of Arizona Press, Tucson, submitted.

EXTRA-SOLAR PLANETS: ATMOSPHERES

J. KASTING
Pennsylvania State University
525 Davey Lab., University Park, PA 16802
USA

Manuscript not provided.

J.-M. Mariotti and D. Alloin (eds.), Planets Outside the Solar System: Theory and Observations, 173.

Part II : Observational Methods

Out of the debris of dying stars,
this rain of particles
that waters the waste with brightness;
the sea-wave of atoms hurrying home,
collapse of the giant,
unstable guest who cannot stay;

the sun's heart reddens and expands,
his mighty aspiration is lasting,
as the shell of his substance
one day will be white with frost.

In the radiant field of Orion
great hordes of stars are forming,
just as we see every night,
fiery and faithful to the end.
(...)

John Haines
Little Cosmic Dust Poem

J.-M. Mariotti and D. Alloin (eds.), Planets Outside the Solar System: Theory and Observations, 175.

ASTROMETRIC TECHNIQUES

M. M. COLAVITA
Jet Propulsion Laboratory, California Institute of Technology
4800 Oak Grove Dr., Pasadena, CA 91109 USA

1. Astrometric Signatures

An exoplanet, while perhaps not visible from its direct radiation, can still make its presence known indirectly through its gravitational interaction with its parent star. As the planet and star revolve about the center of mass of the system, there will be a component of the star's motion both along and transverse to the line of sight. Velocity fluctuations along the line of sight can be detected using the radial velocity technique [1]; position fluctuations transverse to the line of sight can be detected using astrometry.

The astrometric signature is straightforward. Let θ be the amplitude of the (sinusoidal) astrometric signature. The angular signature is given by

$$\theta = \frac{m}{M}\frac{r}{L}, \tag{1}$$

where m and M are the planet and star masses, r is the planet's orbital radius, and L is the distance to the star. The amplitude θ is in arcsec for r in AU and L in pc. The standard example is the astrometric motion of the Sun caused by Jupiter($r = 5.2$ AU, $m = 1.0 \times 10^{-3} M$) as viewed from a distance of 10 pc: this gives a signature of 0.5 mas.

One feature of the astrometric technique is that it is sensitive to all orbital inclinations. Assume measurements in an orthogonal coordinate system on the sky. A system in a face-on orbit would yield identical signatures in each axis, while an edge-on system which yields no signature along one axis would provide a full signature along the other. This ability to measure the inclination allows determination of the planet-star mass ratio unambiguously.

Another feature of the astrometric signature is that it is inversely proportional to distance, and is thus is most sensitive to nearby targets. This can be contrasted with the radial velocity technique, which while most sen-

J.-M. Mariotti and D. Alloin (eds.), Planets Outside the Solar System: Theory and Observations, 177–188.

TABLE 1. Astrometric signatures at 10 pc for stars of various spectral types (in μas)

Planet	Spectral class F0	G2	K0	M0
Jupiter-mass, 11.9-y orbit	349	497	586	821
Saturn-mass, 11.9-y orbit	104	149	175	246
Uranus-mass, 11.9-y orbit	16	23	27	38
Earth-mass, 1-y orbit	0.21	0.30	0.36	0.50

sitive to nearby systems strictly from a detection-noise perspective (i.e., more photons), has a signature which is independent of distance [2].

We can also apply Kepler's law to write the astrometric signature in terms of the period P of the planet's orbit as

$$\theta = \frac{m}{L}\left(\frac{P}{M}\right)^{2/3}, \tag{2}$$

which is in arcsec for P in years and M and m in solar masses. Thus the astrometric technique is most sensitive to large orbital radii and long periods. In this sense it is complementary to radial-velocity techniques [1, 2], which are most sensitive to small orbital radii and short periods.

Figure 1 shows the motion of our Sun as viewed from a distance of 10 pc. The fundamental period is caused by Jupiter in its 11.9-y orbit, which introduces a component with a 500 μas amplitude. However, the presence of the other planets makes the signature more interesting, and a suitably accurate technique with an adequate time base could detect multiple planets from such data. Table 1 give the signatures for Jupiter, Saturn, and Uranus-mass objects in assumed 11.9-y orbits at a distance of 10 pc (see [1] for the expected astrometric signatures of the planets discovered through radial velocities). Thus, for the nearby sample, astrometric accuracies of 10–100 μas are required to do a comprehensive search for planets with masses down to that of Uranus. The table also gives the signature of an Earth-mass planet in a 1-y orbit; detection of such planets requires an astrometric accuracy of <1 μas.

Such astrometric measurements are challenging. The current state of the art for astrometry is $\sim$1 mas. This includes the results from the recent Hipparcos space mission [3], which measured thousands of stars to this accuracy, as well as parallaxes measured with ground-based CCD astrometry [4], and positions from the MAP instrument at Allegheny Observatory [5, 6]. Thus, current techniques fall short of what is required for a comprehensive search, and far short of what would be needed to detect an exoearth. However, application of the MAP and special-purpose CCD astrometry [7] on

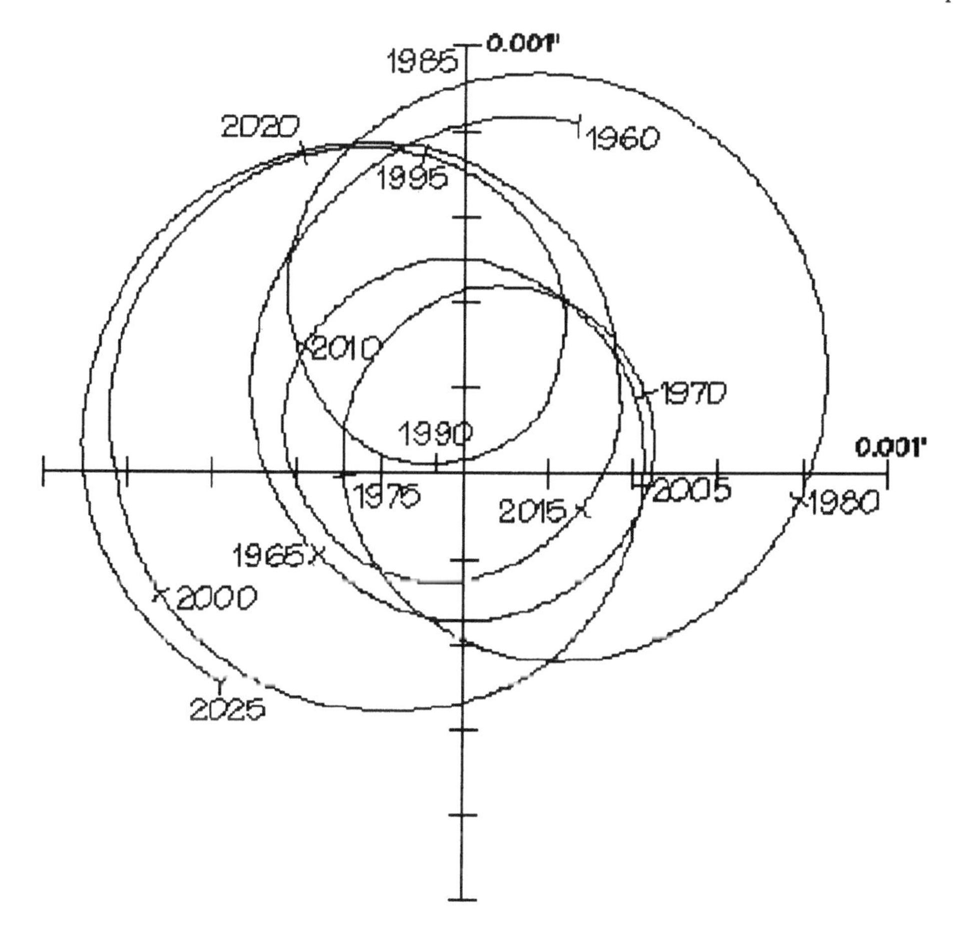

Figure 1. The astrometric motion of the Sun as viewed from 10 pc.

large telescopes offers the promise of 100's of μas accuracy. Interferometry on the ground [8] offers the potential of 10's of μas accuracy. Interferometry in space with a mission such as SIM[9] is needed for accuracies < 10 μas.

2. Astrometric Techniques

Astrometric techniques in the visible and infrared use either single telescopes or interferometers. In principle, both instruments can be used on the ground or in space.

2.1. CCD ASTROMETRY

CCD astrometry for planet detection uses a CCD detector at the focus of a telescope to image the putative parent star plus nearby astrometric references. The field stars are used to calibrate the scale and geometry of the image plane [4]. For ground-based measurements, the atmosphere introduces errors attributable to differential chromatic refraction (DCR), and ultimately fundamental limits attributable to atmospheric turbulence. The USNO parallax program achieves mean accuracies of 1 mas using a CCD detector [4]. Pravdo & Shaklan [7] have examined a CCD astrometry system for planet detection, and estimate that an optimized system for use on a large telescope could achieve accuracies to $\sim$100 μas.

2.2. RONCHI RULING PHOTOMETER

As an alternative to CCD detectors, the Allegheny group [5] has used a Ronchi-ruling photometer to measure stellar positions. The approach uses a precision Ronchi ruling at the focus of the telescope. The ruling is translated in the telescope focal plane, imposing intensity modulation on the various stars in the field. Individual detectors for the target and reference stars collect the light which transmits through the Ronchi ruling; the phase shift of the modulated intensity of the various stars is the astrometric observable. The MAP (Multi-channel Astrometric Photometer) achieves accuracies of $\sim$1 mas at its current site; improved performance with an upgraded MAP to $\sim$100 μasis anticipated for planned observations at the Keck telescope (Gatewood97).

2.3. INTERFEROMETRIC ASTROMETRY

The rest of this paper will concentrate on interferometric techniques [10].

Interferometers have advantages over telescopes in several areas. In particular, resolution and sensitivity, viz. baseline and collecting area, can be set separately. With respect to photon noise (discussed later), the astrometric performance of an interferometer with baseline B and apertures of diameter d is similar to a telescope with diameter $\sim \sqrt{Bd}$. However, with respect to systematic errors, the interferometer has two advantages. The long baselines achievable with an interferometer reduce the angular effect of linear errors. Interferometers also have a simple geometry which can be accurately monitored to minimize systematic errors.

For astrometry, an optical interferometer can be looked at geometrically; the problem is identical to the case of a radio interferometer (see [11]). The delay x measured with the interferometer can be related to the interferometer baseline B and the star unit vector s as $x = B \cdot s$. Thus,

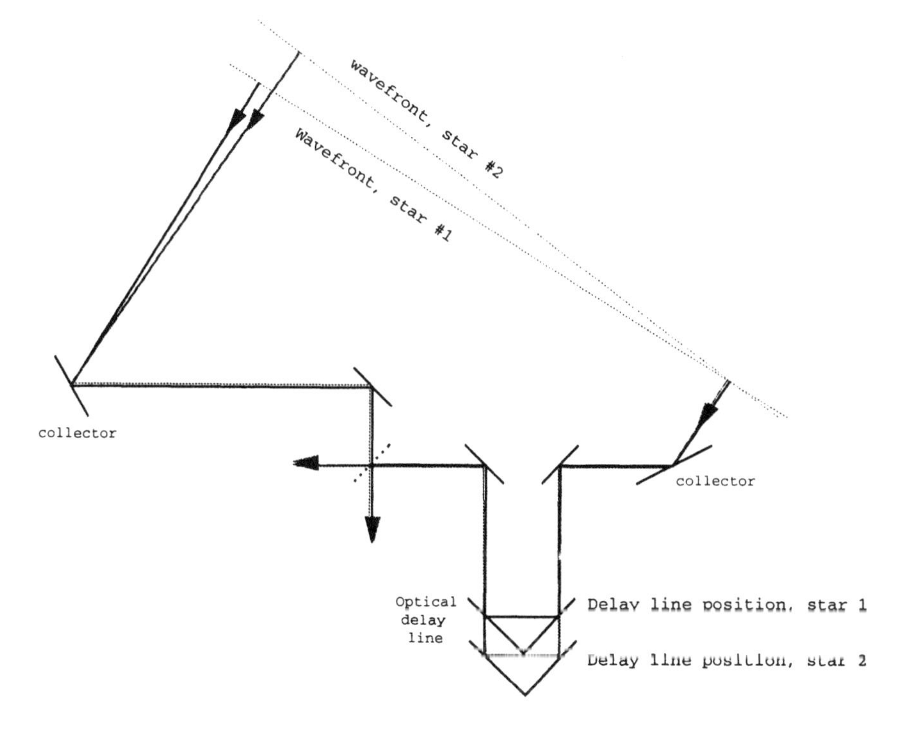

Figure 2. Astrometry with an interferometer.

measurements of delay in conjunction with knowledge of the baseline gives the angle of the star with respect to the baseline vector. This is shown schematically in Fig. 2.3.

The measurement can also be understood perhaps somewhat more intuitively. A stellar wavefront impinging on the interferometer at some angle arrives at one aperture delayed with respect to its arrival at the other aperture. This external delay is what we seek to measure. What the interferometer does is introduce a corresponding internal delay to match the external one; the internal delay is ordinarily introduced using an optical delay line. The internal delay has the advantage that it can be precisely monitored with a laser metrology system. In this case, the interference fringe is used as the measure of equality between the internal and external delays: when the fringe phase is equal to zero, the internal delay is a measure of the external delay.

In practice, the fringe phase can never be made exactly zero, and thus the delay is computed as

$$x = l + k^{-1}\phi, \tag{3}$$

where l is the laser-monitored internal delay, ϕ is the fringe phase, and k is the wavenumber of the interfering light.

3. Sources of Error

We can capture most aspects of the measurement problem by reducing it to 2-d and doing a small-angle approximation for sources near normal to the instrument, viz. $\theta \simeq x/B$. A trivial sensitivity analysis yields

$$\delta\theta = \frac{\delta l}{B} + k^{-1}\frac{\delta\phi}{B} - \frac{\delta B}{B}\theta. \tag{4}$$

The first term incorporates systematic errors in measuring the internal delay; the second term incorporates errors in measuring the fringe phase, including photon and detector noise; the third term incorporates errors in measurement or knowledge of the interferometer baseline.

3.1. SYSTEMATIC ERRORS

Astrometry at high accuracy is to a large extent about the control of systematic errors. The tolerances rapidly become very challenging. For example, the Space Interferometry Mission, SIM, has as one of its goals narrow-angle astrometry with an accuracy of 1 μas. With SIM's 10-m baseline, this corresponds to 50 pm length-measurement accuracy. As these errors are typically allocated among a number of terms, individual tolerances are smaller.

Laser metrology at the picometer level is a major technology area for SIM [12]. The types of effects which must be addressed include errors internal to the laser gauge itself, errors in the endpoints which define the length being measured, and errors attributable to alignment of the metrology beam to the endpoints. In addition, there are errors attributable to how well the path you can measure corresponds to the path you wish to measure. For example, sampling of the position of the center of a mirror with a laser beam is not necessarily representative of the mean position of the surface as would be illuminated by a stellar beam. While these measurements present challenges, the payoff is large, motivated by the high astrometric precision possible outside of the Earth's atmosphere.

Tolerances for ground interferometers are typically more modest, in part because of atmospheric limitations which preclude astrometry beyond a certain accuracy, and because the baseline lengths can be longer (at least with respect to first-generation space interferometers). For Keck Interferometer in astrometry mode using a 100-m baseline, 10 μas systematic accuracy requires a 5 nm total length error, which is closer to the current state-of-practice.

However, for both ground and space interferometers, some themes are common: longer baselines help given a fixed-level of linear metrology accuracy; differential measurements allows certain systematic errors to become common mode and drop out; measurements which can be performed in a switching mode allow a reduction of requirements on long-term thermal stability.

3.2. BASELINE ERRORS AND NARROW ANGLE ASTROMETRY

For general astrometry, we need to know the baseline vector to the same accuracy as the desired astrometry, that is, if we desire 10 μas astrometry with a 100-m baseline, we must know the baseline orientation to 10 μas, corresponding to knowledge of the vector components to ~5 nm. However, planet detection is a special case in that it is fundamentally a narrow-angle problem, i.e., we can make measurements with respect to nearby astrometric references.

Equation 4 illustrates the difference between wide- and narrow-angle astrometry. For wide-angle astrometry, $\theta \sim 1$, leading to the intuitive result that the required fractional accuracy on the baseline is equal to the desired astrometric accuracy. However, for small fields, the requirement on baseline accuracy decreases: essentially, the baseline becomes more common mode to the differential measurement. For example, for a narrow-angle field of 20", the requirements on the baseline are reduced by a factor of 10^4.

3.3. RANDOM ERRORS - DETECTION NOISE

At some level, the ability to measure the fringe phase places a limit on the achievable accuracy. The error $\delta\phi$ in a phase measurement can be written

$$\delta\phi = (\mathrm{SNR}_\phi)^{-1}, \tag{5}$$

where

$$\mathrm{SNR}_\phi^2 \simeq \frac{1}{2}\frac{N^2V^2}{N + B + M\sigma^2}, \tag{6}$$

where N is the total photon count, B is the total background and dark count, σ^2 is the read-noise variance, and M is the number of reads needed to make the phase measurement.

The detection error shows up in the error expression, Eq. 4, reduced by the baseline. Thus, long baselines help by improving sensitivity for a given accuracy, or by reducing astrometric error for a given source brightness.

The SIM mission, observing in the visible, typically uses the photon-noise limit of this expression: $\mathrm{SNR}_\phi^2 \simeq (1/2)NV^2$. For ground-based interferometric astrometry in the infrared, the reference stars will ordinarily be faint and background limited, yielding $\mathrm{SNR}_\phi^2 \simeq (1/2)(N^2V^2/B)$.

3.4. RANDOM ERRORS - ATMOSPHERIC NOISE

For wide-angle optical astrometry on the ground, the ultimate limit to astrometric accuracy appears to be ~1 mas, the goal of the USNO astrometric interferometer [13]. This limit is established by atmospheric turbulence which perturbs the position of a star. While techniques such as the two-color technique can help (and is incorporated into the number above), this limit appears fundamental, and thus wide-angle microarcsec astrometry remains the purview of space missions like SIM.

Narrow-angle astrometry presents a different problem: for a simultaneous measurement between two or more stars, the light rays through the atmosphere follow increasingly common paths as the angle between them is decreased. This effect can be quantified as shown in Fig. 3.4 (from [8]). With respect to narrow-angle astrometry, there are two key features: for small separations between stars, the errors are linearly dependent on the star separation, and nearly inversely dependent on the size of the instrument (the results are identical for either a telescope of diameter B or an interferometer with baseline B). This is in contrast to the case of larger separations, where the dependence on separation and instrument size is weaker. Thus, if it is possible to conduct a measurement with a long-baseline interferometer over a suitably small field, accuracies of tens of microarcsec are possible.

4. Implementing a Narrow-Angle Measurement

Exploiting the tens-of-microarcsec astrometric accuracy possible with a ground-based narrow-angle astrometric measurement requires the ability to utilize nearby reference stars. One approach to this problem uses a dual-star architecture [8], as shown in Fig. 4. It consists of a long-baseline interferometer with dual beam trains. The light at each aperture forms an image of the field containing the target star and the astrometric reference. A dual-star feed separates the light from the two stars into separate beams which feed separate interferometer beam combiners. These beam combiners are referenced with laser metrology to a common fiducial at each collector. The two beam combiners make simultaneous measurements of the delays for the two stars.

Over a small field, reference stars will invariably be faint, and ordinarily would not be usable by the interferometer. However, searching for exoplanets is a special problem in that the target star is bright, and can serve as a phase reference. With phase referencing, the bright target star is used as a probe of the atmospheric turbulence within the isoplanatic patch of the target star. By compensating for the fringe motion of the target star with an optical delay line, the fringe motion of the faint astrometric reference

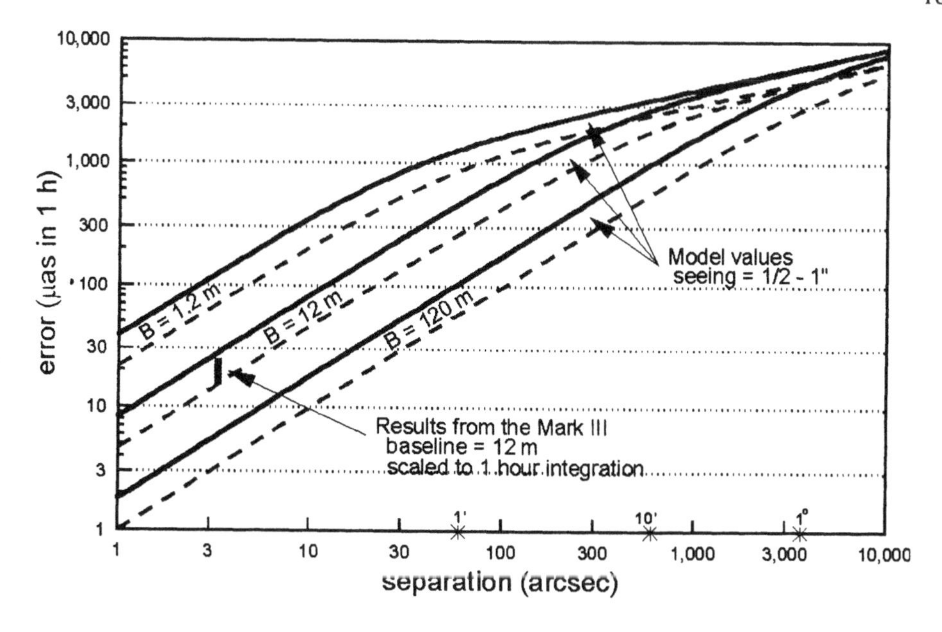

Figure 3. Accuracy of narrow angle astrometry.

star is frozen, allowing for long integration times which greatly increase sensitivity.

The radius of the isoplanatic patch increases with wavelength, and is 20–30 arcsec at 2.2 μm. With phase referencing and 1.5–2.0-m telescopes, astrometric references can be detected around most potential planetary targets.

Conducting a narrow-angle measurement with an architecture like that of Fig. 4 involves two steps. The first step is global astrometry using known reference stars to solve for the interferometer baseline. As discussed above, the required baseline precision for a narrow-angle measurement is much less than for a wide-angle measurement, and the accuracies available from these wide-angle measurements provide sufficient accuracy. (Strictly, if atmospherically-limited accuracy were always proportional to star separation, there would be difficulties with this step; however the break in the error dependence with large angles, as shown in Fig. 3.4, allows for a good baseline solution). There are some subtleties regarding the wide-angle baseline as thus solved and the narrow-angle baseline applicable to the science measurement, and an auxiliary system may be required to tie these two baselines together.

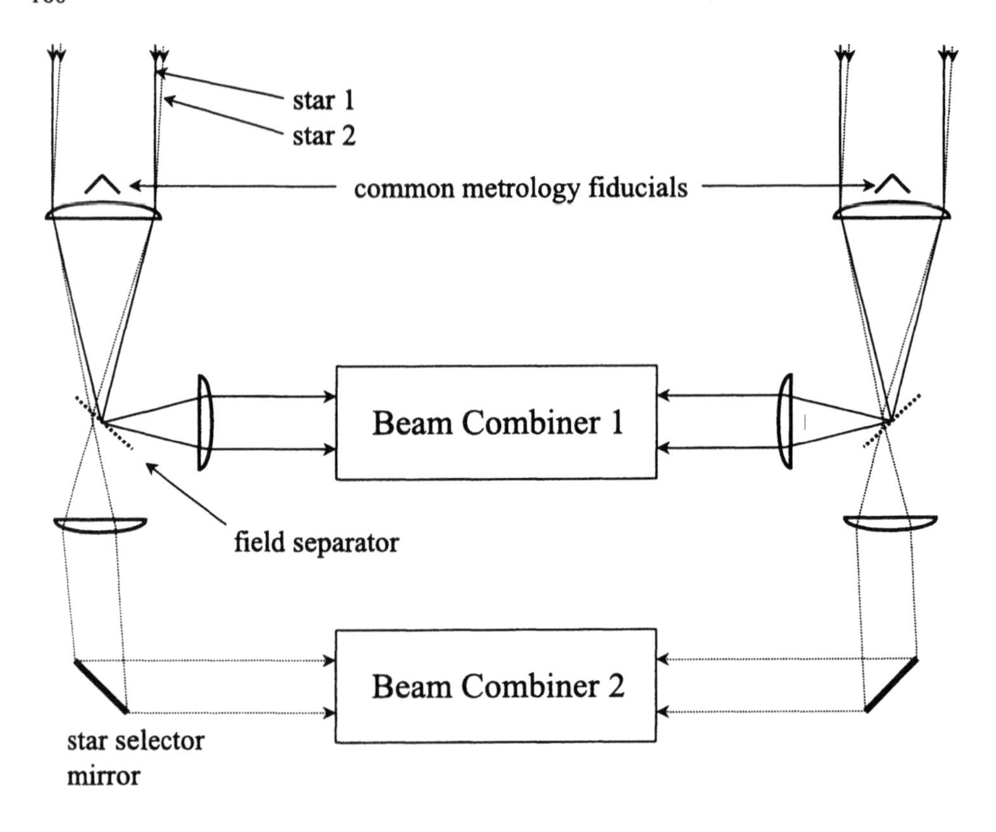

Figure 4. Dual-star architecture.

The second step in the approach is to implement the measurement through chopping. In this approach, one interferometer beam combiner always tracks the target star. The other beam combiner switches repeatedly between the target star and the reference star. This "chopping" approach requires instrument stability only over the chop cycle (a similar approach would also be used for narrow-angle measurements with SIM). The use of even a low-resolution spectrometer in the fringe detector makes the ground-based measurements very insensitive to differential chromatic refraction.

In general, measurements on two orthogonal baselines are needed to detect systems with arbitrary inclinations. Measurements with respect to two reference stars are also desirable. With redundant measurements, astrometric noise in a reference star is in most cases separable from the desired (planetary) signature.

5. Astrometric interferometer projects

5.1. PALOMAR TESTBED INTERFEROMETER

The Palomar Testbed Interferometer (PTI) [14] was designed as a testbed for interferometry techniques applicable to other interferometers, especially the Keck Interferometer. It implements the dual-star architecture described above to develop the dual-star astrometry technique. PTI has been in operation since 1995 at Palomar Observatory. Recent astrometric observations on visual binaries have been used to examine the atmospheric limits predicted above and examine the long-term repeatability of the measurements. While this work is still proceeding, short-term accuracies have been achieved which are consistent with the atmospheric theory.

5.2. KECK INTERFEROMETER

The Keck Interferometer will combine the two 10-m Keck telescopes at Mauna Kea with 4 1.8-m outrigger telescopes as an interferometer [15]. The Keck Interferometer belongs to NASA's Origins [16] program. Science with the Keck Interferometer includes high-angular-resolution imaging [17], direct detection of hot Jupiters via a two-color phase-difference approach, measurement of the quantity of exozodiacal dust around nearby stars, and, apropos to this paper, interferometric astrometry using the 4 outriggers configured to provide orthogonal baselines [18]. Using this mode, a search for planets around several hundred nearby stars is planned. The Keck interferometer draws strongly on lessons learned from PTI. The planned development of the interferometer is over 5 years, starting in 1998, with the astrometric program beginning after installation of the last of the outriggers in 2002.

The Very Large Telescope Interferometer (VLTI) is also considering instrumentation to allow for differential astrometry [19].

5.3. SPACE INTERFEROMETRY MISSION

The Space Interferometry Mission (SIM) [9] is the next major space mission in NASA's Origins Program after SIRTF, and is planned for a 2005 launch. SIM's primary goals are wide-angle astrometry to an accuracy of 4 μas to a limiting brightness of 20 mag, and synthesis imaging with a resolution of 10 mas. SIM uses 4 Michelson interferometers on a single structure with a maximum baseline of approximately 10 m. Several of these interferometers act as high-precision star trackers, stabilizing the interferometer to allow high-precision measurements with the other interferometers.

In addition to its wide-angle capability, SIM will also implement a narrow-angle astrometric capability with an accuracy of 1 μas. This is beyond the capability of ground-based techniques and is directed (through repeated measurements) at detecting planets down to several Earth masses around nearby stars.

6. Acknowledgments

The work reported here was performed at the Jet Propulsion Laboratory, California Institute of Technology, under a contract with the National Aeronautics and Space Administration.

References

1. Queloz, D. 1998, this proceedings
2. Marcy, G. W. & Butler, R. P. 1998, Ann. Rev. Astron. Astrophys. 36, 56
3. Perryman, M. A. C. et al. 1997, A&A, 323, L49
4. Monet, D. 1992, AJ 103, 638
5. Gatewood, G. D. 1987, Astron. J. 94, 213
6. Gatewood, G. D. et al. 1997, in Planets Beyond the Solar System, ASP Conf. Series Vol. 119, D. R. Soderblom, ed., 41
7. Pravdo, S. H. & Shaklan, S. B. 1996, ApJ 465, 264
8. Shao, M. & Colavita, M. M. 1992, A&A, 262, 353
9. Unwin, S., Boden, A., & Shao, M. 1997, in Space Technology and Applications International Forum, AIP Conf. Proc. 387 (AIP Press), 63; see also http://sim.jpl.nasa.gov
10. Shao, M. & Colavita, M. M. 1992, Ann. Rev. Astron. Astrophys, 30, 457
11. Thompson, A. R., Moran, J. M., Swensen, G. W., Jr. 1986, Interferometry and Synthesis in Radio Astronomy, Wiley, New York
12. Laskin, R. A. 1998, Proc. SPIE 3350, 654
13. Hutter, D. J., Elias, N. M., II, Hummel, C. A. 1998, Proc. SPIE 3350, 452
14. Colavita, M. M. et al. 1999, ApJ, in press
15. Colavita, M. M. et al. 1998, Proc. SPIE 3350, 776
16. http://origins.jpl.nasa.gov
17. Vasisht, G., Boden, A. F., Colavita, M. M., Crawford, S. L., Shao, M., van Belle, G. T., Wallace, J. K. 1998, Proc. SPIE 3350, 354
18. van Belle, G. T., Boden, A. F., Colavita, M. M., Shao, M., Vasisht, G., Wallace, J. K. 1998, Proc. SPIE 3350, 362
19. Mariotti, J.-M. et al. 1998, Proc. SPIE 3350, 800

SEARCHING FOR UNSEEN PLANETS VIA OCCULTATION AND MICROLENSING

PENNY D. SACKETT
Kapteyn Astronomical Institute
Postbus 800, 9700 AV Groningen, The Netherlands
psackett@astro.rug.nl

Abstract.
The fields of occultation and microlensing are linked historically. Early this century, occultation of the Sun by the Moon allowed the apparent positions of background stars projected near the limb of the Sun to be measured and compared with their positions six months later when the Sun no longer influenced the light path to Earth. The measured shift in the stellar positions was consistent with lensing by the gravitational field of the Sun during the occultation, as predicted by the theory of general relativity. This series of lectures explores the principles, possibilities and challenges associated with using occultation and microlensing to discover and characterize unseen planets orbiting distant stars. The two techniques are complementary in terms of the information that they provide about planetary systems and the range of system parameters to which they are most sensitive. Although the challenges are large, both microlensing and occultation may provide avenues for the discovery of extra-solar planets as small as Earth.

1. Introduction

Indirect methods to search for extra-solar planets do not measure emission from the planet itself, but instead seek to discover and quantify the tell-tale effects that the planet would have on the position (astrometry) and motion (radial velocity) of its parent star, or on the apparent brightness of its parent star (occultation) or random background sources (gravitational microlensing). All of these indirect signals have a characteristic temporal behavior that aids in the discrimination between planetary effects and other

J.-M. Mariotti and D. Alloin (eds.), Planets Outside the Solar System: Theory and Observations, 189–227.

astrophysical causes. The variability can be due to the changing position of the planet with respect to the parent star (astrometry, radial velocity, occultation), or the changing position of the complete planetary system with respect to background stars (microlensing). The time-variable photometric signals that can be measured using occultation and microlensing techniques are the focus of this small series of lectures.

An occultation is the temporary dimming of the apparent brightness of a parent star that occurs when a planet transits the stellar disk; this can occur only when the orbital plane is nearly perpendicular to the plane of the sky. Because the planet is considerably cooler than its parent star, its surface brightness at optical and infrared wavelengths is less, causing a dip in the stellar light curve whenever the planet (partially) eclipses the star. Since the fractional change in brightness is proportional to the fraction of the stellar surface subtended by the planetary disk, photometric measurements directly yield a measure of the planet's size. For small terrestrial planets, the effect is simply to occult a fraction of the stellar light; the atmospheres of larger gaseous planets may also cause absorption features that can be measured during transit with high resolution, very high S/N spectroscopic monitoring.

The duration of a transit is a function of the size of the stellar disk and the size and inclination of the planetary orbit. Together with an accurate stellar typing of the parent star, measurement of the transit duration and period provides an estimate for the radius and inclination of the planet's orbital plane. Since large planets in tight orbits will create the most significant and frequent occultations, these are the easiest to detect. If hundreds of stars can be monitored with significantly better than 1% photometry, the transit method can be applied from the ground to place statistics on Jupiter-mass planets in tight orbits. Space-based missions, which could search for transits continuously and with higher photometric precision, may be capable of detecting Earth-mass planets in Earth-like environments via the occultation method. Moons or multiple planets may also be detectable, not through their eclipsing effect, but by the periodic change they induce in the timing of successive transits of the primary occulting body.

Microlensing occurs when a foreground compact object (e.g., a star, perhaps with orbiting planets) moves between an observer and a luminous background source (e.g., another star). The gravitational field of the foreground lens alters the path of the light from the background source, creating multiple images with a combined brightness larger than that of the unlensed background source. For stellar or planetary mass lenses, the separation of these images is too small to be resolved, but the combined brightness of the images changes with time in a predictable manner as the lensing system moves across the sky with respect to the background source.

Hundreds of microlensing events have been detected in the Galaxy, a large fraction of which are due to (unseen) stellar lenses. In binary lenses with favorable geometric configurations, the lensing effect of the two lenses combines in a non-linear way to create detectable and rapid variations in the light curve of the background source star. Modeling of these features yields estimates for the mass ratio and normalized projected orbital radius for the binary lens; in general, smaller-mass companions produce weaker and shorter deviations.

Frequent, high-precision photometric monitoring of microlensing events can thus be used to discover and characterize extreme mass-ratio binaries (i.e., planetary systems). With current ground-based technology, microlensing is particularly suited to the detection of Jupiter-mass planets in Jupiter-like environments. Planets smaller than Neptune will resolve the brightest background sources (giants) diluting the planetary signal. For planets above this mass, the planetary detection efficiency of microlensing is a weak function of the planet's mass and includes a rather broad range in orbital radii, making it one of the best techniques for a statistical study of the frequency and nature of planetary systems in the Galaxy. Microlensing can discover planetary systems at distances of the Galactic center and is the only technique that is capable of detecting unseen planets around *unseen parent stars*!

These lectures begin with a discussion of the physical basis of occultation and microlensing, emphasizing their strengths and weaknesses as well as the selection effects and challenges presented by sources of confusion for the planetary signal. The techniques are then placed in the larger context of extra-solar planet detection. Speculative comments about possibilities in the next decade cap the lectures.

2. Principles of Planet Detection via Occultations

Due to their small sizes and low effective temperatures, planets are difficult to detect directly. Compared to stars, their luminosities are reduced by the square of the ratio of their radii (factors of $\sim 10^{-2} - 10^{-6}$ in the Solar System) and the fourth power of the ratio of their effective temperatures (factors of $\sim 10^{-4} - 10^{-9}$ in the Solar System). Such planets may be detected indirectly however if they chance to transit (as viewed by the observer) the face of their parent star and are large enough to occult a sufficient fraction of the star's flux. This method of detecting planets around other stars was discussed as early as mid-century (Sturve 1952), but received serious attention only after the detailed quantification of its possibilities by Rosenblatt (1971) and Borucki and Summers (1984).

Such occultation effects have been observed for many years in the photometry of binary star systems whose orbital planes lie close enough to edge-on as viewed from Earth that the disk of each partner occults the other at some point during the orbit, creating two dips in the combined light curve of the system. The depth of the observed occultation depends on the relative size and temperatures of the stars. For planetary systems, only the dip caused by the occultation of the brighter parent star by the transit of the smaller, cooler planet will be detectable. The detection rate for a given planetary system will depend on several factors: the geometric probability that a transit will occur, the frequency and duration of the observations compared to the frequency and duration of the transit, and the sensitivity of the photometric measurements compared to the fractional deviation in the apparent magnitude of the parent star due to the planetary occultation. We consider each of these in turn.

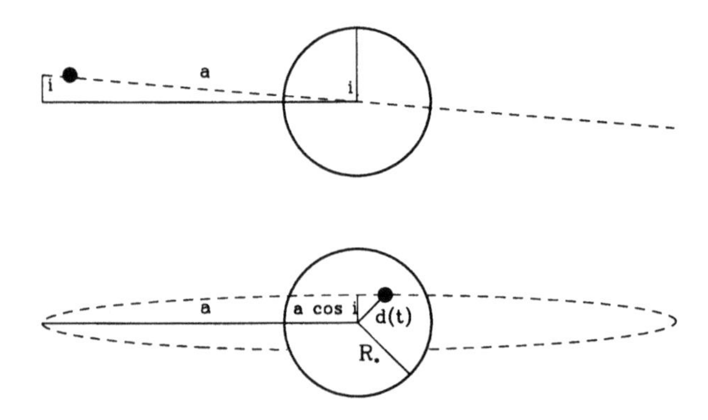

Fig. 1 — Geometry of a transit event of inclination i and orbital radius a as seen from the side (top) and observer's vantage point (bottom) at a moment when the planet lies a projected distance $d(t)$ from the stellar center.

Unless stated otherwise in special cases below, we will assume for the purposes of discussion that planetary orbits are circular and that the surface brightness, mass, and radius of the planet are small compared to that of the parent star. We will also assume that the orbital radius is much larger than the size of the parent star itself.

2.1. GEOMETRIC PROBABILITY OF A TRANSIT

Consider a planet of radius R_p orbiting a star of radius R_* and mass M_* at an orbital radius a. A transit of the stellar disk will be seen by an external observer only if the orbital plane is sufficiently inclined with respect to the sky plane (Fig. 1). In particular, the inclination i must satisfy

$$a \cos i \leq R_* + R_p \quad . \tag{1}$$

Since $\cos i$ is simply the projection of the normal vector (of the orbital plane) onto the sky plane, it is equally likely to take on any random value between 0 and 1. Thus, for an ensemble of planetary systems with arbitrary orientation with respect to the observer, the probability that the inclination satisfies the geometric criterion for a transit is:

$$\text{Geometric Transit Prob} = \frac{\int_0^{(R_*+R_p)/a} d(\cos i)}{\int_0^1 d(\cos i)} = \frac{R_* + R_p}{a} \approx \frac{R_*}{a} \qquad (2)$$

Geometrically speaking, the occultation method favors those planets with small orbital radii in systems with large parent stars. As can be seen in Fig. 2, for planetary systems like the Solar System this probability is small: $\lesssim 1\%$ for inner terrestrial planets and about a factor of 10 smaller for jovian gas giants. This means that unless a method can be found to pre-select stars with ecliptic planes oriented perpendicular to the plane of the sky, thousands of random stars must be monitored in order to detect statistically meaningful numbers of planetary transits due to solar systems like our own.

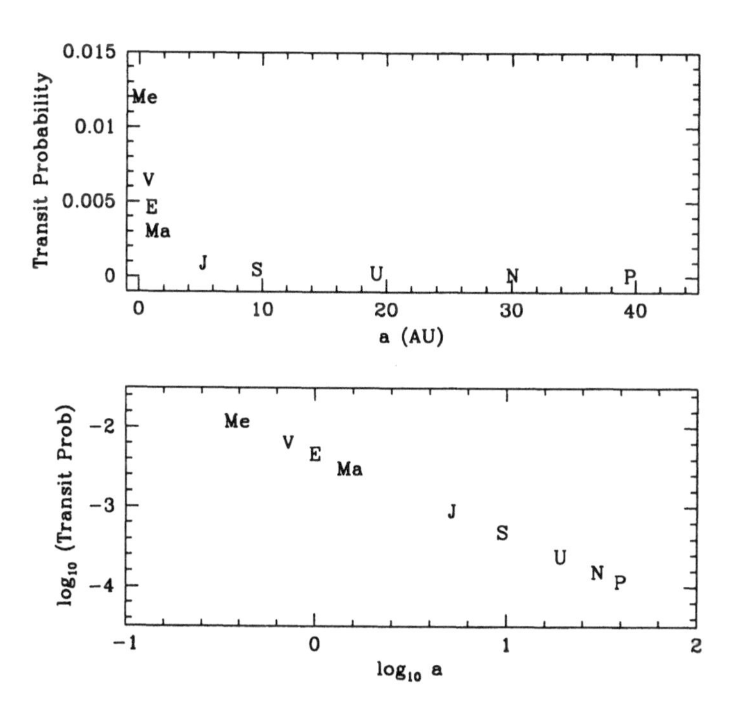

Fig. 2 — Probability of transits by Solar System objects as seen by a random external observer.

2.1.1. *Inclination Pre-selection*

Under the assumption that the orbital angular momentum vector of a planetary system and the rotational angular momentum vector of the parent star share a common origin and thus a common direction, single stars can be pre-selected for transit monitoring programs on the basis of a measurement of their rotational spin. In this way, one may hope to increase the chances of viewing the planetary orbits edge-on. Through spectroscopy, the line-of-sight component of the rotational velocity $v_{*,los}$ of a star's atmosphere can be measured. The period $P_{*,rot}$ of the rotation can be estimated by measuring the periodic photometric signals caused by sunspots, and the radius R_* of the star can be determined through spectral typing and stellar models. An estimate for the inclination of the stellar rotation plane to the plane of the sky can then be made:

$$\sin i_{*,rot} = \frac{v_{*,los}\, P_{*,rot}}{2\pi\, R_*} \quad , \tag{3}$$

and only those stars with high rotational inclinations selected to be monitored for transits.

How much are the probabilities increased by such pre-selection? Fig. 3 shows the probability of the planetary orbital inclination being larger (more edge-on) than a particular value ranging from $89.5° < i < 85°$, if the parent star is pre-selected to have a rotational plane with inclination $i_{*,rot} \geq i_{\mathrm{select}}$.

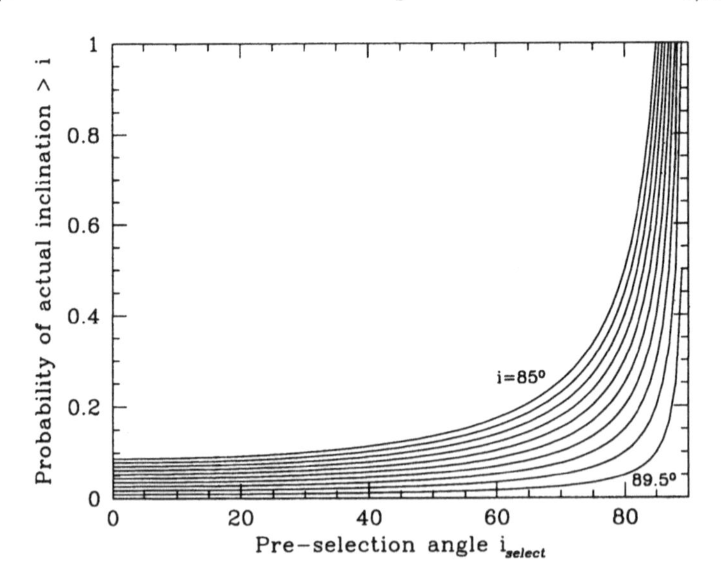

Fig. 3 — Increase of geometric transit probability through pre-selection of the inclination angle to be larger than i_{select}, for example through measurement of the rotational spin of the parent.

In order to produce a detectable transit, most planets will require an orbital inclination $\lesssim 1°$ from edge-on. If planetary systems could be pre-selected to have $i > 85$, the geometric transit probability would be increased by a factor of $\sim$10. Unfortunately, measurement uncertainties in the quantities required to determine $\sin i_{*,rot}$ are likely to remove much of the advantage that pre-selection would otherwise afford. Since $\delta(\cos i) = -\tan i\, \delta(\sin i)$, even small errors in $\sin i_{*,rot}$ translate into large uncertainties in $\cos i_{*,rot}$ and thus the probability that a transit will occur. Furthermore, an accurate measurement of $\cos i_{*,rot}$ does not ensure that $\cos i$ for the planetary orbital plane is known. The planets in our own Solar System are misaligned by about 7° with the Sun's rotational plane, a result that is similar to that found for binaries orbiting solar-type stars (Hale 1994). It is thus reasonable to assume that an accurate measurement of $i_{*,rot}$ will constrain the planetary orbital plane only to within $\sim$10°.

To enhance probabilities, current ground-based attempts to detect transits have taken a different tack by concentrating on known eclipsing binary star systems in which the orbital plane of the binary is known to be close to edge-on. Assuming that any other companions will have similarly aligned angular momentum vectors, it is hoped that such systems will have larger than random chances of producing a transit event. The precession of orbital plane likely to be present in such systems may actually bring the planet across the face of the star more often than in single star systems (Schneider 1994). On the other hand, the evolution and dynamics of single and double star systems is so different that the formation and frequency of their planetary companions is likely to be quite different as well. In particular, it may be difficult for planets in some binary systems to maintain long-lived circular orbits and thus, perhaps, to become the birth place of life of the sort that has evolved on Earth.

Given the uncertainties involved, inclination pre-selection in single stars is unlikely to increase geometric transit probabilities by factors larger than 3 – 5. Ambitious ground-based and space-based initiatives, however, may monitor so many stars that pre-selection is not necessary.

2.2. TRANSIT DURATION

The duration and frequency of the expected transits will determine the observational strategy of an occultation program. The frequency is simply equal to one over the orbital period $P = \sqrt{4\pi^2 a^3/GM_*}$. If two or more transits for a given system can be measured and confirmed to be due to the same planet, the period P and orbital radius a are determined. In principle, the ratio of the transit duration to the total duration can then be used to determine the inclination of the orbital plane, if the stellar radius is known.

The duration of the transit will be equal to the fraction of the orbital period P during which the projected distance d between the centers of the star and planet is less than the sum of their radii $R_* + R_p$. Refering to Fig. 4 we have

$$\text{Duration} \equiv t_T = \frac{2\,P}{2\pi} \arcsin\left(\frac{\sqrt{(R_* + R_p)^2 - a^2\cos^2 i}}{a}\right) \quad , \tag{4}$$

which for $a >> R_* >> R_p$ becomes

$$t_T = \frac{P}{\pi}\sqrt{\left(\frac{R_*}{a}\right)^2 - \cos^2 i} \;\; \leq \;\; \frac{P\,R_*}{\pi\,a} \quad . \tag{5}$$

Note that because the definition of a transit requires that $a\cos i \leq (R_* + R_p)$, the quantity under the square root in Eq. 4 does not become negative.

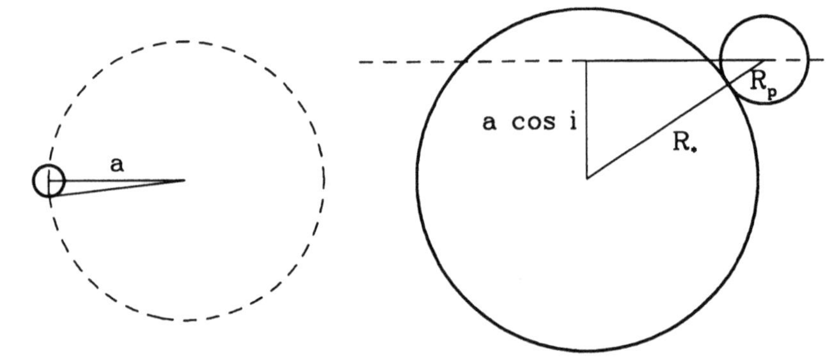

Fig. 4 — Transit duration is set by fraction of total orbit (left) for which a portion of the planet eclipses the stellar disk (right).

Fig. 5 shows the maximum transit duration and period for planets in the Solar System. In order to confirm a planetary detection with one or more additional transits after the discovery of the first eclipse, a 5-year experiment can be sensitive to planets orbiting solar-type stars only if their orbital radius is equal to or smaller than that of Mars. Such planets will have transit durations of less than one day, requiring rapid and continuous sampling to ensure high detection probabilities.

The actual transit duration depends sensitively on the inclination of the planetary orbit with respect to the observer, as shown in Fig. 6. The transit time of Earth as seen by an external observer changes from 0.5 days to zero (no transit) if the observers viewing angle is more than 0.3° from optimal. Since the orbital planes of any two of the inner terrestrial planets in the

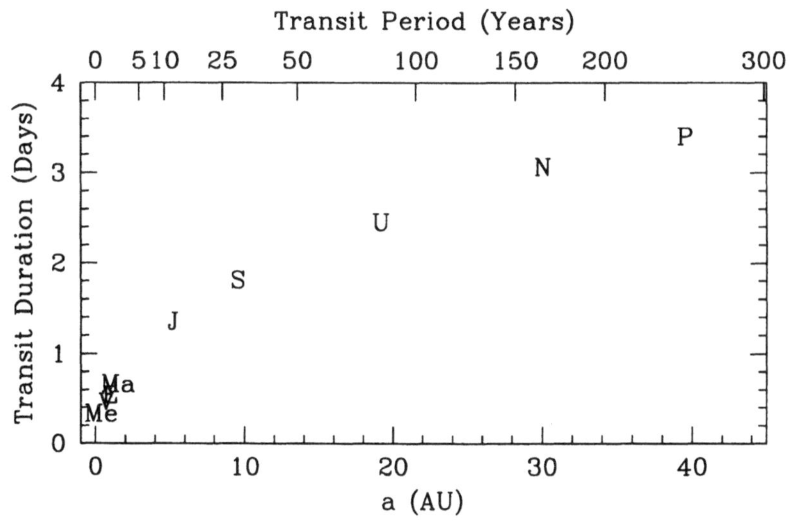

Fig. 5 — Edge-on transit durations and periods for Solar System planets.

Solar System are misaligned by 1.5° or more, if other planetary systems are like our own, a given observer would expect to see transits from only one of the inner planets. This would decrease the detection probabilities for planetary systems, but also the decrease the probability of incorrectly attributing transits from different planets to successive transits of one (mythical) shorter period object.

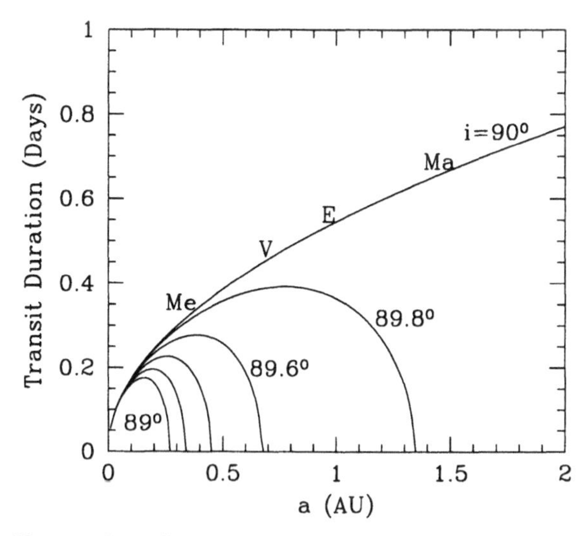

Fig. 6 — "Inner planet" transit durations for different inclinations ($R_* = R_\odot$).

If the parent star can be typed spectroscopically, stellar models can provide an estimate for the stellar radius R_* in the waveband in which the photometric partial eclipse was measured. (It is important to match wavebands since limb-darkening can make the star look larger at redder wavelengths which are more sensitive to the cooler outer atmosphere of the star.) The temporal resolution of a single transit then places a lower limit on the orbital radius a of the planet, but a full determination of a requires knowledge of the period from multiple transit timings which remove the degeneracy due to the otherwise unknown orbital inclination. In principle, if the limb darkening of the parent star is sufficiently well-understood, measurements in multiple wavebands can allow an estimate for the inclination, and thus for a from a single transit; this is discussed more fully in §2.3.1.

2.3. AMPLITUDE AND SHAPE OF THE PHOTOMETRIC SIGNATURE

Planets with orbital radii of 2 AU or less orbiting stars even as close as 10 parsec will subtend angles $\lesssim 50$ microarcseconds; any reflected or thermal radiation that they might emit thus will be confused by normal photometric techniques with that from the parent star. Only exceedingly large and close companions of high albedo would be capable of creating a significant modulated signal throughout their orbit as the viewer sees a different fraction of the starlit side; we will not consider such planets here. All other planets will alter the total observed flux only during an actual transit of the stellar face, during which the amplitude and shape of the photometric dip will be determined by the fraction of the stellar light that is occulted as a function of time.

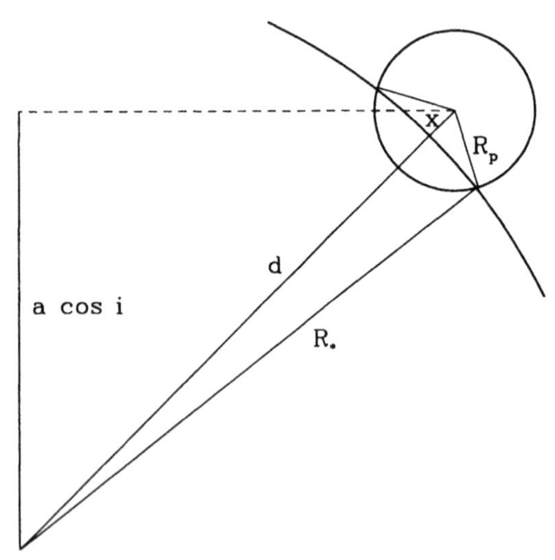

Fig. 7 — The area eclipsed by a planet as it crosses the stellar limb determines the wing shape of the resulting photometric dip.

The maximum fractional change in the observed flux is given by:

$$\text{Maximum} \quad \frac{|\delta\mathcal{F}_\lambda|}{\mathcal{F}_\lambda} = \frac{\pi\mathcal{F}_{\lambda,*}\, R_p^2}{\pi\mathcal{F}_{\lambda,*}\, R_*^2 + \pi\mathcal{F}_{\lambda,p}\, R_p^2} \approx \left(\frac{R_p}{R_*}\right)^2 \equiv \rho^2 \tag{6}$$

The shape of the transit dip will depend on the inclination angle, the ratio of the planet to stellar size, and the degree of limb-darkening in the observational band.

Begin by considering a star of uniform brightness (no limb-darkening) transited by a small planet. The stellar limb will then describe a nearly straight chord across the planet at any time, and integration over planet-centered axial coordinates (see Fig. 7) yields an eclipsing area during ingress and egress of:

$$\mathcal{A}_\mathcal{E} \approx \int_x^{R_p} r_p\, dr_p \int_{-\arccos(x/r_p)}^{+\arccos(x/r_p)} d\phi_p = 2\int_x^{R_p} r_p \arccos\left(\frac{x}{r_p}\right) dr_p \ , \tag{7}$$

where $x \equiv d - R_*$, d is the projected star-planet separation and x is constrained to lie in the region $-R_p < x < R_p$. The last integral can be done analytically to yield,

$$\mathcal{A}_\mathcal{E} \approx R_p^2 \arccos(x/R_p) - R_p x\sqrt{1 - \frac{x^2}{R_p^2}} \ . \tag{8}$$

For larger planets, and to facilitate the introduction of limb-darkened sources, it is more useful to integrate over stellar-centered axial coordinates; the Law of Cosines can then be used to show that

$$\mathcal{A}_\mathcal{E}(t) = 2\int_{\max(0,\, d(t)-R_p)}^{\min(R_*,\, d(t)+R_p)} r_* \arccos[\Theta(t)]\, dr_* \tag{9}$$

$$\text{where} \quad \Theta(t) \equiv \frac{d^2(t) + r_*^2 - R_p^2}{2r_* d(t)} \quad \text{for } r_* > R_p + d(t), \text{ and } \pi \text{ otherwise.} \tag{10}$$

The light curve resulting from the occultation of a uniform brightness source by a planet of arbitrary size, orbital radius and orbital inclination can now be constructed by substituting into Eq. 9 the time dependence of the projected planet-star separation, $d(t) = a\sqrt{\sin^2 \omega t + \cos^2 i \cos^2 \omega t}$, where $\omega \equiv 2\pi/P$. The *differential* light curve is then given by:

$$\frac{\mathcal{F}(t)}{\mathcal{F}_0} = 1 - \frac{\mathcal{A}_\mathcal{E}(t)}{\pi R_*^2} \tag{11}$$

For spherical stars and planets, the light curve will be symmetric and have a minimum at the closest projected approach of planet to star center, where the fractional decrease in the total brightness will be less than or equal to $(R_p/R_*)^2$. For Jupiter-sized planets orbiting solar-type stars, this is a signal of ~1%; for Earth-sized planets the fractional change is $\lesssim 0.01\%$ (Fig. 8).

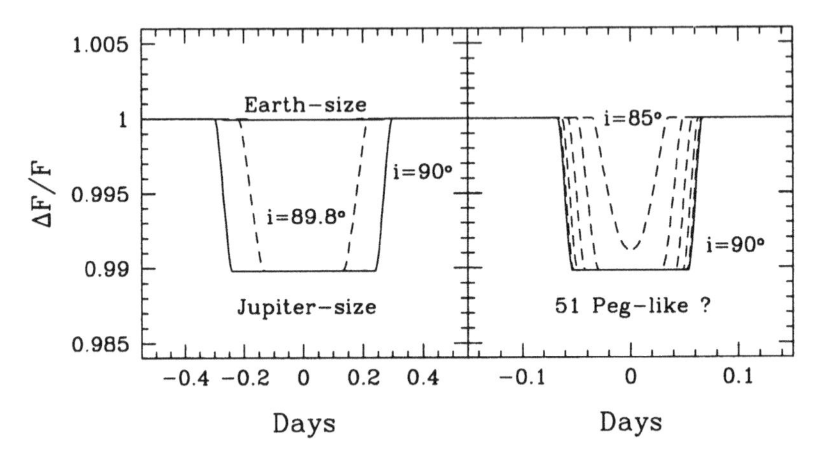

Fig. 8 — **Left:** Photometric light curves for Earth-sized and Jupiter-sized planets orbiting a solar-type star at 1 AU. **Right:** A Jupiter-sized planet orbiting a solar-type star at an orbital radius of 0.05 AU (e.g., 51 Peg) with inclinations ranging from 85° to 90°. The parent star is assumed here to have constant surface brightness. Note change in time scale between two panels.

If proper care is taken, photometry of bright, uncrowded stars can be performed to ~0.1% precision from the ground (Henry *et al.* 1997), so that ground-based transit searches can in principle be sensitive to Jupiter-sized planets at $\lesssim$1 AU — planets perhaps similar to those being found by the radial velocity technique (e.g., Mayor & Queloz 1995, Butler & Marcy 1996). Transit detections of terrestrial planets like those in our own Solar System must await space observations in order to achieve the required photometric precision.

2.3.1. *Effects of Limb Darkening*

Because observations at different wavelengths probe material at different depths in stellar atmospheres, a stellar disk is differentially limb-darkened: the radial surface brightness profile $B_\lambda(r_*)$ of a star is wavelength dependent. In redder bands, which probe the cooler outer regions of the star, the stellar disk will appear larger and less limb-darkened. Limb darkening is important to transit techniques for two reasons: it changes the shape of the photometric signal and it does so in a wavelength-dependent way.

Since a given planet can produce dips of varying strength depending on the inclination i of the orbit, the inclination must be known in order to estimate the planet's radius R_p accurately. In principle, if the parent star has been typed so that its mass and stellar radius R_* are known, Kepler's Law together with Eq. 5 will yield i once the transit time t_T and period P have been measured. Ignoring the effects of limb darkening, however, will result in an underestimate of t_T, and thus an underestimate for the inclination i as well. In order to produce the required amplitude at minimum, the size of the planet R_p will then be overestimated. Furthermore, the sloping shape of the limb-darkened profile might be attributed to the smaller inclination i, reinforcing misinterpretation.

This difficulty will be removed if the limb darkening can be properly modeled. In addition, transit monitoring in more than one waveband could confirm the occultation hypothesis by measuring the characteristic color signature associated with limb darkening. In principle this signature can be used to determine the orbital inclination from a single transit, in which case Eq. 5 can be inverted to solve for the period P without waiting for a second transit.

How strong is the effect of limb darkening? To incorporate its effect, the integral in Eq. 9 used to determine the eclipsing area must be weighted by the surface brightness as a function of stellar radius, yielding the differential light curve:

$$\frac{\mathcal{F}_\lambda(t)}{\mathcal{F}_{\lambda,0}} = 1 - \frac{\int_{\max(0,\, d(t)-R_p)}^{\min(R_*,\, d(t)+R_p)} r_*\, B_\lambda(r_*)\, \arccos\left[\Theta(t)\right] dr_*}{\pi \int_0^{R_*} r_*\, B_\lambda(r_*)\, dr_*} \tag{12}$$

A commonly-used functional form for the surface brightness profile is $B_\lambda(\mu) = [1 - c_\lambda(1-\mu)]$, where $\mu \equiv \cos\gamma$ and γ is the angle between the normal to the stellar surface and the line-of-sight. In terms of the projected radius r_* from the stellar center this can be written as $B_\lambda(r_*) = [1 - c_\lambda(1 - \sqrt{1-(r_*/R_*)^2})]$. Using this form and constants c_λ appropriate for the Sun, light curves and color curves are shown in Fig. 9 for a Jupiter-sized planet orbiting 1 AU from a solar-type star at inclinations of 90° and 89.8°.

As expected, the bluer band shows more limb darkening, which rounds the sharp edges of the occultation profile making it qualitatively degenerate with a larger planet at somewhat smaller inclination. The color curves for different inclinations, however, are qualitatively different and can thus be used to break this degeneracy. During ingress and egress the color curve becomes bluer as the differentially redder limb is occulted; at maximum occultation the color curve is redder than the unocculted star for transits with inclination $i = 90°$ since the relative blue central regions are then occulted. For smaller inclinations, the planet grazes the limb blocking preferentially red light only, and the color curve stays blue through the event. Since the size of the color signal is ~10% of the deviation in total flux, excellent photometry is required to measure this effect and use it to estimate the orbital inclination; even for jovian giants it remains at or just beyond the current limits of photometric precision.

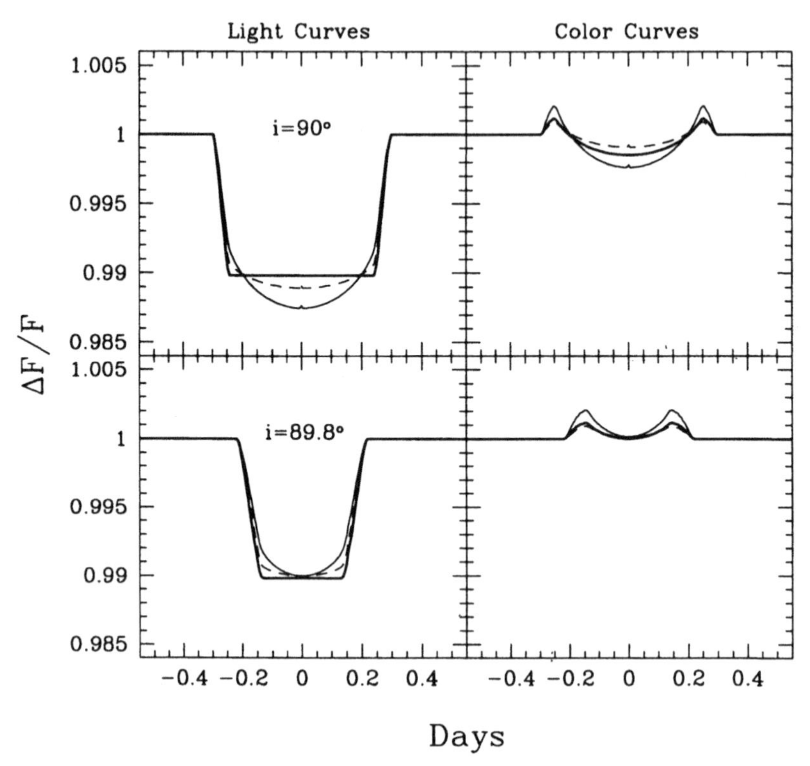

Fig. 9 — **Left:** Light curves for a planet with $R_p = 11R_\oplus$ orbiting a solar-type star with orbital inclinations of 90° (top) and 89.8° (bottom) normalized to the total (unocculted) flux in the indicated band. The thick line shows a uniformly bright stellar disk; thin solid and dashed lines indicate observations in the R and K bands respectively. **Right:** Color curves indicating the flux ratios at any given time between R (thin solid) and K-band (dashed) limb-darkened curves and a uniformly bright target star, and the observed limb-darkened R/K flux ratio (thick).

2.4. OBSERVATIONAL REWARDS AND CHALLENGES

In sum, what can be learned by observing a planetary object transiting the face of its parent star? The amplitude of the photometric signal places a lower limit on the ratio of the planetary radius to stellar radius $\rho \equiv R_p/R_*$, while the duration of the event places a lower limit on the orbital period

P and thus on the orbital radius a as well. If the inclination i is known, these lower limits become measurements. In principle i could be determined by fitting the wings of the transit profile in different wavebands using the known limb-darkening of the star, but in practice this will probably prove too difficult. Instead, multiple transits will be required to time the transits and thus measure the period P of the planet, from which the inclination can be determined from the known transit duration (Eq. 5). This makes the transit method most appropriate for large planets orbiting their parent stars at (relatively) small radii a. The primary challenge then reduces to performing the very precise photometry required on a large enough sample of stars to place meaningful statistics on the numbers of planets at small a.

What limits the photometric accuracy and clear detection of a transit signal? The dwarf stars that have suitably small stellar radii R_* must have apparent magnitudes bright enough (ie, be close enough) that enough photons can be captured in short exposures so that a sub-day transit event can be well-sampled. This will limit the depth of the sample to only nearby stars. Fields with very high stellar densities (like globular clusters or the Galactic Center) or very wide fields that can capture hundreds of candidate stars simultaneously will be required in order to maintain the required temporal sampling on a large enough sample. Regions of high stellar density, however, will be hampered by the additional challenges associated with precision photometry in confused fields.

The use of reference constant stars in the field can reduce the effects of varying extinction to produce the best current photometry in uncrowded fields, precise to the $\sim$0.1% level. Ultimately, scintillation, the rapidly-varying turbulent refocusing of rays passing through the atmosphere, limits Earth-bound photometry to 0.01%. Detection of Earth-mass transits is thus probably restricted to space-borne missions, although in special circumstances, periodicity analyses may be used to search for very short-period Earth-sized transits from the ground (e.g.,Henry *et al.* 1997).

For larger, jovian gas-giants, the signal can be measured from the ground, but must be distinguished from intrinsic effects that could be confused with transits. Late-type dwarf stars often undergo pulsations that cause their brightness to vary on the order of a few hours, but due to their cyclic nature these pulsations should be distinguished easily from all but very short period transits corresponding to $a \lesssim 0.02$ AU or so.

Solar flares produce excess of flux at the $\lesssim 0.001\%$ level, and thus would not confuse a typical transit signal. Later-type dwarfs tend to have more surface activity, however, and thus produce flares that contain a larger fraction of the star's total flux. Since the flares are generally blue, the primary problem will be in confusing the chromatic signal expected due to limb-darkening effects during a transit.

More troublesome will be separating transits from irregular stellar variability due to star spots. Star spots are cool regions on the stellar surface that remain for a few rotations before disappearing. They could mock a transit event and thus are probably the most important non-instrumental source of noise. Although the power spectrum of the Solar flux does show variations on day and sub-day time scales, most of the power during periods of sunspot maximum occurs at the approximate 1-month time scale of the Sun's rotation. Even during sunspot maximum, variations on day and sub-day scales are at or below the 0.001% level (Borucki, Scargle & Hudson 1985). Star spots on solar-type stars will therefore not be confused with the transit signal of a gas giant, but spots might be a source of additional noise for terrestrial-sized planets of small orbital radius ($a \lesssim 0.3$AU) and for parent stars that are significantly more spotted than the Sun.

2.4.1. *Pushing the Limits: Rings, Moons and Multiple Planets*

If the parent star can be well-characterized, the transit method involves quite simple physical principles that can perhaps be exploited further to learn more about planetary systems. For example, if a system is discovered to contain large transiting inner planets, it can be assumed to have a favorable inclination angle that would make it a good target for more sensitive searches for smaller radius or larger a planets in the same system.

If the inner giants are large enough, differential spectroscopy with a very large telescope before and during transits could reveal additional spectral lines that could be attributed to absorption of stellar light by the atmosphere of the giant (presumably gaseous) planet (see Laurent & Schneider, this proceedings). A large occulting ring inclined to the observer's line-of-sight would create a transit profile of a different shape than that of a planet (Schneider 1997), though the signal could be confused with limb-darkening effects and would likely be important only for outer gas giants where icy rings can form more easily.

Finally, variations in the ingress timing of inner planets can be used to search for cyclic variations that could betray the presence of moons (Schneider 1997) or — in principle — massive (inner or outer) planets that are nearly coplanar but too misaligned to cause a detectable transit themselves. Transit timing shifts would be caused by the slight orbital motion of the planet around the planet-moon barycenter or that of the star around the system barycenter. (The latter is unobservable for a single-planet system since the star's motion is always phase-locked with the planet.)

3. Principles of Planet Detection via Microlensing

Microlensing occurs when a foreground compact object (e.g., a star) moves between an observer and a luminous background object (e.g., another star). The gravitational field of the foreground lens alters the path of the light from the background source, bending it more severely the closer the ray passes to the lens. This results in an image as seen by a distant observer that is altered both in position and shape from that of the unlensed source. Indeed since light from either side of a lens can now be bent to reach the observer, multiple images are possible (Fig. 10). Since the total flux reaching the observer from these two images is larger than that from the unlensed source alone, the lens (and any planets that may encircle it) betrays its presence not through its own luminous emission, but by its gravitational magnification of the flux of background objects. Einstein (1936) recognized microlensing in principle, but thought that it was undetectable in practice.

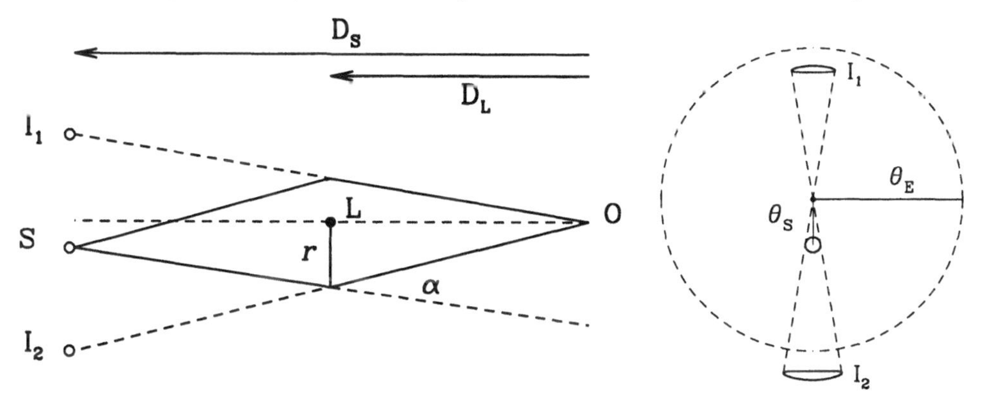

Fig. 10 — **Left:** A compact lens (L) located a distance D_L nearly along the line-of-sight to a background source (S) located at a distance D_S will bend incoming light rays by differing amounts α to create two images (I_1 and I_2) on either side of the line-of-sight. **Right:** An observer O does not see the microlensed source at its true angular sky position θ_S, but rather two images at positions θ_1 and θ_2.

Ray tracing, together with the use of general relatively to relate the bending angle α with the lens mass distribution, produces a mapping from the source positions (ξ, η) to the image positions (x,y) for a given mass distribution. For "point" masses, the angle α is just given by the mass of the lens M and the distance of closest approach r as:

$$\alpha = \frac{4GM}{c^2 r} = \frac{2R_S}{r} \quad , \tag{13}$$

as long as r is well outside the Schwarzschild radius R_S of the lens. Simple geometry alone then requires

$$\theta_S \, D_S = r \, \frac{D_S}{D_L} - (D_S - D_L) \, \alpha(r) \quad , \tag{14}$$

which can be rewritten to yield the lens equation

$$\theta_S = \theta - \frac{D_{LS}}{D_S} \, \alpha(r) \quad , \tag{15}$$

giving the (angular) vector image positions θ for a source at the angular position θ_S as measured from the observer-lens line-of-sight. D_S and D_L are the source and lens distances from the observer, respectively, and $D_{LS} \equiv D_S - D_L$.

For convenience, the characteristic angular size scale is defined as

$$\theta_E \equiv \sqrt{\frac{2R_S D_{LS}}{D_L \, D_S}} = \sqrt{\frac{4GMD_{LS}}{c^2 \, D_L \, D_S}} \quad . \tag{16}$$

Since $r = D_L \, \theta$, Eq. 15 can now be rewritten to yield a quadratic equation in θ

$$\theta^2 - \theta_S \, \theta \; - \theta_E^2 = 0 \quad , \tag{17}$$

with two solutions $\theta_{1,2} = \frac{1}{2}\left(\theta_S \pm \sqrt{4\theta_E^2 + \theta_S^2}\right)$ giving the positions of images I_1 and I_2. When the source lies directly behind the lens as seen from the observer, $\theta_S = 0$ and the two images merge into a ring of radius θ_E, the so-called "Einstein ring." For all other source positions, one image will lie inside θ_E and one outside. The flux observed from each image is the integral of the image surface brightness over the solid angle subtended by the (distorted) image. Since the specific intensity of each ray is unchanged in the bending process, so is the surface brightness. The magnification $A_{1,2}$ for each image is then just the ratio of the image area to the source area, and is found formally by evaluating at the image positions the determinant of the Jacobian mapping J that describes the lensing coordinate transformation from image to source plane:

$$A_{1,2} = \left. \frac{1}{|det \, J|} \right|_{\theta=\theta_{1,2}} = \left| \frac{\partial \, \theta_S}{\partial \, \theta} \right|^{-1}_{\theta=\theta_{1,2}} \quad , \tag{18}$$

where θ_S and θ are (angular) position vectors for the source and image, respectively.

What is most important for detection of extra-solar planets around lenses is not the position of the images but their magnification. For stellar lenses and typical source and lens distances within the Milky Way, the typical image separation ($\gtrsim 2\theta_E$) is ~1 milliarcsecond, too small to be resolved with current optical telescopes. The observer sees one image with a combined magnification $A \equiv A_1 + A_2$ that can be quite large. In order to distinguish intrinsically bright background sources from fainter ones that appear bright due to microlensing, the observer relies on the characteristic brightening and dimming that occurs as motions within the Galaxy sweep the source (nearly) behind the lens-observer line-of-sight. The unresolved images also sweep across the sky (Fig. 11); their combined brightness reaches its maximum when the source has its closest projected distance to the lens.

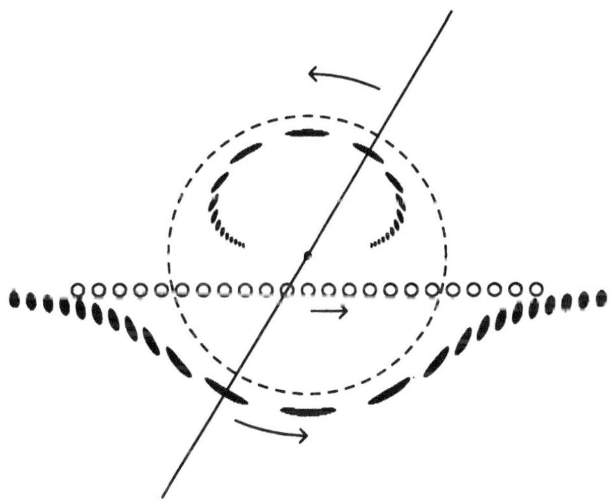

Fig. 11 — As a background source (open circle) moves nearly behind a foreground lens (central dot), the two microimages remain at every moment colinear with the lens and source. (Adapted from Paczyński 1996.)

For a single lens, the combined magnification can be shown from Eqs. 17 and 18 to be:

$$A = \frac{u^2 + 2}{u\sqrt{u^2 + 4}} \quad , \tag{19}$$

where $u \equiv \theta_S/\theta_E$ is the angular source-lens separation in units of the Einstein ring radius. For rectilinear motion, $u(t) = \sqrt{(t - t_0)^2/t_E^2 + u_{min}^2}$, where t_0 is the time at which u is minimum and the magnification is maximum, and $t_E \equiv \theta_E D_L/v_\perp$ is the characteristic time scale defined as the time required for the lens to travel a projected distance across the observer-source sightline equal to the Einstein radius r_E. The result is a symmetric light curve that has a magnification of 1.34 as it cross the Einstein ring radius and a peak amplification that is approximately inversely proportional to the source impact parameter u_{min}. Since the u_{min} are distributed randomly, all of the light curves shown in Fig. 12 are equally probable.

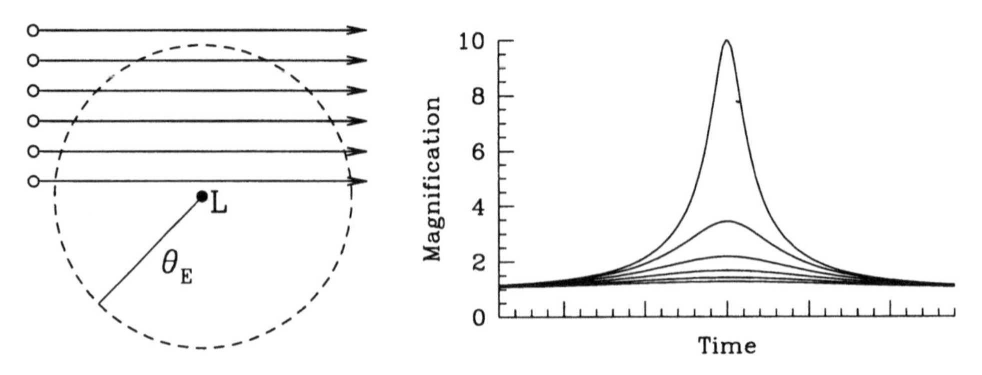

Fig. 12 — **Left:** Equally-probable source trajectories. **Right:** The corresponding single microlens light curves.

Typical event durations $\hat{t} = 2t_E$ for microlensing events detected in the direction of the Galactic Bulge are on the order of a few weeks to a few months, generally matching expectations for stellar lenses distributed in the Galactic disk and bulge.

3.1. MICROLENSING BY BINARY LENSES

Microlensing was proposed as a method to detect compact baryonic dark matter in the Milky Way by Paczyński in 1986. In 1991, Mao and Paczyński suggested that not only dark lenses, but possible dark planets orbiting them may be detected through their microlensing influence on background stars.

The magnification patterns of a single lens are axially symmetric and centered on the lens; the Einstein ring radius, for example, describes the position of the $A \equiv A_1 + A_2 = 1.34$ magnification contour. Binary lens structure destroys this symmetry: the magnification patterns become distorted and are symmetric only upon reflection about the binary axis. Positions in the source place for which the determinant of the Jacobian (Eq. 18) is zero represent potential source positions for which the magnification is formally infinite. The locus of these positions is called a "caustic." For a single point-lens, the only such position is the point caustic at $\theta_S = 0$, but the caustics of binary lenses are extended and complicated in shape. In the lens plane, the condition $|\det J| = 0$ defines a locus of points known as the critical curve; when the source crosses a caustic a pair of new images of high amplification appear with image positions θ on the critical curve.

A static lens configuration has a fixed magnification pattern relative to the lens; the observed light curve is one-dimensional cut through this pattern that depends on the source path. As Fig. 13 illustrates, the exact path of the source trajectory behind a binary lens will determine how much its light curve deviates from the simple symmetric form characterizing a single lens. Due to the finite size of the source, the magnification during a caustic crossing is not infinite, but will be quite large for sources that are small compared to the size of the caustic structure. Several binary-lens light curves have already been observed and characterized (Udalski *et al.* 1994, Alard, Mao & Guibert 1995, Alcock *et al.* 1997, Albrow *et al.* 1998b, Albrow *et al.* 1999).

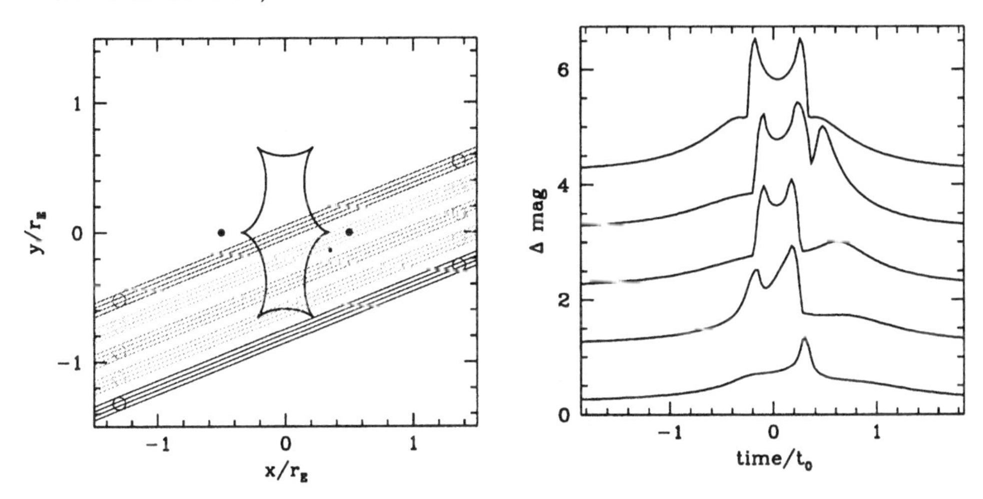

Fig. 13 — **Left:** The caustic (thick closed line) for two equal mass lenses (dots) is shown with several possible source trajectories. Angular distances are scaled to the Einstein ring radius of the combined lens mass. **Right:** The light curves resulting from the source trajectories shown at left; the temporal axis is normalized to the Einstein time t_E for the combined lens. (Adapted from Paczyński 1996.)

A single lens light curve is described by four parameters: the Einstein crossing time t_E, the impact parameter u_{min}, the time of maximum amplification t_0, and the unlensed flux of the source F_0. Only the first of these contains information about the lens itself. Three additional parameters are introduced for binary lenses: the projected separation b of the lenses in units of θ_E, the mass ratio q of the two lenses, and the angle ϕ that the source trajectory makes with respect to the binary axis. Given the large number of free parameters and the variety of complicated forms that binary light curves can exhibit, it may seem quite difficult to characterize the binary lens with any degree of certainty on the basis of a single 1-D cut through its magnification pattern. In fact, with good data the fitting procedure is unique enough that the *future* behavior of the complicated light curve — including the timing of future caustic crossings — can be predicted in real time. This is important since the ability to characterize extra-solar planets via microlensing requires proper determination of the planetary system parameters b and q through modeling of light curve anomalies.

3.2. PLANETARY MICROLENSING

The simplest planetary system is a binary consisting of a stellar lens of mass M_* orbited by a planet of mass m_p at an orbital separation a. The parameter range of interest is therefore $q \equiv M_*/m_p \approx 10^{-3}$ for jovian-mass planets and $q \approx 10^{-5}$ for terrestrial-mass planets. The normalized projected angular separation $b \lesssim a/(\theta_E D_L)$ depends at any moment on the inclination and phase of the planetary orbit. The light curve of a source passing behind a lensing planetary system will depend on the form of the magnification pattern of the lensing system, which is influenced by the size and position of the caustics. How do the magnification patterns vary with b and q?

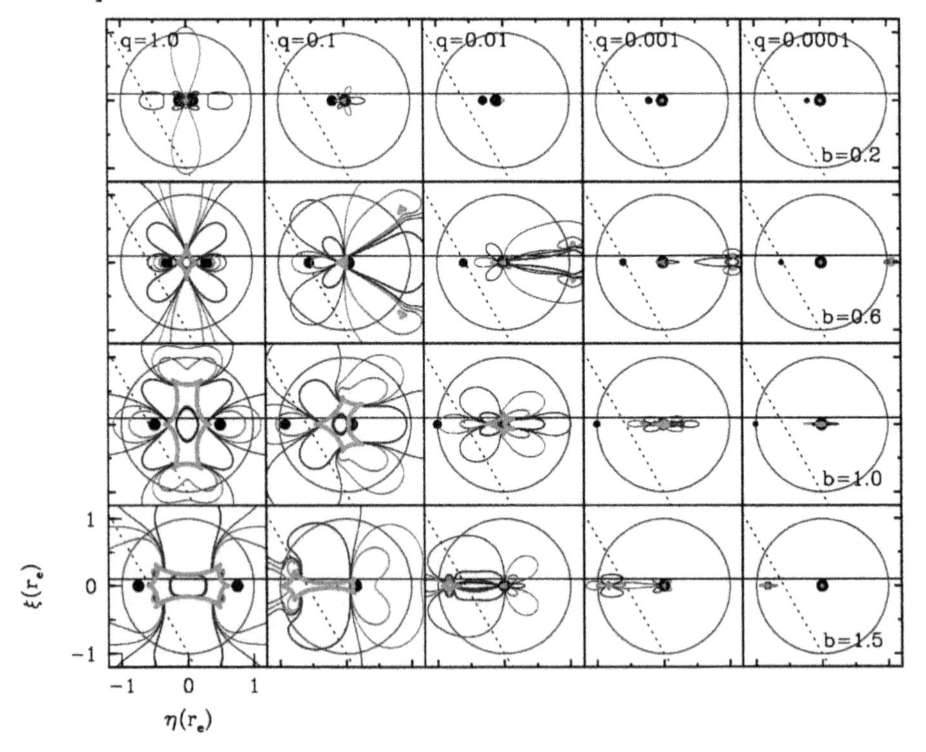

Fig. 14 — Positive (magenta) and negative (blue) 1% and 5% *excess* magnification contours for binary lenses (black dots) of different projected separations b and mass ratios q. Caustics are shown in red. Dimensions are normalized to the Einstein ring radius of combined system (green circle). Dashed and solid lines are two possible source trajectories. (Adapted from Gaudi & Sackett 1998.)

Shown in Fig. 14 is the *excess* magnification pattern of a binary over that of a single lens for different separations b and mass ratios q. The deviations can be positive or negative. High-mass ratio binaries (i.e., q not too much less than 1) are easier to detect since their excess magnification contours cover a larger sky area making it more likely that a source trajectory will cross an "anomalous" region. For a given mass ratio q, the 1% and 5% excess magnification contours also cover more sky when the binary separation is comparable to the Einstein ring radius of the system, i.e., whenever $b \approx 1$.

The symmetric caustic structure centered between equal mass ($q = 1$) binaries becomes elongated along the binary axis for smaller mass ratios, eventually splitting the caustic into a central caustic centered on the primary lens and outer "planetary" caustics. For planetary separations larger than the Einstein ring radius $b > 1$, the planetary caustic is situated on the binary axis between the lens and planet. For $b < 1$, the planetary caustics are two "tips" that are symmetrically positioned above and below the binary axis on the opposite side of the lens from the planet. As the mass ratio decreases, all the caustics shrink in size and the two "tips" approach the binary axis, nearly — but not quite — merging.

3.2.1. *The "Lensing Zone"*

For the planetary (small q) regime, a source that crosses the central caustic will generate new images near the Einstein ring of the primary lens; a source crossing a planetary caustic will generate new images near the Einstein ring of the planet, i.e., near the position of the planet itself. Planets with separations $0.6 \lesssim b \lesssim 1.6$ create planetary caustics inside the Einstein ring radius of the parent lensing star; this is the region in which the source must be in order to be alerted by the microlensing survey teams. For this reason, planets with projected separations $0.6 \lesssim b \lesssim 1.6$ are said to lie in the *"lensing zone."* Since the separation b is normalized to the size of the Einstein ring, the physical size of the lensing zone will depend on the lens mass and on the lens and source distances. Most of the microlensing events in the Milky Way are detected in the direction of the Galactic bulge where, at least for the bright red clump sources, it is reasonable to assume that the sources lie at $D_S \approx 8\,\mathrm{kpc}$. Table I shows the size of the lensing zone for foreground lenses located in the disk ($D_L = 4\,\mathrm{kpc}$) and bulge ($D_L = 6\,\mathrm{kpc}$) for typical stellar masses, assuming that $D_S = 8\,\mathrm{kpc}$.

One of the reasons that microlensing is such an attractive method to search for extra-solar planets is that the typical lensing zone corresponds to projected separations of a few times the Earth-Sun distance (AU) — a good match to many planets in the Solar System. Planets orbiting at a radius a in a plane inclined by i with respect to the plane of the sky will traverse a range of projected separations $a \cos i/(\theta_E\, D_L) < b < a/(\theta_E\, D_L)$, and can thus be brought into the lensing zone of their primary even if their orbital radius is larger than the values given in Table I.

TABLE I. Typical Lensing Zones for Galactic Lenses

	disk lens (4 kpc)	bulge lens (6 kpc)
1.0 $M_\odot$ solar-type	2.4 - 6.4 AU	2.1 - 5.5 AU
0.3 $M_\odot$ dwarf	1.3 - 3.5 AU	1.1 - 3.0 AU

Planets that are seldom or never brought into the lensing zone of their primary can still be detected by microlensing in one of two ways. Either the light curve must be monitored for source positions outside the Einstein ring radius of the primary (i.e., for magnifications $A < 1.34$) in order to have sensitivity to the isolated, outer planetary caustics (DiStefano & Scalzo 1999), or very high amplification events must be monitored in order to sense the deviations that are caused any planet on the central primary caustic (Griest & Safizadeh 1998).

3.2.2. *Determining the Planet-Star Mass Ratio and Projected Separation*
The generation of caustic structure and the anomalous magnification pattern associated with it makes planetary masses orbiting stellar lenses easier to detect than isolated lensing planets. Even so, most planetary light curves will be anomalous because the source passed near, but not across a caustic (Fig. 15). How is the projected planet-star separation b and the planet-star

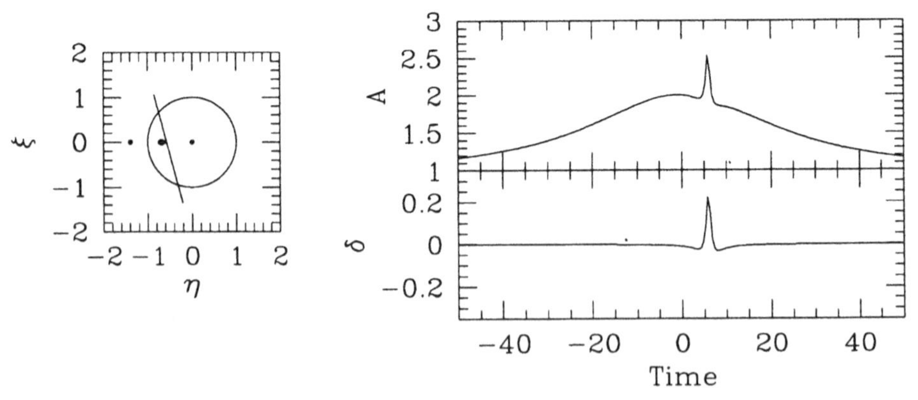

Fig. 15 — **Left:** A background point source travels along a trajectory that just misses the caustic structure caused by a "Jupiter" with mass ratio $q = 0.001$ located at 1.3 Einstein ring radii (several AU) from its parent stellar lens. **Right:** The resulting light curve is shown in the top panel; the excess magnification δ associated with the planetary anomaly is shown in the bottom panel; time scale is in days.

mass ratio $q = m_p/M_*$ extracted from a planetary anomaly atop an otherwise normal microlensing light curve? In practice, the morphology of planetary light curve anomalies is quite complex, and detailed modeling of the excess magnification pattern (the anomalous change in the pattern due to the planet) is required, but the general principles can be easily understood.

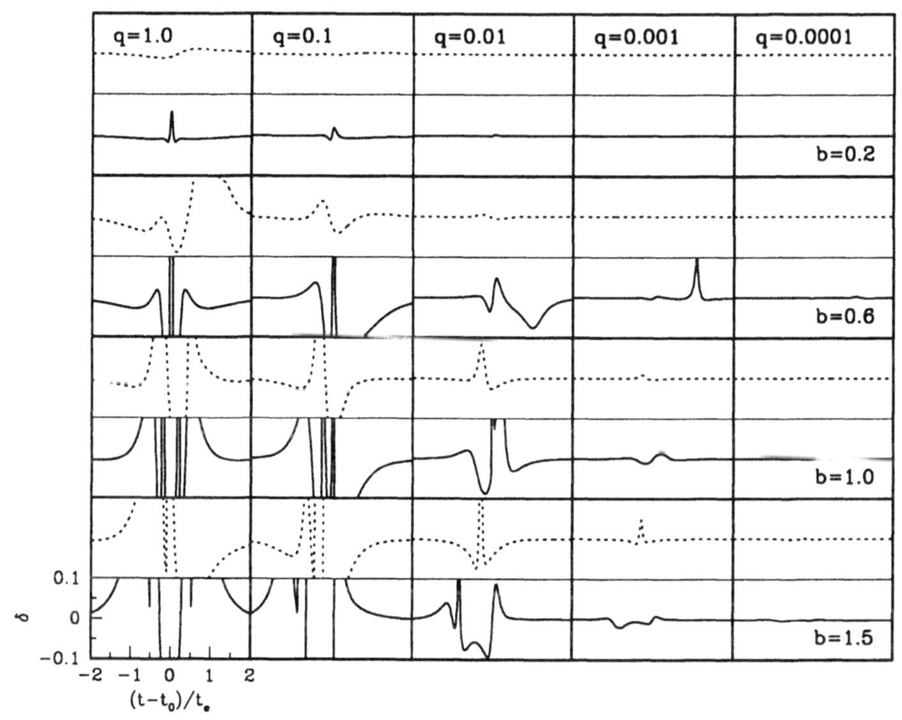

Fig. 16 — Excess magnifications δ for the (solid and dotted) trajectories of Fig. 14 are shown for (the same) range of planetary mass ratios and projected separations. "Super-jupiters" with $q \sim 0.01$ should create detectable anomalies for a significant fraction of source trajectories in high quality light curves. (Adapted from Gaudi & Sackett 1998.)

Since the planet and parent star lenses are at the same distance D_L and travel across the line of sight with the same velocity $v_\perp$ (ignoring the planet's orbital motion), Eq. 16 shows that the mass ratio q is equal to the square of the ratio of the Einstein ring radii $(\theta_p/\theta_E)^2$. Observationally this can be estimated very roughly by the square of the ratio of the planetary anomaly duration to the primary event duration, $(t_p/t_E)^2$. The time difference between the primary and anomalous peaks (normalized to the Einstein time) gives an indication of the placement of the caustic structure within the Einstein ring and thus the position of the planet relative to the primary lens, b. The amplitude of the anomaly $\delta \equiv (A - A_0)/A_0$, where A_0

is the unperturbed amplitude, indicates the closest approach to the caustic structure and, together with the temporal placement of the anomaly, yields the source trajectory angle through the magnification pattern.

Since the magnification pattern associated with planetary caustics for $b > 1$ and $b < 1$ planets is qualitatively different, detailed dense monitoring should resolve any ambiguity in the planetary position. Light curve anomalies associated with $b > 1$ planets, like the one in Fig. 15, will have relatively large central peaks in δ surrounded by shallow valleys; $b < 1$ anomalies will generally have more rapidly varying alterations of positive and negative excess magnification, though individual exceptions can certainly be found.

From the shape of light curve anomalies alone, the mass of the planet is determined as a fraction of the primary lens mass; its instantaneous projected separation is determined as a fraction of the primary Einstein radius. Reasonable assumptions about the kinematics, distribution, and masses of the primary stellar lenses, together with measurements of the primary event duration $2t_E$ and fraction of blended light from the lens should allow r_E and M_* to be determined to within a factor $\sim 3-5$. Detailed measurements of the planetary anomaly would then yield the absolute projected separation and planetary mass to about the same precision.

3.2.3. *Durations and Amplitudes of Planetary Anomalies*

It is clear from Figs. 14 and 16 that, depending on the source trajectory, a range of anomaly durations t_p and amplitudes δ are possible for a planetary system of given q and b (see also Wambsganss 1977). Nevertheless, rough scaling relations can be developed to estimate the time scales and amplitudes that will determine the photometric sampling rate and precision required for reasonable detection efficiencies to microlensing planetary systems.

For small mass ratios q, the region of excess magnification associated with the planetary caustic is a long, roughly linear region with a width approximately equal to the Einstein ring of the planet, θ_p, and a length along the planet-lens axis several times larger. Since $\theta_p = \sqrt{q}\,\theta_E$, both the time scale of the duration and the cross section presented to a (small) source vary linearly with θ_p/θ_E and thus with $\sqrt{q}$. Assuming a typical $t_E = 20$ days, the duration of the planetary anomaly is given roughly by the time to cross the planetary Einstein diameter, $2\,\theta_p$,

$$\text{planet anomaly duration } = 2\,t_p \approx 1.7\,\text{hrs}\,(m/\mathrm{M}_\oplus)^{1/2}(\mathrm{M}/\mathrm{M}_\odot)^{-1/2}. \quad (20)$$

Caustic crossings can occur for any planetary mass ratio and should be easy to detect as long as the temporal sampling is well matched to the time scales above. Most anomalies, however, will be more gentle perturbations associated with crossing lower amplitude excess magnification contours. At the most favorable projected lens-planet separation of $b = 1.3$, and the most ideal lens location (halfway to the Galactic Center), well-sampled observations able to detect 5% perturbations in the light curve would have planet sensitivities given roughly by (Gould & Loeb 1992):

$$\text{ideal detection sensitivity } \approx 1\% \; (m/\mathrm{M}_\oplus)^{1/2}(\mathrm{M}/\mathrm{M}_\odot)^{-1/2} \qquad (21)$$

This ideal sensitivity is relevant only for planets at $b = 1.3$; at the edges of the lensing zone the probabilities are about halved. Detection with this sensitivity requires photometry at the 1% level, well-sampled over the duration of the planetary event.

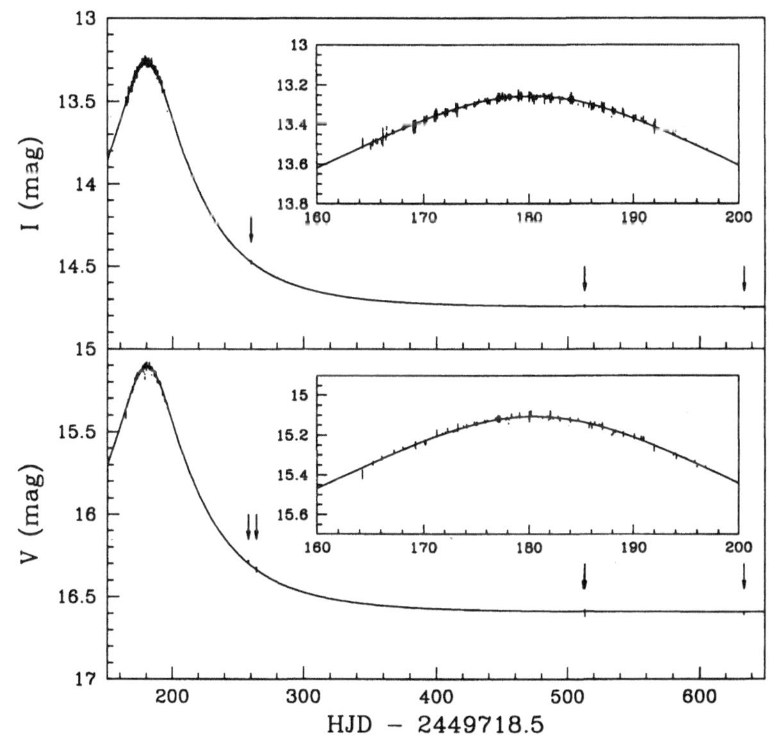

Fig. 17 — PLANET collaboration monitoring of MACHO-BLG-95-13 in the I (upper) and V (lower) bands. Insets show a zoom around the peak of the event; arrows indicate points taken many months to more than a year later. Vertical scale is magnitudes; horizontal scale is days (Albrow *et al.* 1998a).

Can such photometric precision and temporal sampling be obtained in the crowded fields of the Galactic bulge where nearly all microlensing events are discovered? Fig. 17 shows observations of one bright microlensing event monitored by the PLANET collaboration during its month long pilot season in 1995 (Albrow *et al.* 1998a). The residuals from the single point-lens/point source light curve are less than 1% for this event, and the typical sampling rate is on the order of once every 1-2 hours, even accounting for unavoidable longitudinal and weather gaps.

A true calculation of detection probabilities must integrate over the projected orbital separations and the distribution of lenses along the line of sight, must take into account the actual distribution of source trajectories probed by a particular set of observations, and the effect of uneven temporal sampling and photometric precision (Gaudi & Sackett 1998). In the following section we discuss the additional complication of finite source effects that is encountered for very small mass planets for which the size of the planetary Einstein ring is comparable to or smaller than the source size, $\theta_p \lesssim \theta_*$.

3.3. OBSERVATIONAL REWARDS AND CHALLENGES

What can be learned by observing a planetary anomaly in a microlensing light curve? The duration, temporal placement relative to the event peak, and relative amplitude of the anomaly can be used to determine the mass ratio q of the planetary companion to the primary (presumably stellar) lens and their projected angular separation b in units of the Einstein ring radius θ_E. Since in general the lens will be far too distant to type spectrally against the bright background source (except possibly with very large apertures, see Mao, Reetz & Lennon 1998), the absolute mass and separation must be determined statistically by fitting the properties of an ensemble of events with reasonable Galactic assumptions. Measurements of other sorts of microlensing anomalies associated with source resolution, observer motion, or lens blending can produce additional constraints on the lens properties and thus on the absolute planetary characteristics.

Except for very large a planets orbiting in nearly face-on ($i \approx 0$) orbits, cooperative lensing effects between the lens and companion boost the detectability of lensing planets over that expected for planets in isolation. Since current detection and monitoring schemes focus on those events with an essentially random distribution of impact parameters u_{min} for $u_{min} < 1$, the anomaly sensitivity is primarily restricted to planets in the "lensing zone" with projected separations of 0.6 — 1.6 times the Einstein ring radii of the primary lens. For typical distributions of lens masses, and lens and source distances, this translates into the highest probabilities for planets with instantaneous orbital separations *projected onto the sky plane* of $a_p \approx 5\,\mathrm{AU}$, with a zone of reduced detectability extending to higher a_p. Since the efficiency of planetary detection in these zones is likely to be a few or a few tens of percent (Eq. 21), many microlensing events must be monitored with ~1% photometry on the ~hourly time scales (Eq. 20) to achieve statistically meaningful constraints on the number planets in the Milky Way, and their distribution in mass and orbital radius.

What limits the photometric precision and temporal sampling? Round-the-clock monitoring of events, necessary for maximum sensitivity to the 1 – 2 day durations of Jupiter-mass anomalies requires telescopes scattered in geographical longitude, at the South Pole, or in space. Temporal sampling is limited by the number of events monitored at any given time, their brightness, and the desired level of photometric precision. Higher signal-to-noise can generally be obtained for brighter stars in less exposure time, but ultimately, in the crowded fields that typify any microlensing experiment, photometric precision is limited by confusion from neighboring sources, not photon statistics. Pushing below ~1% relative photometry with current techniques has proven very difficult in crowded fields.

If an anomaly is detected, it must be distinguished from other intrinsic effects that could be confused with a lensing planet. Stellar pulsation on daily to sub-daily time scales in giant and sub-giant bulge stars is unlikely, but this and any form of regular variability would easily be recognized as periodic (i.e., non-microlensing) with the dense and precise sampling that is required in microlensing monitoring designed to detect planets. Star-spot activity may be non-negligible in giants, but will have a time scale characteristic of the rotation period of giants, and thus much longer than typical planetary anomalies. In faint dwarf stars spotting activity produces flux changes below the photometric precision of current experiments. Flare activity should not be significant for giants and is expected to be chromatic, whereas the microlensing signal will always be achromatic (except in the case of source resolution by exceedingly low-mass planets).

Blending (complete photometric confusion) by random, unrelated stars along the line-of-sight can dilute the apparent amplitude of the primary lensing event. This will have some effect on the detection efficiencies, but most significantly — with data of insufficient sampling and photometric precision — will lead to underestimates for the time scale t_E and impact parameter u_{min} of the primary event, and thus also to mis-estimates of the planetary mass ratio q and projected separation b.

3.3.1. *Pushing the Limits: Earth-mass and Multiple Planets*

Planets with very small mass ratio will have caustic structure smaller than the angular size of typical bulge giants. The ensuing dilution of the anomaly by finite source effects will present a large, but perhaps not insurmountable, challenge to pushing the microlensing planet detection technique into the regime of terrestrial-mass planets (Peale 1997, Sackett 1997).

Near small-mass planetary caustics, different parts of the source simultaneously cross regions of significantly different excess magnification; an integration over source area is required in order to derive the total magnification. The severity of the effect can be seen in Fig. 18. Earth-mass caustic crossings against even the smaller-radii bulge sources will present a challenge to current photometry in crowded fields, which is generally limited by seeing to stars above the turn-off mass.

The most numerous, bright microlensing sources in the bulge are clump giants with radii about 13 times that of the Sun (13 $R_\odot$), and thus angular radii of 7.6 microarcseconds (μas) at 8 kpc. Since a Jupiter-mass planet with $m_p = 10^{-3}$M$_\odot$ has an angular Einstein ring radius of 32 μas at 4 kpc and 19 μas at 6 kpc, its caustic structure is large compared to the size of the source. An Earth-mass planet with $m = 3 \times 10^{-6}$M$_\odot$, on the other hand, has an angular Einstein ring radius of 1.7 μas at 4 kpc and 1 μas at 6 kpc, and will thus suffer slight finite source effects even against turn-off stars (1.7 μas), though the effect will be greatly reduced compared to giant sources (Bennett & Rhie 1996).

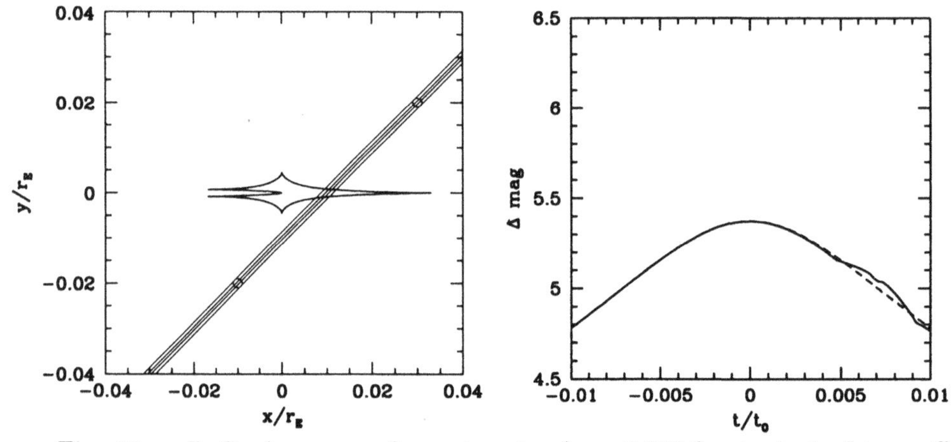

Fig. 18 — **Left:** A source of angular size $\theta_* = 0.001\,\theta_E$, typical of turn-off stars in the bulge, crosses the central caustic caused by a terrestrial-mass planet with mass ratio $q = 10^{-5}$. *Right:* Due to source resolution effects, the resulting anomaly differs from single-lens microlensing only at the ~1% – 3% level. Note the expanded spatial and temporal scales. (Adapted from Paczyński 1996.)

For extreme source resolution, in which the entire planetary caustic lies inside the projected source radius, the *excess fractional* magnification associated with the planetary anomaly scales with the square of the ratio of the planetary Einstein ring radius to the angular source size, $\delta \propto (\theta_p/\theta_*)^2$. On the other hand, source-resolved small q anomalies will have longer durations than implied by Eq. 20, since the characteristic time scale is the time to cross the source θ_* (not θ_p). Furthermore, the cross section for magnification at a given threshold now roughly scales with θ_*/θ_E (not θ_p/θ_E), and is thus approximately independent of planetary mass.

Because the anomaly amplitude is suppressed by source resolution, unless the photometry is excellent and continuous, small-mass planetary caustic crossings can be confused with large impact parameter large-mass planetary anomalies. This degeneracy can be removed by performing multi-band observations (Gaudi & Gould 1997): large impact parameter events will be achromatic, but sources resolved by small-mass caustics will have a chromatic signal due to source limb-darkening that is similar to (but of opposite sign from) that expected for planetary transits (§2.3.1). Source limb-darkening and chromaticity have now been observed during a caustic crossing of a stellar binary (Albrow *et al.* 1998b).

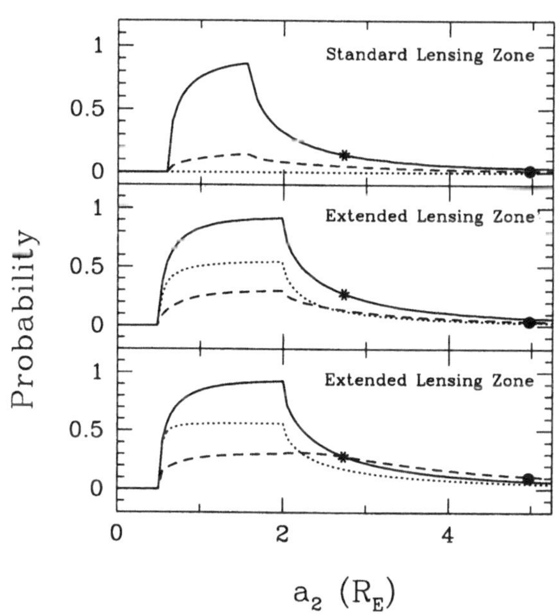

Fig. 19 — **Top:** Probability that two planets with true orbital radii a_1 and a_2 (in units of the Einstein ring r_E) simultaneously have projected separations, b_1 and b_2, in the standard "lensing zone," defined as $0.6 < b < 1.6$. The probability is shown as function of a_2, for fixed $a_1 = 1.5$ (solid), $a_1 = 0.6$ (dotted) and $a_1 = 2.7$ (dashed). The probability for two planets with orbital radii of Jupiter and Saturn around solar-mass primary (star) and a $0.3 M_\odot$ primary (dot) are shown. **Middle:** Same, but for the extended "lensing zone," $0.5 < b < 2.0$. **Bottom:** The conditional probability that both b_1 and b_2 lie in the extended "lensing zone," given that either b_1 or b_2 satisfies this criterion. (Gaudi, Naber & Sackett 1998.)

Finally, since all planetary lenses create a central caustic, low-impact parameter (high magnification) microlensing events that project the source close to the central caustic are especially efficient in producing detectable planetary anomalies (Griest & Safizadeh 1998). For the same reason, however, the central caustic is affected by *all* planets in the system, and so — if possible degeneracies due to the increased caustic complexity can be removed — rare, low impact parameter events offer a promising way of simultaneously detecting multiple planets brought into or near the lensing zone by their orbital motion around the primary lens (Gaudi, Naber & Sackett 1998). As Fig. 19 demonstrates, the statistical probabilities are large that a Jupiter or 47UMa orbiting a solar-mass star (solid and dotted lines, respectively) will instantaneously share the lensing zone with other planets of orbital radii of several AU. However, the light curves resulting from crossing a multiple-planet central caustic may be difficult to interpret since the caustic structure is so complicated (Fig. 20).

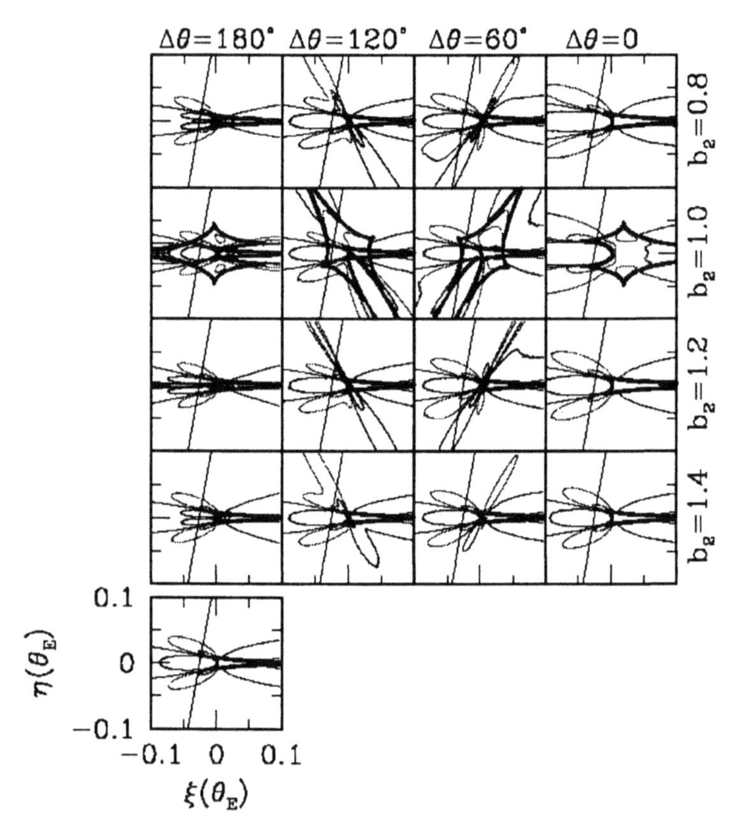

Fig. 20 — Contours of 5% and 20% fractional deviation δ, as a function of source position in units of the θ_E. The parameters of planet 1 are held fixed at $q_1 = 0.003$, $b_1 = 1.2$; the projected separation b_2 and the angle between the axes, $\Delta\theta$, are varied for a second planet with $q_2 = 0.001$. Only planet 1 is present in the bottom offset panel. Positive, negative, and caustic ($\delta = \infty$, thick line) contours are shown. (Gaudi, Naber & Sackett 1998.)

4. Photometric Mapping of Unseen Planetary Systems: Matching the Tool to the Task

Both the transit and microlensing techniques use frequent, high-precision monitoring of many stars to discover the presence of the unseen extra-solar planets. The transit method monitors the light from the parent star in search of occultation by an unseen planet; the microlensing technique monitors light from an unrelated background source in search of lensing by an unseen planet orbiting an unseen parent star. Indeed, microlensing is the only extra-solar planetary search technique that requires *no photons from either the planet or the parent star* and for this reason is the method most sensitive to the distant planetary systems in our Galaxy.

The two techniques are complementary, both in terms of the information that they provide about discovered systems, and in terms of the range of planetary system parameters to which they are most sensitive. Multiple transit measurements of the same planet will yield its planetary radius R_p and orbital radius a. Characterization of a microlensing planetary anomaly gives the mass ratio $q \equiv m_p/M_*$ of the planet to lensing star and their projected separation b at the time of the anomaly in units of θ_E.

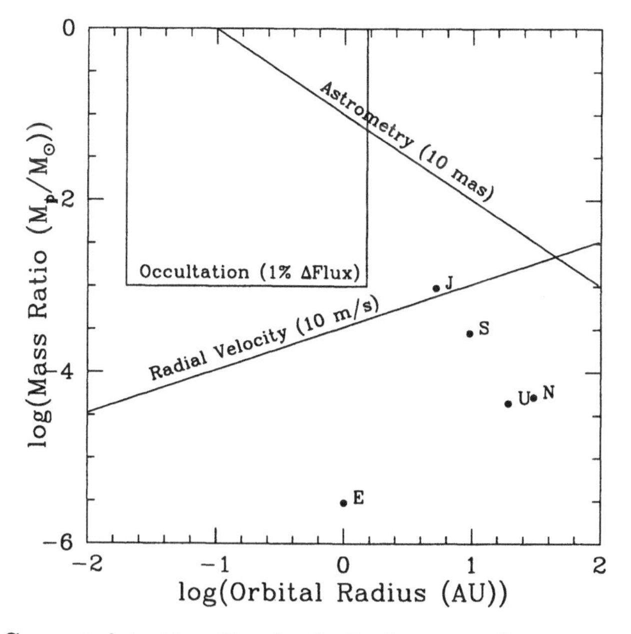

Fig. 21 — Current detection thresholds for long-running programs that rely on planet orbital motion, shown as a function of planetary mass ratio and orbital radius. The occultation threshold must be multiplied by the appropriate geometric probability of a transit to derive detection efficiencies. Selected Solar System planets are shown.

Current ground-based photometry is sensitive to jovian-size occultations; space-based missions may extend this into the terrestrial regime. The transit method is sensitive to planets with small a because they create a detectable transit over a wider range of inclinations of their orbital planes, and because they transit more often within a typical 5-year experiment. These constraints limit the range of orbital radii to about $0.02 \lesssim a \lesssim 1.5\,\mathrm{AU}$ for jovian-size planets. If improvements in photometric precision would allow the detection of Earth-size planetary transits, this range would still be possible, but would suffer from noise due to star spot activity on time scales that could be confused with transits by small-mass planets with $a \lesssim 0.3\,\mathrm{AU}$.

Since the planets in own solar system fall roughly on the same $\log m_p$-$\log R_p$ relationship, it is reasonable to assume that jovian-size planets may also have jovian masses. This assumption was used to place the current transit detection capability on the same plot (Fig. 21) with the current detection thresholds for radial velocity and astrometric techniques. All three of these techniques require long-term projects to detect long-period (large a) planets since the measurements of velocity, position, or flux must be collected over at least one full orbital period. The occultation threshold must be convolved with the geometric transit probability to derive efficiencies *per observed star*.

Photometric precision in crowded fields together with source resolution effects limit current microlensing planet searches to Neptune masses and above. The actual efficiency with which a given planetary system can be detected depends on its mass ratio and projected orbital separation. Fig. 22 shows estimates of microlensing detection efficiency contours for planets of a given mass ratio and true orbital separation (in units of the Einstein radius). The contours are based on the work of Gaudi & Gould (1997) for high-mass ratios, Gould & Loeb (1992) and Gaudi & Sackett (1998) for intermediate mass ratios, and Bennett & Rhie (1996) for small ratios. Integrations have been performed over the unknown but presumably randomly oriented inclinations and orbital phases. Although planets with a in the lensing zone and orbiting in the sky plane are the easiest to detect, a tail of sensitivity extends to larger a as well because inclination effects will bring large-a planets into the projected lensing zone for some phases of their orbits. The efficiencies assume $\approx$1% photometry well-sampled over the post-alert part of the microlensing light curve.

Examination of Fig. 22 makes it clear that different indirect planetary search techniques will are sensitive to different portions of the $\log m_p$-$\log a$ domain. Current ground-based capabilities favor the radial velocity method for short-period ($a \lesssim 3$ AU) planets (see also Queloz, this proceedings). The occultation method will help populate the short-period part of the diagram, and if the programs are carried into space, will begin to probe the regime of terrestrial-sized planets in terrestrial environments. Ground-based astrometry is favored for very long-period ($a \gtrsim 40$ AU) planets, although the time scales for detection and confirmation are then on the order of decades. Space-based astrometry promises to make this method substantially more efficient, perhaps by a factor of 100 (see also Colavita, this proceedings). Microlensing is the only technique capable of detecting in the near term substantial numbers of intermediate-period ($3 \lesssim a \lesssim 12$ AU) planets. Somewhat longer period planets may also be discovered by microlensing survey projects as independent "repeating" events in which the primary lens and distant planet act as independent lenses (DiStefano & Scalzo 1999). Very short-period planets interior to 0.1 AU may be detectable using the light echo technique (Bromley 1992, Seager & Sasselov 1998, Charbonneau, Jhu & Noyes 1998), at least for parent stars with substantial flare activity, such as late M dwarfs.

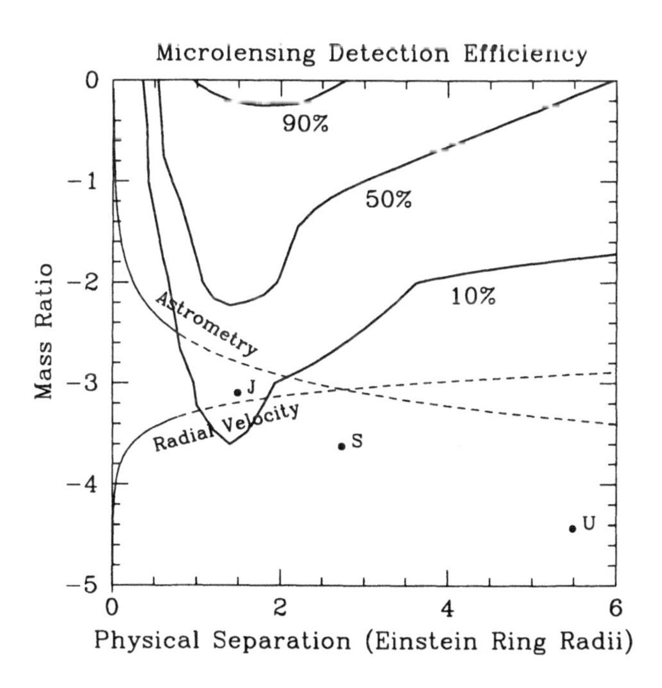

Fig. 22 — Estimated detection efficiency contours for microlensing planet searches as a function of the logarithm of the planetary mass ratio $q \equiv m_p/M_*$ and the true orbital separation a in units of the Einstein ring radius. Efficiencies have been integrated over the phase and inclination of the orbits, under the assumption that they are circular. To make comparisons with other techniques, the Einstein ring radius is taken to be 3.5 AU. Solid lines indicate what can be achieved in an observational program of 5-years duration or less. Note that the vertical scale remains logarithmic, but the horizontal scale is now linear.

The techniques are complementary in another sense as well. Those that rely on light from the parent star will be limited to nearby planetary systems, but will benefit from the ability to do later follow-up studies, including spectroscopy and interferometry. Microlensing, on the other hand, will see the evidence of a given planetary system only once, but can probe planetary frequency in distant parts of the Milky Way, and can collect statistics over a wide range of orbital separations in a relatively short time.

TABLE II. Comparison of Current Ground-Based Capabilities

	OCCULTATION	MICROLENSING
Parameters Determined	R_p, a, i	$q \equiv m_p/M_*$, $b \equiv a_p/R_E$
Photometric Precision of	0.1%	1%
Limits R_p or m_p to	Neptune	Neptune
Orbital Radius Sensitivity	$\sim 0.02 - 1.5\,\mathrm{AU}$	$\sim 1 - 12\,\mathrm{AU}$
Typical Distance of Systems	$< 1\,\mathrm{kpc}$	$4 - 7\,\mathrm{kpc}$
Number of Stars to be Monitored	few 10^3	few 10^2
for Meaningful Jovian Sensitivity at	$\lesssim$1 AU	$\sim$5 AU
In Principle Possible to Detect:		
Multiple Planets	yes	yes
Planets around Binary Parent Stars	yes	yes
Earth-mass Planets in Future	yes (space)	yes

4.1. TOWARD THE FUTURE

The field of extra-solar planets is evolving rapidly. The number of groups conducting transit and microlensing planet searches, planning future programs, and providing theoretical support is growing at an ever-increasing rate. For that reason, this series of lectures has centered on the principles of the techniques rather than reviewing the current players. In order to help the reader keep pace with this accelerating activity, however, a list of relevant Internet Resources with links to occultation and microlensing planet search groups is provided at the end of this section. What can we expect in the next decade from these research teams?

Several ground-based transit searches are already underway (Villanova University, TEP, WASP, Vulcan, and EXPORT). Some focus on high-quality photometry of known eclipsing binaries. This is likely to increase the probability of transits — if planets exist in such binary systems. Two transit-search teams recently issued (apparently contradictory) claims for a possible planet detection in the eclipsing M-dwarf system known as CM Draconis (IAU circulars 6864 and 6875, see also Deeg *et al.* 1998), but no clear, undisputed planetary signal has seen. One class of planets known to exist in reasonable numbers and also relatively easy to detect via occultation is the "hot jupiter;" the planet in 51 Peg is a prototype of this class. If such a planet is the size of Jupiter, its orbital plane would have a $\sim$10% chance of being sufficiently inclined to produce a detectable eclipse of a solar-type parent as seen from Earth. An aggressive ground-based program should be able to detect large numbers of such planets in the next decade — planets that could be studied with the radial velocity technique thereby yielding both planetary mass and radius. Space-based missions (COROT and KEPLER) planned for launch within this decade should have the sensitivity to detect transits from terrestrial-mass objects, but in order to detect Earth-like planets in Earth-like environments (i.e., orbiting solar-type stars at 1 AU) they will need long mission times.

Microlensing planet searches are being conducted or aided by international collaborations (PLANET, GMAN, MPS, MOA, and EXPORT) that intensely monitor the events discovered by microlensing search teams (EROS, MACHO, OGLE, and MOA). MACHO and OGLE electronically alert on-going microlensing in the direction of the Galactic Bulge on a regular basis: at given time during the bulge observing season several events are in progress and available for monitoring. Both the PLANET and GMAN collaborations have issued real-time secondary alerts of anomalous behavior (including binary lenses, source resolution, " lensing parallax"), but to date no clear detection of a lensing planet has been announced. Especially if caustics are crossed, it may be possible to obtain additional information on microlensing planets from the sky motion of the caustics during the event that is induced by planetary motion (Dominik 1998). The number of high-quality microlensing light curves monitored by the PLANET collaboration is already beginning to approach that required for reasonable jovian detection sensitivities (Albrow *et al.* 1998a), so meaningful results on Jupiter look-alikes can be expected within the next few years.

As more telescopes, more telescope time, and wider-field detectors are dedicated to dense, precise photometric monitoring capable of detecting planetary transits and planetary microlensing, we can feel certain that — if jovian planets with orbital radii less than $\sim$6 AU exist in sufficient numbers — they will be detected in the next few years by these techniques.

ACKNOWLEDGMENTS

I am grateful to NATO for financial support and to the Institute's efficient and gracious scientific organizers, Danielle Alloin and (the sorely missed) Jean-Marie Mariotti, for a productive and pleasant school. It is also a pleasure to thank B. Scott Gaudi for assistance in the preparation of some of the figures in the microlensing section and for permission to show the results of our work before publication.

INTERNET RESOURCES

General Extra-Solar Planet News:

Extrasolar Planets Encyclopedia (maintained by J. Schneider):
http://www.obspm.fr/departement/darc/planets/encycl.html

and the mirror site in the U.S.A.:
http://cfa-www.harvard.edu/planets/

Occultation:

EXPORT: http://pollux.ft.uam.es/export/

TEP: http://www.iac.es/proyect/tep/tephome.html

Villanova University: http://www.phy.vill.edu/astro/index.htm

VULCAN: http://www.iac.es/proyect/tep/tephome.html

WASP: http://www.psi.edu/ esquerdo/wasp/wasp.html

Microlensing:

EROS: http://www.lal.in2p3.fr/EROS

MACHO: http://wwwmacho.anu.macho.edu

MACHO Alert Page: http://darkstar.astro.washington.edu

OGLE: http://www.astrouw.edu.pl/~ftp/ogle

MOA: http://www.phys.vuw.ac.nz/dept/projects/moa/index.html

MPS: http://bustard.phys.nd.edu/MPS/

PLANET: http://www.astro.rug.nl/~planet

References

1. Alard, C., Mao, S. & Guibert, J. 1995, *Astron. Astrophysics*, 300, L17
2. Albrow et al. 1998a, *Astrophys. J.*, 509, 000, astro-ph/9807299
3. Albrow et al. 1998b, *Astrophys. J.*, in preparation
4. Albrow et al. 1999, *Astrophys. J.*, 512, 000, astro-ph/9807086
5. Alcock, C. *et al.* 1997, *Astrophys. J.*, 479, 119
6. Bennett, D. & Rhie, S. H. 1996, *Astrophys. J.*, 472, 660
7. Borucki, W.J. & Summers, A. L. 1984, *Icarus*, 58, 121
8. Borucki, W.J., Scargle, J.D., Hudson, H.S. 1985, *Astrophys. J.*, 291, 852
9. Bromley, B.C. 1992, *Proc. Astron. Soc. Pacific*, 104, 1049
10. Butler, P., & Marcy, G. 1996, *Astrophys. J.*, 464, L15
11. Charbonneau, D., Jha, S. & Noyes, R.W. 1998, *Astrophys. J.*, 507, L153
12. Colavita M. 1998, this proceedings.
13. Deeg, H.J. *et al.* 1998, *Astron. Astrophysics*, 338, 479
14. Dominik, M. 1998, *Astron. Astrophysics*, 329, 361
15. Di Stefano, R., & Scalzo, R. 1999, *Astrophys. J.*, 512, 000, astro-ph/9810147
16. Einstein, A. 1936, *Science*, 84, 506
17. Gaudi, B. S., & Gould, A. 1997, *Astrophys. J.*, 486, 85
18. Gaudi, B.S., Naber, R.M. & Sackett, P.D. 1998, *Astrophys. J.*, 502, L33 37
19. Gaudi, B.S. & Sackett, P.D. 1998, in preparation
20. Hale, A., & Doyle, L.R. 1994, *Astro. and Space Sci.*, 212, 335
21. Henry, G.W., Baliunas, S.L., Donahue, R.A., Soon, W.H. & Saar, S.H. 1997, *Astrophys. J.*, 474, 503
22. Gould, A., & Loeb, A. 1992, *Astrophys. J.*, 396, 104
23. Griest, K., & Safizadeh, N. 1998, *Astrophys. J.*, 500, 37
24. Laurent, E. & Schneider J. 1998, this proceedings.
25. Mao, S., Reetz, J. & Lennon, D.J. 1998, *Astron. Astrophysics*, 338, 56
26. Mao, S., & Paczyński, B. 1991, *Astrophys. J.*, 374, 37
27. Mayor, M., & Quelóz, D. 1995, *Nature*, 378, 355
28. Paczyński, B. 1986, *Astrophys. J.*, 304, 1
29. Paczyński, B. 1996, *Ann. Rev. Astron. & Astrophysics*, 34, 419
30. Queloz, D. 1998, this proceedings.
31. Rosenblatt, F. 1971, *Icarus*, 14, 71
32. Peale, S. J. 1997, *Icarus*, 127, 269
33. Sackett, P.D. 1997, Final Report of the *ESO Working Group on the Detection of Extrasolar Planets*, astro-ph/9709269
34. Schneider, J. 1994, *Planet. Space Sci.*, 42, 539
35. Schneider, J. 1997, Final Report of the *ESO Working Group on the Detection of Extrasolar Planets*
36. Seager, S. & Sasselov, D., 1998, *Astrophys. J.*, 502, L157
37. Sturve, O. 1952, *The Observatory*, 72, 199
38. Udalski, A. *et al.* 1994, *Astrophys. J.*, 436, L103
39. Wambsganss, J. 1997, *Mon. Not. Royal Astron. Soc.*, 284, 172

INDIRECT SEARCHES: DOPPLER SPECTROSCOPY AND PULSAR TIMING

D. QUELOZ
Observatoire de Genève
51 ch des Maillettes
1290 Sauverny
Switzerland †

Abstract. The search for planetary systems by monitoring the motion of stellar objects along the line-of-sight is, up to now, the most efficient and successful way to detect planets. For millisec-pulsars, the measurements of pulse chronometry is accurate enough to detect orbiting Earth-like planets. For Sun-like stars, with specific instrumentation and fine-tuned reduction techniques, giant planets can be detected. A general description of the iodine cell and the simultaneous thorium techniques are given as examples.

The interpretation of the observed radial velocity changes as a gravity effect is not unique. Many other phenomena, intrinsic to stellar atmospheres, may mimic a planetary signature. Stellar activity and stellar pulsations effects are discussed in details in that context.

Finally, a summary of the main planetary searches is presented. A prospective is made about expected results for the next millenium.

1. Measurements of orbit parameters from the primary star motion along the line of sight.

In the plane of the orbital motion, an elliptic orbit may be described by:

$$r(\nu) = \frac{a(1-e^2)}{1+e\cos\nu}, \tag{1}$$

where a is the semi–major axis, e the eccentricity, ν the position angle of the object measured from periastron.

†Now Visiting Scientist at JPL, 4800 Oak Grove Drive, Pasadena, CA 91109, USA

J.-M. Mariotti and D. Alloin (eds.), Planets Outside the Solar System: Theory and Observations, 229–247.

We are only interested in the z motion (along the line of sight) of the object. With i the angle between the normal to the orbital plane and the line of sight, $z(t)$ can be written:

$$z(t) = r\sin(\nu(t) + \omega)\sin i, \tag{2}$$

where ω is the position of the periastron measured in the orbital plane from the intersection between the perpendicular plan to the line of sight and the orbital plane (line of nodes) (see on Fig. 1).

Figure 1. Orbital motion and key parameters for the orbit description.

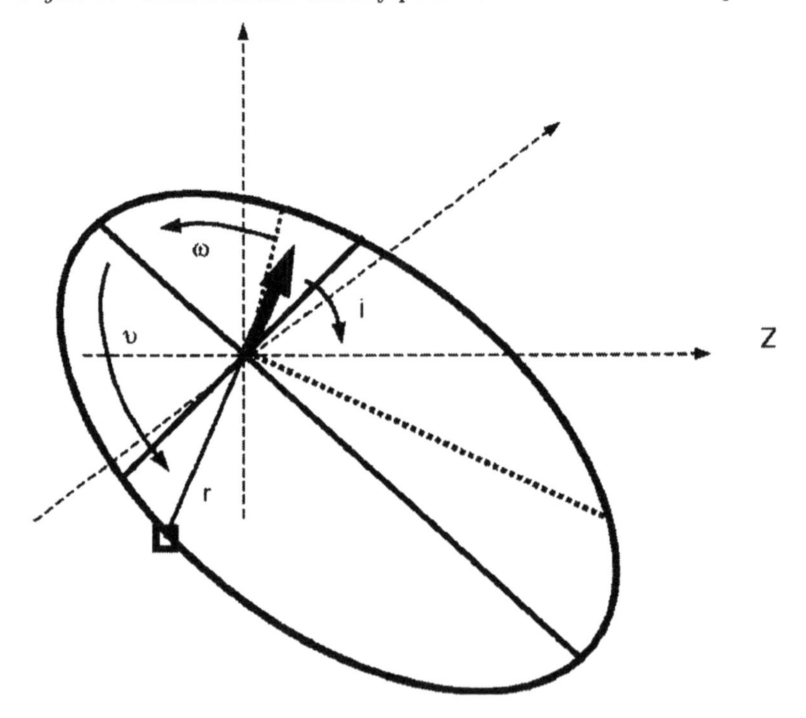

Using Kepler's second law: $r^2 \mathrm{d}\nu/\mathrm{dt} = 2\pi \mathrm{P}^{-1}\mathrm{a}^2\sqrt{1-\mathrm{e}^2}$, we compute the radial velocity $V_r = \mathrm{dz/dt}$ and one gets:

$$V_r = K_j(\cos(\nu + \omega) + e\cos\omega), \tag{3}$$

where K_j is the radial velocity amplitude of the j-th component of the system:

$$K_j = \frac{2\pi}{P}\frac{a_j \sin i}{\sqrt{1-e^2}}. \tag{4}$$

For a two-body system, we have $a^3(m_1+m_2)^{-1} = P^2G(4\pi^2)^{-1}$, $a = a_1 + a_2$ and $m_1a_1 = m_2a_2$, thus we can relate the mass of the secondary with the amplitude of the primary radial velocity orbit. With $a_1{}^3 = P^2G(4\pi^2)^{-1}m_2{}^3(m_1+m_2)^{-2}$, we get the expression:

$$\frac{(m_2 \sin i)^3}{(m_1+m_2)^2} = \frac{P}{2\pi G}K_1{}^3(1-e^2)^{3/2}. \quad (5)$$

The left-hand side of this equation is known as the mass function of the system. From the sole radial velocity orbit one cannot measure the individual masses of objects but only an expression combining the two. However when $m_2 \ll m_1$ and when m_1 can be estimated one gets the measurement of the mass of the secondary times a projection factor ($\sin i$):

$$m_2 \sin i \approx \left(\frac{P}{2\pi G}\right)^{1/3} K_1 m_1^{2/3}\sqrt{1-e^2}. \quad (6)$$

In this low mass approximation regime, we also have an expression linking the period with the semi-major axis of the secondary orbit:

$$a_2 = \left(\frac{G}{4\pi^2}\right)^{1/3} P^{2/3}m_1{}^{1/3}. \quad (7)$$

Rewriting Equ. (6) in more straightforward units, for a circular orbit ($e = 0$), one has:

$$m_2 \sin i \; [\mathrm{M_J}] \approx 3.5 \cdot 10^{-2} K \; [\mathrm{m\,s^{-1}}] \, P^{1/3} \; [\mathrm{yr}]. \quad (8)$$

For a Jupiter mass companion detected with an edge-on orbit ($\sin i = 1$) the amplitude of the V_r effect is $K = 12.5\,\mathrm{m\,s^{-1}}$.

In the low mass approximation, the $\sin i$ uncertainty on m_2 measurement is not as sever as one may think. It turns out that it is far more likely to see an edge-one orbit rather than a face-on system. By looking at all possible configurations, one sees that the 2π azimuthal possible range is weighted by the aera of the solid angle traced by the normal vector to the orbital plan. The solid angle is much larger for a given $\mathrm{d}i$ when i is close to 90° than when i is close to 0. From simple geometric arguments, one easily gets the angle probability $P(i, i+\mathrm{d}i) = \int_i^{i+\mathrm{d}i} P(i)\mathrm{d}i = \cos(i) - \cos(i+\mathrm{d}i)$. Therefore, considering random orientations, 87% configurations have $\sin i$ larger than 1/2. This is the reason why we often consider the $m_2 \sin i$ measurement as a good approximation of the real mass of objects.

2. Pulsar timing

In the Universe, the most stable clocks known so far are Millisecond-pulsars (Davis et al., 1985). They are believed to be old"dead" pulsars revived by matter accretion coming from final pulsation-ejection stages of a companion AGB star. This addition of angular momentum speeds up the pulsar up to ms rotation rates. Since the pulsar is old, it behaves like an isolated solid body sphere and it provides an extraordinarily accurate clock.

The Doppler effect on pulses can be detected by frequency changes in pulse arrivals: $\Delta V/c = \Delta\nu/\nu$. In $\mathrm{m\,s^{-1}}$ unit: $\Delta V_r = 3 \cdot 10^8(\Delta\nu/\nu)$. Given the current precision reached by radio-astronomers and the intrinsic stability of Millisecond-pulsars, a frequency measurement precision of about $(\Delta\nu/\nu) \sim 10^{-11}$ is possible ($\mathrm{cm\,s^{-1}}$ detection level), which is enough to search for Earth mass bodies in orbit. (See Konacki in this issue for a more detailed description of the analysis on the radio data from pulsars.)

On PSR 1257+12, a planetary system of 3 bodies (Earth and Moon-like masses) has been detected (Wolszczan et al., 1992), (Wolszczan, 1994). An orbital 3:2 resonance has even been observed between the B and C component which definitely proves the gravity perturbation interpretation for the observed frequency changes.

The detection of planetary bodies orbiting a pulsar has been quite surprising and challenges theoretical explanations. However, it is believed that the planetary formation occurred after the pulsar formation. Such a system gives us an interesting and unique possibility to study the formation of Earth-mass bodies and may help understanding the formation of telluric planets.

3. Line Doppler spectroscopy

To measure the radial velocity changes of stars, spectral line shift measurements are used. In order to detect planets, a 10^{-4} [Å] precision level in the visible ($\Delta V_r[\mathrm{km\,s^{-1}}] = 60\Delta\lambda[\text{Å}]$ at 5000Å) must be reached on the measurement of the mean position of spectral lines. This is about 1000 times smaller than the best available resolution in this wavelength range. High signal-to-noise data and multi-lines approaches are required to reach this level of precision.

The strongest radial velocity signature in a spectral line is obtained when $|\partial F/\partial\lambda|\sqrt{F}$ is maximum (Connes, 1985). This condition is met on both sides of the line profile (see on Fig. 2). Stars having lots of thin lines are the easiest targets for planet searches by Doppler spectroscopy .

In the next sections the precision issues are addressed. First, we look at all possible intrinsic limitations from the star surface itself and we discuss how surface instabilities affect the accuracy of the measurement of the star radial velocity (V_r). Second, the instrument and measurements issues are considered. Some state-of-the-art techniques currently used to measure radial velocities from stellar spectra at few $\mathrm{m\,s^{-1}}$ precision level are described.

Figure 2. Illustration of the amount of Doppler information and its location in line profile. (**a-c**), a set of thin lines and its relative amount of Doppler information on bottom (fig c.). (**b-d**) same but for a broad line. The comparison between the two sets of lines clearly shows that numerous thin lines contain much more Doppler information than broad ones. By comparison between a. and c. one can see that most of the Doppler information is located on the profile sides but not in the line core.

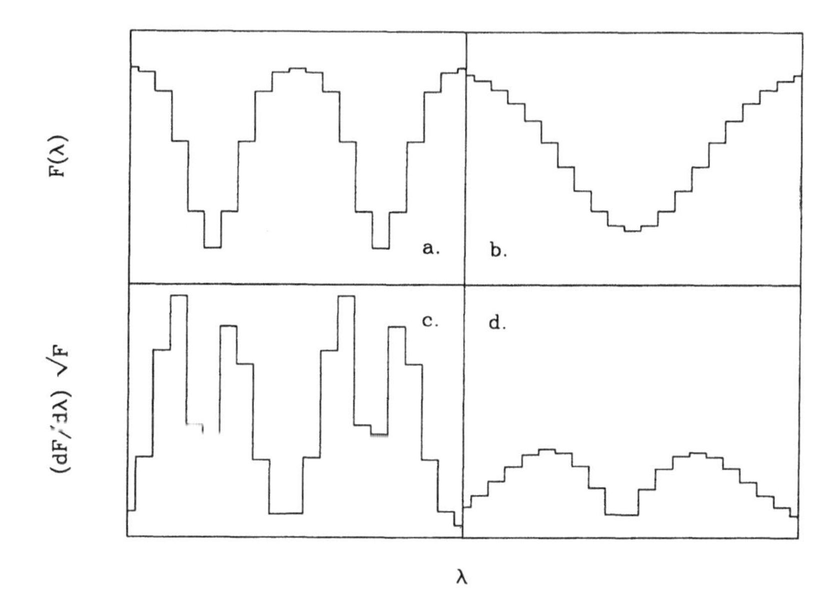

Finally, the problem of the data interpretation is addressed. How sure can we be that any tiny observed radial velocity changes are indeed the result from the whole motion of the star rather than the results of subtle motions at the star surface?

3.1. LIMITATIONS FROM THE STAR ITSELF

The V_r–amplitude of a planetary signature is small enough to make us worry about all possible intrinsic stellar phenomena affecting the stability of spectral lines and more precisely the line profiles. Astrophysical effects such as accretion, stellar wind (mass losses), magnetic activity, non-uniform convection and pulsation have direct effects on line profiles. Any non-symmetric variations of the line profile would be seen as a change of the mean radial velocity of the star. It may be easy to avoid very active or accreting young objects or stars with a strong wind but it becomes more tricky when one has to deal with very tiny changes of line profiles of quiet stars like our Sun.

Their is a tight link between the rotation and the magnetic activity of cool stars (see e.g. Fig 9. in (Queloz, 1998)). Up to a threshold (usually called saturation limit), the more a star rotates the more magnetic related activities are observed (spots, X-ray , chromospheric lines,...). It is believed that the amount of internal differential rotation that generates the magnetic field ($\alpha - \omega$ dynamo model) scales with the surface rotation of stars. After their initial contraction, Main Sequence stars lose their angular momentum through the interaction between their wind and the magnetic field ((Schatzmann, 1962), (Weber & Davis, 1967)) and gradually slow down. The breaking efficiency is also related to the mass of stars. Very low-mass stars (M stars) take more time to slow down than solar-type stars. Stars without convective envelopes (earlier than F2-5) never break down their rotation speed (see (Bouvier, 1998) for more details). Therefore, a younger star will display more magnetic related activity than an older one and for a given age and we shall expect more stellar activity from M stars than from G dwarfs.

Figure 3. Illustration of spot effects on stellar line profiles of a rotating star. When the spot is on the side of the star going towards the observer, a part of the approaching velocity field is hidden by the spot. In the integrated spectrum from the whole star, it looks like as if the star was going away from the observer. When the spot is on the other side, the opposite effect takes place and the star looks as if it was moving towards us. As long as the spot is present, the velocity change is synchronized with the rotation rate of the star and may be misinterpreted as the effect of a short period orbit of a low mass companion. A careful analysis of the line profile helps prevent such a misinterpretation.

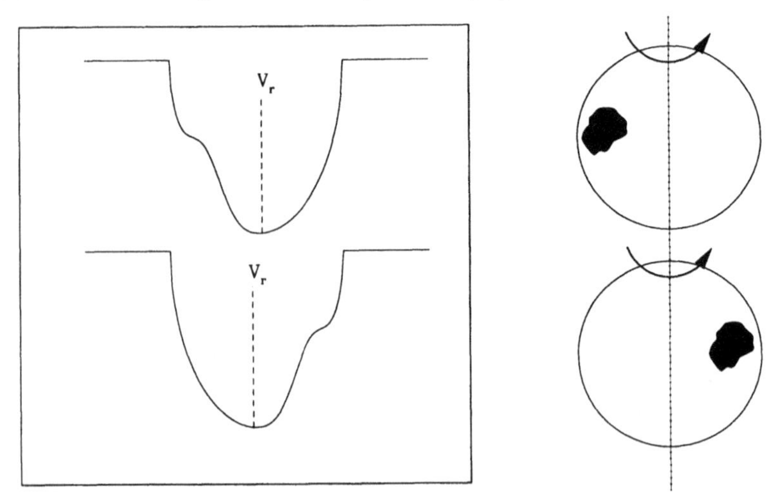

Active stars are covered by cool spots. For very young ones, the surface area ratio covered can easily reach several percents (compare with 0.15% for the Sun). The spots are usually detected photometrically as a time variable signal in phase with the star rotation rate. In spectroscopy, the cool spots act like a screen which alternatively hides a portion of star surface and hence distorts the rotation line profile (see Fig. 3). If not detected, this effect shifts the center of gravity of spectral lines and produces a V_r periodic signal which mimics a short period orbit system. A spot has a typical lifetime of a few months: therefore, we don't expect the phased signal to last long enough to be misinterpreted as a short period planet but only to add a V_r intrinsic noise to the data. However, it has been observed on the Sun that active regions tend to reform in the same region of past outbursts (Brouwer & Zwaan, 1990) leading to "active longitude" zone. This phenomena may sometimes increases the apparent coherence of the phase of the periodic signal.

Effects of inhomogeneous convection (on star surface) may also affect the line profiles. On Sun magnetic regions, the granulation is suppressed and the line profiles change with the magnetic areas (Brandt & Solanki, 1990). Extrapolating to young active stars covered with more active regions, we may expect to see a variable non-axisymmetric velocity field, rotating on top of the mean convection field. This would affect the line profiles on short and long time scales.

The first results from extra-solar planet surveys, regarding the effects of the stellar activity on V_r scattering, suggest a correlation between the rotation speed of the G and K dwarfs stars and their V_r variability (Mayor et al., 1997), (Saar et al., 1998) (see on Fig. 4). All G and K dwarf stars with rotation rates lower or equal than 10 days have intrinsic V_r noises above the $10\,\mathrm{m\,s^{-1}}$ level. For late F stars, the relation between their rotation rate and their V_r scattering is not so obvious, probably because of their very thin convective envelope and their lack of magnetic activity.

On long time scale, the Sun has an observed intrinsic variability less than $4\,\mathrm{m\,s^{-1}}$ (McMillan et al., 1993). The level of long term stability of the integrated Sun stellar profile is still an open issue. With the regular increase in precision of Doppler measurements, detailed analysis of lines profiles at a few $\mathrm{m\,s^{-1}}$ level will soon be necessary to disentangle tiny profile changes from real star motion for Sun-like stars.

All evolved stars exhibit intrinsic variability. The K2 giant αBoo ($R \sim 25\,\mathrm{R_\odot}$, $\log g \sim 2$) is a very illustrative example. This star has a multi-periodic and multi-amplitude V_r variability pattern ranging from a few days to about a year ((Hatzes & Cochran, 1993), (Hatzes & Cochran, 1994)). It is interpreted as a combination of rotation signatures from surface features and pulsation phenomena of its loose atmosphere. The evolved stars are difficult targets for radial velocity searches of planetary companions.

Figure 4. Observed radial velocity scattering (corrected from the mean internal errors) for G and K dwarfs of the Lick planetary search sample [data from (Saar et al., 1998)]. The Sun V_r stability upper limit is displayed with an open dot and stars with planets corrected for the planetary motion with open star symbols. For rotation periods less than about 10 days all stars have intrinsic scatterings larger than $10\,\mathrm{m\,s^{-1}}$. For longer periods it is difficult to disentangle the intrinsic scattering from the star with the one coming from instrumental effects (about $5\,\mathrm{m\,s^{-1}}$).

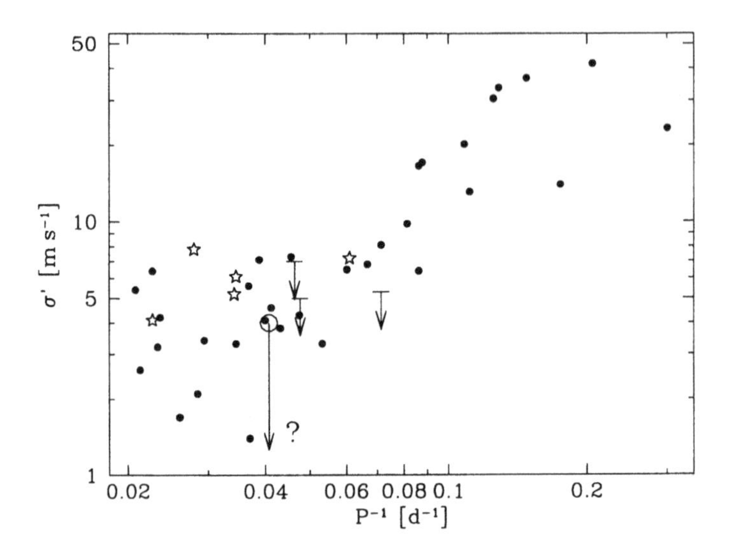

3.2. INSTRUMENT PERFORMANCES

Very schematically, a high resolution spectrograph is composed of an entrance slit, a collimator, some dispersive elements, an objective and a detector. The main effect of the instrument is to project the image of the entrance slit into multi-wavelength images spatially spread onto the detector. At first approximation, a spectrograph behaves like an imaging device. If one moves a spot on the entrance slit, the motion will be seen on the detector. But, since on the detector spatial and wavelength information are mixed up, the main effect of any changes in slit illumination is to introduce systematic offsets between the calibrated frame and the science observations (see Fig. 5). Moreover, a change of the image size also has an impact on the wavelength resolution.

To achieve very high precision in V_r measurements, the most important issue is to find out a calibration process that really matches the recorded stellar data without any systematic offsets (time or spatial) between the two. As pointed on earlier, the slit illumination is one of the very critical issues. Taking into account tracking and centering errors and variable seeing

Figure 5. Schematic illustration of a spectrograph. The gray rectangle on the entrance slit represents a non-uniform input slit illumination. Its effect on the detector is a slight shift of the image (spectrum) with respect to the mean position of each slit images which defines the zero-V_r-calibration of the spectrum. This offset between the calibrating frame and the observed spectrum leads to systematic instrumental errors on the measurement of spectral line position (or V_r of the star).

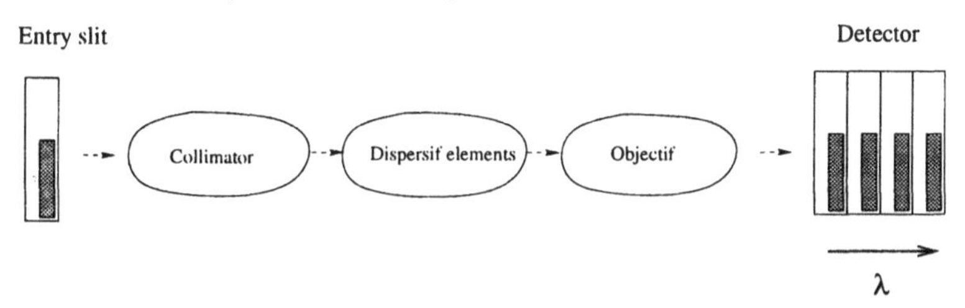

conditions, it is usually impossible to reach a stable image better than 1/50 of the slit width, corresponding to about 50-100 $\mathrm{m\,s^{-1}}$ for a 100,000 ($\lambda/\Delta\lambda$) resolution spectrograph. An internal calibration or an input beam scrambling device must be used to avoid this undesirable effect.

The intrinsic stellar profile for a cool star without rotational broadening is about 2 $\mathrm{km\,s^{-1}}$ (FWHM at 5000 Å). Typically, the best resolution reached by (echelle) spectrographs is about 50,000-100,000 (3–6 $\mathrm{km\,s^{-1}}$ at 5000 Å). Therefore, the observed line profile is mostly the image of the instrument point spread function (PSF) rather than the star stellar profile itself. We have already seen in the previous chapter how much the computation of the V_r of a line depends on the stability of the line profile. Here, we point out that the PSF stability of the instrument is even more crucial if one considers typical resolutions of current high-resolution spectrographs. If the PSF of the instrument varies during the night or from night to night because of mechanical bending or thermal effects (expansion, contraction), it will introduce systematic and random V_r changes in the same way that stellar activity adds noise to the V_r data.

The physical conditions of the atmosphere affects the stability of the instrument and may change its zero-point. The variation of the air index from a pressure change (100 $\mathrm{m\,s^{-1}\,mmHg^{-1}}$) or a temperature change (200 $\mathrm{m\,s^{-1}\,K^{-1}}$) during an exposure requires an instantaneous calibration of the data. Another solution consists in isolating the instrument from any external pressure or any temperature changes.

The observed spectrum will move on the detector during the year because of the Earth's motion effect. The instantaneous projected velocity change may be as high as $\partial V_r/\partial t = 2\cos\delta\,[\mathrm{m\,s^{-1}\,min^{-1}}]{+}0.5\,[\mathrm{m\,s^{-1}\,min^{-1}}]$. To correct for this effect, at the 1 $\mathrm{m\,s^{-1}}$ precision level, requires a good knowledge of the true mean exposure time ($\int F t\mathrm{d}t/\int \mathrm{d}t$). Moreover, the spectrum won't be on the same pixels of the detectors along the year. The simultaneous use of many lines and therefore many different pixels is a way to smear out any systematic pattern that may appear from the intrinsic sensitivity differences or structure of detector pixels

A spectrum of a non-rotating cool star is very rich in Doppler information (see on Fig. 6). A very efficient way to compute the radial velocity of a star is to make use of the systematic behavior of all the spectral lines by using all of them to compute the Doppler shift. The echelle spectrograph is probably the best kind of instrument for conducting a systematic V_r-survey of planet orbiting nearby stars. With their large wavelength range, many thousand lines can be simultaneously observed and a photon limited precision better than 1000 times the spectral resolution of the spectrograph can easily be achieved without the need of very high signal-to-noise data and therefore without using a prohibitive amount of exposure time (see e.g. in (Baranne et al., 1996)).

Figure 6. Blue part of the 51 Peg spectrum observed at R=40'000 with the echelle spectrograph Elodie. One clearly sees that thousand lines are available in this wavelength range for measuring the Doppler shift of the star.

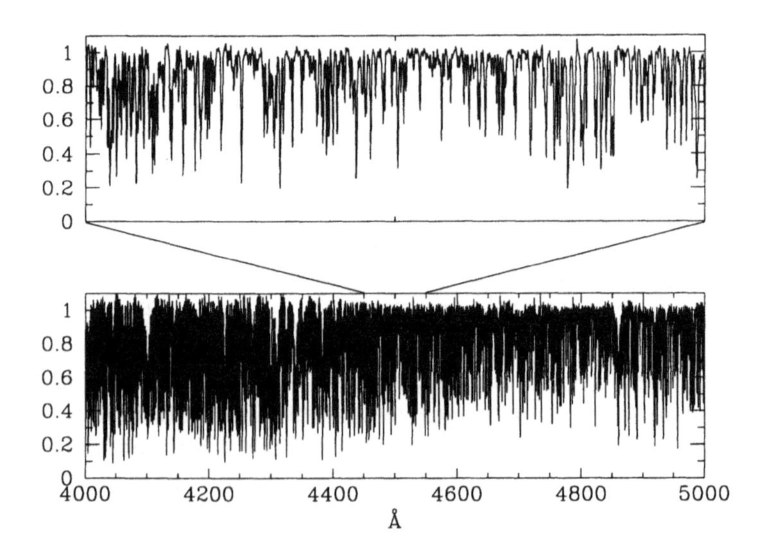

3.2.1. *The simultaneous thorium technique*

The "simultaneous thorium" technique has been successfully used by the ELODIE echelle spectrograph (Baranne et al., 1996) to detect the planet orbiting the star 51 Peg by Doppler shift measurements (Mayor & Queloz, 1995). This technique is based on the assumption that one can build an instrument stable enough to have a constant PSF. As a matter of fact, ELODIE is located in a room, isolated from any external mechanical or thermal effects and is linked to the telescope via an optical fiber, which ensures a very stable input illumination.

A perfect fiber has the property of "scrambling" any input beam in order to get an uniform illumination at the output. More specifically, the scrambling at the output is perfect on the azimuthal axis of the fiber section. An extra "scrambling device" (e. g. double scrambler (Brown, 1990)) may also be used to increase the scrambling effect in the radial direction –which is not so good– and get a very stable and flat illumination of the spectrograph entrance slit (Casse & Vieira, 1997), (Queloz & Mayor, 1999) (see on Fig. 7).

Figure 7. Illustration of the natural scrambling effect by a "perfect" fiber and when a double scrambler is added (simulated data from (Casse, 1997)). **Left** illustration of the way the fiber is fed by the beam (large filled dot). **Right-top**: the toroidal shape visible of the output image of the fiber is the result of a very good azimuthal scrambling combined with a poor radial one. **Right-bottom**: with the addition of a double scrambler device, one see that the radial scrambling is improved. [fig. orig. from (Queloz & Mayor, 1999)]

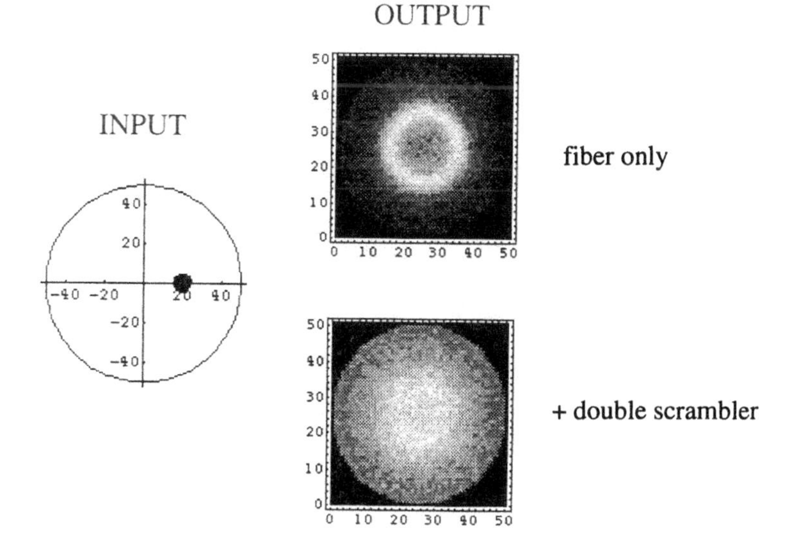

A thorium spectrum is recorded with a second beam at the same time as the stellar spectrum. This second beam follows the same optical path than the primary one but hits the detector just above the science spectrum (see on Fig. 8). This beam is used to correct from any zero point variations which may occur during the science exposure.

Figure 8. Illustration of the"simultaneous thorium" concept. **Top:** Illustration of the two different illuminations of the entrance slit used first for calibrating the zero velocity point and second to measure radial velocity at very hich accuracy. **Bottom:** CCD frame of a "simultaneous thorium"exposure, where one fiber is fed by the star and the other illuminated by a thorium lamp (see text for details).

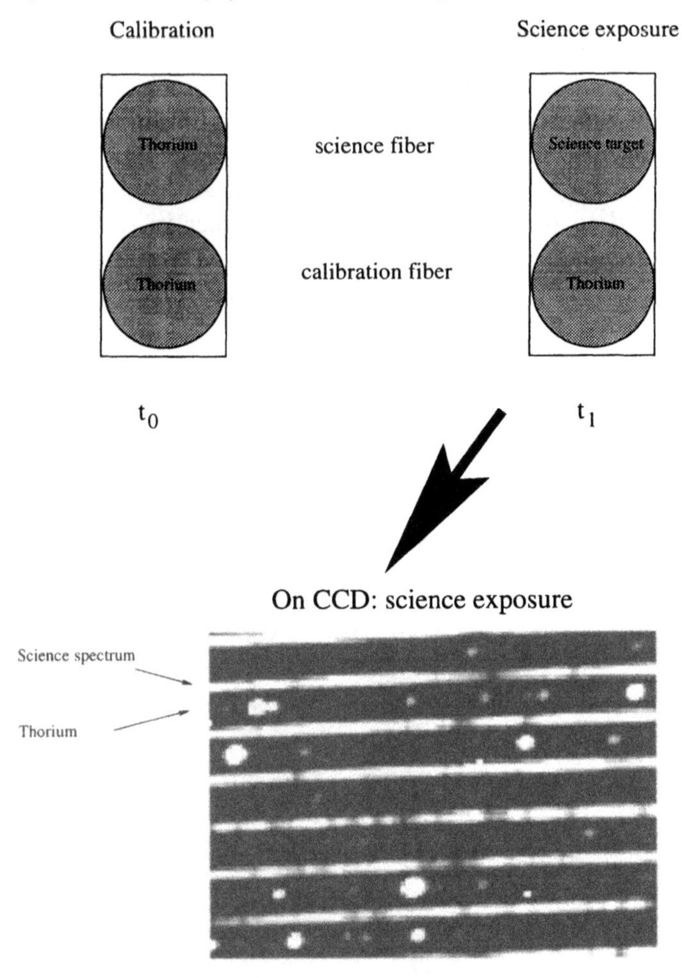

The calibration process is carried out in two steps: first a wavelength solution is computed for both optical beams (science and monitoring thorium). Second, the science exposure is carried out with a thorium lamp illuminating the second optical path. The observed Doppler shift between the two thorium spectra $(\lambda^c(t_1) - \lambda^c(t_0))$ is used to correct for the instrument calibration drifts during the science exposure compared to its previous wavelength calibration: $\lambda(\text{science}) = \lambda(t_0) + (\lambda^c(t_1) - \lambda^c(t_0))$, where λ^c refers to the wavelength calibration from the thorium monitoring spectrum.

This technique is used on ELODIE, CORALIE (Baranne et al., 1996), and AFOE (Brown et al., 1994) echelle spectrographs. A precision of 10 m s$^-1$ has been reached on long term stability V_r measurements with ELODIE (Mayor *et. al.*, 1999). The simultaneous thorium has no strong wavelength limitation. Many thorium lines are available on the whole visible range. The use of fibers does not significantly reduce the instrument brightness. A well mounted fiber has about 80% transmission efficiency. The only weak point in this technique is the assumption that the PSF is stable. At first approximation this assumption is observed to be true but given the level of precision that one is aiming at, this might be wrong and may be the main reason for the observed 10 m s^{-1} limitation on long term accuracy.

3.2.2. *The Iodine cell*

In the Iodine cell technique a reference spectrum is superimposed on the stellar one to calibrate it. A cell filled with I_2 gas sits right in front the spectrograph slit. No efforts are made to keep the instrument stable. The stellar beam goes through it and a forest of I_2 lines are superimposed on the stellar spectrum providing us with a zero-reference. The intrinsic resolution of the I_2 spectrum is above 500'000 but most of the lines range only from 5000 to 6000 Å. A model of the I_2 spectrum and a star template is used to adjust the PSF and the relative Doppler shift of the observed spectrum compared to the template (see details in (Butler et al., 1996)): $I_s(\lambda + \Delta\lambda)T_{I_2} \otimes \mathrm{PSF} = I(\mathrm{observed})$, where I_s is the stellar template, T_{I_2} is the iode spectrum. (see on Fig. 9). If a good enough PSF model can be found and if the star template is accurate, this technique should mostly only be limited by the CCD sampling and photon noise. With large aperture telescopes, V_r precisions, up to 3-5 m s^{-1} are routinely reached on relative Doppler shift measurements for nearby Solar-type stars (Butler & Marcy, 1997).

3.3. DATA MIS–INTERPRETATIONS

The planetary interpretation of the observed Doppler shift relies on the hypothesis that the star atmosphere is not moving (pulsating). In general the geometry of a pulsation is described by spherical harmonics Y_l^m and n the number of nodes of the mode eigenfunction along the stellar radius.

In spherical coordinate the velocity from any pulsation can be written:

$$V_r = V_p Y_l^m \quad V_\theta = V_p\, k\, \frac{\partial Y_l^m}{\partial \theta} \quad V_r = V_p\, k\, \frac{\partial Y_l^m}{\partial \phi}\frac{1}{\sin\theta},$$

Figure 9. From top to bottom: I_2 spectrum, star template, fit of the observed data with a PSF model and the residuals from the fit. [from (Butler et al., 1996). Copyright ©1996 by Astronomical Society of the Pacific]

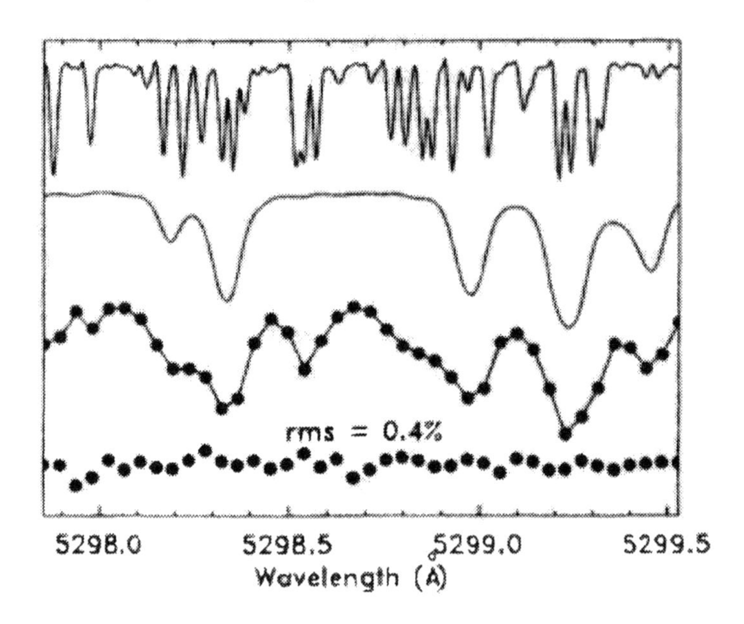

where V_p is the amplitude of the pulsation and k the ratio between the horizontal and the vertical velocity amplitude. Since $\frac{\partial Y_l^m}{\partial \phi} \sim l$ and $\frac{\partial Y_l^m}{\partial \theta} \sim l$ one can see that when $kl \gg 1$, the vertical speeds V_ϕ and V_θ are larger than V_r and therefore most of the "fluid" is in horizontal motion. On the contrary, if kl is small, the pulsation is mostly vertical (radial).

For a radial pulsation, there is a direct relation between the observed radial velocity amplitude and the motion of the atmosphere of the star:

$$\mathrm{d}R = \int_0^{T/4} V_r \mathrm{d}t. \tag{9}$$

Since the area of the star surface changes, the luminosity of the star changes ($L = 4\pi R^2 T_{\rm eff}^4$) and dL/L=2dR/R + 4dT/T.

For 51Peg, if one interprets the observed amplitude of the radial velocity curve ($59\,\mathrm{m\,s^{-1}}$ (Mayor & Queloz, 1995)) as a sole pulsation effect, we get $\mathrm{d}R/R = 0.5\%$. If $\mathrm{d}T/T \sim 0$ we should observe $\mathrm{d}L/L \sim 1\%$. Since the photometry is measured to be constant at better than 0.1% (Henry et al., 1997) a radial pulsation interpretation can be ruled out for that star. Similar conclusions may be reached for all other detected short period planets (see in (Marcy & Butler, 1998)). A pulsation scheme where $\mathrm{d}T$ would compensate $\mathrm{d}R$ such that $\mathrm{d}L/L < 0.1\%$ would be rather far fetched.

A radial pulsation cannot account for the observed radial velocity changes of any of the short period planetary candidates recently detected.

In the case of a non-radial pulsation, the V_r amplitude is not related to the radius change and therefore to the star luminosity. The star may well be pulsating and having very few luminosity changes. The two typical non-radial pulsation modes are p-modes (pressure driven) and g-modes (gravity driven).

On the Sun, p-mode pulsations are well observed. This "acoustic" pulsation mode is like a sound wave propagating at sound speed with typical periods less than one hour. The largest solar p-modes oscillations have amplitudes of $15\,\mathrm{cm\,s^{-1}}$. It is too small to account for any observed radial velocity amplitudes on recently detected star with planets. Moreover the periods would be too small.

For gravity modes (g-modes) the restoring force origins from an *horizontal* density gradient (buoyancy phenomena). By definition these modes have $l \neq 0$. Their propagation depends on the restoring force what is absent or even negative in convective regions. Therefore gravity pulsation modes can not propagate within stellar convection zones. They have never been detected at the Sun stellar surface. Typical time scales depend on the detailed density stratification of internal star. However for Sun-like stars with $n \gg l$:

$$T \approx \frac{(n + l/2 + \delta)}{l(l+1)}\,[\mathrm{h}]. \tag{10}$$

In order to get 4 day g-mode pulsations, n has to be close to 100. For such a high order one may expect a whole family of modes close to that value to be excited as well.

A non-radial pulsation may be detected by careful analysis of line profiles. The understanding of the behavior of a line during various phases of the pulsation cycle can be calculated by using disk integration of a synthetic star (see (Hatzes, 1996) for more details). Detail analysis of line profiles of 51 Peg has been carried out (Mayor & Queloz, 1995), (Brown et al., 1998a), (Brown et al., 1998b), (Hatzes et al., 1999) and any change of the line bisectors can be ruled out down to a few $\mathrm{m\,s^{-1}}$. Given the current data on profile stability from the observations of stars with short period planetary candidates, there is no reason to believe that any of these stars may be pulsating.

4. Summary of current researches

All recent announcements of planet detections are based on data from surveys started years ago. Our current knowledge about all "the new planetary systems" (see Queloz in this issue) stems from a relatively small sample of

TABLE 1. Summary of the "old" high precision Doppler surveys which led to all recent planet detections.

Teams, Instruments and telescopes	Samples	Beg./duration [yr]
Cochran & Hatzes (Cochran & Hatzes, 1994) at Mc Donald Obs.	~ 30	1989/9
Delfosses et al. (Delfosses et al., 1999) with ELODIE at OHP	~ 120	1995/3+
Marcy & Butler (Marcy & Butler, 1992) with Hamilton at Lick Obs.	~ 120	1988/10+
Mayor & Queloz (Mayor & Queloz, 1995) with ELODIE at OHP	~ 140	1994/4+
Noyes at al. (Noyes *et al.*, 1998)with AFOE at Whipple Obs.	~ 100	1994/4+
Walker et al. (Walker et al., 1995) with coudé spectro at CFHT	21	1980/12

stars monitored (about 300) with precision from 7 to $20\,\mathrm{m\,s^{-1}}$ (see in Table 1). If one considers all detected companions with $m_s \sin i < 5\,\mathrm{M}_J$ as a planet, a detection rate of about 2-3% planets per star has been achieved. This is a first and very rough estimate of the occurrence of giant planets amongst stars in the solar neighborhood.

In recent ongoing surveys the total amount of stars being monitored to detect small Doppler shifts reaches almost 3000. Moreover, the accuracy has been improved by almost a factor 2 (Butler et al., 1996), (Queloz *et al.*, 1998). From the increase in both the amount of stars and the measurement accuracy one may expect more than 100 stars with a planetary companion to be detected in the next coming years. Moreover, the precision gain should definitely lead to the detection of a set of twins to our own Jupiter and maybe to the detection of a first planetary system with two giant planets orbiting a star (see on Fig. 10).

With the increasing number of planets detection, the amount of detected 51 Peg-like systems known should soon be large enough to find one with a transit. For such close-by system, the geometrical probability to transit its star is only 1/20. Such a discovery would give us access for the first time to some basic structure parameters of an extrasolar planet (radius, gravity) and would yield the definitive confirmation that all these new and "weird" planetary systems are real.

References

Baranne, A., Queloz, D., Mayor, M., Adriansyk, G., Knispel, G., et al., 1996, "ELODIE: A spectrograph for Precise Radial Velocity Measurements", A&A Suppl. Ser 119, 1.

Bouvier, J., 1998, "The surface and internal rotation of low-mass stars in young clusters", in "Cool Stars in Cluster and Associations: Magnetic Activity and Age Indicators", eds G. Micela, R. Pallavicini and S. Sciortino, Memorie della Societa Astronomia Italiana, vol. 68, 881.

Brandt, P. N., Solanki, S. K., 1990, "Solar line asymmetries and the magnetic filling

Figure 10. Simulation of the effect of the orbit of Jupiter and Jupiter & Saturn together on radial velocity of a solar type star observed at various precision level. Above 10 m s^{-1} accuracy there is very little chance to detect a Jupiter. With a 5 m s^{-1} accuracy the detection of a giant planet like our own Jupiter is within reach of current survey. The dotted line on left diagrams displays for comparison the effect of a single Jupiter. One sees that a precision of 1 m s^{-1} is needed to detect any second planetary body like Saturn.

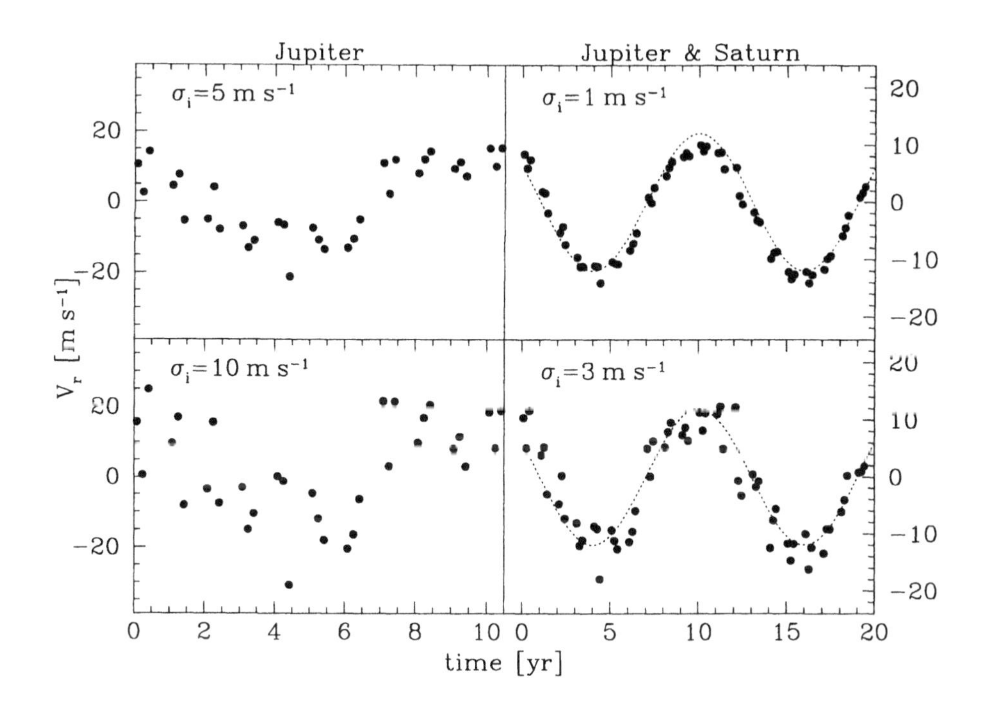

factor" A&A, 231, 221.

Brouwer, M. P., Zwaan, C., 1990. Sol. Phys., 129, 221.

Brown, T. M., 1990, in CCds in Astronomy, ed. G. Jacoby, ASP Conf. Ser., 8, 335.

Brown, T. M., Noyes R. W., Nisenson P., Korzennik S. G., Horner S., 1994, "The AFOE: A Spectrograph for precise Doppler Studies", PASP 106, 1285.

Brown, T. M., Kotak, R., Horner, S. et al., 1998, "A search for line shape and depth variations in 51 Peagsi and τ Bootis", ApJ, 494, L85.

Brown, T. M., Kotak, R., Horner, S. et al., 1998, "Exoplanet or dynamic atmosphere? The radial velocity and line shape variations of 51 Pegasi and τ Bootis", ApJ Suppl. Ser., 117, 563.

Butler, R. P., Marcy G. W., Williams E., et al., 1996, "Attaining Doppler Precision of 3 m s^{-1}", PASP, 108, 500.

Butler, R. P., Marcy G. W., 1997, "The Lick Observatory Planet Search", IAU Colloquium 161, 5th International Conference on Bioastronomy, "Astronomical and Biochemical Origins and the Search for Life in the Universe", eds. C. Cosmovici, S. Bowyer, and D. Werthimer, 331.

Casse, M, Vieira, F., 1996, "Comparison of the scrambling properties of bare optical fibers with microlens coupled fibers", in "Optical telescopes of today and tomorrow",

SPIE proceedings, 2871, 1187.
Casse M., 1997, private comm. from "Solid Scrambler design" Report.
Cochran, W. D., Hatzes, A. P., 1994, "A high-precision radial-velocity survey for other planetary systems", Astrophysics and Space Science, vol. 212, 281.
Connes, P., 1985, "Absolute astronomical accelerometry", ApSS, 110, 211.
Davis, M.M., Taylor, J.H., Weisbergg, J.M., Backer, D.C., 1985, "High-precision timing observations of the millisecond pulsar PSR 1937+21", Nature, 315, 547.
Delfosses, X., Forveille, T., Udry, S., et al., 1999, "The multiplicity of the field M dwarfs", in IAU Colloquium 170, "Precise stellar radial velocities", eds J.B.Hearnshaw and C.D.Scarfe, ASP Conference Series, in press.
Hatzes, A. P., Cochran, W. D., "Long-period radial velocity variations in three K giants", ApJ, 413, 339.
Hatzes, A. P., Cochran, W. D., 1994,"Short-period radial velocity variations of Alpha Boo: evidence for radial pulsations", ApJ, 422, 366.
Hatzes, A. P., 1996, "Simulations of Stellar radial-Velocity and spectral line bisector variations. I. Nonradial pulsations", PASP, 108, 839.
Hatzes, A. P., Cochran, W. D., Bakker, E. J., 1998, "Further evidence for the planet around 51 Pegasi.", Nature, 391, 154.
Henry, G. W., Baliunas, S. L., Donahue, R. A., Soon, W. H., Saar, S. H, 1997, "Properties of Sun-like Stars with Planets: 51 Pegasi, 47 Ursae Majoris, 70 Virginis, and HD 114762", ApJ, 474, 503.
Marcy, G. W., Butler, R. P., 1998, "Extrasolar Planets Detected by the Doppler Technique",in "Brown Dwarfs and Extrasolar Planets", ASP Conf, eds. R. Rebolo, E. L. MArtin, M. R. Zapatero, 134, 128.
Marcy, G. W., Butler R. P., 1992, "Precision radial velocities with an iodine absorption cell", PASP, vol. 104, 270.
Mayor, M., Beuzit, J.-L., Mariotti, J.-M., et al., 1999, "Searching for giant planets at the Haute-Provence Observatory", in IAU Colloquium 170, "Precise stellar radial velocities", eds J.B.Hearnshaw and C.D.Scarfe, ASP Conference Series, in press.
Mayor, M., Queloz D., Udry S. 1997, "Mass function and orbital distributions of substellar companions", in "Brown Dwrafs and Extrasolar Planets", ASP Conf. Ser., vol. 134 , Eds. R. Rebolo, E.L. Martin, M.R. Zapatero Osorio, 140.
Mayor, M., Queloz, D., 1995, "A Jupiter-mass companion to a solar-type star", Nature, vol. 378, 355.
Mayor, M., Queloz, D., 1996, "A search for Substellar Companions to Solar-type Stars via precise Doppler Measurements: A first Jupiter Mass Comppanion Detected", in Cool Star, Stellar Systems, and the Sun, 9th Cambridge Workshop, ASP Conf. Ser. vol. 109, R. Pallavicini and K. Dupree (eds.), 35.
McMillan, R. S., Moore, T. L., Perry, M. L., Smith, P. H., 1993, "Radial velocity observations of the Sun at night", ApJ, 403, 801.
Noyes, R., Jha, S., Korzennik, et al., 1997, "The AFOE Program of Extra-Solar Planet Research", in "Planets Beyond the Solar System and the Next Generation of Space Missions", Space Telescope Science Institute Workshop, PASP, vol. 119, D. Soderblom (ed.), 119.
Queloz, D., Mayor, M., Sivan, J. P., et al. 1998, "The Observatoire de Haute-Provence Search for Extrasolar Planets with ELODIE", in "Brown Dwarfs and Extrasolar Planets", ASP Conf. Ser., vol. 134, R. Rebolo, E. L. Martín and R. Zapatero Osorio (eds.), 324.
Queloz, D., Allain, S., Mermillod, J.-C., Bouvier, J., Mayor, M., 1998, "The rotation velocity of low-mass stars in the Pleiades cluster", A&A, 335, 183.
Queloz, D., Casse, M., Mayor, M., 1999, "The fiber-fed spectrographs: a tool to search for planets", in IAU Colloquium 170, "Precise stellar radial velocities", eds J.B.Hearnshaw and C.D.Scarfe, ASP Conference Series, in press.
Sarr, S. H., Butler, R. P., Marcy, G. W., 1998, "Magnetic activity-related radial velocity variations in cool stars: first results from the Lick extrasolar planet survey", ApJ,

498, L153.
Saar, S. H., Donahue R. A., 1997, "Activity-related radial velocity variation in cool stars", ApJ, 485, 319.
Schatzmann, E., Ann. d'Astrophys., 25, 18.
Walker, G. A. H., Walker, A. R., Irwin, A. W. et a. 1995, "A search for Jupiter-mass companions to nearby stars", Icarus, 116, 359.
Weber, E. J., Davis, L., 1967, ApJ, 148, 217.
Wolszczan, A., Frail, D. A. 1992, "A planetary system around the millisecond pulsar PSR1257+12", Nature, vol. 355, 145.
Wolszczan, A., 1994, "Confirmation of earth-mass planets orbiting the millisecond pulsar PSR B 1257+12", Science, vol. 264, 5.

FREQUENCY ANALYSIS AND EXTRASOLAR PLANETS

MACIEJ KONACKI AND ANDRZEJ J. MACIEJEWSKI
Torun Centre for Astronomy
Nicolaus Copernicus University
Gagarina 11, 87-100 Torun, Poland

1. Introduction

The discovery [13] (and confirmation [14]) of the first extrasolar planetary system by Wolszczan & Frail (1991) around the pulsar B1257+12 began a new era in this exciting field of astronomy. A few years later, in 1995, the first extrasolar planet around a normal star was found by Queloz & Mayor [10]. This was followed by other such discoveries [9]. The common point of all these findings is that the planets were indirectly deduced from the observations of their parent stars. In the case of the PSR B1257+12 the planets were found in the timing data of the pulsar and in the case of normal stars in their Doppler velocity measuerements. The usual approach to analyzing such data is to fit by meanas of the least-squares technique a Keplerian orbit (or orbits) that would account for the observed planetary effects. In most situations this is perfectly satisfactory. However, it is possible to show examples in which this technique fails. In this paper we show a new approach to the analysis of the timing observations of pulsars, as well as the Doppler velocity measurements of normal stars, when one suspects to have planetary signatures in the data. We disscus how this method, frequency analysis, can give more insight into to the data or even properly solve the problem when the least-squares fit of Keplerian orbits is not capable to do so.

2. Quasi-periodic functions

Many mechanical, or more generally, physical processes are quasi-periodic in their nature. For example, motion of planetary systems, precession of planets, pulsations of stars, are in most cases quasi-periodic. This is why it is important to know basic properties of quasi-periodic signals.

J.-M. Mariotti and D. Alloin (eds.), Planets Outside the Solar System: Theory and Observations, 249–260.

Omiting technical details by a quasi-periodic function we understand a function which can be represented by a multiple Fourier series of the form

$$f(t) = \sum_{k_1,\dots,k_n \in \mathbb{Z}} f_{k_1 \dots k_n} e^{i(k_1\omega_1 + \dots + k_n\omega_n)t}, \tag{1}$$

where sum is taken over all integer values of indices k_i (the set of integers we denote $\mathbb{Z}$). Let us observe that in this case $f(t)$ has a countable set of frequencies and a finite number of basic frequencies. Because of this the assumption that an observed signal is quasi-periodic put strong constraints on its properties which, when taken into account, make the process of data analysis more powerful and reliable.

3. Pulsar timing observations

The pulsar timing observations are basically very precise measurements of the times of arrival (TOAs) of pulsar pulses. Since we can predicted these TOAs with high accuracy, any advances or delays of TOAs due to the spatial motion of the pulsar are easily detectable. In particular, the TOA variations, Δt, due to the orbital motion of the pulsar may be written in a very simple form:

$$\Delta t(t) = -c^{-1} Z(t), \tag{2}$$

where Z is the z-th coordinate of the pulsar in such a system that the Z-th axis is directed to the observer; c is the speed of light. In additon, since the motion of the pulsar is described in the center of mass reference frame, assuming that there are N planets in the system, we have the following formulae:

$$\mathcal{M}_\star Z + \sum_{j=1}^{N} m_j Z_j = 0, \tag{3}$$

where $\mathcal{M}_\star$ is the mass of the pulsar and m_j, Z_j are masses and z-th coordinates of planets respectively. Thus, the formulae for TOA variations may be written in the form:

$$\Delta t(t) = (c\mathcal{M}_\star)^{-1} \sum_{j=1}^{N} Z_j(t). \tag{4}$$

3.1. KEPLERIAN MOTION

Let us assume, for simplicity, that we have a planetary system consisting of a pulsar and one planet. We can calculate the Z-th coordinate of the pulsar from the following equation

$$Z_\star(t) = -\frac{m}{\mathcal{M}_\star} Z(t), \tag{5}$$

where $\mathcal{M}_\star$ is the mass of the pulsar, m is the mass of the planet and Z is its Z-th coordinate. Next, because of the Keplerian motion of the system, we have

$$Z(t) = a \sin i \left(\sin\omega\,(\cos E - e) + \cos\omega\sqrt{1-e^2}\sin E \right), \tag{6}$$

where a, i, ω, e are Keplerian elements of the planet (semi-major axis, inclination, longitude of periastron and eccentricity) and E is its eccentric anomaly that can be calculated from the Kepler equation

$$E - e\sin E = M, \qquad M = n(t - T_p), \qquad n = \frac{2\pi}{P}, \tag{7}$$

where M is the mean anomaly, P is the orbital period of the planet and T_p is the time of periastron. The function can be expanded in the following series (for all details see [3])

$$Z_\star(t) = -\frac{m}{\mathcal{M}_\star}\left(Z_0 + \sum_{k=-\infty}^{k=+\infty}{}' Z_k \mathrm{e}^{ikn(t-t_0)} \right), \tag{8}$$

where amplitudes Z_k are functions of Keplerian elements of the planets.

The expansion (8) has a very important property—terms corresponding to successive harmonics have monotonicaly decreasing amplitudes. These considerations allow us to state the following. In general, assuming that there are N planets in the system moving in Keplerian orbits, in the spectra of the TOA variations, we should see fundamental frequencies of planets equal to the mean motions $n_j, j = 1,\ldots,N$ and a few of their higher harmonics with frequencies equal $k_j n_j, j = 1,..,N$ where $k_j = 1,2,\ldots$ if eccentricities are not too small. Succesive harmonics (with frequencies $n_j, 2n_j, 3n_j, \ldots$) should have decreasing amplitudes.

3.2. REAL MOTION

In this situation, except for frequencies similar to those from the Keplerian case, we should see some additional frequencies, namely we should see a few of the following frequencies:

$$f_{k_1 k_2 \ldots k_N} = k_1 n_1 + k_2 n_2 + \ldots + k_N n_N, \tag{9}$$

where $k_1, \ldots, k_N$ are integer numbers. This is true for realistic planetary systems with weak interactions between planets. Since, in such a case, as stated by the KAM theorem [1], coordinates and velocities of planets are almost always quasi-periodic functions of time.

4. Doppler velocity measurements

The expansion of the Keplerian motion shown above can be easily applied to calculate a star's radial velocity resulting from its motion in the system. Namely, having the expansion of the Z-th coordinate of a star, we derive its radial-velocity by differentiating (8) over time

$$v_r(t) = -\frac{\mathrm{d}Z_\star(t)}{\mathrm{d}t} = \frac{\mathrm{d}}{\mathrm{d}t}\left(\frac{m}{\mathcal{M}_\star}Z_0 + \sum_{k=-\infty}^{k=+\infty}{}' \frac{m}{\mathcal{M}_\star}Z_k \mathrm{e}^{\mathrm{i}kn(t-t_0)}\right). \tag{10}$$

Thus we have

$$v_r(t) = \sum_{k=-\infty}^{k=+\infty}{}' \frac{m}{\mathcal{M}_\star}Z_k \mathrm{i}kn \mathrm{e}^{\mathrm{i}kn(t-t_0)}. \tag{11}$$

As is easily noticed, the above expansion has the same feature as the expansion of the Z-th coordinate. In fact, it is possible to show that the ratio of two successive harmonics is always less than 1. Thus, all useful properties of the radial velocity expansion can be derived from that of the Z-th coordinate. In addition, in a similar way one can examine characteristic features of the spectra of other processes (e.g. stellar pulsations) and compare them to the expansion of Keplerian induced radial-velocity variations.

In summary, if we observe radial-velocity variations that are of planetary origin then in their spectra we will detect a basic frequency corresponding to the planet orbital period and its harmonics with decreasing amplitudes (depending on the value of eccentricity). Finally, we should mention that, because of the finite accuracy of our observations, we can only detect a few of higher harmonics. Thus, from the observational point of view, expansion (11) is always finite and includes the basic term (corresponding to the planet orbital period) and a few of its harmonics.

5. Model-dependent least-squares approach

The usual way of proving that there are planets around a star (or a pulsar) consists of a direct least-square fit of Keplerian model to the observations. This procedure always gives certain values of orbital parameters and their formal errors. In the case of 'good' data it is the best and the fastest way to obtain reliable results. However, when we have spare data with large errors it is necessary to prove that the least-squares method can be used and that the found parameters' values and their errors are good estimates of the real values. The other, conceptional problem connected with a direct least-square fit of any physical model is such that it is not always certain that a chosen model is the correct one. In other words, it is possible to obtain a good and formally accepted fit of a model that in fact is not the real one (and thus responsible for the observed effects). For example the observer can, in some situations, misunderstand a system with two planets as a system with one planet moving in an eccentric orbit. This can happen when these two planets have orbital periods corresponding to a certain basic frequency and its first harmonic. One such example is presented in Figure 1. The radial velocity changes in Figure 1 (c) are the sum of velocities from (a) and (b) of two planets in circular orbits with orbital periods of about 800 and 400 days. For such a system we then calculate fake data at the times of real observations of 16 Cygni B with the observational errors of 16 Cygni B. Next we fit an eccentric orbit to the fake data and we get a very good fit wit the eccentricity about 0.51 (Figure 1, d). In this way we get an eccentric, entirely false orbit for a system that in fact consists of two planets while the least-squares fit is as good as possible. Even the χ^2 map [11] on (a, e) plane shows that the model is good and acceptable (see Figure 2) (for all details see [4]). These facts are especially siginficant for planets' hunters as their most crucial problem is whether the observed effects are really of planetary origin and what are the correct values of the orbital elements.

In the next section, we propose an alternative approach to the analysis of planet search observations that may have advantages over a direct least-square fit of Keplerian model. The approach that may properly go over the problems described above.

6. Frequency analysis

Let us consider a time series of observations $\{\psi_i\}_{i=1}^N$, representing a quasi-periodic process. In practice, in order to properly analyze the real data one must account for the following facts: (I) data are unevenly sampled and contaminated by noise, (II) a possible range of signal amplitudes can be very large, (II) periods corresponding to some harmonics present in the signal may be longer than the data span.

We approach the above problem by means of the following algorithm. Using the least squares method, we fit a function $F_{(1)}(t) = a_1 \cos(2\pi f_1 t) + b_1 \sin(2\pi f_1 t)$ to the observations $\{\psi_l\}_{l=1}^N$. As the first approximation of f_1, we take a frequency corresponding to the maximum in the periodogram of $\{\psi_i\}_{i=1}^N$. For this purpose we use Lomb-Scargle periodogram [7, 12, 11].

After the k-th iteration, we have assembled a set of residuals $R_k =$

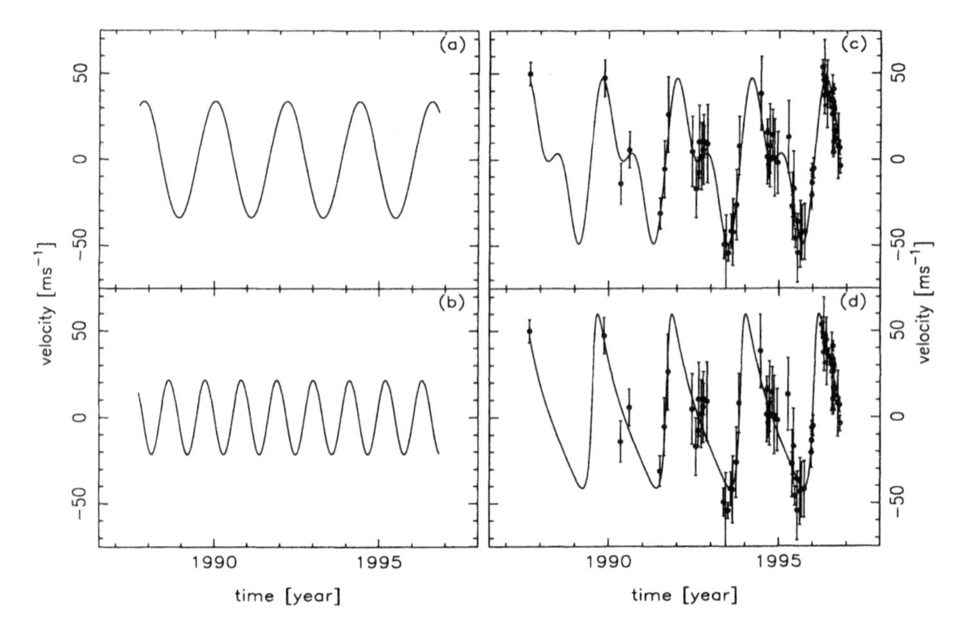

Figure 1. The fake radial velocities calculated for a planetary system consisting of two planets on circular orbits (a,b) with orbital periods of about 800 and 400 days. The set of 70 radial velocity measurements at the moments of real observations of 16 Cygni B (c) with the the errors of 16 Cygni B for the planetary system with planets from a and b. (d) The fit of an eccentric orbit ($e = 0.51$) to the data from c.

$\{r_l^{(k)}\}_{l=1}^N$ defined as

$$r_l^{(k)} = \psi_l - F_{(k)}(t_l), \qquad l = 1, \ldots, N, \tag{12}$$

where

$$F_{(k)}(t) = F_{(k-1)}(t) + a_k \cos(2\pi f_k t) + b_k \sin(2\pi f_k t),$$

and t_l, $l = 1, \ldots, N$ denote observation times. Using the periodogram of R_k, we estimate f_{k+1} and fit $F_{(k+1)}$ to the *original data* $\{\psi_i\}_{i=1}^N$. The whole process can be continued until a desired number of terms is obtained, or the final residuals fall below a predefined limit.

We call the above algorithm *unconstrained frequency analysis*, as we assume that all frequencies we find are independent. However, there exists a modification of the algorithm to a constrained form. For description see [6].

The general idea of the frequency analysis is that one can analyze any data (i.e. extract all the significant components present) without assuming any specific physical model. This is possible under the only assumption that the observations have quasi-periodic nature. This assumption is weak enough to include practically all these processes that can mimic planets.

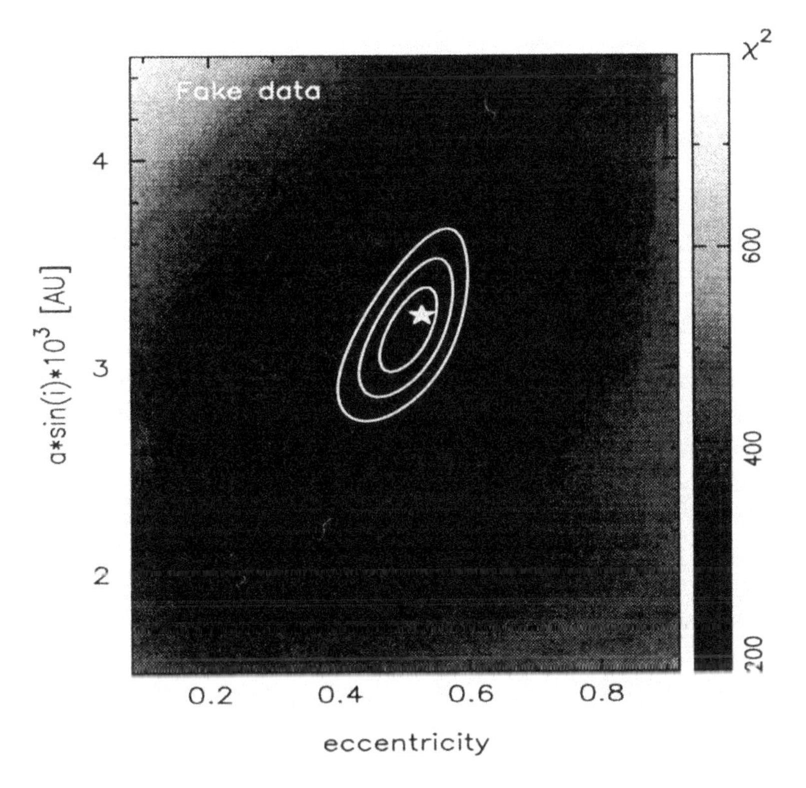

Figure 2. χ^2 map on the plane (e, a) of the fake data from Figure 1. The filled star indicates the global minimum found.

And even when we are sure that our observations' variations are of planetary orgin, the frequency analysis gives an independent solution that can be compared with that from the least-squares fit of Keplerian orbits.

7. Tests of the method

In order to present a typical run of the frequency analysis, below we perform two tests: one for the simulated timing observations of a pulsar and one for the simulated radial velocity measurements. For tests we have chosen the orbital parameters of the planets around PSR B1257+12 [14] and 70 Virgins [8]. Other tests of the method one can find in [3, 5, 6].

7.1. PSR B1257+12

For purposes of this test we computed about 400 unevenly sampled topocentric TOAs and added $3\mu s$ Gaussian noise. We assumed the parameters of the pulsar B1257+12 together with its planets [14]. As there are three planets in this system we expected to find three basic frequencies and possibly their first harmonics (as for planets $\mathcal{B}$, $\mathcal{C}$ the eccentricities are small but not zero). Additionaly, in the PSR1257+12 planetary system we have a 3:2 resonance between planets $\mathcal{B}$ and $\mathcal{C}$ with the resonance frequency $f_r = |2n_{\mathcal{B}} - 3n_{\mathcal{C}}| \approx 1/5.56$ years where n are mean motions of planets. Thus, in the TOA residuals of PSR1257+12, besides Keplerian frequencies, some additional frequencies should be seen. Precisely, we should see frequencies:

$$f_{1/2} = f_{\mathcal{B}} \pm f_r \quad \text{and} \quad f_{3/4} = f_{\mathcal{C}} \pm f_r, \tag{13}$$

with small but noticeable amplitudes. The detection of the above described combination of frequencies together with their specific values of amplitudes is the most significant signature that the observed variations are of planetary origin. Obviously, the frequency analysis, see Figure 3, finds the frequencies as predicted by the theory but in general this method can be used to analyze any quasi-periodic TOA variations. And the findings of the frquency analysis (i.e. frequencies, amplitudes, phases) can be subsequently used to accept (or reject) a hypothesis about the origin of these variations.

7.2. 70 VIRGINIS

For this test we simulated data consisting of 39 unevenly sampled radial velocity measuerements from 1988 to 1996.25 with 10 m/s Gaussian noise. In this way we had a data set resambling the real one [8]. The results of the frequency analysis are presented in Figure 4. From this analysis we can learn that the eccentricity of the orbit of 70 Virginis is large since we are able to detect three periodic terms in the data (main term and its two harmonics). From the ratio of the amplitudes of the basic term to its first harmonic, we precisely calculate the eccentricity and obtain that it is as assumed in the simulation. Thus the frequency analysis can be used to confirm the results from the orbital fit (for further details see [4]).

7.3. APPLICATION TO THE REAL DATA

The application of the frequency analysis to the real data has been desrcibed in the papers [4] (in case of the radial velocity measurements of 16 Cygni B) and [6] (in case of the timing observations of the pulsar B1257+12). In summary, in case of the pulsar B1257+12, the frequency analysis allowed to detect not only the Keplerian signatures but also the presence of the resonanse in the timing observations of this pulsar. Thus, the existence of this planetary system has been confirmed in this "model-independent" way. It is also possible to show how, by means of the frequency analysis, one can reject other explanations for the observed PSR B1257+12 TOA residual

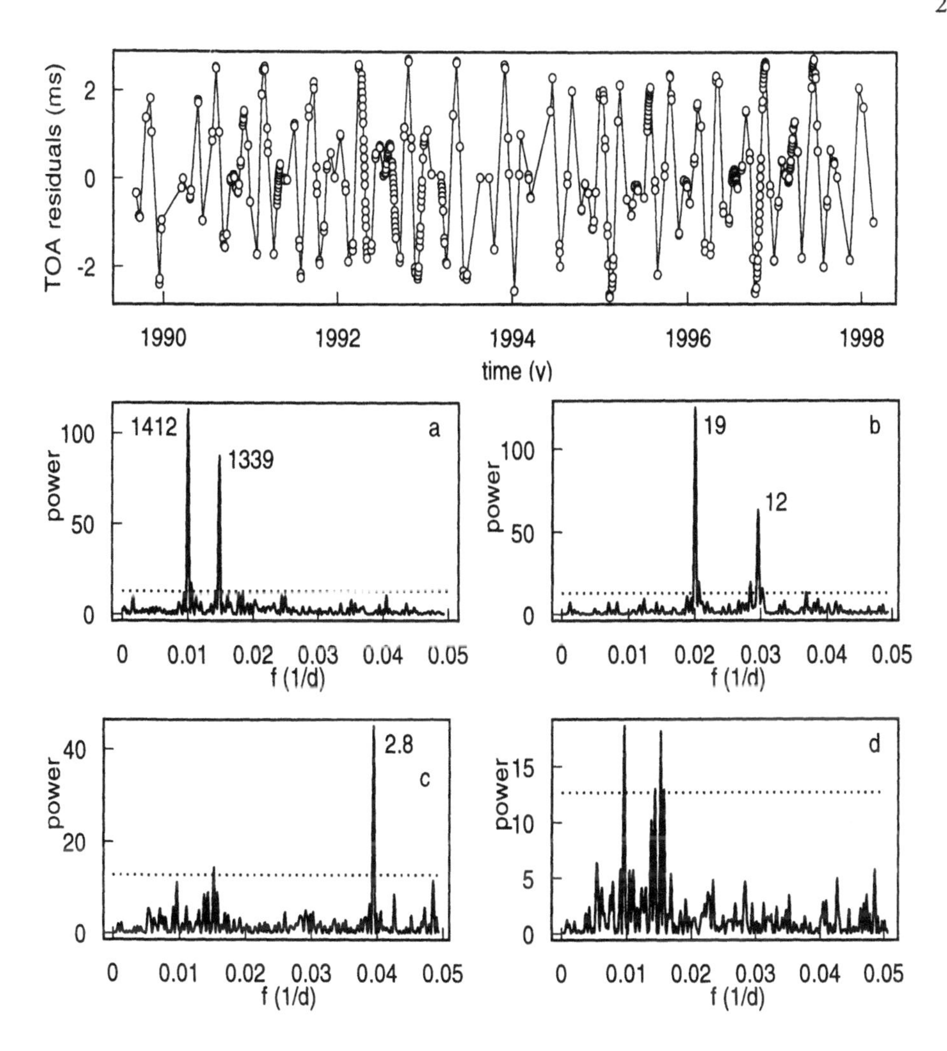

Figure 3. Frequency analysis of the simulated timing observations of PSR 1257+12. The top graph show TOA residuals after fitting the standard timing model. Tho other pictures are Lomb-Scargle normalized periodograms (LSNP) for which on the x-axis we have frequencies (in 1/day) and on the y-axis we have power units of LSNP. A number near a pick in each periodogram is equal to the amplitude (in microseconds) of the component of the signal. Dotted lines correspond to 3σ confidence levels of LNSP. In (a) we can see only fundamental frequencies of planets $\mathcal{B}$ and $\mathcal{C}$. In the next step, in (b) we can see frequencies of second harmonics of planets $\mathcal{B}$ and $\mathcal{C}$. Having subtracted them all from the signal we can see the fundamental frequency of the planet $\mathcal{A}$ (c). And finally, after additionaly subtracting the fundamental term of the planet $\mathcal{A}$, we can see some of the peaks corresponding to the resonance frequencies $f_{\mathcal{B}} \pm f_r$ and $f_{\mathcal{C}} \pm f_r$ in LSNP of the fake data (d).

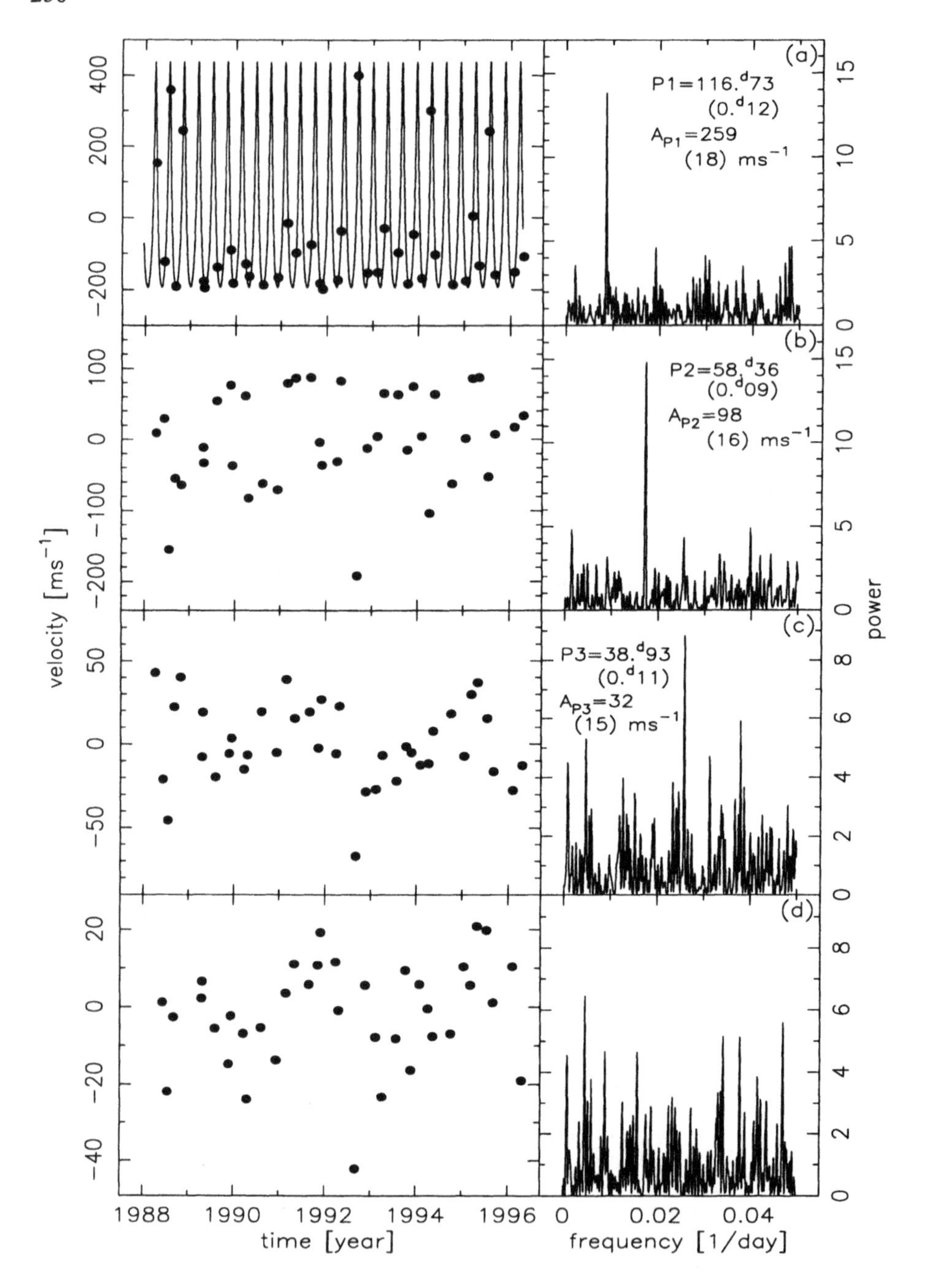

Figure 4. Frequency analysis of the simulated reflex velocity of 70 Virginis. *a,b,c,d:* Subsequent steps of the frequency analysis. Found periods and their amplitudes are presented. Numbers in parentheses are 3σ uncertainties of the parameters. There are three detectable periodicities in the data.

variations [5]. Whereas for the Doppler velocity measurements of 16 Cygni B, the frequency analysis allowed to show that with the current accuracy of the observations it is not possible to obtain the eccentricty of the orbit which on the basis of the least-squares orbital fit has been claimed to be as high as 0.6 [2].

What conerns our artificial example of the system at 1:2 resonance, let us underline that in this case also the frequency analysis does not rule out one planet solution with an eccentric orbit, however using this method we can only claim that two frequencies are detectable and one can interpret them as a signal caused by one planet with significant eccentricity or two planets in resonant circular orbits. Thus, using the frequency analysis one can make more justified hypotheses.

8. Final conclusion

The main idea of the freqeuncy analysis is to analyze the observations without apriori assuming any physical model. The only assumption is that the observations have quasi-periodic nature (in the sense of the definition from Section 2). This assumption is weak enough to include practically all the processes that can mimic planetary signatures in our data. In case of the pulsar timing observations, the frequency analysis has the potential to detect the resonance in the system as well as reject other hypothesis about the origin of the observed effects. In case of the radial velocity measurements, the frequency analysis gives us more insight into the data and allows us to verify the results from the orbital least-squares fit. Thus, it is suitable to use the frequency analysis together with the usual model-depenedent least-squares approach especially that the frequency analysis is numerically not demanding and easy to apply.

References

1. V. I. Arnold. *Mathematical Methods of Classical Mechanics*. Graduate Texts in Mathematics. Springer-Verlag, New York, 1978.
2. W. D. Cochran, A. P. Hatzes, R. P. Butler, and G. W. Marcy. The discovery of a planetary companion to 16 Cygni B. *ApJ*, 483:457–463, 1997.
3. M. Konacki and A. J. Maciejewski. A method of verification of the existence of planets around pulsars. *Icarus*, 122:347–358, 1996.
4. M. Konacki and A. J. Maciejewski. Frequency analysis of radial velocities of stars with planets. *ApJ*, 1998. to appear.
5. M. Konacki and A. J. Maciejewski. Methods of searching planets around pulsars. *MNRAS*, 1998. to appear.
6. M. Konacki, A. J. Maciejewski, and A. Wolszczan. Resonance in PSR B1257+12 planetary system. *ApJ*, 1998. to appear.
7. N. R. Lomb. Least-squares frequency analysis of unequally spaced data. *Ap&SS*, 39:447–462, 1976.

8. G. W. Marcy and R. P. Butler. A planetary companion to 70 Virginis. *ApJL*, 464:L147–L151, 1996.
9. G. W. Marcy and R. P. Butler. Characteristics of observed extrasolar planets. 1997. to appear in the Proceedings of The Tenth Cambridge Workshop on Cool Stars, Stellar Systems and the Sun.
10. M. Mayor and D. Queloz. A jupiter-mass companion to a solar-type star. *Nature*, 378:355+, 1995.
11. W. H. Press, S. A. Teukolsky, W. T. Vetterling, and B. P. Flannery. *Numerical Recipes in C. The art of Scientific Computing.* Cambridge University Press, New York, second edition, 1992.
12. J. D. Scargle. Studies in astronomical time series analysis. II. Statistical aspects of spectral analysis of unevenly spaced data. *ApJ*, 263:835–853, 1982.
13. A. Wolszczan and D. A. Frail. A planetary system around the millisecond pulsar PSR 1257+12. *Nature*, 355:145–147, 1992.
14. A. Wolszczan. Confirmation of Earth-mass planets orbiting the millisecond pulsar PSR B1257+12. *Science*, 264:538–542, 1994.

DIRECT SEARCHES: IMAGING, DARK SPECKLE AND CORONAGRAPHY

A. LABEYRIE
Collège de France
& Observatoire de Haute-Provence (CNRS)
F-04870 Saint Michel l'Observatoire
labeyrie@obs-hp.fr

1. Introduction

Large telescopes in space and the much larger interferometric arrays of telescopes currently studied will presumably provide images, spectra and resolved images of exoplanets. There is a huge potential for progress in this direction, and detailed information will be obtainable on planetary objects including possibly on their life-bearing characteristics.

2. The contrast problem in star/planet images

As seen from a distance of 10 parsecs, the solar system shows the Sun as a 5th magnitude star, with Earth appearing 0.1 arc-second away as a 30th magnitude source. Jupiter is 0.5 arc-second away and magnitude 27.5. Both planets are bright enough to be imageable with a Hubble-like Telescope, and spaced far enough from the Sun to be angularly resolved, but the contamination of the focal image with scattered light from the Sun would prevent the detection of the planets unless special precautions are taken.

If we expect a neighbouring star to have such planets (and one case was announced by M. Mayor after the Cargèse school [1]), we face a similar situation. In the 10 micron infra-red region, the planet's contrast is improved owing to its thermal emission, with 10^{-6} rather than 10^{-10} relative luminosity in the Earth's case, but the angular resolution is degraded 20 times and the zodiacal light emission, both in the observer's planetary system and in the target system, affect the planet's contrast.

3. Coronagraphy with opaque and phase masks

A similar problem of contamination by diffracted light long prevented the observation of the solar corona, which could be seen only during rare eclipses. A solution was found by astronomer Bernard Lyot in the 1930's [2], in the form of his coronagraph which provided excellent images of the corona, and films showing its evolution during hours of observation.

In addition to building a very clean lens, free from scratches and maintained free from dust, he tried to reduce the diffractive spreading of light from the disk into the image of the corona. In the telescope's focal plane, he installed an opaque mask blocking the disc's geometric image. Looking at the pupil through the mask (smaller than the pupil of his eye), he could locate the brighter areas which contributed most diffracted light around the mask. In the absence of dust, he could see a double bright ring circling the pupil, and had the idea of removing it with a mask slightly smaller than

J.-M. Mariotti and D. Alloin (eds.), Planets Outside the Solar System: Theory and Observations, 261–279.

the pupil's image obtained with a relaying lens. With a second lens relaying the focal plane, he could thus obtain a cleaned image of the source. The corona, little affected by the field and pupil masks, became visible.

This empirically discovered method was subsequently justified analytically by a number of authors, using the powerful tool of the Fourier transformation to model diffraction effects. In coherent light, i.e. from a point source with monochromatic spectrum, the Fourier transform relates the vibration amplitude distributions in a pupil and the image. Bonneau et al. [3], Mauron [4] and Malbet [5], Ftaclas [6], Brown [7]) Clampin et al. [8], among others, have studied variants of the basic scheme. When attempting to use the method with large ground-based telescopes for observing stellar environments, it became necessary to implement adaptative corrections of atmospheric turbulence.

When the Hubble Space Telescope was being conceived by NASA in 1976, I proposed a Lyot coronagraphic camera to look for star's planets. I made laboratory tests on a new 1.5m spherical mirror which J. Texereau had just completed at Observatoire de Paris for Calern's Schmidt telescope; and our calculations [3] had indicated that it would be feasible, in a few hours of exposure, to detect exo-planets with the 3 m aperture then considered. NASA invited me to contribute to the project study as a member of its instrument definition teams, although it did not express much interest in exo-planets in those days.

I therefore suggested to NASA that the instrument be built by the European Space Agency in exchange for a share of telescope time. This was agreed between NASA and ESA. I eventually resigned from my involvement with NASA, following its decision that the telescope would not be tested optically before launch. Given the absence of testing, the spherical aberration problem found after launch was no surprise to optical experts, but the corrective COSTAR optics subsequently added changed the pupil size and made the Lyot mask inoperant. The mirror size reduction, from 3 to 2.4 m, with respect to the initial plans had also made impossible the coronagraphic detection of a Jupiter-like planet, but some of the fast orbiting planets recently discovered are expected to be much brighter and may prove detectable with a 2.4 m space telescope and Lyot coronagraph, using the refinements described below.

Several ideas for improved coronagraphy emerged since. Claude and François Roddier [9] described the use of a phase-shifting mask replacing the occulting mask of Lyot. Suitably sized within the Airy peak, it can provide a destructive interference which cancels most of the star light in the relayed pupil. The light is rejected outside of the geometric pupil, and occulted with the same kind of aperture mask used by Lyot, as shown in figure 1. Its advantage, in theory, is that a planet as close as the first Airy ring can be imaged with a phase mask, whereas the Lyot mask cannot be made smaller than the third or fifth ring without losing much of its effect.

The classical representation of diffraction by Fourier transforms, if applied to the Lyot or the Roddier coronagraph, indicates that the residual starlight in the output image has an amplitude distribution which is a convolution of two complex distributions: the amplitude distribution just downstream from the occulting or phase-shifting mask, and the diffraction pattern from the Lyot diaphragm in the relayed pupil. This latter pattern is typically identical to the pattern just up-stream from the mask, but slightly wider in the usual case where the pupil diaphragm is made slightly smaller than the geometrical pupil to remove diffracted light.

This convolution description also applies to the case of a multiple aperture such as discussed in section 5.

Making the phase mask achromatic, for usability across a wide spectral band, requires a mask diameter and optical path shift proportional to wavelength. Reflective

Bragg holograms (Denysiuk [10], Stroke & Labeyrie [11]) can in principle achieve the required non-interacting superposition of many color-selective masks. Equivalent structures can in principle be built through the vacuum deposition of multi-layer dielectric films, each layer having a radially graded refractive index.

Little practical testing has yet been reported for phase mask coronagraphy, and it is unclear whether the theoretical advantages can be materialized. It may be remarked that the micro-dips used to encode data bits in CD-ROM systems are also phase masks. Used in reflective fashion, they make the dips appear dark although the disc is uniformaly coated with relective aluminum.

Another idea which emerged in the way of improved coronagraphic devices is the Achromatic Interference Coronagraph of Gay [12]. It uses a beam splitter, according to a modified Michelson interferometer scheme, to cancel the star's wavefront. It however produces a double image of the planet.

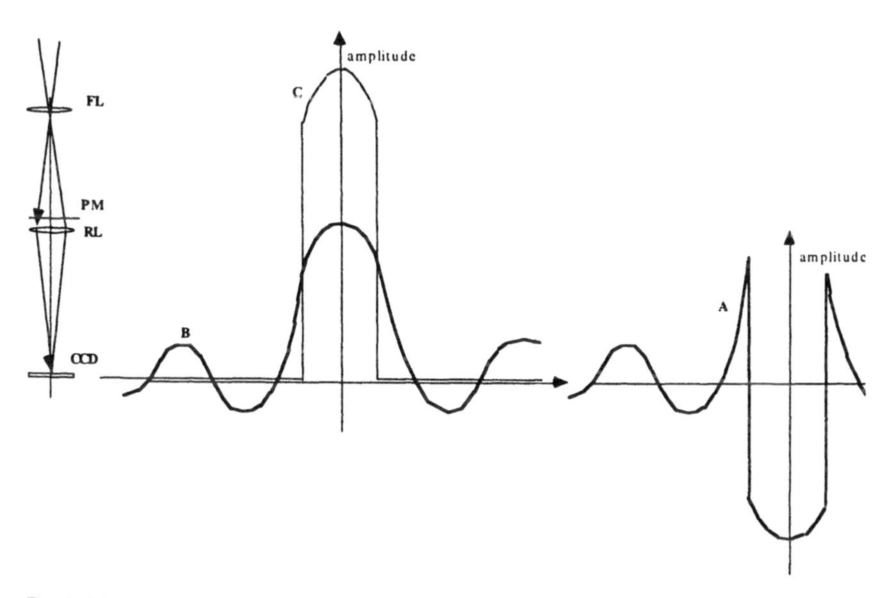

Fig. 1. Principle of the Roddier & Roddier phase mask coronagraph. FL field lens with phase mask in the image, PM opaque pupil diaphragm, RL, relay lens, CCD : detector in the relayed image. As seen in the vibration amplitude profile across the star's Airy pattern, the phase mask changes the vibration sign in the central part of the peak (A). The transmitted vibration distribution may be considered as the difference of the unmasked distribution (B) and twice the masked part of it (C). Diffraction being a linear transformation, in terms of vibration amplitudes, the corresponding difference is also found in the relayed pupil. The first term is the unaffected pupil image and the second is a broadened version of it, the amplitude of which can be balanced to cancel the star light as much as possible within the geometrical pupil. The balancing is achieved by adjusting the diameter of the phase mask, and possibly by attenuating the field outside of the phase mask. An aperture mask and re-imaging lens similar to those used by Lyot produce a cleaned field image where the star's diffraction rings are attenuated 10 or 100 times.

4. Dark hole and dark speckle imaging

Additional levels of cleaning are required to bring out planet images. Once basic coronagraphic cleaning is applied, most of the organized diffracted light is removed, i.e. the diffraction rings and the spikes caused by the structural spiders, but a speckled halo of residual starlight inevitably remains. It results mostly from residual bumpiness on the wavefront, and can be attenuated, in a selected region, by applying weak corrections to the shape of the main mirror, using actuators and the algorithm developped by Malbet, Yu and Shao [13].

The performance of this second cleaning step is limited by the accuracy of the wave mapping. Phase-contrast techniques may be considered for accurate wave mapping in this case, and they should be applied up-stream from the Roddier phase mask, where the pupil is uniformly illuminated. Phase-contrast is perhaps not applicable for wave sensing on ground-based telescopes since the high-altitude turbulence creates a shadow pattern which tends to confuse the phase contrast pattern.

Such optimization being made, which is easier said than done, there is again a residual speckled halo which still affects the detection of any faint Airy peak contributed by a planet. It cannot be further attenuated with actuators on the mirrors, but slight re-adjustments of the actuators, as caused by servo noise in the feedback loop, generate "boiling" of the speckle pattern.

Speckles are caused by interference of many vibrations having random phases, received from all parts of the mirror. The superposition can be more or less destructive or constructive, as one guesses from a vector representation of vibrations, where the addition amounts to a classical random walk. The speckle's scale size matches the width of the Airy peak. The two initial steps of darkening produce a highly destructive interference in the field zone considered. This is a random walk situation where the last step brings the walker very close to its starting point, and one guesses that very small phase shifts affecting randomly the numerous component vibrations suffice to degrade significantly the intensity cancellation, making the system exquisitely sensitive to low-level disturbances, such as residual "seeing" on Earth or fixed mirror bumpiness in space.

A long exposure, relative to the speckle lifetime, produces an image where the speckled halo becomes smoothed. This improves the visibility of underlying planet peaks, but photon noise ultimately limitates the detectable planet level: the planet should provide at least $n^{1/2}$ photons if n photons are detected from the star in a speckle-sized area of the field [24].

Rather than long exposures, the "dark speckle" analysis method [14], [15] instead uses thousands of exposures shorter than the speckle life-time. In each exposure, the speckle pattern is different and dark speckles appear at different locations. A dark speckle appearing at the planet's location improves its detectability since the contaminating photon count n is decreased.

The reduction algorithm builds a map of dark speckle appearances, and the planet tends to stand out as a location where no dark speckles ever appears, owing to the addition of intensities from the star and the planet, their light being mutually incoherent.

The method therefore provides a third level of cleaning. It can be applied when observing on Earth using a telescope equipped with an adaptive coronagraph, where residual turbulence achieves the speckle "boiling". It has been recently proposed to NASA for the Hubble Space Telescope, in the form of a "Faint Source Coronagraphic Camera", but the instrument was not selected.

Another opportunity arises with the Next Generation Space Telescope, for which a "dark-speckle coronagraph" is again being proposed. Suitable adaptive optics can make it usable in the infra-red and in the visible as well. Calculations indicate that exo-planets will be imageable.

Once a planet is detected, the contrast of its image can be improved by creating a permanent dark speckle in the star light at the planet's location. This allows in principle to obtain low-resolution spectra of the planet: the above-mentioned planet detection condition may be written $n_p > n^{1/2}$, n_p being the number of photons received from the planet, and implies that $n_p / k > (n / k^2)^{1/2}$. This indicates that the detection condition is still satisfied in each among k spectral channels, carrying k times fewer planet photons, if the star's light is attenuated k times in the white band, equivalent to k^2 times in the spectral channel considered. If, following the planet's discovery with "boiling speckles", a permanent dark speckle is thus created to attenuate the local stellar contamination k=100 times, a 100-element spectrum of the planet becomes obtainable in the same time which it took to discover it.

This challenging prospect announces attempts to detect spectroscopic signatures of life, and particularly analogs of the broad spectral bands caused by chlorophyll on Earth. Assuming vegetated planets, the contrast of the broad photosynthetic bands to be expected depends highly on the average vegetation density, and can reach high values in forested areas. There are reasons to believe however that a cloud cover approaching 50% must be present on densely vegetated planets. Whether diurnal and seasonal variations of the cloud and vegetation spectra can help discriminating photosynthetic spectral features from mineral features remains to be investigated. On Earth, a diversity of photosynthetic pigments providing broad absorbtion bands in varied parts of the visible spectrum have evolved since the onset of photosynthetic life. Whether analogous photochemistry, or completely different photosynthetic systems, may have evolved on other planets is unclear, but much insight will be gained if corresponding spectral features can be detected.

Ground-based observations with dark-speckle techniques have been initiated on the 1.52m telescope at Haute Provence, using 80-element adaptive optics developped by the Office National d'Etudes et de Recherches Aérospatiales [15]. These are yet far from achieving the expected dynamic range, but improved photon-counting detectors and adaptive correction should allow significant progress.

5. Extension to interferometric arrays: towards an Exo-Earth Discoverer

The option of using several free-flying elements for interferometry in space was first proposed in the early 1980's [16] and is now actively studied by the space agencies. Following a feasibility study by ESA [17], the agency is studying the proposal of Léger and Mariotti for DARWIN [18] [19], an infra-red array dedicated to the search for extra-solar planets. As described in the lectures of J.M. Mariotti, the coronagraphic cancelling of the star's light which he proposed extends and improves markedly the Bracewell scheme, where a beamsplitter recombines the plane waves from two apertures in opposite phases. The Terrestrial Planet Finder concept studied by american groups involves a similar optical principle, although initially ivolving a rigid beam-shaped structure to carry the optical elements.

A different coronagraphic approach is possible, also using a number of free-flying telescopes as collecting elements, according to the principle of the proposed "Exo-Earth Discoverer" [20]. As shown in figures 2,3,4 and 5 it involves a recombination scheme [21] [22] which violates the "golden rule" of W.Traub and J.Beckers . It causes a densification of the exit pupil, in comparison with the entrance pupil: the size/spacing ratio of the sub-pupils is considerably increased.

If certain conditions are met, and the wavefront elements properly phased, a directly usable recombined image is generated (figures 3, 4, 5). The conditions are: 1- all sub-apertures must be identical in size and shape; 2- the pattern of sub-aperture centers must be identical in the entrance and exit pupils. 3- no differential rotation of sub-pupils is allowed.

The pupil densification avoids the problem encountered with highly diluted Fizeau interferometers. For the case of many apertures, mainly considered in this section, these have a spread function which has a sharp interference peak surrounded by a wide halo of side-lobes. The peak-to-halo ratio, in terms of intensity levels, is equal to N, the number of apertures, but most of the energy goes in the halo owing to its large diameter, thousands of times larger than the interference peak.

A multiple star or complicated object gives an image which is a convolution of this spread function (intensities) with the object's distribution of intensities. The halo makes the image useless, since its convolution tends to create a dominating uniform background, with little energy left in the convolved peak (figures 2, 3, 4).

With a densified pupil instead, the halo, being the sub-aperture's diffraction pattern, is shrunk and the energy concentrated. The central interference peak still dominates the halo by a factor N, but now carries a much larger fraction of the energy since it is wider compared to the size of the shrunk halo.

The displacement of the diffraction function sketched in figure 3 is negligible when the pupil is highly densified, in which case the image of an extended object is a convolution of the intensity distributions in the object and the interference function, followed by a multiplication with the diffraction function.

The multiplication cancels most of the light in the feet of the latter function, and this causes a field limitation for the high resolution image, as seen in the simulations of figure 4 (bottom) where one of the stars seen at the edge of the simulated cluster in the Fizeau case vanishes from the densified-pupil image. The multiplication also darkens the vast halo of sidelobes found in Fizeau images, and since the total energy is conserved, the consequence is a strong intensification of the interference function appearing within the window.

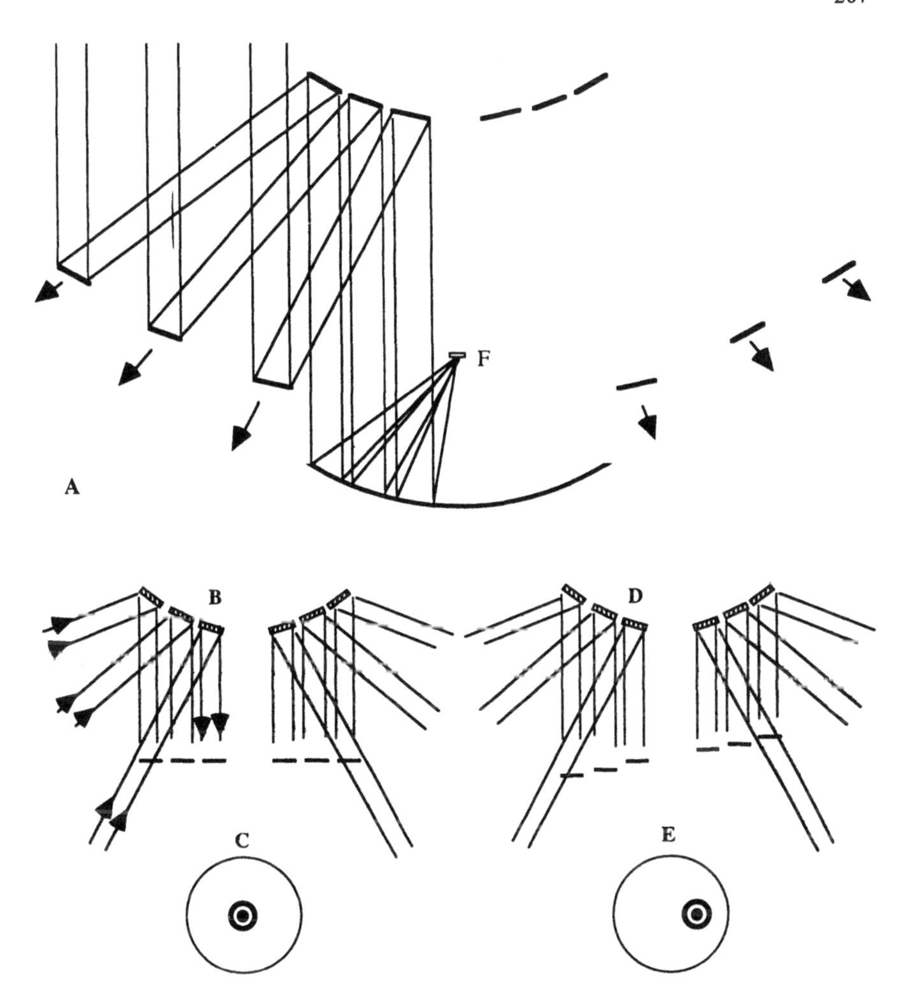

Fig. 2. Basic scheme of a Densified-Pupil Imaging Interferometer, here shown for clarity with flat collecting mirrors rather than telescopes (**A**). Collimated beams received from several flat mirrors (or from afocal telescopes) on flat mirrors Mi become parallel. The entrance aperture can be made thousands of times larger than the sub-apertures by translating (arrows) the collecting optics to enlarge their paraboloidal locus. The size ratio of sub-apertures and the aperture is considerably increased in the exit pupil, where the sub-apertures can be nearly adjacent. With a point source on axis (**B**), the wavefront segments (fat lines) can be phased to reconstruct a flat wave. The resulting image (**C**) has a peaked interference pattern (fat peak and ring), multiplied by a wider "window" (within the thin ring), the diffraction function, which is the Airy pattern of the sub-apertures. An off-axis point source provides wave-front segments (**D**)which are slightly tilted, but the comparatively large optical path differences occurring in the interferometer arms cause a much stronger global tilt of the reconstructed wave. In the image (**E**), the stair-shaped reconstructed wave generates a slightly displaced diffraction function, and a more markedly displaced interference function. Assuming many point sources or an extended source, the classical convolution holds for the interference function, but the diffraction function causes a field limitation. The number of resolved elements is limited to N^2 with N apertures, in accordance with the general rule stated by Lannes [25] for snapshot imaging.

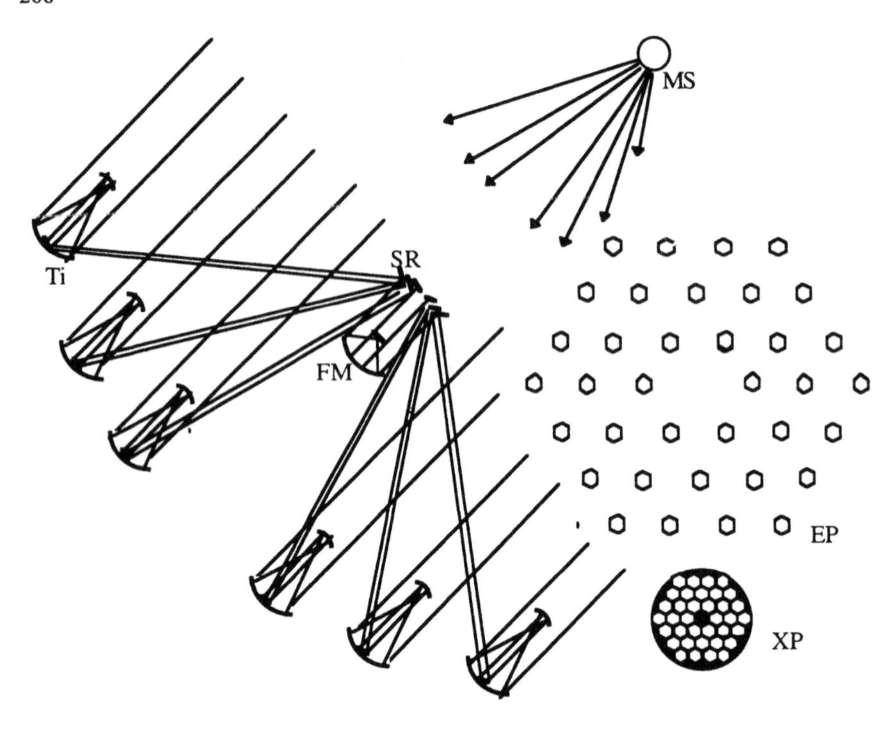

Fig. 3. Proposed Exo Earth Discoverer, a 1-to-100km interferometric array in space. With the 36 free-flying telescopes shown here the exit pupil can be densified for efficient imaging and coronagraphy. Collimated coudé beams from telescopes Ti are reflected from flat mirrors SR towards the concave image recombination mirror FM. For optical path equality, the telescopes are located on a paraboloïdal surface, with focus at SR and curvature center at the metrology satellite MS. The arrangement provides a compact exit pupil XP which is a densified version of the diluted entrance pupil EP. The full densification achievable with hexagonal sub-pupils favors coronagraphic masking (Lyot or Roddier type) in the recombined image and a relayed pupil downstream, at visible and infra-red wavelengths.

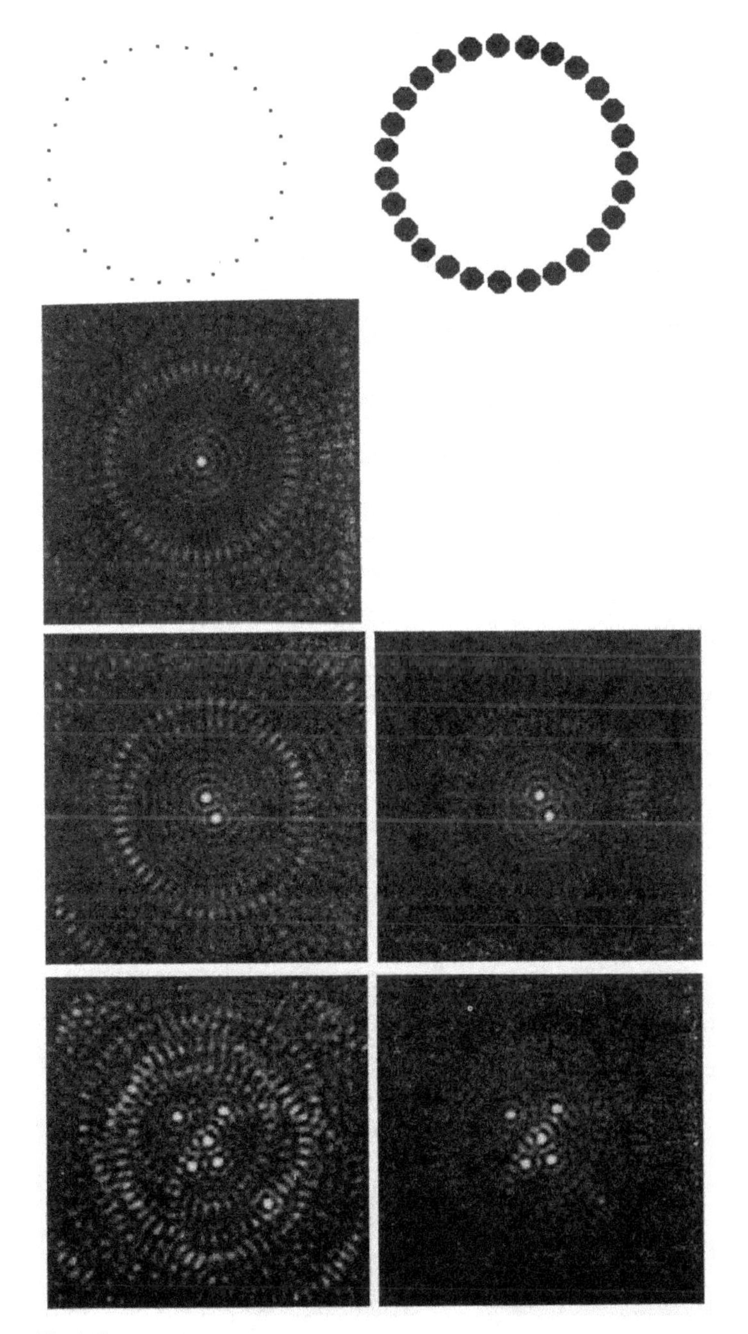

Fig. 4. Comparison of image formation at the combined focus of 27 phased apertures with Fizeau (left) and densified (right) pupils. Top: entrance pupil with 27 sub-apertures 100 times smaller than the ring (left), and the densified exit pupil obtained from the same entrance pupil (right) to attenuate the side lobes of the spread function (left, second line). The following rows show corresponding images for a binary and a sextuple star. The densification intensifies the images at right 10 000 times since the shrunk diffraction envelope concentrates the energy from the halo of sidelobes, but the field extent is decreased.

5.1. Expanding EED into an "Exo-planet imager"

Figure 5 illustrates the Earth and Moon, as they would appear in yellow light at the combined focus of the EED from a distance of 3 parsecs. A real pictures of these bodies, produced by the Galileo probe, was convolved with the interference function of a 150-element EED and subsequently multiplied by the diffraction pattern of the sub-pupil. Further post-detection image processing is normally required for better rendering but these simulations verify the theoretical analysis[21] previously mentioned. Its equation (6) gives the photon count in the planet image, and equation (9) the signal to noise ratio.

Depending on the aperture configurations, it requires at least 30 telescopes of 3 meter size, arrayed across 100km, to produce images such as these. The exposure time should not exceed 30mn to avoid rotational blurring, although stroboscopy with repeated short exposures at diurnal intervals can be attempted in the absence of fluctuating cloud cover. Light from the parent star must be removed as much as possible to obtain such images, and this requires coronagraphic masking, plus active dark-speckle implementation, in each subpupil up-stream from the combining optics.

The method is also applicable in the infra-red, but baselines of several thousand kilometers become needed, requiring rather large optics to retransmit the beams from the collecting stations to the combiner. This difficulty is avoided by the Moth-Eye Interferometer concept discussed below.

Such images can be recorded through an integral field spectrograph to obtain independant spectra in each of their pixels. If spectral features analogous to the broad absorbtion bands of terrestrial photosynthetic pigments are present, their distribution on the planet can be mapped.

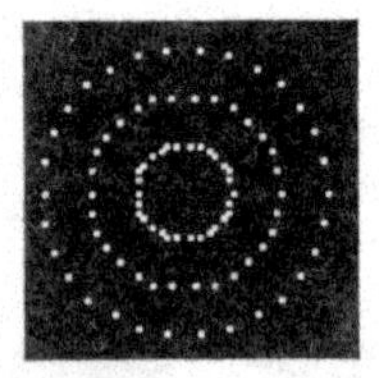

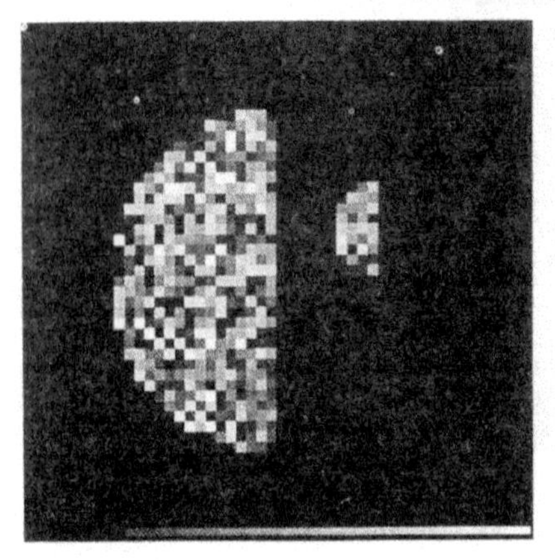

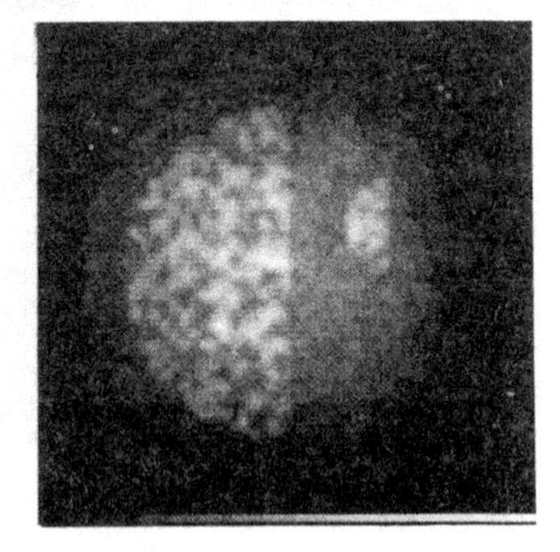

A

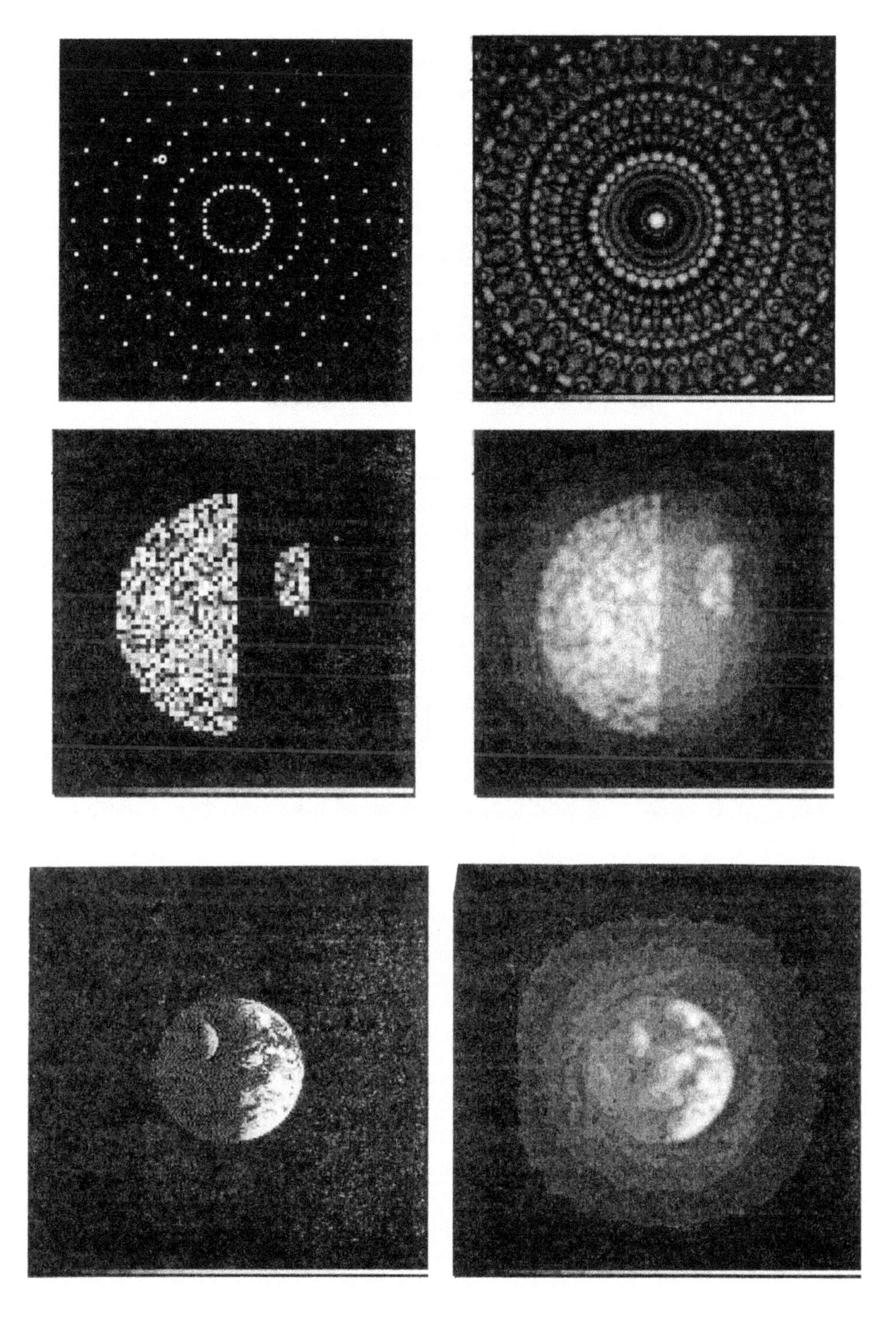

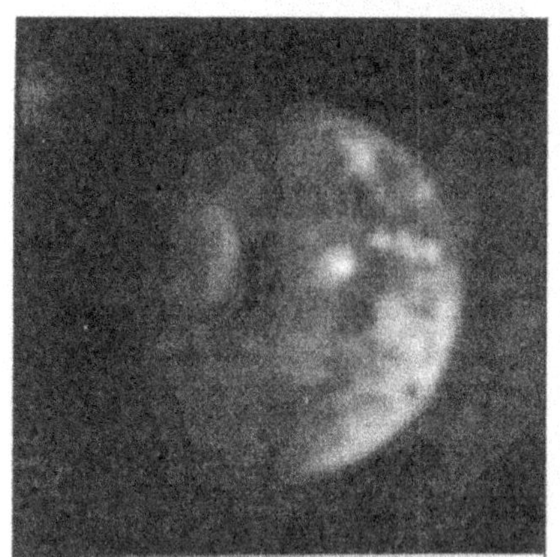

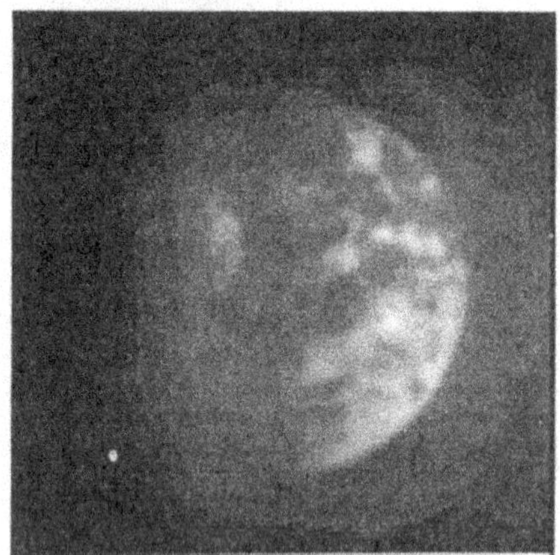

Fig. 5. Numerical simulations of exo-planet imaging with the EEI. **A**- densified pupil of a 90-aperture EED, fake planet with random dots and its raw image obtained at the combined focus. **B**- Same with a 150-aperture EED, where more reconstructed details are seen, and (bottom) case of the Earth with Moon as seen from 3 parsecs with a 150 km EEI in visible light. **C**- In the absence of post-detection deconvolution to remove the feet of the interference function, improved resolution (bottom right) degrades the contrast. If taken through a green filter, such images should evidence vegetated areas such as the Amazon.

5.2. Coronagraphy with the EED

The very small field extent obtained in the observing situation just discussed, tens or hundreds of micro-arc-seconds, requires that the planet's exact location be previously known from observations with smaller intruments. Relative to its parent star, the contrast of a planet is typically improved to 10^{-6} in the 10-micron infra-red, and coronagraphic observations in this range facilitate the detection. The 8m aperture of the planned NGST is however too small for separating planets correctly at 10 microns, although it should be attempted in the visible as mentioned above, hence the larger size of the DARWIN and the Terrestrial Planet Finder projects.

The EED can also provide efficient coronagraphy (figures 6, 7). Configurations with 100m-1km size are adapted to the situation just mentioned. Much larger sizes are in principle also efficient for coronagraphy on neutron stars, active galactic nuclei and other luminous compact sources.

Classical coronagraphy involved a mask in the image plane, and a second mask to remove diffracted light in a relayed pupil. When EED has its exit pupil completely densified, i.e. with sub-pupils made exactly adjacent and possibly hexagonal in shape for full "paving", the coronagraph's effect on an unresolved source is the same as it would be in a conventional filled telescope with the same pupil pattern. Indeed, the densified pieces of the wavefront collected in the entrance pupil cannot be distinguished, in terms of diffraction, from the continuous wavefront in the exit pupil of a conventional telescope. Nearby sources such as exoplanets can be imaged without being significantly affected by the pair of masks.

Either the Lyot or the Roddier techniques can be used, the latter being more tolerant of gaps between the sub-pupils and providing higher resolution.

The "Exo Earth Discoverer" being proposed to the space agencies [20] follows these principles, and the concept can be made compatible with DARWIN, in terms of using a single set of free-flying telescopes to feed interchangeable focal optics so that both recombination principles be usable.

For a large usable field, the number of elements N should be as large as possible, and several kinds of ring, hexagonal or square paving geometries appear of interest. For coronagraphic uses, a periodic hexagonal diluted paving (figures 6A and B) may be preferable, but the option of using a ring with a Roddier phase mask (figure 7) in the recombined image is also appealing and being studied. The amplitude convolution description of the cleaned stellar image, mentioned in section 3, indicates that a multi-aperture diaphragm matching the geometrical pupil, located downstream from the recombined field where a phase mask is inserted in the interference peak, causes a convolution of the amplitude distributions found just up-stream and downstream from this mask. The size of the phase mask is optimal when the convolution has a minimal intensity in the annular field of interest.

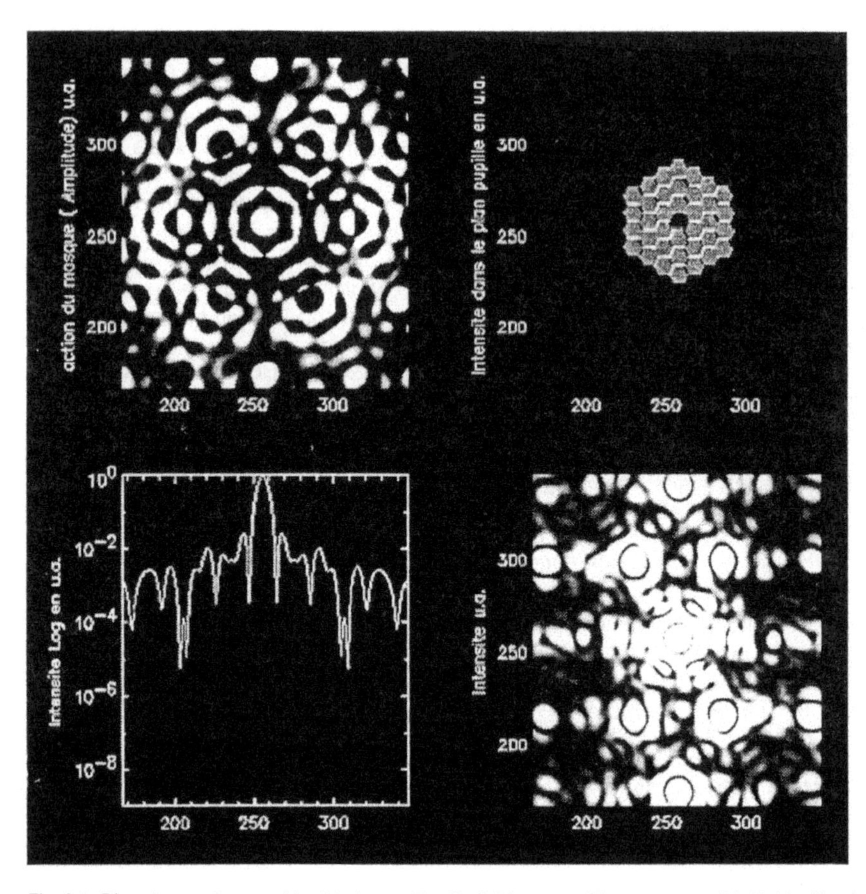

Fig. 6 A: Direct image of a star with a 36-element Exo-Earth Discoverer. The entrance pupil is highly diluted, and the exit pupil is fully densified, with adjacent hexagonal elements. Top left: real part of the image amplitude in the combined focal plane, where negative areas appear dark and positive areas appear bright; top right: exit pupil, bottom right: image intensity; bottom left: intensity profile.

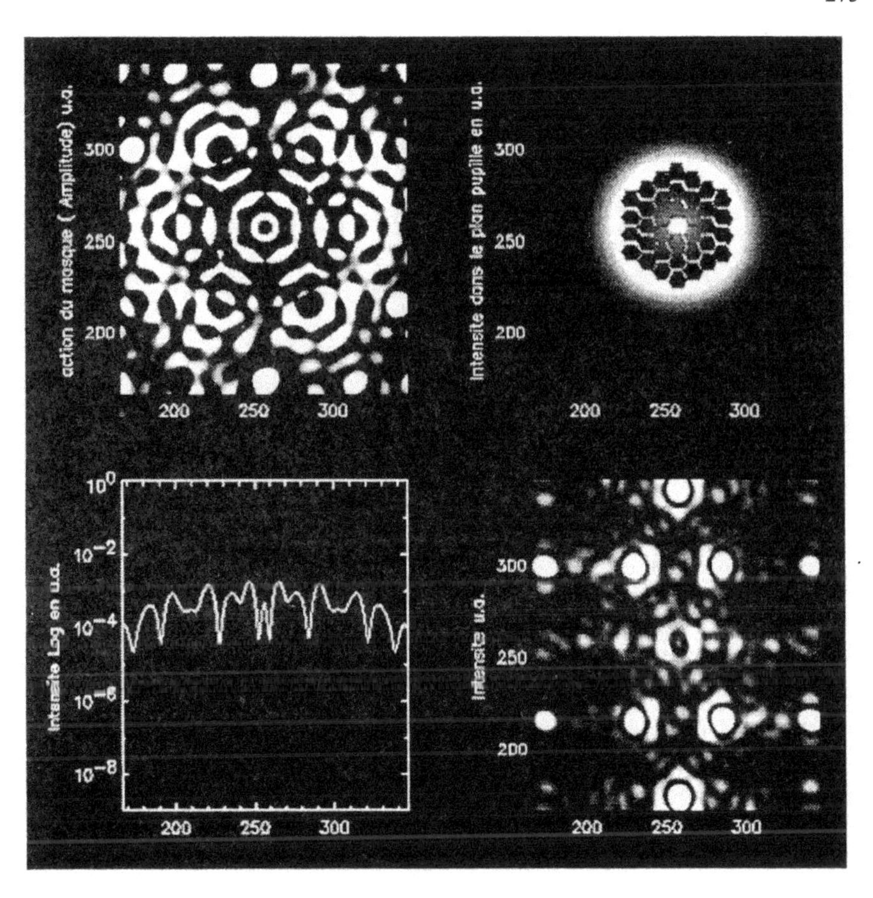

Fig. 6 B: Same as figure 6A with a Roddier phase-mask coronagraph now in operation. The central dark disk now seen in the amplitude image (top left) just downstream from the phase mask indicates its effect. The pupil intensity (top right) is now shown just up-stream from the pupil diaphragm, the pattern of which matches the geometric pupil image formed after the focal plane. The darkened hexagons and brighter halo now apparent illustrates the effectiveness of the phase mask. The cleaned image (bottom right) has an attenuated central peak and sidelobes. The image of a planet, 12 magnitudes fainter than the star, becomes visible (arrow). A white circle indicates the size of the high resolution field. Spurious deviations from the expected ternary symetry in the star's diffraction pattern result from small position errors in the pupil pattern. Dark hole and dark speckle techniques applied to the residual star light are expected to gain the extra 3 to 5 magnitudes needed to detect a typical Earth-like planet, expected to be a million times (15 magnitudes) fainter than its star in the 10 micron infra-red. (from Riaud, Moutou, Boccaletti &Labeyrie, in preparation).

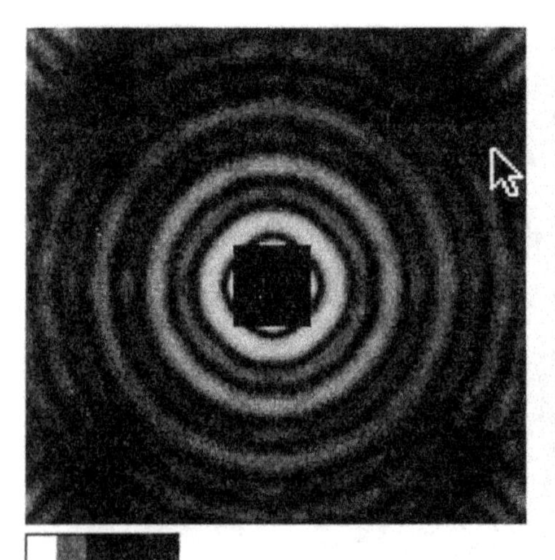

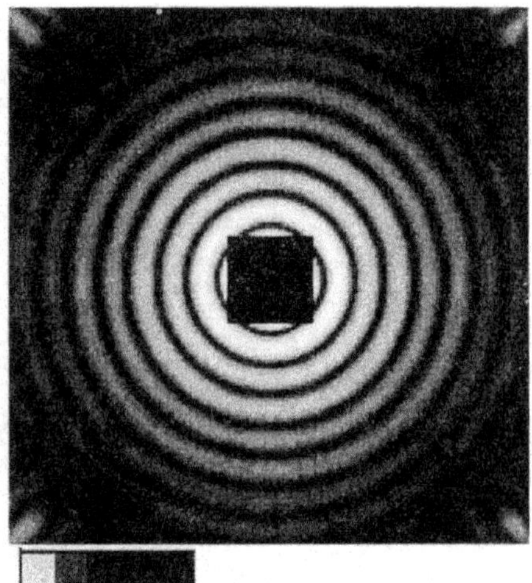

Fig. 7: Coronagraphy with an EED having 36 circular apertures arrayed along a circle. The simulated star image is shown with a phase mask (left) and without (right). The central black square is a software mask added in the display for better visibility of the faint feet, and identical intensity scales (steps of 10 in the calibration marks at bottom left) are used in both images, with 0.5 contrast for better visibility. A faint horizontal band, 2,000 times fainter than the star's peak intensity obtained at right with the software mask removed, is added to both images. It is barely detectable at left (arrow) but invisible at right. A planet with 10^{-6} relative level in the infra-red would be similarly detectable as a faint peak, with additional cleaning techniques such as dark hole apodization and dark speckle analysis. The imaging field is here wider than with the hexagonal arrangement of figures 6, but the image is fainter, with equal collecting areas, since the pupil is not so dense.

Also achievable is a progressive up-grading of the array with additional telescopes, as operating experience is gained. The array size can also be extended for observing compact objects having a high surface brightness. Low-brightness objects such as an exo-planet are in principle observable with 27 apertures of 8m, arrayed as a 20km circle. At 5 pc, a Jupiter is resolvable in 20x20 elements. Brigther objects are observable with smaller (1.5m) sub-apertures using 100 km baselines.

6- Extension of EED towards a Moth Eye Interferometer

With their spherical symetry, Schmidt-type telescopes provide a wide field coverage. Similarly, a giant interferometer in space can provide high-resolution images with unlimited field if arranged as sketched in figure 8.

The collecting telescopes of EED are replaced by simple spherical mirrors, all part of a single giant sphere. The instrument requires no global pointing and keeps a fixed attitude, with only the focal stations being moved to produce high-resolution images of selected parts of the sky. Densified-pupil imaging is an essential part of the concept since the collectors are necessarily diluted, if only to avoid masking each other.

Each focal station contains small mirrors and delay lines arranged to correct the huge spherical aberration of the large spherical reflector, together with the corresponding optical path differences. It also achieves the pupil densification, required to produce images such as those of figure 5. Coronagraphic masks can also be used in the focal station to clean either the sub-images or the combined image. The former case arises when attempting to image details, clouds or continents, of an exo-planet which is well separated from its parent star by the sub-apertures. The latter case occurs typically when attempting to find an exo-planet with short baselines (100m) in the infra-red, as discussed for EED.

The Moth Eye Interferometer will be of obvious interest after the initial operation of a pointable imaging interferometer such as the EED. Later, many EEDs may become needed for observing many fields simultaneously, as currently done on Earth with the multitude of existing telescopes, but the Moth Eye Interferometer will provide a much more economical integrated approach since each of its mirrors will contribute to simultaneous observing in many directions, given as many focal stations. Whether a first-generation EED should already be conceived as a "bare-bone" Moth-Eye Interferometer, suitable for later up-grading towards wider and wider sky coverage, remains to be investigated.

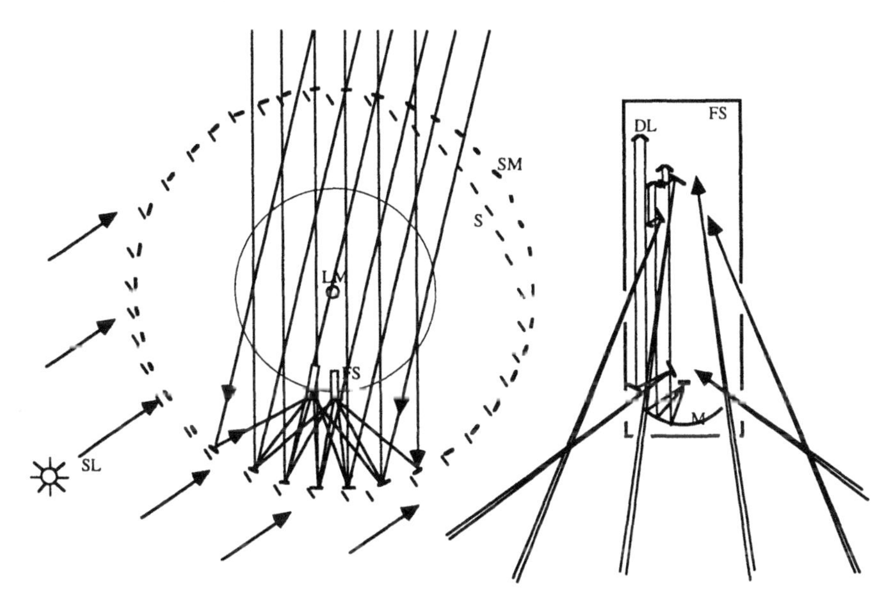

Fig. 8. Proposed "Moth Eye Interferometer" using many free-flying mirrors. The mirrors SM are elements of a single giant sphere. Focal stations FS are mobile along the focal surface. Each collects light beams (inset at right) from multiple elements SM and corrects the spherical aberration of the giant sphere, using delay lines DL. The focal optics also provides a densified pupil, for efficient direct imaging and coronagraphy in the recombined image. The free-flying mirrors SM are maintained at fixed positions by small solar sails S receiving sun light SL, under control from a centrally located laser metrology system LM. Since each mirror SM can feed many focal stations, the Moth Eye concept allows efficient parallel observing in many independant directions, with fewer mirrors than would be required if many narrow field interferometers were to be used. Dimensions from kilometers to thousands of kilometers appear feasible, and the number of mirror segments may be progressively increased from dozens to thousands.

7. Feasibility of giant baselines

Objects of extremely high surface luminosity, such as the few known neutron stars emitting visible light, can in principle be observed with extremely long baselines. 100,000 km baselines thus appear usable on the few optical pulsars known [20]. As shown in figure 4, auxiliary optics in the form of out-rigger mirrors could be added at some stage to the EED for this purpose. The fringe acquisition will be difficult, but probably not impossible: in the absence of a better method, one can always increase progressively the baselines while monitoring the fringes.

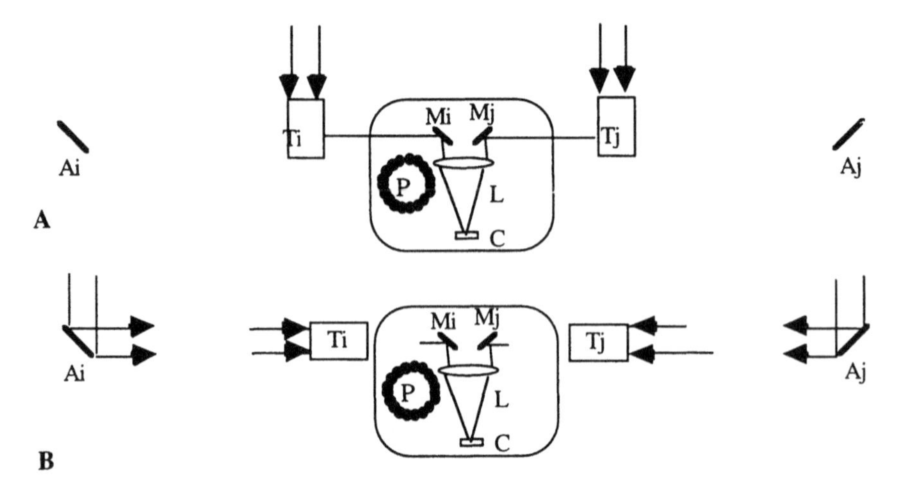

Fig. 9. Space interferometer and outrigger mirrors for giant baselines of 1,000 to 100,000 kilometers, usable for neutron star imaging. **A-** Normal interferometer mode, with free-flying telescopes Ti, Tj pointed axially. The nearly flat out-rigger mirrors Ai, Aj are unused, but travel towards their assigned location for the subsequent observing session with the giant baseline; **B-** Giant baseline mode, with Ti, Tj pointed towards the out-rigger mirrors. The neutron star image is focused by Ai, Aj onto the telescope apertures. Background stars appearing around Ai, Aj provide information on the array attitude, while laser beams (not shown) define its internal geometry.

8. Conclusion

Coronagraphic and interferometric approaches to the detection of exo-planets in images are actively developped by a number of groups. There is a potential for tremendous gains in this area. The goal of searching spectral signatures of photosynthetic life on exo-planets is quite challenging but does not appear unrealistic. Nor does it appear btain images showing surface details on an exo-planet. With free-flyer arrays, the required multi-kilometric sizes may be quickly achievable once smaller baselines become mastered.

9. References

1. Mayor, M. and al. (1998) *in press*
2. Lyot, B. (1931) Photogaphie de la couronne solaire en dehors des éclipses, *CRAS* (Paris) **193**, 1169
3. Bonneau, D., Josse, M. and Labeyrie, A (1975) in *Image processing techniques in Astronomy.* In: de Jager C., Nieuwenhuijzen H. (eds), Dordrecht, Reidel, 403-409
4. Mauron, N. (1980) Haute résolution angulaire et coronographie en astronomie spatiale. Etude theorique et experimentale du mode coronographique de la "faint object camera", *thesis University Aix-Marseille*
5. Malbet, F. (1996) High angular resolution coronagraphy for adaptative optics, *A&AS*, **115**, 161-174
6. Ftaclas, C. (1995) *Proceedings of the 15th NSO/Sac Peak Summer Workshop*, ed. J.R. Kuhn and M.J. Penn (Singapore, World Scientific), p. 181
7. Brown, R.A. (1990), in *Astrophysics from the Moon.* In: Mumma M.J., Smith H.J. (eds) *AIP Conf. Proc.* **207**, New York, AIP, 87-94 (19)

8. Clampin, M., Notta, A., Golimowski, D.A., Leitherer, C. and Durrance S.T., (1993) Coronographic imaging of the bipolar nebula around the luminous blue variable R127, *ApJ* **410**, L35
9. Roddier, F. and Roddier, C. (1997) Stellar Coronagraph with Phase Mask, *PASP* **109**, 815-820
10. Denysiuk, Yu. N. and Ganzherli, N.M. (1992) Opt. Eng., 31, 731, Opt. Eng., 32, 958
11. Stroke, G.W. and Labeyrie, A. (1966) White light reconstruction of holographic images using the Lippmann-Bragg diffraction effect, *Phys. Letters* **20**, 4
12. Gay, J. and Rabbia, Y. (1996) Principe d'un coronographe interférentiel", *CRAS* **322 (IIb)**, 265
13. Malbet, F., Yu, J.W. and and Shao, M. (1995) High-Dynamic-Range Imaging Using a Deformable Mirror for Space Coronagraphy, *PASP* **107**, 386-398
14. Labeyrie, A. (1995) Images of exo-planets obtainable from dark speckles in adaptative telescopes, *A&A* **298**, 544-546
15. Boccaletti, A., Moutou, C., Labeyrie, A., Kohler, D. and Vakili, F. (1998) Present performance of the dark-speckle coronagraph, *A&A, in press*
16. Labeyrie, A., Authier, B., Boit, J.B., De Graauw, T., Kibblewhite, E., Koechlin, L., Rabout, P. and Weigelt, G. (1984) TRIO, a kilometric optical array controlled by solar sails, *Bull. Am. As. Soc.* **16**, 828
17. ESA/Space Interferometry Study Team (1996). *Kilometric Baseline Space Interferometry*. ESA SCI(96)7.
18. Leger, A., Mariotti, J.M., Mennesson, B., Ollivier, M., Puget, J.L., Rouan, D. and Schneider, J. (1996) The Darwin Project, *Ap&SS* **241**, 135
19. Mennesson, B., Léger, A., Mariotti, J.M. and Ollivier, M. (1997) A Space Infrared Interferometer to Detect and Characterize Earth-like Extrasolar Planets; the DARWIN project, *JENA conf.* E, Thessaloniki, Greece, 84
20. Labeyrie, A. (1998) Exo-Earth Discoverer, a free-flyer interferometer for snapshot imaging and coronagraphy, proc. conf. "Extrasolar planets: Formation Detection and Modelling, Lisbon 27 April-1 May 1998, Kluwer.
21.Labeyrie, A. (1996), Resolved imaging of extra-solar planets with future 10-100 km optical interferometric arrays, Astron. Astrophys. Supp. **118**, 517-524
22. Labeyrie, A. (1997) La percée attendue des télescopes géants interférométriques, CRAS 325, Ser. II, p. 45-50
23. Labeyrie, A. (1998) Kilometric arrays of 27 telescopes: studies and prototyping for elements of 0.2m, 1.5m and 12-25m size, *SPIE* **3350**, *Astronomical interferometry*, Kona, 20-24 march
24 Angel, J.R.P., (1994) Ground-based imaging of extrasolar planets using adaptative optics, Nature **368**, 203-207.
25 Lannes, A., Anterrieu, E. & Marechal, P. CLEAN and WIPE, Astron. Astrophys. Supp. **123**, pp. 183-198 (1997).

DIRECT SEARCHES: INTERFEROMETRIC METHODS

J.-M. MARIOTTI
European Southern Observatory
Karl-Schwarzschild-Str.2, D-85748 Garching b. Muenchen
(notes written by B. Mennesson and A. Léger, November 10, 1998)

Abstract. After giving a brief introduction to optical and infrared interferometry, its observables and theoretical capabilities, we focus on the specific problem of imaging faint point-like companions close to very bright sources with large interferometric arrays. Low resolution spectroscopy of Earth-like planets is feasible from space, preferably with large infrared nulling interferometers. We put the emphasis on recent concepts that could make such projects realistic in the near future.

1. Interferometry: an introduction

We recall here a few basic principles about interferometry and concentrate on aperture synthesis imaging. We deliberately chose to give a rather qualitative analysis of interferometric imaging principles and properties. More detailed computations are given in previous courses (Mariotti 1988, Haniff 1994) while exhaustive presentations of the field can be found in dedicated reviews (Shao 1992, Ridgway 1998).

1.1. INTERFEROMETRIC IMAGING

Interferometric imaging can be readily seen as an application of Fourier optics, and linear systems theory, as soon as two fundamental conditions are fulfilled:

- The optical system under consideration can actually be represented using the linear system theory. This means that for some specific variable, the overall optical system can be seen as a black box linear operator S, yielding an output g_1 for any input g_0:

J.-M. Mariotti and D. Alloin (eds.), Planets Outside the Solar System: Theory and Observations, 281–296.

$g_1(x_1, y_1) = S[g_0(x_0, y_0)]$
The input can always be decomposed as a sum of impulse functions:
$g_0(x_0, y_0) = \int\int_{-\infty}^{\infty} g_0(\epsilon, \eta).\delta(x_0 - \epsilon, y_0 - \eta)\, d\epsilon\, d\eta$
The optical system being linear versus parameter g, we get:

$$g_1(x_1, y_1) = \int\int_{-\infty}^{\infty} g_0(\epsilon, \eta).S[\delta(x_0 - \epsilon, y_0 - \eta)]\, d\epsilon\, d\eta \qquad (1)$$

Eq.1 is called the superposition theorem and $S[\delta(x, y)]$ is the impulse response of the system.

- The impulse response should be space invariant, which refers to the well known isoplanetism condition. With this second condition fulfilled, the impulse response does not depend on the object location close to the optical axis, so that (with a magnification of 1):
$S[\delta(x_0 - \epsilon, y_0 - \eta)] = h(x_1 - \epsilon, y_1 - \eta)$
hence,
$g_1(x_1, y_1) = \int\int_{-\infty}^{\infty} g_0(\epsilon, \eta).h(x_1 - \epsilon, y_1 - \eta)\, d\epsilon\, d\eta$
which is a convolution relation,
$g_1(x, y) = g_0(x, y) * h(x, y)$.
In the Fourier plane with conjugate coordinates (u,v), we obtain the simple relation:
$\tilde{g_1}(u, v) = \tilde{g_0}(u, v).\tilde{h}(u, v)$
where $\tilde{h}(u, v)$ is the transfer function of the system.

In the case of image formation, Fourier conjugated coordinates are the angular direction of the source in the object plane (x,y) and the reduced coordinates (u,v) in the image plane (expressed in units of wavelengths). Using the scalar monochromatic theory of electromagnetic waves, the linear variable of the system can be:

1. The electric field complex amplitude in the case of spatially coherent incident light. In that case the transfer function is simply the entrance pupil function P expressed in units of wavelengths. i.e.:
P(u,v)=1 at the pupil location (for an ideal pupil, possibly diluted) and
P(u,v)=0 elsewhere.
2. The irradiance spatial distribution in the incoherent case. This is the case of interest when dealing with astronomical sources at optical wavelengths. The resulting irradiance distribution I(x,y) in the image plane is then linked to the object spatial distribution O(x,y), by a convolution relation, or more simply in the conjugate Fourier plane by:

$$\tilde{I}(u, v) = \tilde{O}(u, v).\tilde{h}(u, v) \qquad (2)$$

where the transfer function $\tilde{h}(u, v)$ (currently designed by MTF(u,v)) is the autocorrelation of the pupil function P(u,v) defined above.

We note here that no assumption has been made on the shape of the entrance pupil which might consist in a monolithic mirror or in several disconnected apertures as in the case of aperture synthesis. This leads to the usual statement that imaging and interferometry both describe the same physical process. Conversely to the case of geometrical optics, Fourier optics also take into account optical aberrations, phase distortions induced by the atmosphere, or scintillation. These effects can indeed be easily included, replacing the ideal pupil function (1 on the whole pupil) by $t(\vec{r}/\lambda).exp(i.\phi(\vec{r}/\lambda))$ where $\vec{r}$ is the spatial coordinate in the pupil plane.

1.2. WHAT IS EXACTLY THE INTERFEROMETRIC OBSERVABLE?

Now let us assume that we sample a stellar wavefront at two disconnected apertures located at $\vec{r_1}$ and $\vec{r_2}$ (a wavefront division interferometer) of small individual dimensions as compared to their mean distance B=$|\vec{r_2} - \vec{r_1}|$. We obtain interference fringes (spatially or temporally encoded, depending on the recombination scheme) characterized by their complex degree of spatial coherence V_{12}, also named complex visibility. This quantity is simply defined as the mean correlation between the complex amplitudes of the electric fields of the two sampled wavefronts.

$V_{12} = < \Phi(\vec{r_1}, t)\Phi(\vec{r_2}, t) >$

where the average is taken around mean time t, on an integration time much longer than the electric field pulsation period.

In interferometry, the observables are the phase and modulus of the coherence factor of the incoming wavefront.

The Zernicke van Cittert theorem tells us that for an incoherent source in quasi-monochromatic conditions, this complex coherence factor is equal to the normalized Fourier transform of the source brigthness distribution at the spatial frequency (u,v)=$\vec{B_p}/\lambda$, where B_p is the apparent projected baseline as seen from the source. We then have $V_{12}(u, v) = \tilde{O}(u, v)$. In theory, measuring the complex degree of correlation between two separate wavefronts (i.e. the complex fringe contrast) directly gives one Fourier component of the source spatial distribution of intensity.

The goal of aperture synthesis is to obtain an image with *a spatial resolution corresponding to the diffraction limit of a telescope whose diameter is the interferometer largest baseline*. Images are theoretically recovered by measuring complex visibilities with various baseline vectors, which gives access to as many different Fourier components of the source spatial distribution. Fidelity of the reconstructed image depends on the number of telescopes (i.e. of baselines), and on the accuracy of visibility measurements both in modulus and phase [5].

Spatio-spectral information on the source can also be retrieved by using the so called double Fourier interferometric techniques [14], scanning around the zero optical path difference and freezing the atmospheric piston.

In this case the quantity of interest is the spatio-temporal correlation between the complex amplitudes of the electric fields of the two sampled wavefronts:

$V_{12}(\delta) =< \Phi(\vec{r_1},t)\bar{\Phi}(\vec{r_2},t+\delta) >$ where δ designates the variable time delay between the two wavefronts at the recombination stage.

1.3. INSTRUMENTAL AND ATMOSPHERIC LIMITATIONS

In the real world, Eq.2 tells us that the object complex visibility $V_{12}(u,v) = \tilde{O}(u,v)$ is degraded by the optical system, and that we actually observe $V_{12}(u,v).MTF(u,v)$. The modulation transfer function MTF(u,v) generally includes a rather steady instrumental transfer function (due to static aberrations) and an atmospheric transfer function highly time dependent, with non stationary statistics evolving with seeing conditions. The visibility information is then degraded on Earth by a rapidly variable amount difficult to calibrate. The influence of atmospheric effects on visibility moduli measurements can be strongly reduced by using adaptive optics and /or by coupling the starlight into single mode waveguides at the recombination stage ([7]), whereas information on visibility phases can be recovered using phase closure techniques ([8]), or off-axis referencing techniques (see Colavita in this volume).

2. A few (frightening!) figures about extrasolar planets

One of the announced ultimate goals of NASA Origins program is to image Earth-like extrasolar planets. This is an extraordinary difficult task for various reasons.

2.1. PLANETS ARE SMALL!

Let us have a look at some orders of magnitude. The naked eye cannot resolve Jupiter: at the closest point of the giant planet's orbit, its resolution limit corresponds to about 1.3 R_J. Galileo's 5 cm diameter telescope triggered spectacular improvements. Many planetary features were observed or measured for the first time such as the phases of Mercury and Venus, the surface features and ice caps on Mars, the clouds structure, giant red spot and flattening of Jupiter, the rings of Saturn, the sizes of Uranus and Neptune.

If we want a comparable revolution to be achieved in the visible by resolving extrasolar planets, say 5 pc away (Gliese 876b [12] is the closest known giant extrasolar planet at 4.72 pc from the solar system), telescope diameter or resolving baseline is now about 10 km !!

2.2. PLANETS ARE FAINT!

Here again very simple calculations lead to frightening figures. The naked eye limit in sensitivity is Uranus ($m_v \sim 6$). Assuming an integration time of 0.1 s, the required collecting area to detect an extrasolar giant planet at 5 pc, in the visible, is then $10^6\,m^2$, which is roughly 10000 Kecks or VLTs...

Now if we take $10\,m^2$ as a more reasonable collecting area, the required integration time to get the same object with a signal to noise ratio of 10 against the sky background (i.e. assuming a free floating object) is 25000 s, i.e. about 7 hours.

Note that the incident flux from the Earth as seen from a 5 pc distance ($m_v \sim 27$) peaks to about 1 photon.$m^{-2}.s^{-1}.\mu m^{-1}$ in the mid infrared.

2.3. PLANETS HAVE STARS!

Although we cannot rule out the possibility of free floating planets, these objects appear to be very uncommon. So far, all planetary discoveries used indirect methods, monitoring stellar gravitational reaction to potential orbiting planets. Starlight actually helps indirect methods and almost kills direct ones: the dynamic range needed is huge. The contrast ratio between an Earthlike planet and a solar type star is roughly 5.10^9 in the visible, and is still 7.10^6 in the thermal infrared.

Before even thinking of resolving the planet continents as in section 2.1, resolving the planet from its parent bright star is a very challenging task. Table 1 gives angular resolutions needed to resolve 51 Peg b, the Earth and Jupiter from their parent stars when observed from various distances.

2.4. WHAT ARE THE MILESTONES IN THE DIRECT CHARACTERIZATION OF EXTRASOLAR PLANETS?

From the various figures of resolution and sensitivity given above, and state of the art technology, direct searches for extrasolar planets can be subdivided in three phases of increasing difficulty:

- The detection and low resolution spectroscopy of extrasolar giant planets seen as unresolved spots around bright closeby stars. It will be performed on the ground in the next decade using either massive adaptive optics in the visible or near infrared [1], or long baseline near infrared interferometry [6].
- The direct detection and low resolution spectroscopy of Earthlike extrasolar planets. It could be carried out in space either using the NGST

TABLE 1. Angular resolutions (in mas) needed to separate 51 Peg b, the Earth and Jupiter from their parent stars when observed from various distances.

D(in pc)	5	20	50	200
51 Peg b	10	25	1	0.25
Earth	200	50	20	5
Jupiter	1000	250	100	25

in the visible, with improved coronographic masks and dark speckles signal processing (cf. Labeyrie in this volume), or with a space infrared nulling interferometer, performing wide *spectral analysis* from 5 to 20 microns searching for CO_2 and possibly biotic features such as H_2O and O_3. This latter concept is developed in section 3.

- The multipixel imaging of extrasolar planets using second generation instruments, with large arrays of space telescopes (cf. Labeyrie in this volume).

3. Nulling methods

3.1. SCIENTIFIC REQUIREMENTS

Rejection of the starlight

At 10 μm, the ratio of the Sun to Earth flux is 7 10^6. For an Earth-like planet around a star, this ratio is in the range (15-0.5) 10^6 for stellar types F5V-M0V.

The central problem is to reduce the stellar flux down to a level that will not prevent the detection and spectral analysis of the planetary flux. This level depends upon the noise spectrum of the stellar leaks.

A low limit estimate is obtained assuming a steady leakage flux. The noise would be the quantum noise ("photon" noise) associated with the flux. For a 5 σ detection of an Earth-size planet located at 10 pc, in 10 hours, using 6 x 1.5 m telescopes and typical numbers, the required rejection is: $\rho < 5\,10^4$.

In real conditions, the stellar leaks will be strongly variable with a mean value given by $\rho\ F_{star}$. The associated noise in the frequency range of the detection of the planetary signal will depend on the power spectrum of the stellar leaks. The later will be driven by the characteristics of the servo systems insuring the interferometer nulling. Its knowledge requires a major study. A *guess* is that

$$\rho \simeq 10^6, \tag{3}$$

is probably needed.

Rejection of the Local Zodiacal Light (LZL)

The LZL is another major source of noise. To lower it, one can:

- send the mission to a large heliocentric distance (e.g. 5 AU) where the Interplanetary Dust temperature is lower than at 1 AU and the 10 μm about 200 times lower (Wien part of the Planck curve). However, this introduces major complications to the mission e.g. scarceness of the solar energy, no possibility to send any spare element if any failure;
- reduce the optical étendue of individual telescopes;
- increase the diameter of telescopes.

Rejection of the Exo Zodiacal Light (EZL)

Observed from distance, the 10 μm emission of the Solar Zodiacal dust is $\simeq$ 300 times larger than the Earth's one. Fortunately, it is mostly symmetrical with respect to the Sun and this can be used to discriminate it from that of a planet. This symmetrical distribution of interplanetary dust is a consequence of the differential angular rotation for a extended object in Keplerian orbit and a similar situation is expected for other solar systems. Therefore, interferometer designs that can distinguish between symmetrical emissions (EZL) and asymmetrical ones (planets) have a major advantage.

3.2. OPTIMISING BRACEWELL'S IDEA

Bracewell (1978) proposed the use of an adequate interference between 2 telescopes to cancel the light from a pointed star and detect the emission of a faint nearby companion. The principle of his design is represented in Fig.1.

If recombined after an equal optical path, but for a π shift in one arm, the electric fields provided from an on axis source, through the 2 telescopes, cancel. If the interferometer points at the star and if its basis size is ade-

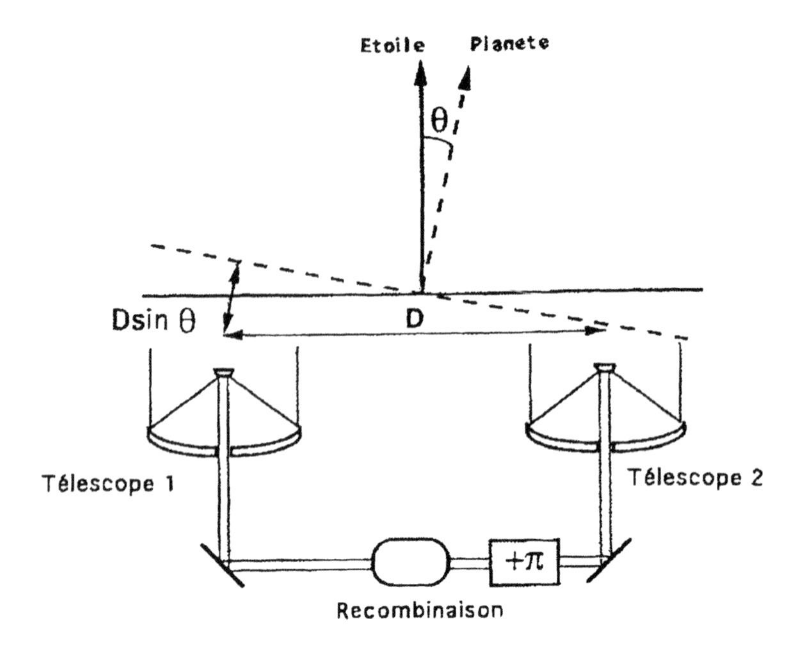

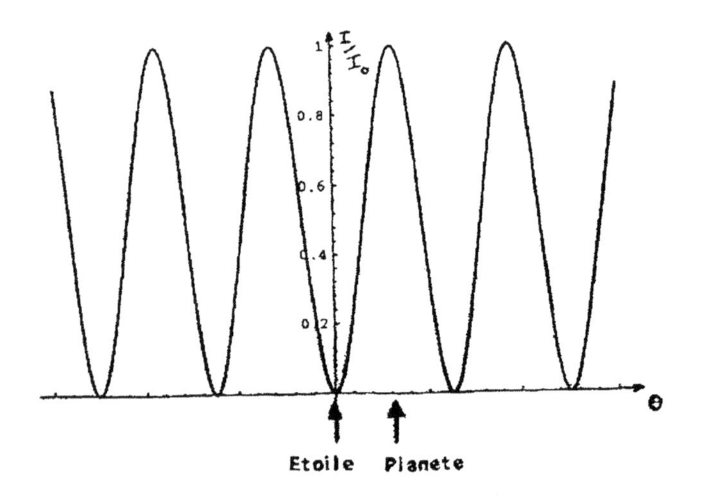

Figure 1. Sketch of a Bracewell (1978) interferometer. For an on axis source, e.g. a star, the waves entering the 2 telescopes are in phase. The subsequent optical paths to the recombining device are equal but for a π shift in one arm. This provides an achromatic destructive interference of its light and a null transmission. A source at angular distance θ provides wave fronts to the telescopes with a phase shift $(D/\lambda)\, sin\theta$. The output of the recombination is now a function of θ as shown in the lower curve. In principle i.e. for a point source star, if the distance between telescopes is adapted to the angular distance of a planet to its star, the transmission of the interferometer could be zero for the star and about unity for the planet, allowing the study of the emission of the latter without been overwhelmed by the former

quately chosen, in principle, the light from the star could cancel out and that of the planet transmitted.

In practice, stellar disks having a finite size, even if the interferometer quality and tuning were perfect, the rejection factor of the stellar light could not be higher than 10^3 when the interferometer size is adapted to the detection of a planet located at 1 AU from a G2V star (Fig.2).

If θ is the source angular offset from the Bracewell interferometer axis, its transmission varies as (1 - $\cos\theta$), i.e. θ^2 for small angle. Angel (1986) showed that a proper combination of 4 identical telescopes can provide a θ^4 central dependence of the transmission, solving, for an optical perfect system, the problem risen by the finite stellar disk. Few years latter, Angel and Woolf (Angel and Woolf, 1997) pushed the limit further (OASES concept), proposing a design providing a θ^6 dependence with a linear array of telescopes whose diameters scale as 1, 2, 2, 1 (Fig.2).

Léger et al (1996) proposed an array of 5 telescopes located on a circle that provides a θ^4 dependence and can distinguish between symmetrical and asymmetrical sources around the star. This design is illustrated in Fig.3 and the corresponding image reconstruction of the Solar System observed from distance is shown in Fig.4.

3.3. SPATIAL FILTERING

The requirement for low optical defects in order to provide a rejection of 10^6 is, a priori, frightening e.g. a standard deviation of micro roughness of 2.5 10^{-4} λ, or 2.5 nm at 10 μm. Ollivier and Mariotti (1997) made *a major step ahead* , showing that optical filtering by a pine hole, or better a monomode optical fiber, reduces this requirement to 5 10^{-3} λ, or 50 nm.

A laboratory interferometer is under construction that should investigate the actual rejection factor that can been obtained and the use of optical filtering ([19])

3.4. INTERNAL MODULATION

3.4.1. *Rotation or internal modulation?*

The physical rotation of the telescopes array, in the plane perpendicular to the line of sight, as initially suggested (Angel, 1996; Léger et al. 1996) is used for 2 functions. It insures the modulation of the signal emitted by an off-axis source, and insures the (u,v) coverage.

In the case of internal modulation, the planetary signal is also modulated (in a very different way though, see thereafter) but the (u,v) coverage is mainly obtained by the multiwavelength operation of the detection: recording signals simultanously at different wavelenghts gives access to different spatial frequencies (but not different azimuths).

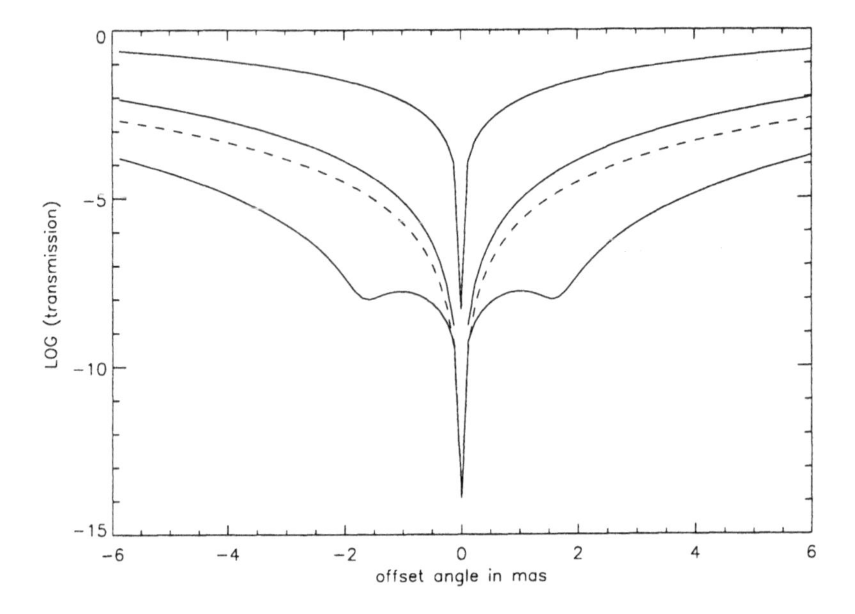

Figure 2. Residual transmission of a nulling interferometer close to its pointing direction. The leaks are calculated at 6μm (worst case) as a function of the separation angle to the pointing direction, expressed in mas. From top to bottom, curves correspond to: 2 telescopes (Bracewell's concept), 4 telescopes in a cross pattern (Angel's cross), 5 telescopes in a circular array (a possible configuration for the DARWIN proposal) and 4 telescopes, in a linear array with diameter ranging as 1, 2, 2, 1, (OASES proposal). For each configuration the largest dimension is 50 m, a dimension adapted to the observation of a planet at 1 AU from a star at 20 pc. For reference, the critical rejection factor is about 10^6 and the solar disc radius, seen from 20 pc, is 0.23 mas. If optically perfect, the last 3 designs meet the requirements for detecting such a planet. The OASES one is the less demanding for the pointing accuracy of the interferometer. Courtesy of Mennesson and Mariotti (1997)

The rotation of the telescopes array can be made only at low frequency. For moderate interferometer size e.g. 50 m, a minimal period of about 3 hours is imposed by fuel consumption requirements. This leads to a demanding requirement on the detector stability. They have to be stable enough to extract the large extrasolar zodiacal light signal from the weak planetary one at frequency of 0.1 mHz, an extreme requirement for *low frequency noise.* Internal modulation can be performed with period of say 10 s, displacing the stability requirement by 3 decades (100 mHz) and making the characteristic much easier to achieve.

Building an interferometer with internal modulation would be *another major step forwards.* Thereafter, we describe a concept that may achieve this performance.

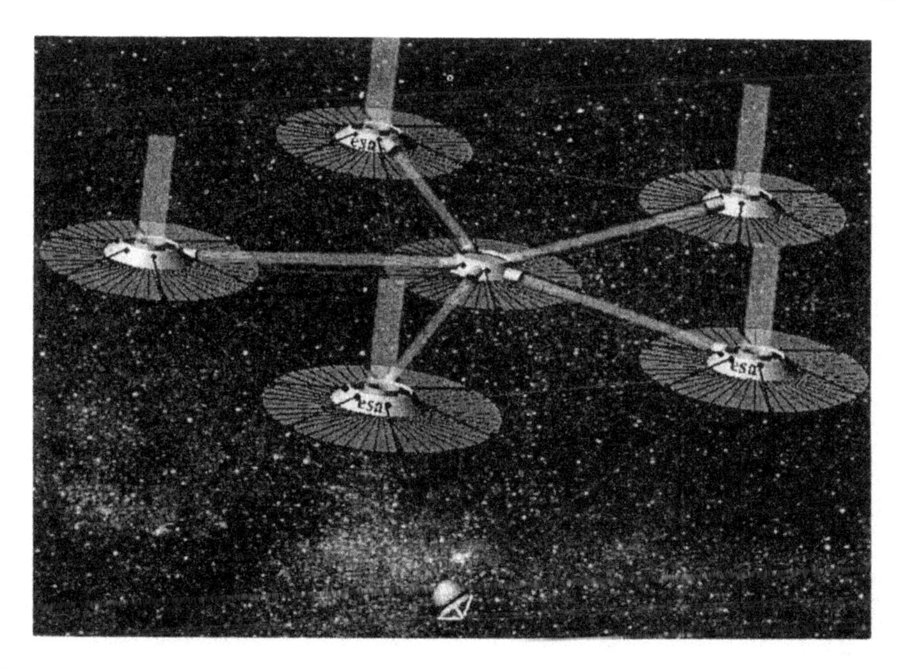

Figure 3. A free flyer concept of the DARWIN proposal. 5 telescopes are displayed on a circle and send their light to a central recombining central laboratory. The interferometer structure is dynamically maintained through laser metrology implying an out-of-plane auxiliary vessel, and accurate ionic thrusters (FEEPs) to a level of about 1 mm. A second stage metrology, in charge of the equality of optical paths to the central recombining hub, is provided by delay lines, ~ 1 cm stoke, to the required level (~ 3 nm)

3.4.2. *The 3DAC or "Mariotti" concept*

- *Geometry of the entrance pupil*

The concept uses 3 Degenerated Angel Cross (DAC) i.e. a 4 telescope Angel Cross interferometer where 2 telescopes displayed on one diagonal are merged, and is illustrated by Fig.5. Six telescopes of equal diameters are arranged on an equilateral baseline. Each side of the triangle represents a DAC. Each beam coming from a telescope located on a summit is "shared" with the adjacent DAC through a 50/50 beam splitter. Telescopes of each DAC are recombined to produce a nulled output for the on-axis star (usual mode for a DAC). These nulled outputs are then recombined by pairs. In a pair, a variable phase shift, f(t), is added to one member before recombination.
As an illustration, Fig.6 shows how beams of DAC1 and DAC2 are recombined.

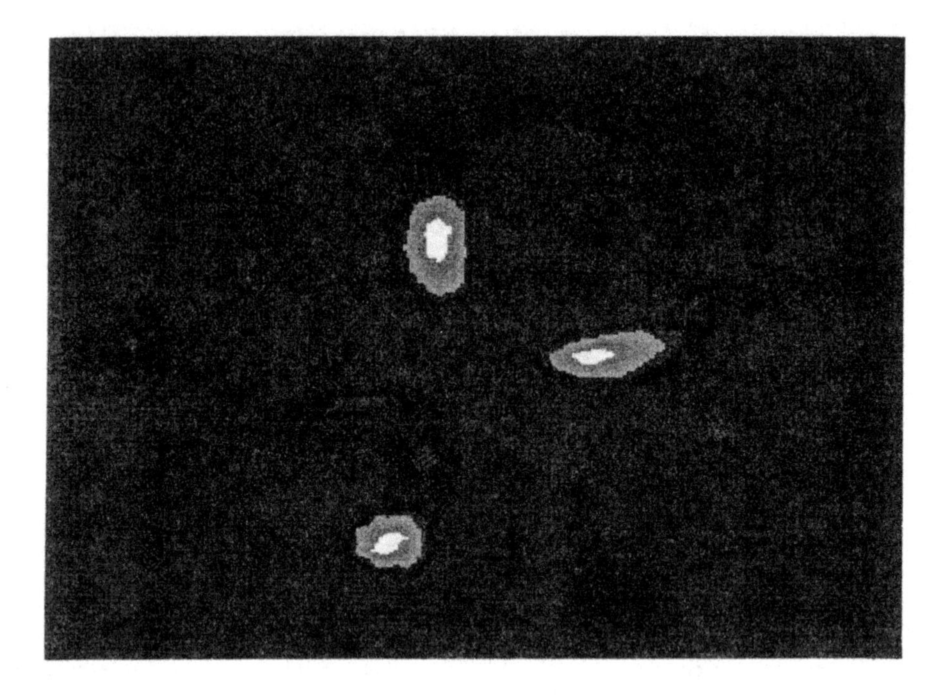

Figure 4. Simulation of the image recovery of three Earth size planets orbiting around a Sun-like star, 10 pc away, by the proposed DARWIN mission. The integration time is 30 hours. The exo-system is seen under a 30 degree inclination and the *projected* distances of the planets from their star are 0.5 AU, 0.7 AU and 1.2 AU, respectively, corresponding to that of Venus, Earth and Mars at the 1st of January 2001. The planets are the 3 bright areas and the star, located in the center, is supposed to be fully nulled out by the interferometer. The other features are artefacts. Courtesy of Mennesson and Mariotti (1997)

- *Internally modulated transmission provided by the recombination of DAC's*

  Six outputs will be usable for science (two for each recombination of 2 DAC's, as presented in fig.6).
  Phase shift, f(t), between 2 outputs can be provided by a delay line and can be arbitrary chosen. The (symmetrical) exozodiacal light and the (asymmetrical) emission of a planet lead to different dependence of the ouput signal upon the phase shift f(t) which allows their discrimination.

- *Internal modulation maximum frequency*

  The integration time between two consecutive samplings of the internally modulated signal must be large enough so that detector readout

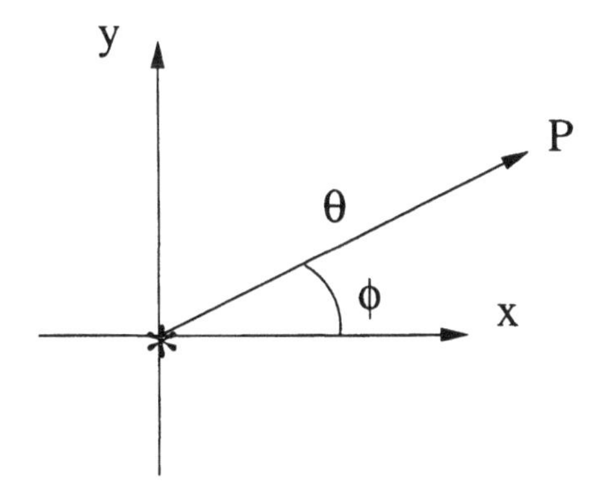

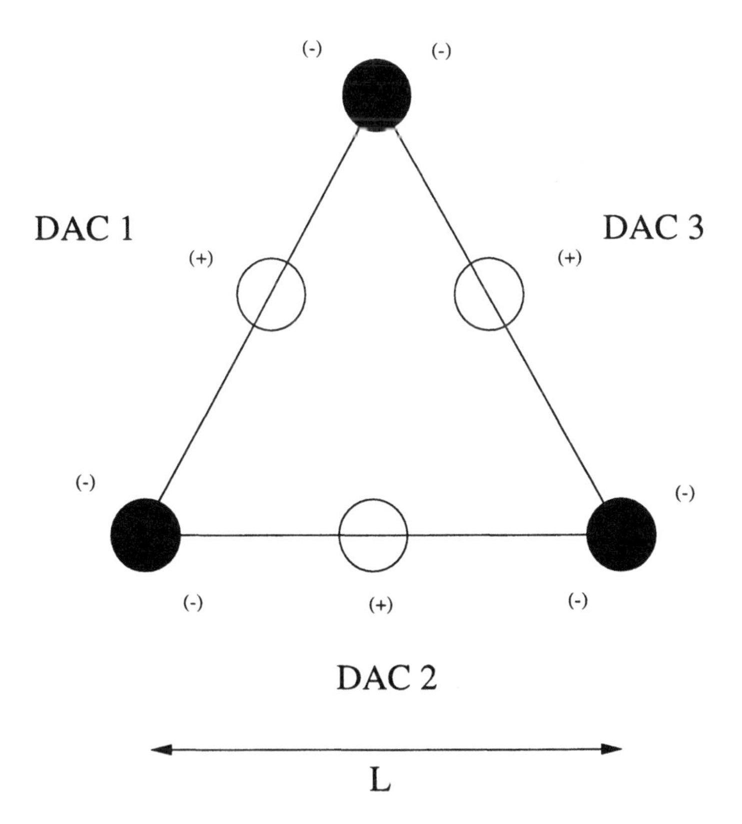

Figure 5. Upper sketch gives notations adopted to locate a point source P on the sky. θ represents its angular distance from the central star, and ϕ its azimuth. Lower figure gives the geometry of the Mariotti concept. (+) and (-) stand for 0 and π phase shifts respectively (after recombination).

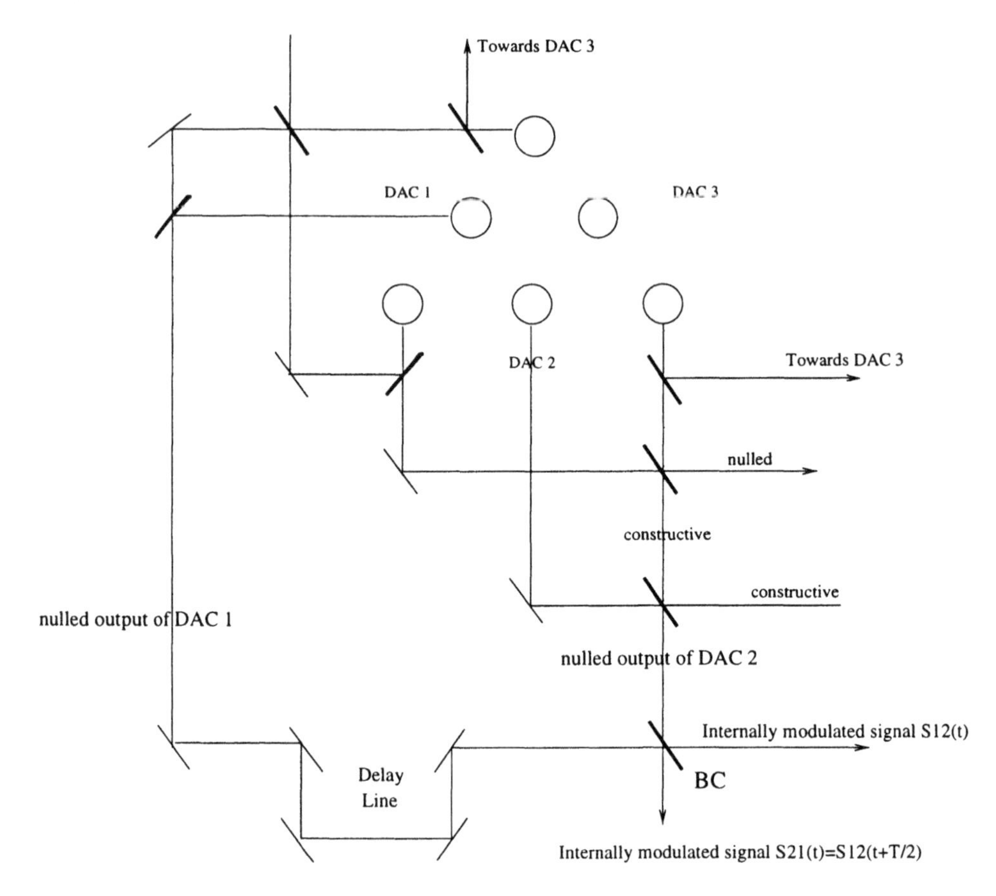

Figure 6. Recombination scheme between the nulled outputs of DAC 1 and 2. For the sake of simplicity all delay lines have been omitted, except the one providing internal modulation (phase shift f(t)) between the 2 outputs, which will probably use a short stroke piezo mirror. Beam splitters are all 50/50 and represented in bold. Note that both final outputs S12 and S21 are usable for science, after the final beam combiner BC.

noise does not dominate unavoidable noises (e.g. photon noise). Using a readout noise of 13e- rms, and assuming that the modulated signal is sampled 50 times per period T, we then have the simple condition:

$$Nd.T/50 > 170 \text{ photoelectrons},$$

where Nd is the flux (in phe-/s) of the dominant source of astrophysical background.

For a mission in Earth orbit, the solar ZL noise is dominant, and regardless the collecting aperture, we get:

Nd=1100phe-/s and T$\geq$ $8s$ ($\nu leq 120\, mHz$).

- *Conclusion*

The Mariotti concept has the major advantage of an internal modulation which can operate at much higher frequency (typically 100 mHz) than when modulation is achieved through the interferometer rotation. This should drastically reduce 1/f type noises.
In order to insure a better coverage of the azimuths in the (u,v) plane, it is mandatory to turn the array by 20 degrees twice. For spectroscopy, an homothetic expansion of the baselines is required. Moving the free flyers is then still needed, but energy consumption is expected to be much lower than in the case of the continuous rotation of the network. The impact of chromatic internal modulation (such as the one given by a delay line) has not been studied yet, but since we are recombining achromatically nulled outputs, it should not be a problem.
The case of several planets has not yet been looked at either. Imaging capabilities of extended objects look poorer than for 2D continuously rotating configurations, although further studies still need to be carried out on this point.

References

1. Angel, J.R.P., 1994, Nature, 368, 203
2. Angel, J.R.P. and Woolf, N.J., 1997, ApJ, 475, 373
3. Baudoz, P. et al, 1997, First results with the Achromatic Interfero-Coronograph, Proc. SPIE 3350, in press
4. Bracewell, R.N., 1978: Nature 274, 780
5. Cornwell, T.J., 1988, Image restoration, in "Diffraction limited imaging with very large telescopes", NATO ASI Series, Alloin and Mariotti eds.
6. Coudé du Foresto, V., Mariotti, J.-M., and Perrin, G. Direct observation of extrasolar planets with an infrared interferometer, in "Science with the VLT Interferometer", F. Paresce ed, ESO Garching 1996.
7. Coudé du Foresto V., Perrin G., Ruilier C., Mennesson B., Traub W. and Lacasse M., 1998, Proc SPIE, 3350, in press
8. Haniff, C.A., 1988, Phase closure imaging - Theory and Practice, in "Diffraction limited imaging with very large telescopes", NATO ASI Series, Alloin and Mariotti eds.
9. Haniff, C.A., 1994, Introduction to wave optics, in "Adaptive optics for astronomy", NATO ASI Series, Alloin and Mariotti eds.
10. Lannes, A. et al., 1996, Journal of Modern Optics, 43, 105-138.
11. Léger, A., Mariotti, J.-M., Mennesson, B. et al., 1996, Icarus 123, 249
12. Marcy, G., Butler, P. Vogt, S. et al., 1998, ApJL, 505, L147
13. Mariotti, J.M., 1988, Introduction to Fourier optics and coherence, in "Diffraction limited imaging with very large telescopes", NATO ASI Series, Alloin and Mariotti eds.
14. Mariotti, J.-M. and Ridgway, S.T., 1988, A&A 195, 350

15. Mennesson, B., and Mariotti, J.-M., 1997, Icarus 128, 202
16. Mennesson, B., Ruilier, C., Ollivier, M. in preparation.
17. Neumann E.G., 1988, Single Mode Fibers, Springer-Verlag
18. Ollivier M. and Mariotti J.-M., 1997, Applied Optics 36, 5340
19. Ollivier M. et al, 1997, International Conference on Space Optics 1997, edited by the Centre National d'études Spatiales, Paris
20. Ridgway S., 1998, Optical interferometry today and tomorrow, presented on the occasion of the "Journée scientifique à la mémoire de Jean -Marie Mariotti", September 14, 1998.
21. Shao, M. and Colavita,M.M., 1992, Ann. Rev. Astron. Astrophys. 30, 457
22. Voit G. Mark, 1997, ApJ, 487, L109
23. Woolf N.J. and Angel J.R.P., 1997, Planet Finder Options I: New Linear Nulling Array Configurations, in "Planets beyond the solar system and the next generation of space missions", ASP Conference Series, ed. David Soderblom.
24. Woolf et al., 1998, PDI, A potential precursor to Terrestrial Planet Finder, Proc SPIE, 3350, in press.

LARGE GROUND-BASED TELESCOPES WITH HIGH ORDER ADAPTIVE OPTICS FOR IMAGING FAINT OBJECTS AND EXTRA-SOLAR PLANETS

M. LANGLOIS
Steward Observatory,
933 N Cherry Ave., Tucson, USA

D. SANDLER
Steward Observatory,
933 N Cherry Ave., Tucson, USA,
ThermoTrex Corporation,
9550 Distribution Ave., San Diego, CA, USA

AND

D. MCCARTHY
Steward Observatory,
933 N Cherry Ave., Tucson, USA

The new 6-8 m class ground based telescopes equiped with very high-resolution adaptive optics have the potential to detect Jupiter-like planets around nearby stars. Direct detection will allow discoveries of planets, beyond the angular radius where Doppler spectroscopy achieves maximum sensitivity. In addition, direct imaging (and spectroscopy) will allow confirmation for those indirect detections which lie within 0.3-2 arcseconds in orbital radius. However, the technical requirements for direct imaging using high order adaptive optics are at the theoretical limits of performance and hence very challenging. Here we review the limiting performance of such systems. We give the exposure time required to detect such companions, and we point out the improvements required in order to accomplish exo-planets detection.

J.-M. Mariotti and D. Alloin (eds.), Planets Outside the Solar System: Theory and Observations, 297–305.

1. Introduction

The challenge in direct detection of extra-solar planets comes from their faintness (10^9 fainter than their sun for a jupiter twin) and their location very close to their hosting star. In consequence imaging as well as spectroscopy of these faint companions require high resolution and high dynamical range particularly very close to the target star. The former can be obtained with ground based telescopes equiped with high-order adaptive optics system to reduce the light scattered from atmospherical disturbances, the latter can be obtained by using in addition to the adaptive optics system a coronograph which reduces the scattered light from the diffraction.

2. Adaptive optics systems

The atmospheric turbulence disturbes the incoming wavefront from the star-companion system and in consequence degradate the resolution. Adaptive optics (AO) systems have the power to image and compensate the wavefront disturbances to restore the diffraction limit. Further information about Adaptive optics can be found for example in (Beckers 93), (Léna 94), (Roddier 94), (Rousset 94),(Sandler et *al.* 94). The AO corrected images are not perfectly diffraction limited but they consist of two components: a diffraction limited core and a broad halo. The diffraction limited core appears when the low spatial frequences wavefront errors are corrected by the system. There are already many AO systems operating at this level by using less than 10 correcting elements per diameter and they perform extremely good correction for $D/r_0 < 10$. The intensity of the halo starts to decrease once the higher spatial frequencies are reduced. There are already three routinely operating AO systems operating at this level: Mt Wilson 2.5m (Shelton et *al.*), Starfire optical range (SOR) 1.5m (Fugate et *al.*) and 3.5m (31 correcting elements per diameter). Figure 1 shows the efficiency of the AO correction at SOR 3.5m in terms of resolution and in terms of peak amplitude. The corrected peak is 7 times higher (which correspond to a strehl of 35%) and is $\frac{1}{5}$ wider than the uncorrected one. The residual irregularities in phase left behind after the AO correction will lead to a loss in central intensity which is redistribuated in the halo. The halo intensity level generates background photon noise at the planet location which needs

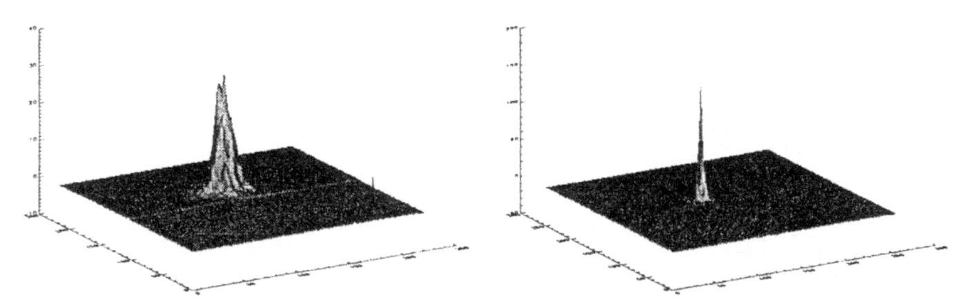

Figure 1. Comparison of the uncorrected (right) and corrected (left) image profile from SOR 3.5 m

to be decreased in order to increase the signal to noise of the detection. This background level is usually expressed in terms of gain, G, defined as the ratio of the peak intensity divided by the halo intensity at a given radius. For a uniform halo the mean gain is given by (Angel 94):

$$G = (\frac{D}{\sigma d})^2 \tag{1}$$

where D is the pupil diameter, σ the residual error on the wavefront (related to the strehl ratio (ratio of the observed peak intensity to the airy maximum intensity) by $s = e^{-\sigma^2}$) and d is the correcting element size so $\frac{D}{d}$ is the number of correcting elements per diameter. In order to improve the gain and in consequence to reduce the halo level the need is to increase the number of correcting elements per telescope diameter. (Angel 94) has shown that a gain of 500000 is needed to detect planets which correspond for the MMT new 6.5m telescope to 10000 correcting elements in total or about 120 across the primary mirror diameter.

3. coronography devices

When the halo level is reduced by the AO correction the airy pattern becomes a predominant contribution to the halo at small radii and is needed to be supressed by using an appropiate coronographic device. In the case of a Lyot coronograph composed of an occulting mask at the focal plane associated with a aperture stop to remove the diffraction from the edge of the mask before reimaging the pupil, the critical parameter to achieve good suppression is the relative size of the occulting mask to the point source image width. Because the AO correction greatly reduces the image width a much smaller mask can be used and in consequence a closer view

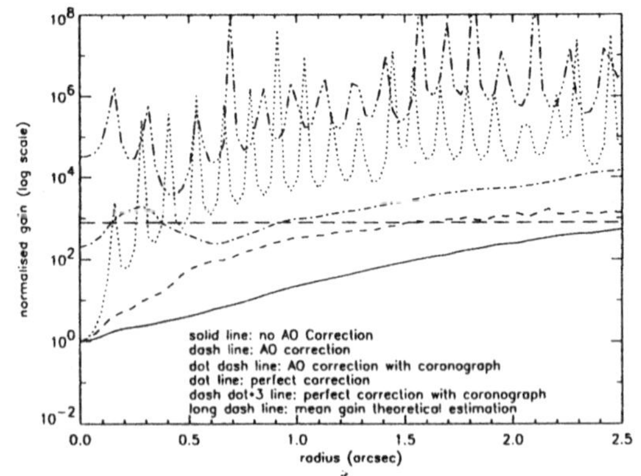

Figure 2. Gain as a function of angular radius. The two upper curves are theoretical, the three bottom ones are from SOR 3.5m

of the star can be achieved. Figure 2 shows that the use of a coronograph (.943" occulting mask) leads to a simulated gain of G=25000 (1000 without coronograph) at 1" in the case of a perfectly flat wavefront, 300 for AO corrected images (SOR 3.5 m), and G=20 in the case of no correction. Theoretical and experimental performances caracterisations for the Lyot coronograph already have been performed by many authors (Malbet 96), (Ftaclas et *al.*),(Beuzit et *al.*) mainly to work in imaging mode but also to recover the image by using the dark speckle technique (Boccaletti et *al.*). Other approachs toward substraction the light from the central star have also been studied. Here are 3 examples of interferometric methods (Roddier, Roddier 93), (Hinz 98), (Baudoz 98). In order to image planets the coronographic gain needs to be 10^6 at the planet location and that has not been achieved yet altough we will consider in the following that this could be achieved.

4. speckle pattern

To what extend the halo fluctuations will average out in long exposure is a critical question in searching for faint companions. The noise level comes from the halo level, the spatial stucture which can be removed but also from the speckle pattern which in the case of AO corrected images may have longer lifetime than without AO correction. Figure 3 is a 70 ms AO corrected exposure taken at the SOR 3.5m facility with a Lyot coronograph. It can be seen that waffle pattern and speckle pattern are present on the image, and they can be mistaken for faint companions if they are not average out on the long exposure. The more actuators the AO system has the more of these structures would be present but at a fainter level. (Ryan et *al.*) have shown that this speckle noise does not decrease with the $\sqrt(Texposure)$ because the AO residual errors are correlated. They also estimated the correlation time which is the time required to obtained independant realisation of the residual speckle in the halo.

Given the correlation time and the AO gain we can calculate the exposure time, T required to detect a brown dwarf companion by using (Angel 94),(Stahl 95): $T = \tau(\frac{R}{G}(SNR))^2$. R is the ratio of star brightness to companion brightness and is about 10^4 for a brown dwarf 1Gyr old with a mass 30 times the jovian one, located at 10 parsecs and orbiting a star with aparent K magnitude of 4. Decreasing the correlation time as well as increasing the gain will reduce the needed exposure time. At .75", a correlation time of 11 milliseconds and a gain of 300 were measured at the SOR 3.5 m facility. These measurements lead to an exposure time of 2 hours to detect a brown dwarf at that radius with a signal to noise of 5. But with this system it would still take 10^5 hours to image a Super Jupiter. With the very high order MMT 6.5 m AO system a gain of 500000 can theoretically be achieved with a correlation time of 24 ms that leads to a 12 hours exposure to detect a companion 10^7 times fainter than its sun. By adding temporal prediction to the wavefront reconstruction (Stahl 95) predicted that the speckle correlation time can be reduce to less than 1 ms and this would lead to detection of even fainter companions.

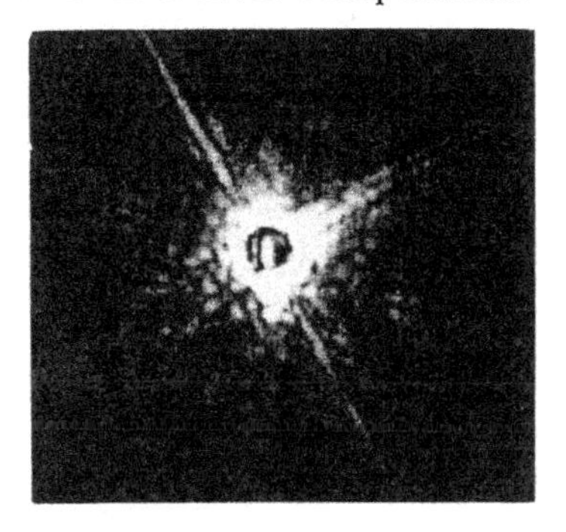

Figure 3. 71 milliseconds frame of HR1925 in K band with AO and coronograph

5. mirror surface roughtness

Since the primary mirror is not a perfect optical surface, Its surface roughtness will also scatter light in the halo. The purpose of this section is to estimate what is the gain due to micro ripples for an actual technology mirror and to compare it to the high order AO gain. In order to get an estimation of this gain we used an uncorrected image from the actively supported WIYN telescope and estimated its profile from 0 to 400 $\frac{\lambda}{D}$. The analytical as well as simulated Kolmogorv profiles give a good fit of the data up yo 2" as seen on Figure 4 (*Left*). At radii larger the Kolmogorov and the Airy contributions become negligeable and the halo, called aureole, is assumed to be formed by the small scales error on the mirror surface. The fit to the observed profile is given by

$$f(\theta) = \frac{(\frac{\theta}{HW})^{-2}}{29} \tag{2}$$

Using this estimation and assuming that at a given radius θ the amount of light in the halo comes mainly from ripples at scale $\frac{\lambda}{\theta}$ but also, in a smaller amount, from scales smaller than that, we can estimate the halo level due to ripples at various scales for the MMT 6.5m telescope which is similar to the WYIN mirror with an rms surface error of the order of 16nm. The AO system would not correct for these high order mirror structure (smaller than 5 cm which is approximately the interactuator spacing for the proposed high order MMT AO system). Thus it is important to compare the amplitude of the halo formed from the mirror irregularities to the AO residual halo given by the equation (1). Figure 4 (*Right*) shows that when the AO system corrects every wavefront irregulatity larger than 5 cm with a strehl of 96% the uncorrected ripples structure form an halo smaller than the AO residual halo. In consequence the mirror ripples should not prevent direct imaging of extra-solar planets using large telescopes.

6. detection limits

We will consider here the detectability of three type of extrasolar objects: a brown dwarf and two giant planets using actual and futur AO system. The actual system (case (a)) is taken to be the SOR 3.5 m telescope equiped with AO (941 actuators) with a strehl of 70%. The theoretical AO system (case (b)) is the one wich was studied in (Angel 94), (Stahl 95), (Sandler 97) for

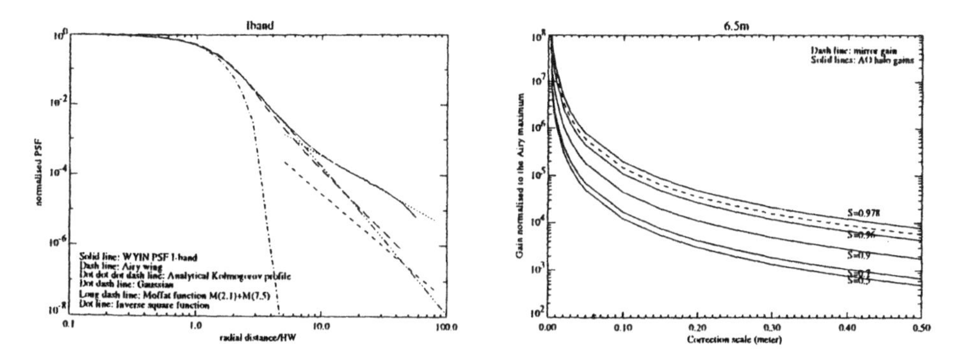

Figure 4. (*Left*) The observed PSF profile is compared to a gaussian profile, and an analytical Kolmogorov profile with D/r0=14.5, (*Right*) Halo gain for various strehs (from .5 to .978) and 6.5 m primary compared to the mirror gain

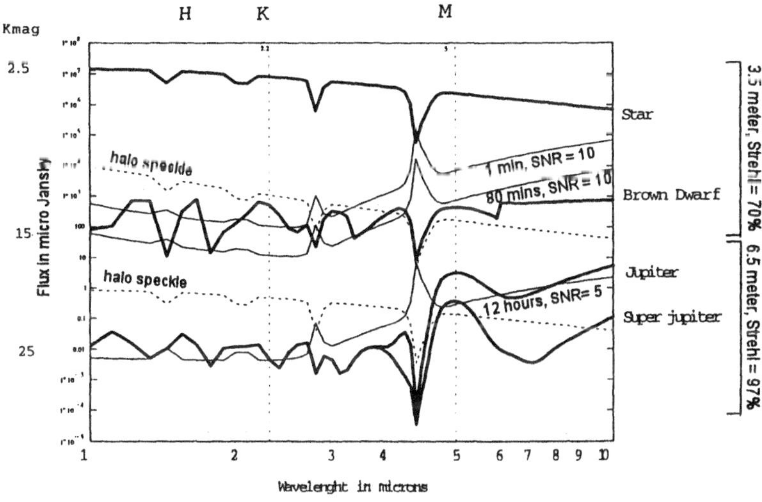

Figure 5. detection limits

the MMT 6.5 meter using light from a relatively bright central star (K<5) and assuming that it would correct larger scales than 5 cm with a strehl of 97%. Figure 5 shows the schematic spectra for a brown dwarf with 25 jovian mass, a super jupiter like planet with 5 jovian mass and a jupiter like planet with 1 jovian mass. These extrasolar objets are 1 Gyr old, located at 10 parsec and were extrapolated from models by A. Burrows models (this volume). The flux for these giant planets comes from thermal emission and does not depend on their orbits unlike the flux from small planet which is mostly reflected starlight. We have include here the atmospheric and telescope transmissions. The former is responsible for the absorption lines between the atmospheric windows. We have also plotted in Figure 5 the average flux from individual speckles given in (Angel and Woolf 97) by the total flux distributed in the halo devided by the number of speckles $(\frac{D}{d})^2$ in the halo. In both cases the flux of individual speckle is about the same than the target objects and it is needed to reduce their britghness by averaging them out either by increasing the exposure time or by decorrelating the speckle pattern (section 4) either by rotating the telescope amoung the line of sight. Assuming good coronographic suppression and no residual speckle in the halo, the noise level of the background at the planet location on a long exposure image comes from the photons scattered in a smooth halo by the AO residual errors. We estimated the background count level by using the equations (1) for the AO residual halo and (2) for the mirror halo. Increasing the order of AO correction up to 5 cm scale will make the residual errors comparable to the mirror surface error. We also include the contribution of the telescope thermal emission (which becomes comparable to the halo contribution at 3 microns in the case (a) and at 2 microns in case (b)), sky brightness (which becomes comparable to the halo contribution at 4 microns in the case (b))). The detector noise assuming to have the Nicmos detector caracteristics is negligeable in case (a) but is comparable to the halo in case (b)).

Given the background count C_b, the flux limit wich can be detected can be expressed by (Angel and Woolf 97):

$$F_\lambda(S) = \frac{A}{tD^2}(\frac{S^2}{2} + \sqrt{\frac{S^4}{4} + S^2 C_b}) \tag{3}$$

where S is the signal to noise ratio, t is the integration time and A is a constant depending on the bandwith, the pixel scale, and the quantum efficiency of the system. This flux is compared to the extra-solar objects flux on figure 5 and it shows that brown dwarfs can be detected with the SOR 3.5m telescope in the near infrared in about an hour if the diffraction pattern has been substracted well enough at the object location. But It would take a 6.5 meter telescope equiped with an extremely good AO system (strehl = 97%) to detect a giant planet in 12 hours in the J, H, K bands or at 5 microns where its spectra present a peak due to the predicted very low atmospheric opacity (Gillett). Taking the advantages of the increasing quality and number of AO systems, the large ground-based telescopes can now be prepared to detect brown dwarfs but also in the future detect extra-solar giant planets around bright stars within 10 parsecs from earth.

Acknowledgements

We thank P. Ryan, B. Fugate, and the SOR team for their support with the new 3.5 m AO system. We thank R. Angel, P. Léna, G. Rousset for stimulating discussions. We thank to T. von Hippel and J. Jurcevic for the WIYN images. This work was supported by grant F49620-96-1-0366 from the Air Force Office Research to the CAAO at Steward Observatory and the IR camera was founded from NSF grant AST 9203336.

References

Angel, J. R. P., 1994, *Nature*, vol. 368, pp. 203–208.

Angel, J. R. P., and Woolf, N. 1998, *Science with the Next Generation Telescope in PASP Conf. Series*, Vol. 133, pp. 172-187

Baudoz P., Rabbia Y., and Gay J., 1998, *Astronomical Telescopes and Instrumentation, Proc. SPIE.*

Beckers, J., 1993, *Annual Review A&A*, vol. 31, pp. 13–62.

Beuzit, J.-L. et *al.*, 1997, *A&A Supp. Ser*, Vol. 125, pp. 175-182.

Boccaletti A. et *al.*, 1998, Accepted in Jan 98 by *A&A*.

Ftaclas, C. et *al.*, 1997, *A&A Supp. Ser*, Vol. 125, pp. 175-182.

Fugate, R. Q et *al.*, 1991, *Nature*, vol. 353, pp. 144-146.

Gillett, F. C., Low, F. J., Stein, W. A., 1969, ApJ, Vol. 157, pp. 925.

Hinz P., Angel R., McCarthy D., Hoffmann W., and Woolf N., 1998, *Astronomical Telescopes and Instrumentation, Proc. SPIE.*

Langlois M., Sandler D., Ryan P., McCarthy D., 1998 in *Astronomical Telescopes and Instrumentation, Proc. SPIE.*

Léna ,P., 1994, *Nato ASI Series:"Adaptive Optics for Astronomy"*, Vol 423, pp 321-333.

Malbet, F. 1996, *A&A Supp. Ser*, Vol. 115, pp. 161-174.

Roddier C., Roddier F., 1997, *PASP*, Vol. 109 ,pp. 815

Roddier F., 1994, *Nato ASI Series:" Adaptive Optics for Astronomy"*, Vol 423 , pp 89-111.

Rousset G., 1994, *Nato ASI Series:" Adaptive Optics for Astronomy"*, Vol 423 , pp 115-138.

Ryan P., Fugate R., Langlois M., and Sandler D., 1998, *Astronomical Telescopes and Instrumentation,* Proc. SPIE.

Ryan, P., 1996, *PhD dissertation*, University of Arizona, Tucson.

Sandler D. G.,Stahl S., Angel J. R. P., Llyod-Hart M., and McCarthy D., 1994, *J. Opt. Soc. Am.*, Vol. 11 No. 2, pp. 925.

Sandler D. G., 1997, *Planets seyond the solar system and the next generation of space missions, PASP Conference Series.*

Sandler D. G.,Stahl S., 1995, *Ap. J.* , 454, L156.

Shelton J. C. et *al.*, 1995, *SPIE*, Vol. 2534, pp. 72.

Stahl, S. and Sandler, D., 1995, *ApJ*, vol. 454, pp. L153–L156.

REFLECTED LIGHT FROM CLOSE-IN EXTRASOLAR GIANT PLANETS

D. CHARBONNEAU
Harvard-Smithsonian Center for Astrophysics
60 Garden St., Cambridge, MA 02138, USA

Abstract. A Jupiter size planet at a distance of 0.05 AU could produce a reflected light component at the level of $\sim 7 \times 10^{-5}$ relative to its host star. Although the star and planet are at a very small angular separation, they would be well resolved spectroscopically, due to the large orbital velocity of the companion. The distinctive signature produced by the addition of this secondary reflected light spectrum should be observable given sufficient spectral resolution and signal-to-noise. A detection would be extremely significant as it would be the first direct detection of a planet around another star. It would yield the orbital inclination and hence the mass of the companion. Furthermore, it would measure a combination of the planetary radius and albedo, from which a minimum radius may be deduced.

1. Reasonable Reflected Light

There are currently four known close-in extrasolar giant planets (CEGPs) inferred indirectly through the radial velocity variations induced on their host stars. The semi-major axes of these systems, τ Boo, 51 Peg, υ And and ρ^1 Cnc, are 0.046, 0.051, 0.057 and 0.11 AU respectively (Mayor & Queloz [17], Butler *et al.* [5]). These CEGPs constitute half the population of known radial velocity companions with minimum masses less than 13 M_J orbiting G and F type stars. From this early vantage point, these objects appear to constitute an exceptionally interesting sub-population of extrasolar giant planets, and the motivation to know more about them than the information available through radial velocity surveys is strong.

What can we aspire to given the current observational limits? The small semi-major axes of these systems translate into tiny angular separations ($\sim$ 3 milliarcsec for τ Boo) which makes prospects for direct imaging bleak.

J.-M. Mariotti and D. Alloin (eds.), Planets Outside the Solar System: Theory and Observations, 307–318.

The predicted orbital motion of the star on the sky ($\sim 9\,\mu$arcsec in the case of τ Boo for a minimum mass companion) is also well below current astrometric capabilities. However, while the star and planet are poorly spatially separated, the difference in their orbital velocities is large. The amplitude of the observed radial velocity of the planet (K_p) is simply the observed radial velocity amplitude of the star (K_s) multiplied by the mass ratio of the star to the planet. Since the orbital inclination i is unknown, we have only a lower limit on the planetary mass. In the case of τ Boo,

$$K_p = -\frac{M_s}{M_p} K_s \leq 152 \text{ km s}^{-1}, \tag{1}$$

which is orders of magnitude larger than current high resolution spectroscopic capabilities.

In the visible, emission from these CEGPs might be dominated by reflected light. At first, the idea of detecting planets through their reflected light might seem absurd. Indeed, in the case of the solar system, Jupiter reflects a mere 2×10^{-9} of the solar output. However, these systems are roughly a factor of 100 closer to their stars than Jupiter is to the Sun, and hence the reflected light may be $\sim 10^4$ brighter. An estimate of the flux ratio from a CEGP system can be achieved as follows. Assume that the planet is a diffusing sphere of radius R_p and reflective efficiency η, ie. a surface element on the planet re-radiates a fraction η of the incident flux uniformly back into the local half-sky. Denote by D the distance from the star to the planet, and let α denote the angle between the star and the Earth as seen from the planet. Neglecting occultation by the star and assuming $R_p \ll R_s \ll D$, integration over the sphere yields the ratio of the observed flux of the planet to that of the star at $\alpha = 0$,

$$\epsilon = \frac{2\eta}{3}\left(\frac{R_p}{D}\right)^2. \tag{2}$$

For 51 Peg b, using $D = 0.051$ AU, $R_p = 1.3R_J$ (Guillot *et al.* [12]) and $\eta = 0.7$, the result is $\epsilon = 7 \times 10^{-5}$.

The observed flux ratio will vary with the angle α, which is a combination of the orbital inclination i and the orbital phase ϕ. Here I define the orbital phase to be the one conventionally given by radial velocity measurements, $\phi = \frac{t-t_0}{P}$, where t_0 is the time of maximum recessional velocity of the star and P is the orbital period. The angle $\alpha \in [0, \pi]$ is then given by

$$\cos\alpha = -\sin i \, \sin 2\pi\phi. \tag{3}$$

Integration of the intensity over the surface of the sphere viewed at an angle α yields the phase dependent observed flux ratio,

$$f(\phi, i) = \epsilon\left(\frac{\sin\alpha + (\pi - \alpha)\cos\alpha}{\pi}\right). \tag{4}$$

This estimate of the flux ratio does not consider the underlying source of the albedo, nor any angle dependent scattering effects associated with it. Estimates of the albedos of these CEGPs are currently uncertain as the chemistry of their atmospheres is poorly understood. Recent models by Seager & Sasselov [19] which explicitly solve the radiative transfer through the atmospheres of these CEGPs indicate that a large enhancement to the reflectivity in the visible may occur should the conditions allow for the formation of dust high in the atmosphere (Figure 1). Detailed modelling of the atmospheric chemistry of these CEGPs indicates that condensates such as $MgSiO_3$ may indeed form at the required height (Burrows *et al.* [4]).

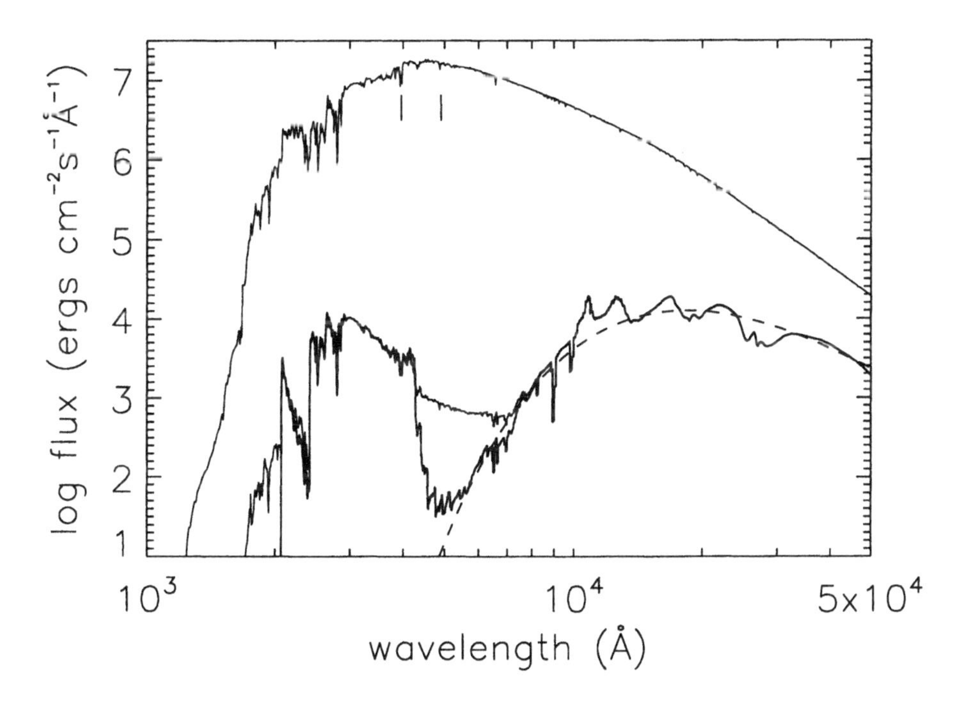

Figure 1. Low resolution spectrum (reflected + thermal) of τ Boo b (lower thick solid curve), compared to that of τ Boo A (upper thin curve). The lower thin curve is τ Boo b with silicate dust formation, where the reflected spectral lines of the primary can be seen. The right tickmark is Hβ, and the left one is the Ca II H + K doublet. Both are seen as reflected features in the planetary spectrum. Here τ Boo b has $R_p = 1.2R_J$ and $T_{eff} = 1580$K; τ Boo A has a radius $R_s = 1.2R_{sun}$. The dashed line is a 1580 K blackbody. This figure is from Seager & Sasselov (1998).

2. Photometric Variations

For a given orbital inclination, Equation 4 describes the time varying component that could be observed as a photometric modulation of the system. For a nearly edge-on configuration (but ignoring occultation effects) and $\epsilon = 10^{-4}$, there will be $\sim$ 0.1 millimag photometric difference between $\phi = \frac{3}{4}$ (when the planet is fully illuminated) and $\phi = \frac{1}{4}$ (when it does not contribute any observed reflected light). High precision photometric monitoring (Henry *et al.* [15], Baliunas *et al.* [1]) of three CEGP systems (51 Peg, τ Boo, and ρ^1 Cnc) has been conducted to detect transits which would result in a reduced apparent stellar brightness as the planet occults the star. None of these systems is observed to transit. However, these observations also constrain the reflected light signal. In the case of τ Boo, Baliunas *et al.* [1] can exclude a 0.4 millimag peak-to-peak sinusoidal variation at the 3.3 day radial velocity period, which is approaching the predicted $\sim$ 0.1 millimag reflected light variation. If the companion to τ Boo were of radius R_p and possessed a reflective efficiency η, then the expected photometric variation would be

$$\Delta m = -2.5 \log_{10}(1 + f(\phi, i)) \quad (5)$$

$$\approx -70\,\eta \left(\frac{R_p}{R_J}\right)^2 \left(\frac{\sin\alpha + (\pi - \alpha)\cos\alpha}{\pi}\right) \mu\text{mag}. \quad (6)$$

Observations from space should be capable of photometric precision at the level of a few μmag and stability over timescales corresponding to typical CEGP orbital periods of several days. This would enable a direct detection for the current list of CEGPs inferred from radial velocity surveys, and would provide a survey technique for new objects. Such photometric precision is the goal of at least two satellite missions scheduled for launch within the next five years, MOST (Canadian Space Agency) and COROT (French Space Agency, CNES). Prior to the launch of these missions, direct detection of CEGPs simply through photometric modulation of the starlight is not feasible. We must make use of the large spectroscopic separation of the reflected light component from the stellar light to attempt a direct detection.

3. Direct Separation of the Secondary Spectrum

If the reflective efficiency η does not vary markedly over some spectral range, then the observed spectrum would consist of the stellar absorption lines and a Doppler shifted copy dependent upon the inclination (and hence reflex velocity) of the planetary orbit. The following conceptual algorithm could be used to achieve a direct detection of the companion: Obtain many high resolution, high signal-to-noise spectra at a variety of orbital phases. Obtain a series of spectra when the planet is near to inferior conjuction, and hence minimally illuminated as viewed by the observer. Although the orbital inclination is unknown, the radial velocity data provides the phase of the planet as it must be $\phi_0 = \frac{1}{2}$ out of phase with the measured stellar motion. Coadding this series, one would obtain a reference stellar spectrum at very high signal-to-noise. This can then be subtracted from any one observation, yielding residuals which should constitute the spectrum of the secondary, yet dominated by the photon noise from the primary. Given an assumed value of $\sin i$, these residuals can then be shifted by the negative of the expected Doppler shift of the planet for each observation, and then co-added to increase the signal relative to the noise. Finally, the resulting summed residuals can be correlated with the stellar template (or an alternate model of the reflected light spectrum) to confirm or exclude the presence of a secondary reflected light component. As the inclination of the orbit is unknown, this method must be carried out for a variety of assumed values of $\sin i$. When the correct orbit is assumed, the planetary features add constructively and the correlation is maximized, yielding the inclination. The amplitude of the correlation peak measures ϵ, which in turn yields the product of the albedo and the square of the planetary radius.

There are several reasons for the choice of τ Boo as the optimal candidate for this experiment: Firstly, of the four CEGP systems, τ Boo b is the closest to it parent star. Since the reflected light amplitude decreases with the square of the planet-star separation, clearly a small orbital separation is desirable. Secondly, the visual brightness of τ Boo is greater than that of 51 Peg or ρ^1 Cnc, and hence more favorable as it is the photon noise of the star which must be overcome. Finally, there exists evidence from observations of Ca II H + K lines (Baliunas *et al.* [1]) that the star is rotating with with the same period as that of the planetary orbit, implying that the star and planet are tidally locked. Marcy *et al.* [16] demonstrate that, in the case of τ Boo, a convective envelope of mass $M_{CE} \approx 10^{-2} M_{sun}$ could be synchronized with the orbital period in less than the age of the system. If the star and planet are tidally locked, then the planet does not see the rotationally broadening of the stellar lines, as there exists no motion between any point on the surface of the star relative to any point on the surface of the planet. Thus the planet reflects a rotationally unbroadened spectrum of the primary, with an characteristic width due to the stellar photospheric convective motions. An observer then sees the reflected spectrum broadened only by the planetary rotation, which is small in this tidally locked scenario. The result is to superimpose sharp planetary lines on the much broader stellar lines. This effect may facilitate the identification of the secondary component in the spectrum, although we do not require this assumption in our methodology.

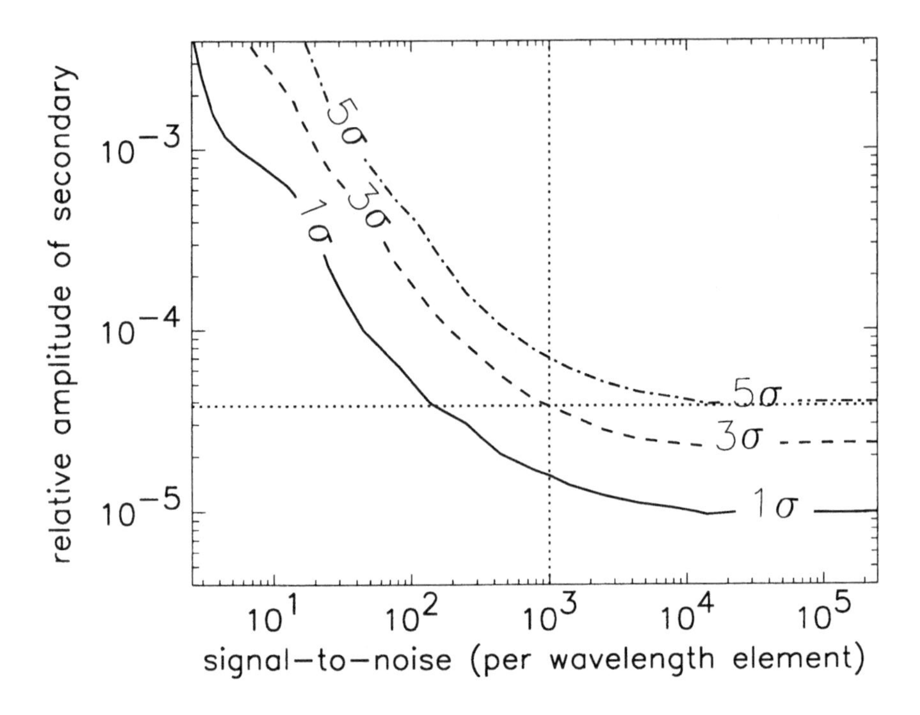

Figure 2. The photon noise limited detectability threshold of a faint secondary using the detection methodology prescribed in the text, applied to simulated spectra, computed to match our τ Boo observations in which a secondary signal and the appropriate Poisson noise are injected. The 3σ detection limit is at 3.8×10^{-5}.

In order to attempt a direct separation of the reflected light spectrum of τ Boo b from that of the primary using the methodology described above, 3 nights of data were obtained on the Keck telescope, utilizing the HIRES echelle spectrograph (Noyes *et al* [18]). Figure 2 illustrates the predicted 3σ detection limit given the statistical significance of the achieved data set. To achieve the photon noise limit, a plethora of sources of systematic errors must first be surpassed. Among these are pixel-pixel sensitivity variations in the CCD, scattered light and spurious movement of the CCD in the focal plane. All of these effects must be quantified and subsequently corrected in order to approach the photon noise limit. A detailed description of the results of this investigation will be presented in Charbonneau *et al.* [7].

4. Distortions to Line Profile Bisectors

The line profile bisector is defined to be the locus of midpoints of a stellar absorption line from the core up to the continuum. Typical bisectors for Sun-like stars show a distinctive **C** shape caused by granulation in the photosphere (Gray [10]). Recently, the interpretation of the observed periodic radial velocity variations from the star 51 Peg (Mayor & Queloz, [17]) were put into question when Gray [8] and Gray & Hatzes [11] claimed to detect distortions to the line bisector at the 4.23 day radial velocity period. They stated that the hypothesis of an orbiting companion could not account for such variations, and hence was no longer viable. Further observations and analyses (Gray [9], Hatzes *et al.* [13], Brown *et al.* [2, 3]) now exclude the claimed period in the line bisector, and the planet hypothesis has emerged as the most reasonable explanation of the radial velocity observations.

Nevertheless, the question of interpretation of line bisector variations remains. Would the planet explanation necessarily be excluded if intrinsic variations in the line profiles at the claimed orbital period were detected? Charbonneau *et al.* [6] have demonstrated that one effect of the addition of a reflected light spectrum is to cause time varying distortions to the line profiles at the orbital period, and at an amplitude which might be observable.

The observed velocity of the planet relative to the star at phase ϕ is

$$v_p(\phi, i) = -K_s \frac{M_s + M_p}{M_p} \cos 2\pi\phi. \tag{7}$$

We have modelled the stellar and planetary line profiles as a function of velocity (with zero velocity corresponding to the center of the stellar line) by

$$\begin{aligned} s(v) &= 1 - a_s \exp\left(-v^2/\Delta v_s{}^2\right) && (8) \\ p(v, \phi, i) &= 1 - a_p \exp\left(-(v - v_p(\phi, i))^2/\Delta v_p{}^2\right) && (9) \end{aligned}$$

where we have explicitly allowed for a different planetary line width (Δv_p) and depth (a_p) due to the rotation rates of the planet and star (as described above).

The intensity of the reflected light is given by Equation 4, so that the observed spectrum of the system is

$$d(v, \phi, i) = \frac{s(v) + f(\phi, i)p(v, \phi, i)}{1 + f(\phi, i)}, \tag{10}$$

where we have renormalized to the continuum flux level. The effect of the reflected light component on the observed bisector is shown schematically (and greatly exaggerated) in Figure 3.

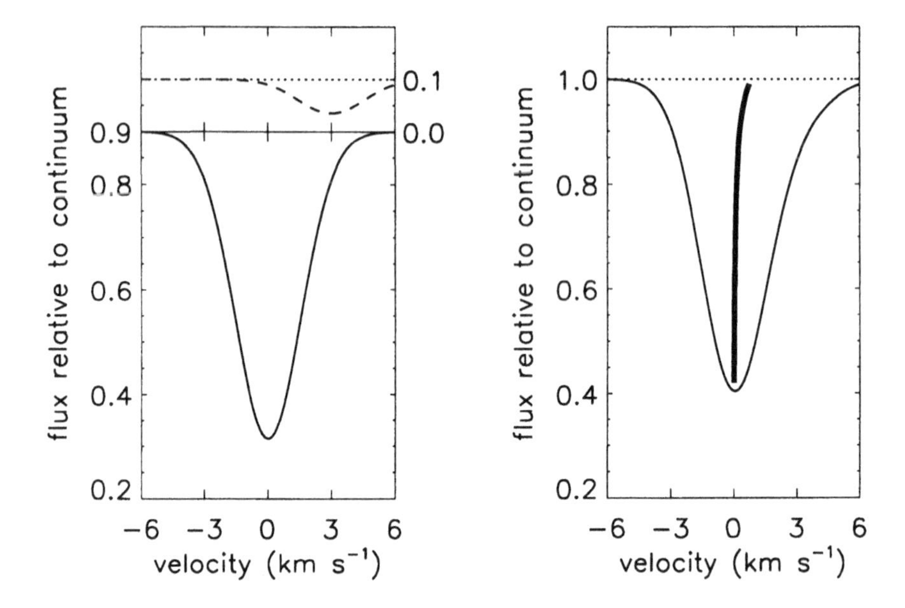

Figure 3. The observed line profile is the sum of the stellar absorption feature and a time varying, Doppler shifted reflected light component. The amplitude of the reflected light has been greatly exaggerated in this plot so as to make the resulting distortion to the bisector visible.

Below are the results for models of two systems, 51 Peg and τ Boo. For 51 Peg ($K_s = 56$ m s^{-1}), we assume that $\epsilon = 10^{-4}$, and that the reflected light is simply an undistorted replica of the stellar line, so that $\Delta v_p = \Delta v_s$ and $a_p = a_s$. For τ Boo ($K_s = 469$ m s^{-1}) we again take $\epsilon = 10^{-4}$, and further assume that the star and planet are tidally locked (as discussed above) and thus the effect is to superimpose relatively narrow planetary lines on the much broader stellar lines.

For each point in the phase, we calculate the bisector, the midpoint between two halves of the absorption feature at all line depths. For comparison with recent observational studies, we characterize the temporal variation in the bisector in terms of the velocity span, a conventional parameter used in line bisector analysis. The velocity span is defined to be the velocity difference in the bisector at two flux levels. Here we choose the flux levels to be within 1% of the continuum and within 1% of the maximum line depth.

The results of our simulations are demonstrated in Figures 4 and 5. In Figure 4, the time varying component of the bisector is plotted for 51 Peg and τ Boo for a selection of phases. The variations in the τ Boo bisector with phase are more complex than those of 51 Peg, owing to the differences

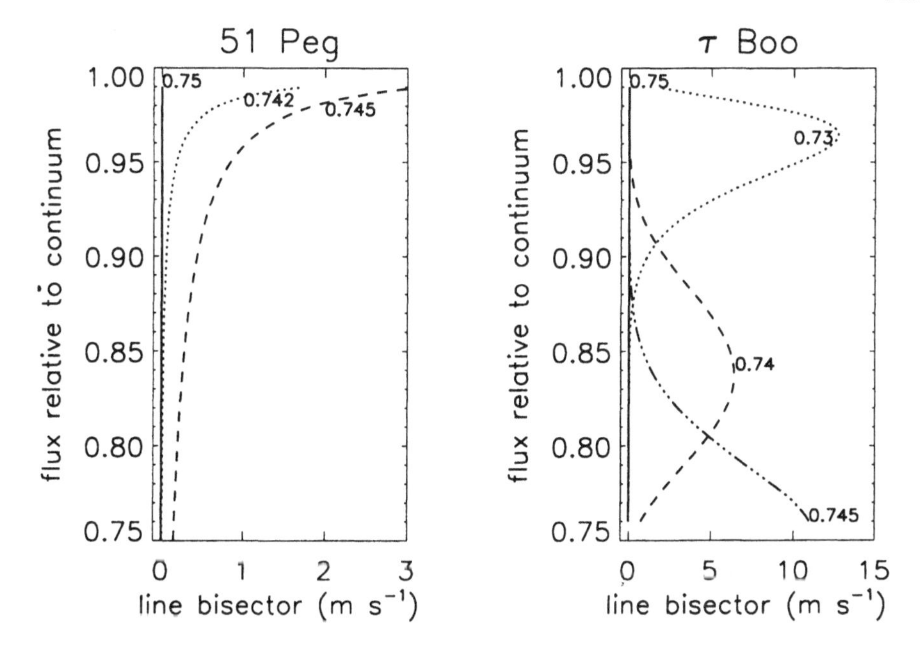

Figure 4. Predicted distortions to the bisector of a typical spectral line of 51 Peg and τ Boo for an orbital inclination of 80°. The bisectors are labelled by radial velocity phase.

in the line widths of emitted light from τ Boo A and reflected light from τ Boo b. The resulting velocity span variations with phase for a selection of orbital inclinations are shown in Figure 5.

The bisector is distorted for the fraction of the phase that the velocity separation of the planet and star projected along the line of sight is less than the stellar line width. For 51 Peg with $i = 80°$, the velocity span is non-zero for roughly 3% of its 4.23 day period, or about 3 hours, centered on $\phi = \frac{3}{4}$. In the case of τ Boo, again with $i = 80°$, the larger stellar line width means that the bisector is distorted for a larger fraction of the phase, roughly 10%, or 8 hours, of the 3.3 day period. The predicted velocity span variations of $\sim 8 \text{ m s}^{-1}$ peak-to-peak in the case of 51 Peg, and $\sim 80 \text{ m s}^{-1}$ peak-to-peak in the case of τ Boo, are commensurate with the current upper limits (Hatzes *et al.* [13], Hatzes & Cochran [14]), though these predictions rely upon measurement of the bisector higher up towards the continuum than current observations realize. Nonetheless, the predicted bisector variations induced by the reflected light from the planet possess a characteristic signature dependent upon the inclination of the orbit and amplitude of the signal which would allow for a direct detection of the companion in these systems should the required measurement precision be achieved.

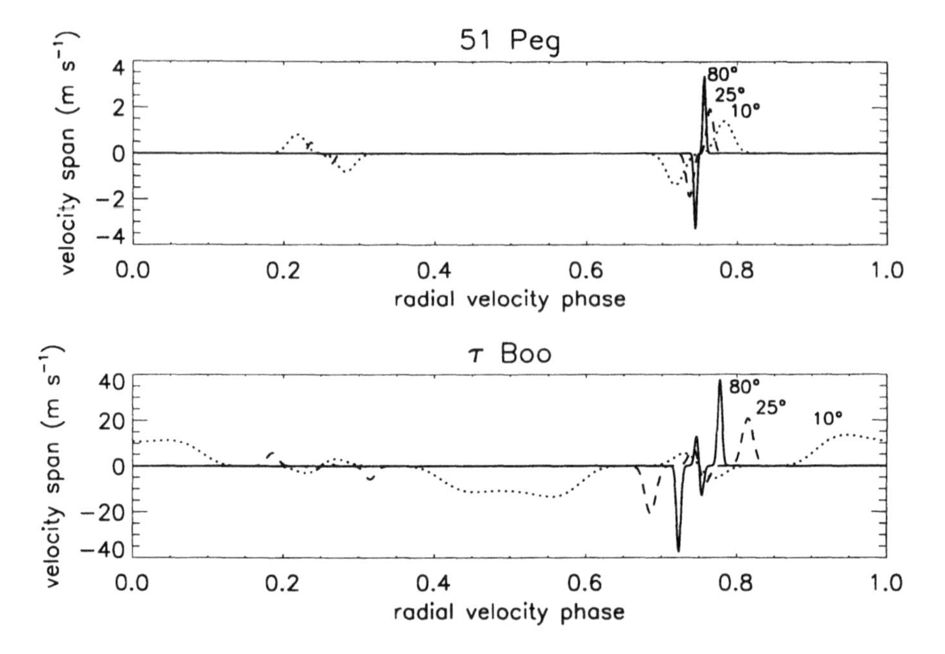

Figure 5. The velocity span variations with radial velocity phase for 51 Peg and τ Boo. The inclinations are 80° (solid), 25° (dashed) and 10° (dotted).

5. Conclusions

Direct detection of close-in extrasolar giant planets through reflected light is facilitated by two factors: the increased amplitude of the reflected light due to the proximity of the planet to the parent star, and the large spectroscopic separation due to the large orbital velocity of the planet. A number of the observational consequences of this reflected light component have been examined: A Jupiter size planet in a close-in orbit such as that of the τ Boo system and possessing a reasonable albedo would produce photometric variations which will be accessible to upcoming satellite missions. We are currently attempting a direct separation of the secondary spectrum from that of the primary. Achieving the photon noise limit will enable a 3σ detection or upper limit of the reflected light from the companion at a level predicted by a Jupiter size planet with an albedo similar to that of Jupiter. Finally, the addition of a reflected, Doppler shifted copy of the stellar spectrum would introduce distinctive temporal variations in the line profile bisector at an amplitude which is commensurate with the current observational limits. A direct detection through the reflected light would be of exceptional interest. It would be the first direct detection of a planet orbiting a star other than our own, and would yield the orbital inclination and hence the mass of the companion. It would also measure a combination of the planetary radius and albedo, from which the radius may be estimated. This would allow for an evaluation of quantities such as the surface gravity and average density, and would provide the first constraints on models of these extrasolar giant planets.

Acknowledgements

I would like to thank Robert Noyes, Saurabh Jha, Sylvain Korzennik and Peter Nisenson for many useful discussions on the topic of reflected light from planets. I would also like to thank Jean-Marie Mariotti and Danielle Alloin for organizing such an enjoyable and profitable summer school.

References

1. Baliunas, S. A., Henry, G. W., Donahue, R. A., Fekel, F. C., Soon, W. H. (1997) Properties of Sun-like Stars with Planets: ρ^1 Cancri, τ Boötis, and υ Andromedae, *Ap. J.* **474**, L119.
2. Brown, T. M., Kotak, R., Horner, S. D., Kennelly, E. J., Korzennik, S., Nisenson, P., & Noyes, R. W. (1998a) A Search for Line Shape and Depth Variations in 51 Pegasi and τ Boötis, *Ap. J. Lett.* **494**, L85.
3. Brown, T. M., Kotak, R., Horner, S. D., Kennelly, E. J., Korzennik, S., Nisenson, P., & Noyes, R. W. (1998b:in press) Exoplanets or Dynamic Atmospheres? The Radial Velocity and Line Shape Variations of 51 Pegasi and τ Boötis, *Ap. J. Supp.*, astro-ph/9712279.
4. Burrows, A., Marley, M., Hubbard, W. B., Sudarsky, D., Sharp, C., Lunine, J. I., Guillot, T., Saumon, D., & Freedman, R. (1997:in press) The Spectral Character of Giant Planets and Brown Dwarfs, in R. A. Donahue & J. A. Bookbinder (eds.), *The Tenth Cambridge Workshop on Cool Stars, Stellar Systems, and the Sun*, ASP, San Francisco, pp. 27-46, astro-ph/9709278.
5. Butler, R. P., Marcy, G. W., Williams, E., Hauser, H., & Shirts, P. (1997) Three New "51 Pegasi-Type" Planets, *Ap. J. Lett.* **474**, L115.
6. Charbonneau, D., Jha, S., & Noyes, R. W. (1998:submitted) Spectral Line Distortions in the Presence of a Close-in Planet, *Ap. J. Lett.*.
7. Charbonneau, D., Korzennik, S., Nisenson, P., Noyes, R. W., & Vogt, S. (1998:in preparation).
8. Gray, D. F. (1997) Absence of a Planetary Signature in the Spectra of the Star 51 Pegasi, *Nature* **385**, 795.
9. Gray, D. F. (1998) A Planetary Companion for 51 Pegasi Implied by Absence of Pulsations in the Stellar Spectra, *Nature* **391**, 153.
10. Gray, D. F. (1992), *The Observation and Analysis of Stellar Photospheres, Second Edition*, Cambridge University Press, Cambridge.
11. Gray, D. F., & Hatzes, A. P. (1997) Non-Radial Oscillation in the Solar Temperature Star 51 Pegasi, *Ap. J.* **490**, 412.
12. Guillot, T., Burrows, A., Hubbard, W. B., Lunine, J. I., Saumon, D. (1996) Giant Planets at Small Orbital Distances, *Ap. J. Lett.* **459**, L35.
13. Hatzes, A. P., Cochran, W. D., & Bakker, E. J. (1998) Further Evidence for the Planet Around the Star 51 Pegasi, *Nature* **391**, 154.
14. Hatzes, A. P., & Cochran, W. D. (1998:in press) A Search for Variability in the Spectral Line Shapes of τ Boötis: Does this Star Really Have a Planet?, *Ap. J.*, astro-ph/9712313.

15. Henry, G. W., Baliunas, S. L., Donahue, R. A., Soon, W. H., Saar, S. H. (1997) Properties of Sun-like Stars with Planets: 51 Pegasi, 47 Ursae Majoris, 70 Virginis, and HD 114762, *Ap. J.* **474**, 503.
16. Marcy, G. W., Butler, R. P., Williams, E., Bildsten, L., Graham, J. R., Ghez, A. M., & Jernigan, J. G. (1997) The Planet Around the Star 51 Pegasi, *Ap. J.* **481**, 926.
17. Mayor, M., & Queloz, D. (1995) A Jupiter-mass Companion to a Solar-type Star, *Nature* **378**, 355.
18. Noyes, R. W., Korzennik, S., Nisenson, P., & Vogt, S. (personal communication).
19. Seager, S., & Sasselov, D. (1998) Extrasolar Giant Planets under Strong Stellar Irradiation, *Ap. J. Lett.* **502**, L157.

STRATEGIES FOR SPACE PROGRAMS

P.Y. BELY
Space Telescope Science Institute
2080 San Martin Drive
Baltimore, MD, 21218

Abstract. The programs currently envisaged by NASA and ESA for the search and analysis of extra-solar planets are presented together with the technical challenges that will be faced.

1. Introduction

Space offers several crucial advantages for astronomical observations and the search for extra-solar planets in particular. On the ground, even at the best sites, atmospheric turbulence typically limits spatial resolution to 0.4 arcseconds in the visible. The atmospheric turbulence effects diminish greatly in the mid-infrared (Fried's parameter of about 5 meters at 10 microns), but the background flux created by the atmosphere, which is approximately that of a black body at 260° K and emissivity of 0.08, and by the telescope optics at 270 – 280° K, is enormous.

In space, given sufficiently good optics, spatial resolution is limited only by diffraction, and the telescope optics can be cooled to minimize self emission. Except for the low Earth orbits, observatories in space also benefit from a benign environment: temperature variations are typically very low (less than 1° K), and the only external disturbance is due to solar pressure. The very low level of external disturbance can be dealt by a low authority attitude control system which in turn minimizes internal disturbances. This is especially important in the case of interferometers used for exo-planet observations because of their high sensitivity to thermal and mechanical disturbances.

In what follows we outline the space missions for exo-planet observations which are currently planned by NASA and ESA, and describe the main technical issues to be faced.

J.-M. Mariotti and D. Alloin (eds.), Planets Outside the Solar System: Theory and Observations, 319–327.

2. NASA's Program

An important portion of NASA scientific research programs has recently been recast under the label of "Origins". The goals of the "Origins" program cover both cosmology and exo-planets and have been defined as follows [1]:

- Goal 1: to understand how galaxies formed in the early universe and to understand the role of galaxies in the appearance of planetary systems and life.
- Goal 2: to understand how stars and planetary systems form and to determine whether life-sustaining planets exist around other stars in the solar neighborhood.
- Goal 3: to understand how life originated on Earth and to determine whether it began and may still exist elsewhere.

A sequence of precursor missions and three generations of space missions is envisaged in order to accomplish these goals (Fig. 1).

The *Precursor Missions* include several missions which should all be operational by the end of the year 2001. Among those, the Hubble Space Telescope (HST) especially with the recently installed NICMOS instrument, the Stratospheric Observatory for Far Infrared Astronomy (SOFIA) and the Space Infrared Telescope Facility (SIRTF), will likely benefit the field of exo-planets because of their infrared capabilities.

The *First Generation Missions* are now in the planning and definition stages, and will be launched in the second half of the first decade in the new millennium. On the scientific side, they will directly contribute to the understanding of planet formation and have the potential for detection of Jupiter-size planets. They will also serve as technological pathfinders for the second generation in the areas of cooled optics, infrared detectors and vibration isolation.

The *Second Generation Mission* will be the first one dedicated to the detection and spectral analysis of earth-like planets.

Finally, the *Third Generation Missions* consist of missions dedicated to the imaging of detected planets. A candidate mission, the Planet Imager (PI), consists of an array of five Planet Finder-class interferometers flying in formation. Each interferometer carries four 8-meter telescopes (a total of 1000 square meters of collecting area for all five interferometers) and one 8-meter telescope to relay the collected light to a combiner spacecraft. The starlight is nulled at each interferometer before the light is relayed to the combining spacecraft. The five interferometers are arranged in a parabola, creating a very large baseline of 6,000 km with the combiner spacecraft at the focal point of the parabola. For now, such a mission remains just a

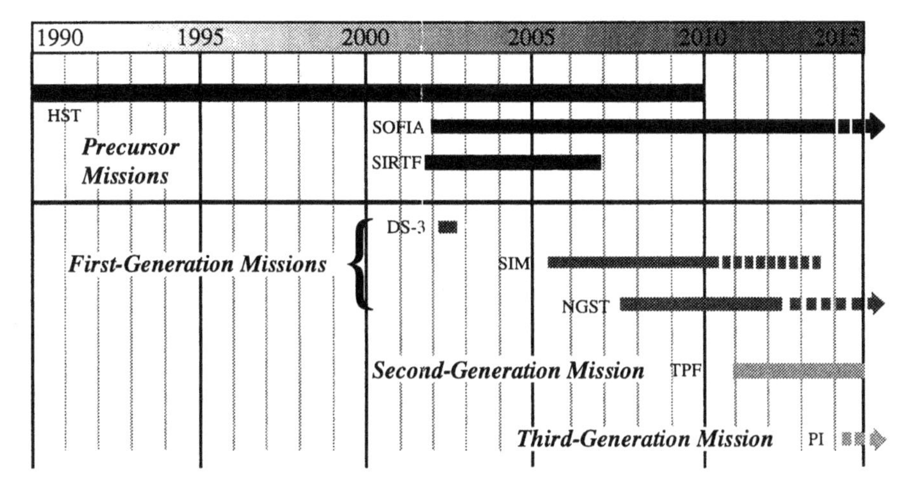

Figure 1. Timeline of the "Origins" missions

vision because the required technology and funding is not on the immediate horizon.

The main characteristics of the first and second generation missions are described below.

2.1. SPACE INTERFEROMETRY MISSION (SIM)

The Space Interferometry Mission (SIM) is dedicated to narrow angle astrometry. Although its main purpose concerns the dynamics of the galaxy, it has the potential to detect massive exo-planets by measuring the motion of their parent stars. This can be done over relatively short periods thanks to its precision of a few millionths of an arc second.

2.2. DEEP SPACE-3 (DS-3)

Prior to SIM, a precursor mission called New Millennium Interferometer, also known as Deep Space-3 (DS-3), will be launched to demonstrate interferometry in space. DS-3 is intended to demonstrate the feasibility of multiple spacecraft precision formation flying and very long baseline interferometry. There is no doubt that the kilometric scale separations in the Planet Imager will require multiple spacecraft flying in formation. But even the TPF which has separations between 75 and 100 meters could possibly benefit from the same technology.

DS-3 is planned to be launched in 2001 in a heliocentric orbit. It is composed of two collector spacecraft directing stellar light to a third combiner spacecraft where the light beams are combined for interferometric measurements. Bright astrophysical objects (14th magnitude and brighter) will be observed in the visible at 0.55 to 0.9 microns, with a 100 to 1000-meter baseline in an equilateral triangle to attain an angular resolution of 1 to 0.1 milliarcsecond.

2.3. NEXT GENERATION SPACE TELESCOPE (NGST)

Although the primary goal of the Next Generation Space Telescope (NGST) is cosmology [2], this 8-meter class infrared telescope (0.6 to 30 microns) will be capable of making significant contributions to the study of planetary discs [3], directly detect Jupiter-sized planets [4], and map zodiacal disks around nearby stars [5]. NGST would be launched in 2007 and its main characteristics are summarized in Table 1.

NGST is optimized for high sensitivity, but not necessarily for exo-planet work. Unlike all previous space telescopes, NGST will have a segmented primary mirror which is deployed after launch. This could generate problems in coronagraphic work because of the diffraction due to notches and gaps between segments and also because of the abrupt phase errors which may occur on each side of these gaps. Another problem for exo-planet work stems from the fact that NGST optics will be diffraction limited only at 2 microns, with a degraded image quality in the visible. But with adequate pupil masking, optical filtering and post correction of the wavefront errors, these effects may be overcome and the very large collecting area of the telescope and the low temperature of the optics could then be used to great advantage.

2.4. THE PLANET FINDER

The Planet Finder is the first mission dedicated to the detection and coarse spectral analysis of Earth-like planets. It will search out planetary systems around the brightest 1000 stars within 13 Pc of our solar system and will characterize the spectra of the brightest 50-100 of the detected planetary systems. The search is best done in the mid-infrared (7 to 17 microns) for two reasons: i) the contrast between the parent star and planet is minimum there (10^6 instead of 10^9 in the visible) and ii) ozone and methane emission lines which would suggest existence of life fall in that waveband [6]. The main characteristics of TPF are summarized in Table 2.

The key problem faced by TPF is the extreme brightness of the parent star compared to faintness of potential surrounding planets. Straight imag-

TABLE 1. NGST main characteristics

Wavelength coverage	0.6 to 30 microns
Aperture	8 meter, quasi-filled
Sensitivity	4 nJy in 10,000s at 2 microns, S/N=10, BP20%
Resolution	0.050" (diffraction limited at 2 microns)
Science instruments	cameras, multiobject spectrograph
Field	up to 4'x4' per instrument
Yearly sky coverage	full sky
Instantaneous sky coverage	> 15%
Optics temperature	< 70° K
Mass	> 3500 kg
Mission lifetime	5 years nominal- 10 years goal
Orbit	Sun-earth Lagrange 2 Halo orbit
Launcher	Medium class launcher

ing is not possible because of the insufficient dynamic range of detectors, therefore the light of the parent star must be somehow eliminated. When working at infrared wavelengths, the potential planet typically falls within the first Airy ring of the image, and traditional coronagraphs would not work because they cancel light up to about the third Airy ring. The solution consists in "interferometric nulling" where the light from one arm of the interferometer is combined with that of the other arm after a π phase shift [7]. The light from the on-axis parent star is cancelled, while that of a potential planet will not be if its angular distance creates a pathlength difference compensating the π phase shift of the system. In practice, a simple two-telescope interferometer is not sufficient, however, because the parent star disk is generally resolved and the null is not broad enough to prevent the star light from spilling over. A wider null can be obtained by combining the light of four or more telescopes.

With the light from the parent star cancelled, the next difficulty arises from the background due to our zodiacal light. Near the Earth, large telescope apertures, about 4 meters, are required to beat down the corresponding noise. An alternate is to operate the interferometer at about 5 astronomical units from the sun where the local zodiacal light is about 300 times less. This would allow smaller telescopes (1.5 to 2 m) to be used, but requires higher launch capability. Table 3 gives the flux of the various background noise components at 1 and 5 AUs [8]. The trade-offs between the two approaches are complex. If, however, the current NGST development program is successful in making large optics in space affordable, a near-Earth orbit which benefits from easier communication, power and better sky coverage

TABLE 2. TPF main characteristics

Parameter	Value
Operating Wavelength	$7-17\mu m$
Spatial Resolution	0.025"
Starlight rejection	10^6
Sensitivity	Detect broad-band (Resolution $\simeq 5$)
	Detect spectral features (Resolution $\simeq 20$)
Number of Targets	100's
Orbit	1 AU (L2, solar, or similar)
	or 3-5 AU (solar)
Telescope Apertures	4-6 m (1 AU)
	1.5-2 m (5 AU)
Interferometer Baseline	75-150 m
Mission Duration	5 years
Detector performance	< 8 electrons/ dark current
	1 electron read noise
Operating temperature	< 35 K

would likely be the most attractive solution.

Finally, a source of background noise that may affect both TPF and future similar missions is starlight scattered by the dust (zodiacal light) which is thought to be present in all planetary systems. If the planetary system has much more dust than our own solar system does, future direct imaging of the planets from space will be more difficult, if not impossible. Characterization of exo-zodiacal disks in nearby star systems will be studied from the ground using the Keck interferometer, from space with SIRTF and possibly later with a dedicated mission such as the Exozodiacal Mapper [9]

3. ESA's Program

The European Space Agency, for its part, has identified Infrared Space Interferometry as a candidate for a "Cornerstone Mission" in the *Horizon 2000+* science plan.

In addition to the study of disks around stars and cores of extra galactic objects such as AGNs and Quasars, a long baseline infrared interferometer mission would be able to detect and characterize potential exo-planets in a manner similar to TPF. A study is currently under way to delineate the scientific objectives, define the main characteristics of the mission of the mission and better understand the problem areas and needed technologies.

TABLE 3. Fluxes at 1 and 5 AU

Distance from sun (Telescope diameter)	5 AU (2 m)	1 AU (4 m)
Earth at 10 parsecs	2.5 10^3	1.0 10^4
Exo zodiacal disk	2.6 10^5	1.0 10^6
Local zodiacal light	1.7 10^4	1.9 10^6
Nulled star	1.2 10^4	4.7 10^4
Detector dark current	5.0 10^4	5.0 10^4
Total counts	3.4 10^5	3.0 10^6
Noise	582	1746
Signal to noise ratio	4.3	5.7

4. Technical Challenges

The suite of missions which are part of NASA's Origins program and ESA's plans is vitally dependent on technologies which are not fully available today. Below we review the main problems to be solved and the solutions which are currently being considered.

4.1. LIGHTWEIGHT MIRRORS

Lightweight primary mirrors are clearly the key to large aperture telescopes such as NGST or the future Planet Imager. But this is also the case for the medium-sized telescopes used in TPF, especially if that mission is to operate near Earth in which case 4-meter class primary mirrors are required.

A lightweight mirror development is currently under way for NGST with an areal density goal of 15 kg/m^2 (mirror blank, supporting structure and actuators). One type being developed by the University of Arizona [10] relies on a 2mm-thick glass meniscus supported on a bed of densely packed actuators (10 cm spacing). The main issues to be solved in this approach are polishing and handling of the very thin glass sheet, supporting and actuating the facesheet at the periphery and supporting the fragile facesheet during launch. Another mirror scheme being developed is a hybrid composed of a glass facesheet bonded to a graphite-epoxy supporting structure [11]. The main problem in this case is avoiding debonding during cooldown due to the coefficient mismatch between the two materials.

Another approach, also under development, uses beryllium which has extremely good stiffness and thermal conduction properties. The blank is made of a facesheet supported by an isogrid structure with very thin webs. The main problem to be solved is reducing the cost of the material and lightweighing process.

4.2. PRECISION DEPLOYMENT

Launcher fairing size limitations (about 4.5 meters in diameter by 11 meters in height) require that the large aperture missions (i.e. NGST, PI) be deployable. This is also the case for long baseline interferometers (TPF) unless a free flyer scheme is used. In both cases, the structure supporting the optics has to be lightweight and must deploy with an accuracy compatible with the range of the optics adjustment system. The required deployment accuracy, a few millimeters for NGST and a few centimeters for the TPF structure, has been achieved by current technology (e.g. Fastmast deployed on Space Shuttle STS-46) but it will be essential to verify that once deployed and locked, these structures remain stable. Long term drifts are correctable by periodic calibration, but short term dimensional changes induced by thermal or dynamic effects would be unacceptable for passive solutions (e.g. NGST) and would require high bandwidth compensation for the active systems such as TPF may have to use.

4.3. FREE FLYING SPACECRAFT

An alternate to using a 75 to 100 meter long deployable boom to support the TPF telescopes is to mount them on separated spacecraft. A preliminary study by ESA [12] suggests that such a solution is feasible with state of the art technology. The main requirements concern spacecraft to spacecraft ranging and powerful but precise and quiet propulsion. If the DS-3 mission is successful, the separated spacecraft solution would certainly be attractive for TPF because the baseline can be varied. This would permit maximizing the signal of a potential planet depending on its distance to the parent star.

4.4. VIBRATION ISOLATION AND SUPPRESSION

Mechanical systems used for observations, such as filter and grating wheels, and for space support, such as reaction wheels, tape recorders, and momentum management devices, emit disturbances which after being amplified by the highly resonant dynamics of the structure may result in vibration amplitudes well into the micrometer range. Passive isolation may in some cases be sufficient [13]; if not, active vibration isolation will have to be used. Such active devices have been successfully developed at JPL over the last few years in the context of the space interferometry mission.

4.5. ACHROMATIC NULLING

As explained above, cancellation of the parent star in a TPF-like concept relies on a π phase shift introduced in one the interferometer arms. This phase shift cannot be obtained by a simple delay line because it would only be good for a given wavelength and detection of faint companions requires a significant bandpass (e.g. 20%). In other words, the phase shift must be "achromatic". Several schemes have been proposed [14], but the most promising one, due to Gay [15, 9] is based on the phase reversal occuring while going through focus. The difficulty will be in achieving the very high cancellation required (10^{-6}).

References

1. NASA (1997), Roadmap for the Office of Space Science Origins Theme, prepared by the Origins Subcommittee of the Space Science Advisory Committee, *NASA publication No JPL 400-700*
2. Next Generation Space Telescope (1997), H. S. Stockman ed., NASA/STScI Publication M-9701.
3. J. C. Mather, E. P. Smith, B. Seery, P. Y. Bély, M. Stiavelli, H. S. Stockman and R. Burg (1998), NGST capabilities and Design Concepts, in *Science with the NGST, ASP Conference Series*, **Vol. 133**, 3-23.
4. R. Angel and N. Woolf (1998), Sensitivity of an Active Space Telescope to Faint Sources and Extra-solar Planets , in *Science with the NGST, ASP Conference Series*, **133**, 172-186.
5. R. Burg, P.Y. Bély, L. Petro, J. Gay, Y. Rabbia, P. Baudoz and D. Redding, (1998), Searching for Exo-Zodiacal Discs with a Rotation Shearing Interferometer for NGST, in *Science with the NGST, ASP Conference Series*, **133**, pp. 221-226.
6. NASA (1996) A roadmap for the exploration of neighboring planetary systems (ExNPS), *NASA publication No JPL 96-22*
7. R. N. Bracewell (1978), *Nature*, **379**, 780.
8. C. Beichman (1998) private communication.
9. P.Y. Bély, L. Petro, R. Burg, L. Wade, C. Beichman, J. Gay, P. Baudoz, Y. Rabbia, and J.M. Perrin, (1998) The Exo-Zodiacal Disk Mapper: a space interferometer to detect and map zodiacal disks around nearby stars, *Experimental Astronomy*, in press.
10. J.H. Burge, J.R.P. Angel, B. Cuerden, H.M. Martin, S.M. Miller, D.G. Sandler (1998) Lightweight mirror technology using a thin facesheet with active rigid support, *SPIE Proc.* **3356**, 690.
11. E.P. Kasl, G.V. Mehle, J.E. Dyer, H.R. Clark, S.J. Connel, D.A. Sheikh (1998) Recent developments in composite-based optics *SPIE Proc.* **3356**, 735.
12. Kilometric baseline interferometry (1996) *ESA Report* SCI-96-7.
13. G.E. Mosier, M. Femiano, K. Ha, P.Y. Bély, R. Burg, D.C. Redding, A. Kissil, J. Rakoczy, L. Craig (1998) Fine pointing control for the Next Generation Space Telescope, *SPIE Proc.* **3356**, 1070.
14. M. Shao and M. Colavita (1992) A&A Reviews, **30** 457.
15. J. Gay and Y. Rabbia (1996) *C.R. Acad. Sci.*, Paris, **322**, série II b, 265.

Part III : Astrobiology

Parmi tous les lits sur terre
où est ton lit?
Parmi les pas qui martèlent le sol par milliers
où sont tes pas?
Parmi la lignée des mystères
où est ta maison?
Faut-il que j'habite une aile pour voir
comment dort la terre?
Les sources coulent-elles en moi
pour que j'oublie tout retour?
Faudra-t-il que je fasse s'incliner l'étoile
pour te voir?

Chawki Abdelamir
Epi des terres paiennes

J.-M. Mariotti and D. Alloin (eds.), Planets Outside the Solar System: Theory and Observations, 329.

BIOLOGICAL FOUNDATIONS OF LIFE

MARIE-CHRISTINE MAUREL

Institut Jacques Monod- Tour 43
2, place Jussieu 75251 Paris Cedex 05
e-mail : maurel@ijm.jussieu.fr

It is possible to simulate, in the laboratory, conditions that may have existed on the prebiotic Earth producing a range of prebiotic compounds. A fundamental question is now to know all kinds of primeval or simple lifes possible and hence what kind of signature of life it is possible to expect on another planets or systems. Entities like-cells might exist (or may have existed) without pedigrees or genealogical history but with a physical history. We have to search all kind of architecture of life in and out the conventional lineages.

1. The emergence of life on Earth.

The emergence of life on Earth is the result of a long chemical evolution which began 4.5 billion years ago (Fig 1).
The oldest known sedimentary rocks containing carbon molecules of biological origin, have been found in the west of Greenland and date from 3.8 billion years ago (Schopf, 1983; Mojzsis et al., 1996). It is thus possible to infere that the oldest organisms of which partially decomposed remains have been found in this carbon rich rocks were living in an environment devoided of dioxygen.
The conditions which existed at this time on the surface of the planet were completely different from those we know today.
As regards to the composition of the primitive atmosphere, different models have been proposed (Chyba and Sagan, 1992). The most famous amongst them, proposes a composition rich in hydrogenated gas like methane, ammonia and water vapour (Oparin, 1938; Kasting, 1993).
In facts, the analyses of trapped gases in ancient rocks and from the existence of sedimentary carbonate deposits dating from this time, suggest that the primitive atmosphere was much less reducing (Braterman et al., 1983). Some people thought that an atmosphere, known as secondary, would have been formed in the course of the cooling of the planet as a result of volcanism and the progressive degassing of the crust and the mantle.

J.-M. Mariotti and D. Alloin (eds.), Planets Outside the Solar System: Theory and Observations, 331–363.

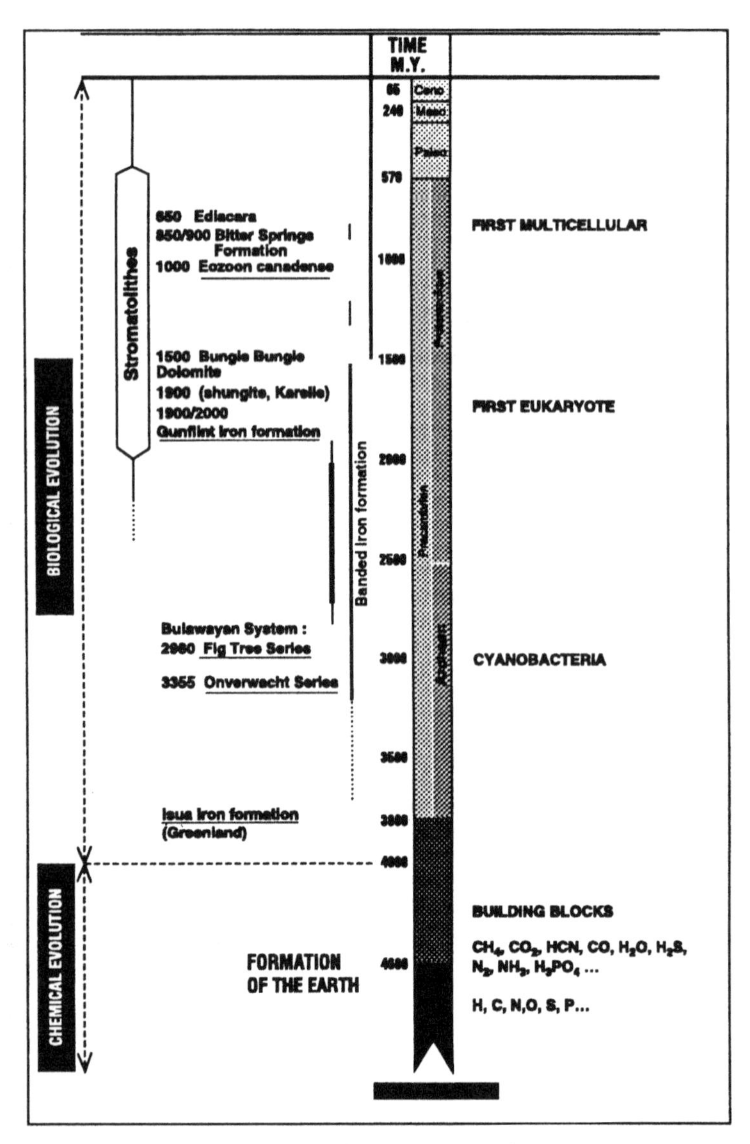

Figure 1. Geological time scale.

The volcanoes must have expelled huge amounts of gas which contributed to the formation of the first atmosphere comprising of carbon dioxide, water vapour, sulphur dioxide, an atmosphere containing small quantities of carbon monoxide, methane and dinitrogen, but not dioxygen (Walker, 1985).

When the temperature of the crust fell below the critical level (of 100°C), large volumes of liquid water accumulated by condensation to form the oceans.

Moreover 4 billion years ago the sun only emitted 75% of the energy it does today. This energy deficit would normally have entailed a glaciation of the Earth unless a compensating greenhouse effect linked to the considerable amounts of carbon dioxide allowed the maintenance of water in its liquid state. The heat given off from an intense radioactive decay, particularly that of potassium-40 at this time, contributed to the maintenance of the relatively elevated temperature. These are, very briefly, the extreme conditions in which life on Earth had to take its first steps.

Today's life forms, such as we know them, are by consequence the result of an evolution process of approximately 4 billion years duration. Paleontological discoveries in Precambrian soils confirm that for 3 billion years the only living systems on our planet were micro-organisms (Schopf and Packer, 1987; Mojzsis et al., 1996). Fossil evidence of microbial life exists in rocks 3.5 billion years old and younger. The simplest Precambrian micro-fossils show striking similarities with present day algae and cyanobacteria. Numerous micro-fossils have been discovered in the silicified parts of stromatolites, which are fossilized microbial mats formed from layers of primarily filamentous procaryotes. These lamellar structures formed by certain bacteria including present day cyanobacteria, producing calcium carbonate and carbon dioxide (by the way of photosynthesis). These fossils, like modern cyanobacteria, were formed from cellular alignments. Although modern stromatolites are composed of filamentous bacteria (oxygenic photoptrophs), this would not have been the case in the oldest stromatolites, because Earth was still anoxic at that time. Stromatolites from this date were probably made by anoxygenic phototrophs.

Today, everyone is in agreement that life almost as we know it today, was already present on the primitive Earth around 800 million years after its formation (Morowitz, 1992; Mojzsis et al., 1996).

2. How is life today.

All living organisms are composed by cells, and they may be classified in terms of the type of cell (Fig 2).

Microbe or bacteria are the procaryotes, organisms in which there is only a single compartment, bounded by membranes that give security against the outside world and insure exchanges. Inside the compartment there is a kind of jelly named cytoplasm, where the chemical reactions of metabolism are taking place. One can also find the chromosome (DNA), that is genetic material, transmitted from one generation to another one, thanks to the reproduction.

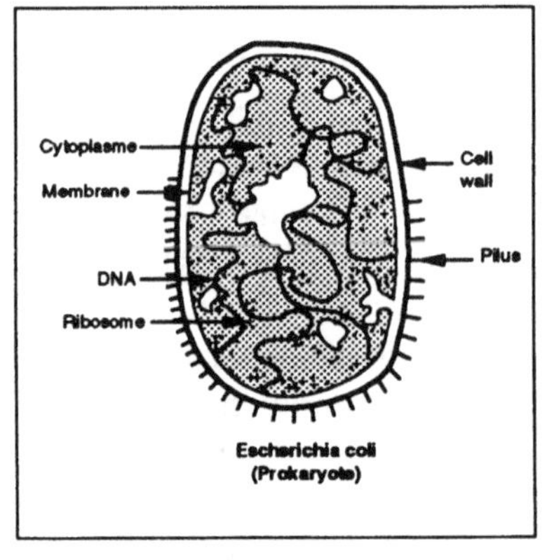

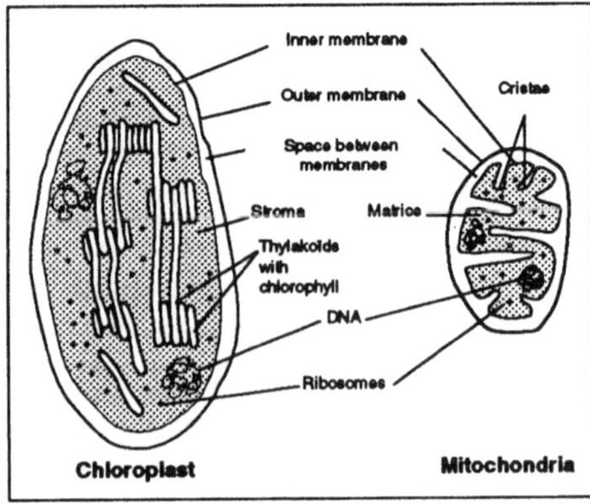

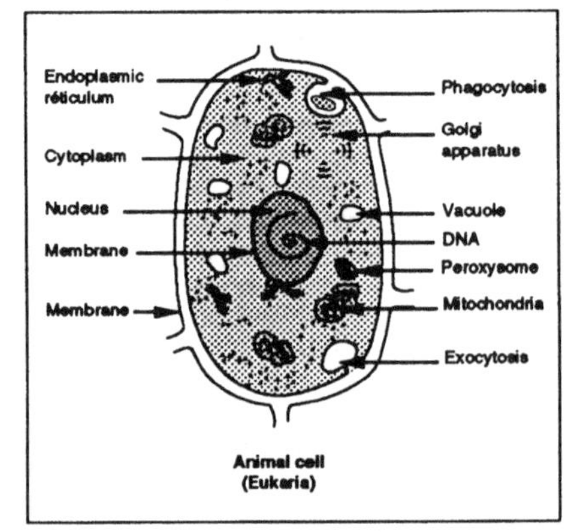

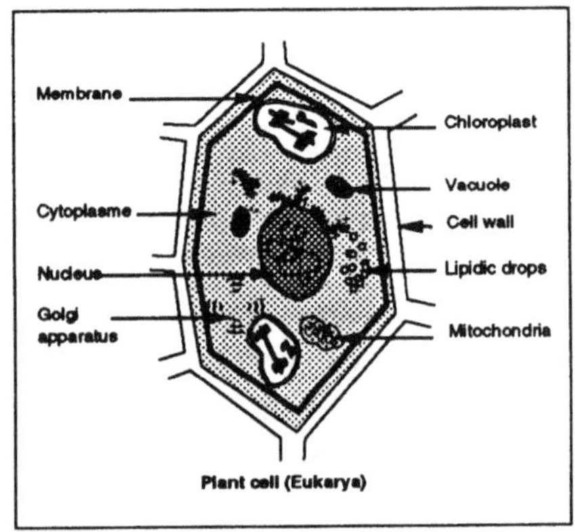

Figure 2. Type of cells (mitochondria and chloroplast are organelles of Eukaria)

Eukaryotes are defined by the division of each cell into a nucleus, containing the genetic material, surrounded by a cytoplasm, which in turn is bounded by the membrane that marks the periphery of the cell. Eukaryotes are also characterized by the presence of numerous organelles such as mitochondrions, plastids, peroxisomes etc... which are specialized parts.

Unicellular organisms (either prokaryotic or eukaryotic) consist of individual cells, able to survive and self-reproduce. Multicellular organisms (usually eukaryotic) exist by virtue of cooperation between many cells, which may have different specialities to contribute to the survival of the individual.

The nature of the genetic information and the basic process of its expression is very similar in all organisms. All organisms use the same or a very similar genetic code and they use the same aminoacids in their proteins.

Although there is a large similarity among organisms, according to morphology, genetic code and metabolism, it is yet possible to notice a great biodiversity.

As a matter of fact, the tree of life is divided in three domains that is eukaryote, eubacteria, and archaebacteria (Fig 3).

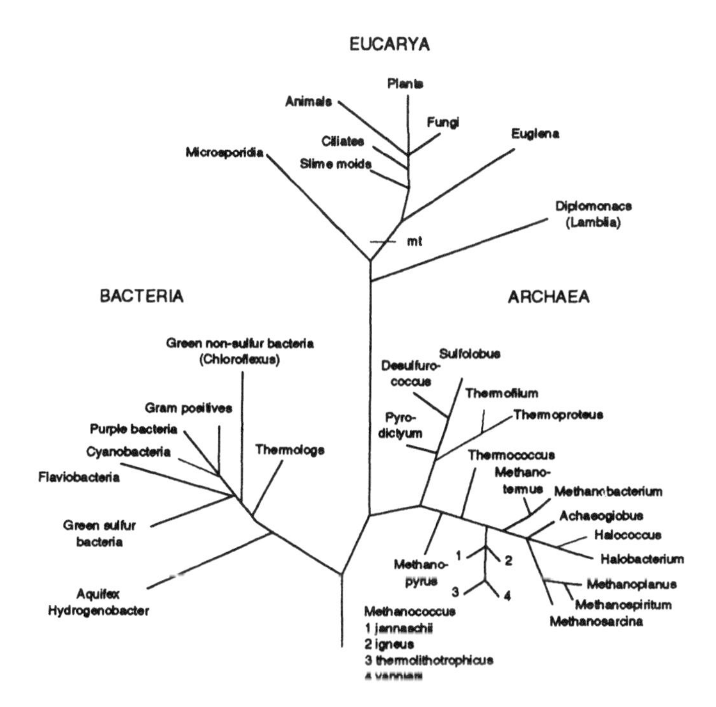

Figure 3. Universal phylogenetic rooted tree, based on the sequence of 16S/18S rRNA, ATPases and protein elongation factor (after Kandler 1993).

Eubacteria include the photosynthetic organisms formerly known as blue-green algae but better known today as cyanobacteria. Archaebacteria are also procaryotic but they are living in extreme conditions.The methanogens live only in oxygen-free milieus such as swamps. These bacteria generate methane also known as "swamp gas" by the reduction of carbon dioxide. Other archaebacteria include the halophiles, which require high concentrations of salt to survive, alkaliphiles growing at 70°C and pH 10-11, and the thermoacidophiles, which grow in hot (80°C) sulfur springs, where a pH of less than 2 is common. At high temperature there are the thermophilic organisms, at cold temperature (below 0°C), there are psychrotrophic cells (Hochachka et Somero, 1984). Some bacteria are living in the sea or in fresh water, others on the earth, or in suspension within the atmosphere, or either inside the crust of the earth. In 1996, Fredickson and Onstott related this amazing discovery of intraterrestrial micro-organisms dwell within the earth's crust inside basalt. The bacterial communities living there include so-called autotroph organisms that synthesize organic compounds from inorganic sources. They use hydrogen gas for energy and derive carbon from inorganic carbon dioxide.

The message of the recent discoveries is now clear : First of all, we know only a few percent of species now living in the biosphere (1.6. 10^6 different species are known, more than 10^7 are estimated to be present on earth). Then, we discover organisms with very surprising ways of life. Everything may be different, the frequency of cell division (which may be once a century!), the metabolisms which might be slowing down etc...

3. Molecules of the cell :

There are two main groups of molecules in the cell as it exists today: nucleic acids which transmit genetic information and participate directly in reproduction; and proteins which are responsible for the metabolic functions and chemical and structural relations within the cell. (Table 1)

. Chromosomes = Nucleic acids **DNA** **RNA**
Nucleic acids = Bases + Sugars + Phosphates **A,T,G,C**
. Proteins = Enzymes = Biological Catalysts **Proteins = Aminoacid 1 + Aminoacid 2 + ...** **Bases and aminoacids are the building bocks of life**

Table 1.

What interests evolutionary biochemistry is how the first molecules related to these one came to exist, their original replication and how the different classes of molecules are related to one another.

What was the first system enabling genetic information to be transfered ? Were the first autoreplicative molecules, the same as those we know today, or did they only have certain similarities to our nucleic acids ?

How did relations between molecules come about so as to construct the first metabolic pathways?

Chemical reactions of metabolism have to be selective, rapid, therefore catalysed, to be efficient. Were the very first catalysts which played a role in the first appearence of metabolism made up of enzymes such as we know them today, that is complex and highly organised proteins ? Or were they quite simply mineral supports or reactive compounds?

From the point of view of molecular evolution, before the appearance of cellular life when prebiotic chemistry furnished the basic building-blocks of the living organisms, we can suggest a number of hypotheses to answer these questions.

First of all, let us examine the different classes of molecules in the contemporary cell. Small molecules are the substrates and products of metabolic pathways, providing the energy needed for cell survival; they fall into four general classes :

sugar, fatty acids, amino acids, and nucleotides. They are formed by individual steps of chemical transformation. Larger molecules, the structural components of the cell are synthesised from the small molecules called monomers. Polysaccharides are assembled from sugars. Lipids are assembled from fatty acids. Proteins are assembled from amino acids. Nucleic acids are assembled from nucleotides. Proteins and nucleic acids are very large molecules known as macromolecules.
A nucleic acid consist of a chemically linked sequence of nucleotides (Fig 4).

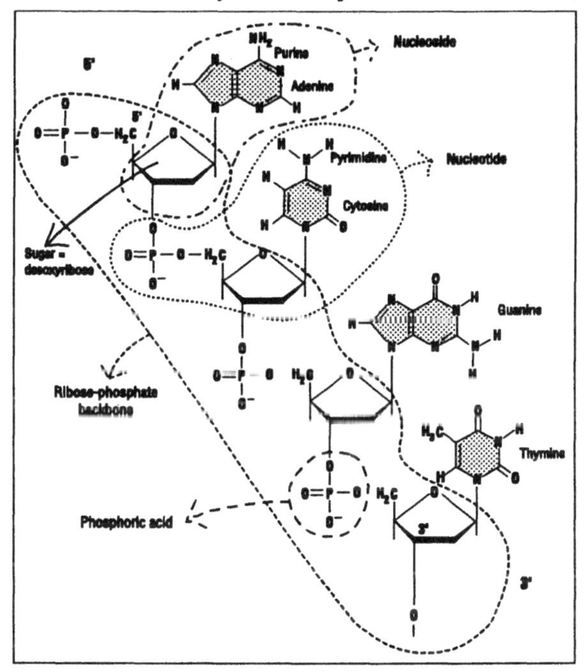

Figure 4. DNA strand.

Each nucleotide contains a heterocyclic ring of carbon and nitrogen atoms (the nitrogenous base), a five-carbon sugar in ring form (a pentose), and a phosphate group. The nitrogenous base fall into the two types shown in the figure, pyrimidines and purines. Each nucleic acid is synthesized from only four types of bases. The sequence of nitrogenous bases is the form in which genetic information is carried. DNA is a long string of four different small molecules arrayed in a sequence that encodes the cell's information. The same two purines, adenine and guanine, are present in both DNA and RNA. The two pyrimidines in DNA are cytosine and thymine; in RNA uracil is found instead of thymine. The bases are usually refered to by their initial letters; so DNA contains A, G, C, T, while RNA contains A, G, C, U.
Also two types of pentose are found in nucleic acids. They distinguish DNA and RNA and give rise to the general names for the two types of nucleic acids. In DNA the pentose is 2-deoxyribose; whereas in RNA it is ribose. The difference lies in the

absence/presence of the hydroxyl group at position 2 of the sugar ring. A base linked to a sugar forms a nucleoside; when a phosphate group is added, the base-sugar-phosphate is called a nucleotide.

Nucleotides provide the building blocks from which nucleic acids are constructed.

The nucleotides are linked together into a polynucleotide chain by a backbone consisting of an alternating series of sugar and phosphate residues. The 5' position of one pentose ring is connected to the 3' position of the next pentose ring via a phosphate group as shown in the figure.

DNA is a double helix in which the two polynucleotide chains are associated by hydrogen bonding between the nitrogenous bases. In the usual forms, G can hydrogen bond specifically only with C, while A can bond specifically only with T. These reactions are described as base pairing, and the paired bases (G with C or A with T) are said to be complementary.

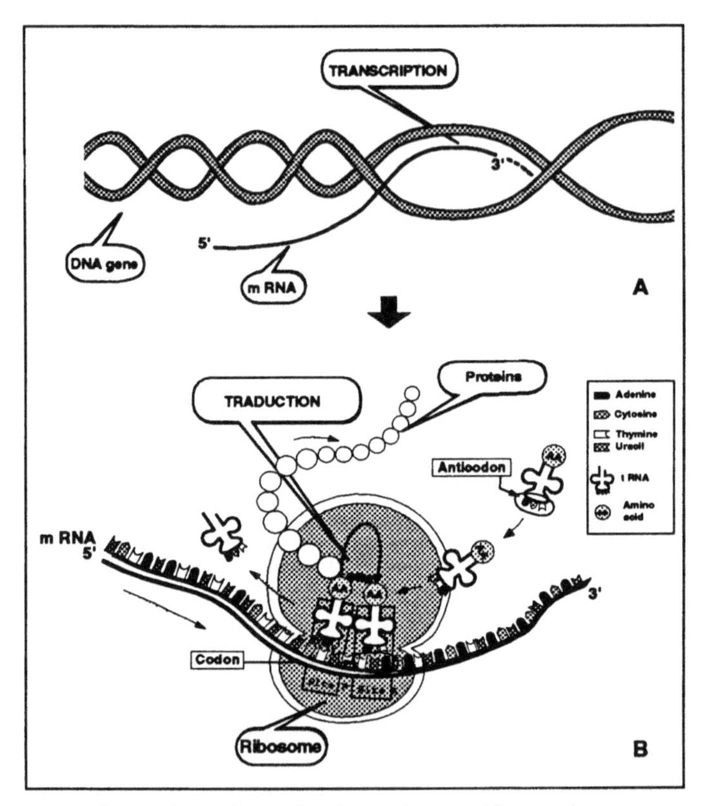

Figure 5. DNA specifies RNA, which specifies proteins.

A gene can be defined as the entire nucleic acid sequence that is necessary for the synthesis of a functional protein or RNA molecule (Fig 5). Genes actually do not specifically direct synthesis of proteins; they do it through an intermediate. Genes are the template for the synthesis of RNA, a molecule chemically very similar to DNA, which serves functions very different from those of DNA. RNA's most important role, however, is to act as a template for the synthesis of specific proteins, leading us to the so-called Central Dogma of biology : DNA specifies RNA, which specifies proteins.

For the inheritance of genetic information, DNA replication is semi- conservative. The two polynucleotide strands are joined only by hydrogen bonds, they are able to separate like a zip (Fig 6). The structure of DNA carries the information needed to perpetuate its sequence, this means that each strand acts as a template for the synthesis of a complementary strand. The parental duplex is replicated to form two daughter duplexes each of which consists of one parental strand and one newly (synthesized) daughter strand. Thus the unit conserved from one generation to the next is one of the two individual strands comprising the parental duplex.

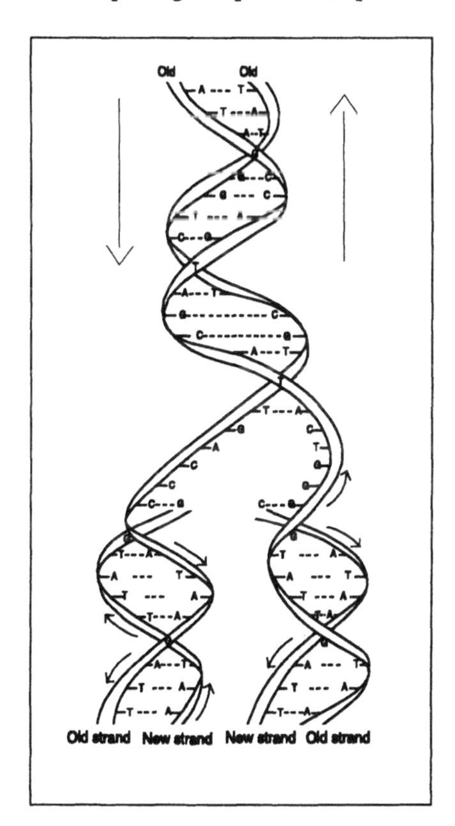

Figure 6. The two polynucleotide strands are able to separate like a zip.

Each protein consists of a unique sequence of aminoacids. 20 different amino acids are joined together to form a protein by peptide bonds which are created by the condensation of the amino (NH_2) group of one amino acid with the carboxyl (COOH) group of the next (Fig 7 a). Proteins show enormous diversity of forms as a result of their ability to generate a huge range of conformations.

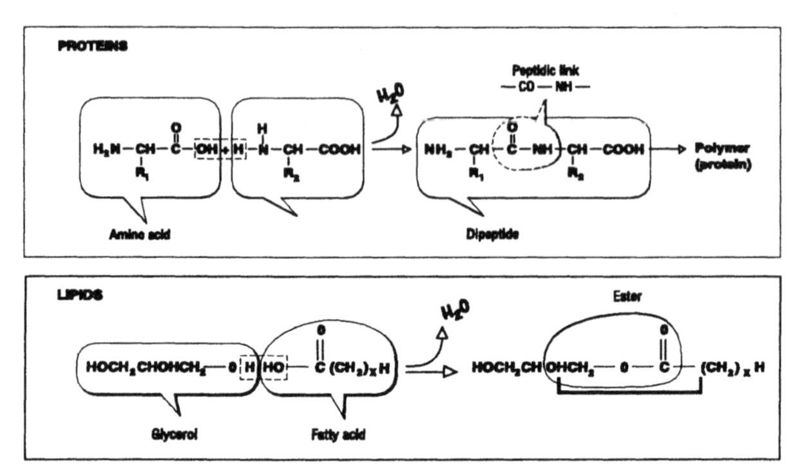

Figures 7a. 7b

Key membrane components are fatty acids, molecules that contains a long hydrocarbon chain attached to a carboxyl group (COOH), that is small hydrophilic groups (Fig 7 b).

Fatty acids are insoluble in water and salt solutions. Consequently membrane lipids orient themselves in sheets to expose their hydrophilic ends to the aqueous environment. The lipidic chains are segregated together, to minimize the disruption of the network of hydrogen bonds present in liquid water.

To emphasize the biodiversity at a molecular scale, it is interesting to notice that in eukarya, these phospholipids are straight-chain esters of a variety of head-groups, most of which contain a phosphate unit. In bacteria, more variety is observed in the acyl chains, which are often branched (iso, anteiso etc...). In archaea, n-acyl chains are normally absent, and are replaced by polyprenyl ones, usually saturated, as ethers. In eucaryotic membranes, cholesterol (or phytosterols) which strengthen the membrane, is present in large amounts. It is absent in bacteria and archaea. The presence of cholesterol in eukaryotic membranes is a sign of modernity.

Concerning metabolism it is noteworthy that the ability of bacteria to grow in a wide variety of environments is a reflection of their metabolic versatility. However, all bacteria have the same fundamental requirements for growth : a source of energy, and a reducing power, of carbon, nitrogen and other elements. There is considerable variation in the sources that can be utilized for these purposes. Energy for bacterial growth may be obtained from light, from oxidation-reduction reactions involving organic compounds. The sources of carbon may be carbon dioxide (for autotrophic organisms) or any one of a large number of organic compounds (for heterotrophic organisms).

A matter of special interest is photosynthesis. Photosynthesis is a series of processes in which electromagnetic energy is converted to chemical free energy which can be uses for biosynthesis. Commonly, photosynthesis is thought of as a process in which bacteria or green plants, with the aid of chlorophyll and light, convert carbon dioxide and water to carbohydrate and molecular oxygen. Variations on this theme occur in which the carbohydrate may be formaldehyde, glucose, or starch.

In the chloroplasts of plant cells and in photosynthetic bacteria, chlorophyll pigments absorb the energy of light, which is then used to synthesize ATP (adenosine triphosphate, that is an energy-rich molecule made of adenine-sugar-triphosphate) from ADP and Pi. Much of the ATP produced in photosynthesis is used to help convert carbon dioxide to polysaccharides that are polymers of six carbon sugars such as glucose ($C_6H_{12}O_6$):

$$6\,CO_2 + 6\,H_20 + ATP \text{ -------hv------> } C_6H_{12}O_6 + 6\,O_2 + ADP + Pi \quad (1)$$

In the cell, through an elaborate set of enzyme-catalyzed reactions, the metabolism of one molecule of glucose is coupled to the synthesis of as many as 38 molecules of ATP from 38 ADP. Thus, one predominant source of energy in cells is the glucose.

$$C_6H_{12}O_6 + 6\,O_2 + 38\,Pi + 38\,ADP \text{ --------> } 6\,CO_2 + 6\,H_20 + 38\,ATP \quad (2)$$

This type of cellular metabolism is termed aerobic because it is dependent on the oxygen in the atmosphere. For instance, if glucose is burned in air, all of this energy is released as heat (almost 686 kcal).

Aerobic degradation of glucose (named catabolism) is found in all higher plant and animal cells and in many bacterial cells. The overall result of glucose respiration, (2), is an exact reversal of the photosynthetic reaction in which polymers of six-carbon sugars are formed, (1), except that the energy of light is essential for the photosynthetic reaction.

Respiration and photosynthesis are the two major processes constituting the carbon cycle in nature. The only net source of energy in this cycle is sunlight. Thus directly or indirectly, photosynthesis is the source of chemical energy for almost all cells.

However, communities of organisms do exist in deep ocean vents and others extreme environments, where sunlight is completely absent. These unusual bacteria derive the energy for converting ADP and Pi into ATP from the oxidation of reduced inorganic compounds, such as H_2S, present in the dissolved vent gas that originates in the center of the earth. The effluent that emerges from beneath the seafloor is rich in reduced ions (principally sulfide S^{--}) which are used as chemical energy by autotrophic bacteria to convert carbon dioxide, water and nitrate (NO^{3-}) into essential organic substances. This process named chemosynthesis may have predated or may be complemented with photosynthesis.

It is important to notice that the advent of oxygen-evolving photosynthesis is one of the central events in the development of life on earth. Before the appearance of this metabolic way of life, the atmosphere was largely anaerobic, and the development of advanced eukaryotic life forms did not take place until the free oxygen in the atmosphere rose to a sufficient level. As a matter of fact anaerobic cells use up 20 times more glucose (per unit of time) than aerobic cells. No other known process, either biogenic or non-biogenic, is capable of producing the large quantities of molecular oxygen that changed the course of life on earth.
It is admitted that the most likely source for the O_2 necessary for the oxidation of Fe^{2+} to the iron oxides that produced the great volumes of iron-formation in the early Proterozoic, would be a biological source. However, for older banded iron-formations that make-up substantially less volume than their early Proterozoic counterparts, the photodissociation of water in the atmosphere can be considered another source for O2 and possibly the principal source 3800 million years ago (Walker et al 1983).
Also, Cairns-Smith (Braterman et al 1983) proposed that iron-formation could form by photochemical reactions that extract oxygen from sea-water. That is the second abiogenic hypothesis for the origin of iron-formation (Walker, 1978). It is possible that both processes operated at different times in different environments in the Archean with biological oxygen becoming the major source by the early Proterozoic.
Several groups of anoxygenic photosynthetic bacteria (performing photosynthesis in which O_2 is not produced) are thought to be more closely related to the earliest photosynthetic organisms. Some of them like purple photosynthetic bacteria oxidize ferrous iron (Fe^{2+}) to the ferric form (Fe^{3+}), others are described which transform hydrogen sulfide (H_2S) in sulfur (S). These are modestly oxidizing species.
An intermediate stage is proposed using hydrogen peroxide, H_2O_2. The early atmosphere has been midly reducing or neutral in overall redox balance (Kasting, 1990?) and water photolysis by U.V light can produce hydrogen peroxide, which then might be concentrated by rain fall in certain protected environments. A metallic complex-like enzyme extracted electrons from H_2O_2 producing the first oxygen-evolving system (Maurel et al 1996). After what, modern photosynthesis with chlorophyll to help transform H_2O in O_2 was ready to evolve (Fig 8).
We have seen life as we know it today, let's see on the geologic time scale (Fig 1) how many time it takes to do it, in other words, how many time is required to go from a protocell like-procaryote to the eucaryotic cell, then to pluricellular organisms?
Almost one billion years was necessary to make procaryote-like organisms. Then these procaryotes complexify themselves to reach more sophisticated morphologies and functions. It needed two billions years for the appearance of the first eucaryotic cell as we know it today. Then, one billion year more was necessary for eucaryotes organized themselves in colony, collectivity or community to give the multicellular organisms and then to reach the exuberant and remarkable biodiversity observed today.

H_2S	S	An intermediate stage is proposed in which hydrogen peroxyde was oxidized.
or	-------------> or	H_2O_2 -----------> $O_2 + 2e^- + 2H^+$
Fe^{2+}	Fe^{3+}	Metal-Enzyme-like extract e- from H_2O_2 producing the first oxygen-evolving complex.
Purple photosynthetic bacteria that can oxidize ferrous iron to the ferric form has been described. There modestly oxidizing species.		
	Then chlorophyll help transform H_2O -----------> O_2	

Figure 8. See text.

4. The origin of organic molecules

What happened in the course of the first billion year of the Earth's history which led to the appearance of organisms similar to present day bacteria?
Can we reconstitute in the laboratory, under the supposed primitive conditions, the early stages of bio-organic molecular synthesis?
Chemical evolution led the simplest elements, hydrogen, carbon, oxygen, sulphur, phosphorus ... to combine to form the organic molecules, methane, carbon monoxide, carbon dioxide and water vapour, molecules, which were present in the atmosphere of the primitive Earth (Fig 9).
If a gaseous mixture of these molecules is violently heated to a high temperature, or submitted to an electrical discharge, complex organic molecules are formed (Miller, 1953 and 1955; Joshi and Pathak, 1975). Such syntheses carried out in the laboratory from simple molecules under conditions believed to represent initial terrestrial conditions are called prebiotic chemistry. One of the most important prebiotic reaction and one of the simplest to carry out, is the formation of hydrogen cyanide (HCN) very soluble in water from atmospheric dinitrogen and methane (Ferris and Hagan, 1984; Ferris, 1991). Another possible prebiotic synthesis, from simple compounds in a gaseous state, is the formation of formaldehyde (HCHO) from methane (CH_4) and water vapour.

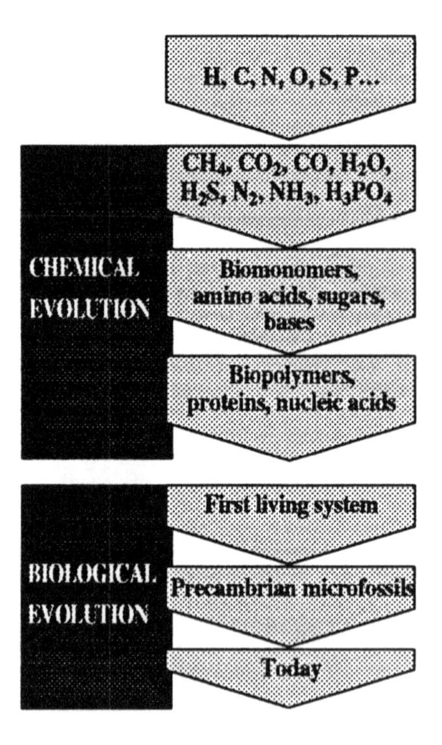

Figure 9.

Hydrogen cyanide and formaldehyde, are probably the atmospheric precursors which, dissolved in the water of the oceans, lagoons or lakes, could have reacted spontaneously to give more complex compounds in order to form the elementary building blocks of life (Eschenmoser and Loewenthal, 1992).

4.1. THE PREBIOTIC 'SOUP'

Oparin and Haldane, in 1924 and 1929 respectively, proposed that these compounds, hydrogen cyanide and formaldehyde, could have formed in water a primitive soup where complex molecules would have formed spontaneously, and would have at an appropriate time given rise to life as we know it today. Protocells or coacervates (Oparin, 1938) initially formed would have drawn from the primitive broth the molecules capable of putting in place the first heterotrophic metabolism .

In 1953, Stanley Miller, working in the laboratory of the chemist Harold Urey reproduced the presumed conditions of the primitive atmosphere (Miller, 1953 and 1955) in an apparatus in which he submitted a gaseous mixture consisting of water, ammonia, methane and molecular hydrogen to an electrical discharge (Fig 10). The results are striking: after a week, Miller found that almost half of the amino acids found in living cells were formed in significant quantities. They were obtained after the initial formation of hydrogen cyanide and formaldehyde.

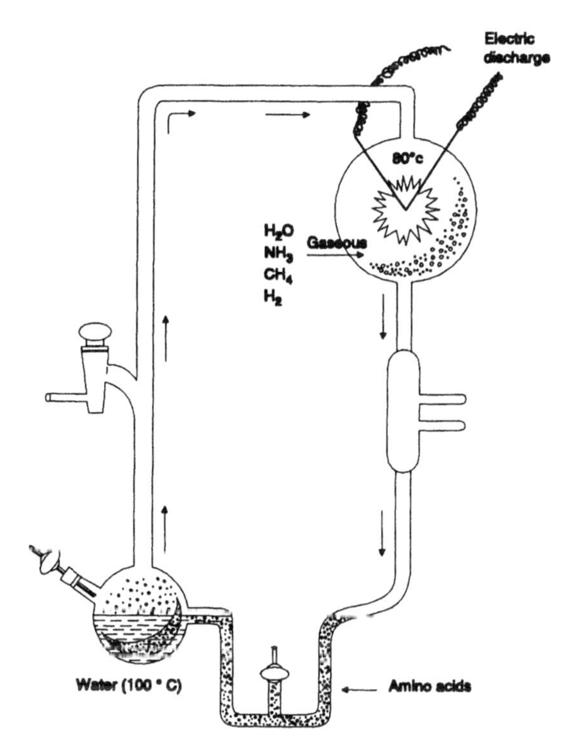

Figure 10 . Miller experiment.

Molecular ratios of some aminoacids formed by abiotic synthesis
From Weber and Miller (1981)

	Molar ratio in aminoacid from	
Amino acid	electric discharge	Murchison meteorite
Glycine	100	100
Alanine	180	36.44
α-amino-n-butyric acid	61	19.18
Norvaline	14	14.3
Valine	4.4	19.18
Norleucine	1.4	2
Leucine	2.6	4
Isoleucine	1.1	4
Alloisoleucine	1.2	4
Proline	0.3	22.16
Aspartic acid	7.7	13.5
Glutamic acid	1.7	20.18
Serine	1.1	

During the sixties, following from this result, it was realised that aqueous hydrogen cyanide could also lead to the nitrogen bases of our nucleic acids. In the 1960s, Orò succeeded in synthesising the purine base adenine (Orò, 1960; Orò and Kimball, 1961) and guanine (Orò, 1965) from aqueous ammonium cyanide) (Fig 11). Adenine was also obtained photochemically from hydrogen cyanide (Ferris and Orgel, 1966).

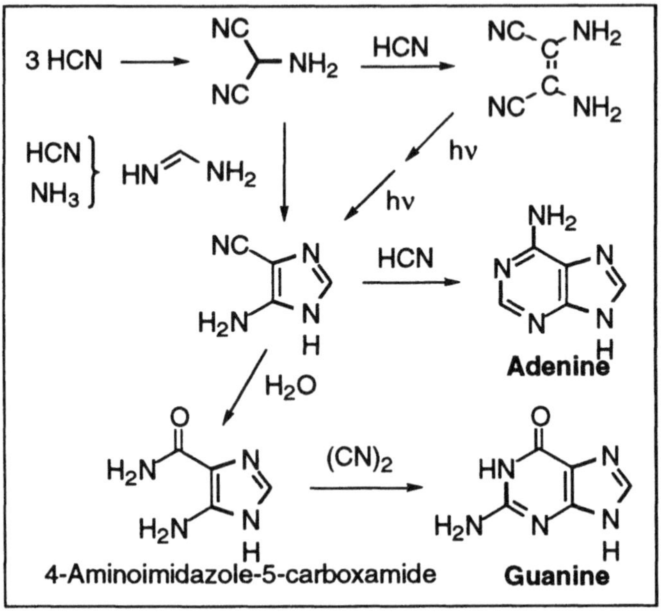

Figure 11. Synthesis of the purine bases adenine and guanine from hydrogen cyanide.

It is not possible to summarise in only a few words the very considerable amount of work carried out in the last forty years. The simplicity with which these reactions occur is a strong argument make them very good candidates as the source of biochemical monomers.

However, this scenario of the prebiotic soup, has a certain number of weaknesses.

The first concerns the low concentration of the precursors; we know that the dilution of organic compounds in the expanse of the aqueous environment is an obstacle to their meeting and hence to the synthesis of more complex compounds.

The second problem, linked to the presence of water as a solvent, is the hydrolysis of the reactants and of the reaction products.

The third problem is that of selection: in a primitive broth one has as much chance of obtaining toxic substances, anti-metabolites as good molecules useful for the purpose. Numerous reactions can interfere with the favorable processes.

Thus we have to search for the biomass, elsewhere.

4.2. INTERSTELLAR ORGANIC CHEMISTRY

The finding of traces of organic matter in meteorites and of polymers of formaldehyde, a famous "prebiotic" molecule allowed evolutionary biologists to become increasingly aware that the course of biological evolution cannot be viewed as operating independently of planetary conditions.

We know today that very significant quantities of organic material from meteoric and cometary sources have been deposited on the primitive Earth (approximately 20 grammes per cm^2); (Allen and Wickramasinghe, 1987; Carle et al., 1988; Anders, E. 1989; Chyba and Sagan, 1992). Organic synthesis is very active in interstellar space. For example, hydrogen cyanide, adenine and polymers of formaldehyde were found in comet Halley. More than ten amino acids, in every way identical to those made in the course of the Miller experiment, have been identified in the Murchison meteorite which fell on Australia in 1969 (Anders, 1989; Chyba et al., 1990) (Fig 10).

The fact that more than 100 tonnes of meteorites now fall on our planet in one year, and in the past, the bombardment was 10 000 times more intense, means that meteorites have brought an enormous quantity of organic molecules to the Earth (Maurette et al., 1987).

4.3. HYDROTHERMAL VENTS AND SULFUR CHEMISTRY

Finally, it is also possible that a portion of compounds necessary for life were expelled by hot water springs. In the 1980's, investigations from the submarine " The Nautilus " led to a completely unexpected discovery, the existence of hot springs at depths greater than 2600 metres. These marine vents resemble small volcanic cones, the sides of which are covered with cracks or fissures through which sea water can infiltrate and come into contact with the hot basalt. When this water emerges, its temperature can be as high as 400°C, it is more acidic and it is enriched with mineral salts and in metallic elements. These marine vents which continuously expel large quantities of hydrogen sulphide and derivatives such as dark metallic sulfides (hence their name "black smokers"), constitute a reducing environment containing all the necessary ingredients for prebiotic chemistry. Today, living communities constituting guenine independent food chains develop around these black smokers without using solar energy. What are the arguments to assume that these hydrothermal springs may have been the place of the origin of life? Corliss has called attention to the discovery in ancient geological sediments dating from 3.5 thousand million years, of micro-fossils resembling forms now found in hydrothermal zones (Corliss et al., 1981). This new area of rescarch probably holds many more surprises.

Recently, Huber and Wächtershäuser (1997) have demonstrated the potential chemical relevance of the "black smoker" environment in prebiotic synthesis. They have shown that mixed iron nickel sulfides act as a catalyst for the conversion of methyl thiol and carbon monoxide, present in vent gases, to methyl thioacetate, a compound containing an activated acetyl group such as that in acetyl-coenzyme A, a central intermediate in the biosynthesis of important biomolecules. The biological C–C bond-forming reaction of Huber and Wächtershäuser can therefore be seen as a preliminary step in the building up of more complex organic molecules in or near the early oceanic vents. Sulfur chemistry may have been essential to the prebiotic synthesis of bio-organic molecules.

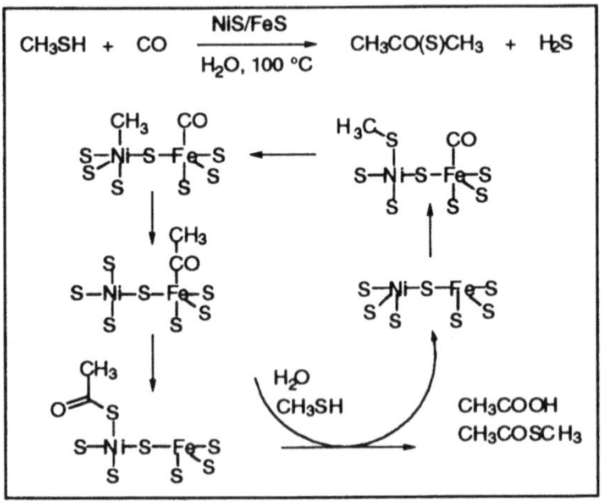

$$CH_3SH + CO \xrightarrow[H_2O,\ 100\ ^\circ C]{NiS/FeS} CH_3CO(S)CH_3 + H_2S$$

Figure 12 . The first evidence of a possible prebiotic chemistry in the "black smoker" environment : C–C bond-forming conversion of methyl thiol and carbon monoxide into methyl thioacetate catalysed by coprecipitated nickel (II) and iron (II) sulfides and representation of a hypothetical mechanism (see Maurel and Décout, 1998).

Lipmann (1971) and de Duve (1991, 1995) proposed that a thioester-dependent mechanism of peptide bond formation may have preceded the RNA-dependent mechanism of protein synthesis in the earliest steps of life.
Recently, Liu and Orgel (1997) described a closely related route to amide formation involving acylation by thioacetic acid and an oxidizing agent. As thioacids are obtained by the hydrolysis of nitrile in presence of H_2S (nitrile are prebiotic compounds easily found in space) this is of interest as it would provide a route to prebiotic peptide synthesis and ligation in aqueous solution.

4.4. A SURFACE METABOLISM?

The English crystallographer Bernal suggested in 1951 that the adsorption of molecules onto a mineral surface, for example clay, could facilitate their meeting and polymerisation. These minerals have a laminar structure characterized by alternate sheets a few angstroms thick, positively or negatively charged and folded like the pages of a book. Amino acids can be fixed between the layers of clay in a structure which favours their condensation.

In the 1970s, Paecht-Horowitz and Katchalsky (1970 and 1978) carried out an experimental demonstration of this theory. They showed that a particular clay, Montmorillonite, acts like a mini reactor: it stores, concentrates and positions the adenylated amino acids between its layers and favours their polymerisation (Fig13)

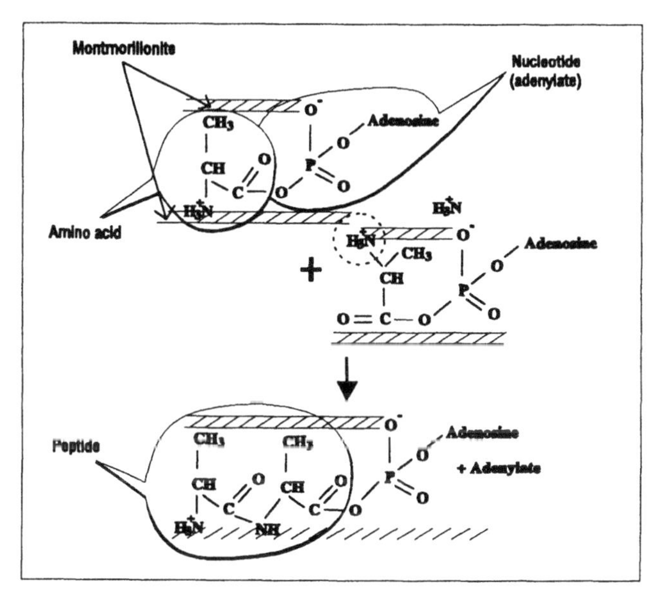

Figure 13. See text.

From this point of view the clay can be considered as a primitive enzyme. Similarly, several researchers succeeded in condensing mononucleotides or amino acids on the clay surface (White and Erickson, 1980; Ferris et al., 1989 ; Ferris et al., 1996).

Hartman from 1975, then Wächtershäuser from 1988 proposed a primordial autotrophic metabolism. Wächtershäuser suggests the possibility of a surface based «organism», in which simple negatively charged organic molecules, become fixed on the positively charged pyrite surface and are able to use atmospheric carbon dioxide directly as a carbon source, exactly as is done by green plants and certain bacteria today. Such an "organism" would have developed a surface metabolism at the time of the origin of life on Earth.

The origin of chirality in living molecules (that is the exclusive involvment of L-amino acids and D-ribose), is at present not satisfactorily explained. We can at the moment only hypothesise, for example that natural chiral minerals may have play a role, and/or have induced a selection by the specificity of certain reactions (Bonner, 1991; Bolli et al., 1997). Cronin and Pizzarello (1997) and Engel and Macko (1997) have demonstrated that the Murchison meteorite contains an enantiomeric excess of L-amino acids (for example, alanine and glutamic acid, over 30 and 50%, respectively) that are, today, almost exclusively present in living systems acids. These results suggest that an extraterrestrial source for an excess of L-enantiomers in the Solar System may predate the origin of life on the Earth. This excess may have resulted from the alteration of initially abiotic racemic mixtures by a process such as preferential decomposition by exposure to circularly polarized light (Kagan et al., 1974; Bonner, 1991).

To conclude this brief overview, it can be seen that there are several plausible hypotheses which can account for the presence of the elementary building blocks of life on the primitive Earth. The average chemical composition of living cell is different from the average chemical composition of the Earth. Unlike inanimate structures, the living organisms are selective chemical system composed essentially of C, H, O, N, S and P. From these elements the first protocells then populations of cells arose from which evolution has selected for improvements and diversification.

5. Molecular Evolution.

In the examination of cellular function how can we trace the key steps in the development of the early life ? What are the respective roles of nucleic acids and of proteins in the genesis of life?

Several arguments lead to the idea that RNA predated DNA in evolution (Orgel, 1968 and 1986; Lazcano et al., 1988; Maurel, 1992). Among these, it is known that the living cell makes the nucleotides of DNA by reduction of those constitutive of RNA (Reichard, 1993). Thymine, a base specific for DNA, is obtained by the transformation of uracil, which is specific for RNA. Furthermore RNAs are indispensable primers in the synthesis of DNA, whereas the synthesis of RNA is carried out without recourse to DNA. In the "RNA world" hypothesis of the origin of life (Gilbert, W., 1986; Gesteland and Atkins, 1993), RNAs were the central macromolecules able to self-replicate by base pairing, conserve information, and catalyse the reactions necessary for a primitive metabolism. Afterwards, DNA has been invented in order to store more efficiently the genetic information.

Many enzymatic cofactors possessing in their structure a ribonucleotidic moiety may be regarded as molecular fossils of the «RNA world» .

The discovery of catalysis by RNAs called ribozymes is another argument for the primacy of RNAs in evolution.

5.1. ORIGIN OF SELF-REPLICATION AND EXPERIMENTAL MODELS

Ancestral RNAs must be able to assemble themselves from a nucleotide soup and evolve in self-replicating patterns, using recombination and mutation to explore new functions (Gilbert, 1986).

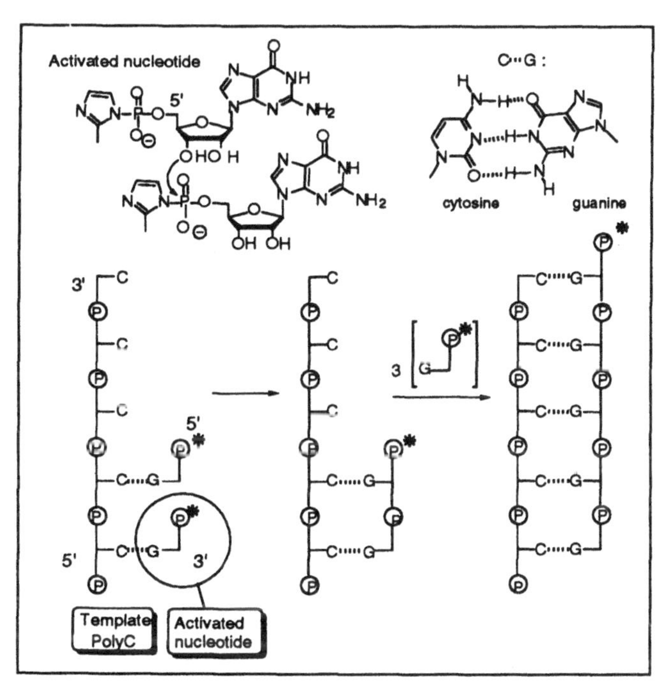

Figure 14. Template-directed synthesis of oligonucleotides (from Orgel, 1987).

In extant organisms, for replication the two nucleic acid strands separate, and by complementarity each one of them serves to regenerate the missing strand. In the cell there are very specialised enzymes which carry out this task; under primitive conditions, we have to suppose that replication may have occurred without the intervention of enzymes by the simplest possible template-directed synthesis.

The principle of this system, studied since the early 1970s by Orgel, is now well known (Orgel and Lohrmann, 1974; Orgel 1987, 1989 and 1992) (Fig14).

Mononucleotides activated in the form of 5'-phosphoro(2-methyl)imidazolides are positioned according to the pairing rules of Watson and Crick on the surface of a preformed poly-pyrimidine template and, because they are activated, they are able to link to one another to form the complementary strand. In this way, many defined oligonucleotide sequences are copied faithfully.

However, when we come to build the first autoreplicative molecule from start to finish and under the primitive conditions, we come up against a certain number of problems (Orgel, 1987; Joyce, 1989). In other words it is very difficult to obtain all the pieces of an RNA molecule under the original conditions.

For instance in the case of bases synthesis Orò (1960, 1965) showed that, under prebiotic conditions, purines could easily be obtained with the evolution of hydrogen cyanide (HCN) in water, but very little pyrimidines (Robertson and Miller, 1995).

When one tries to produce ribose, the sugar of nucleic acids, from formaldehyde by the so-called "formose reaction" (Butlerow, 1861; Decker et al., 1982), a very complex mixture is obtained, in which ribose is a very minor component. Moreover, ribose is unstable on a geological time scale (Larralde et al., 1995).

Contemporary research is moving towards the search for polymers simpler than the well-known strands of RNA.

In theory, the ribose-phosphate backbone may not be required for the transfer of genetic information and a simpler system of replication appearing before RNA can be envisaged. Furthermore, it is postulated that this pre-RNA molecule needed to be more robust than the sensitive and complex components of the RNA world. Work has been directed towards the synthesis of molecules which can self-replicate more simply than contemporary RNA by replacing the ribose by acyclic compounds such as glycerol, acrolein or erythritol (Spach, 1984; Joyce et al., 1987; Joyce, 1989).

Danish chemists, Egholm and Nielsen, replaced the ribophosphate backbone by amide links similar to those in proteins (Nielsen et al, 1991; Egholm et al, 1992). These new nucleic acid analogues called PNA (for peptide nucleic acids) are capable of strongly pairing with oligodeoxyribonucleotides according to the rules of Watson and Crick. PNA can also act as a template that can catalyse the non-enzymatic synthesis of their RNA complements from activated mononucleotides and *vice versa* (Böhler et al., 1995) (Fig 15). This suggests that a transition between different genetic systems is possible without loss information.

Eschenmoser (1994) proposed an alternative structure, an isomer of RNA, p-RNA, in which the sugar is in the six ring pyranose form (Maurel et Convert, 1990). This proposal came from studies run in the presence of formaldehyde, which gives a ribopyranose derivative. This p-RNA was found to have a stronger and a more selective pairing system than natural RNA. Recently, Bolli et al. (1997) demonstrated that base-sequences of p-RNA can be copied by template-controlled replicative ligation of short activated oligomers under mild conditions and that this ligation is highly chiroselective.

Finally, we proposed a model, a polyallylamine chain linked with nitrogenous bases (Maurel et Décout, 1992, Décout et Maurel 1993, Décout et al, 1995, Ricard et al, 1996). This structure (Palad and PaladThy) like RNA catalysts present remarkable catalytic activities thanks to the imidazole moiety of adenine (Maurel et Ninio, 1987).

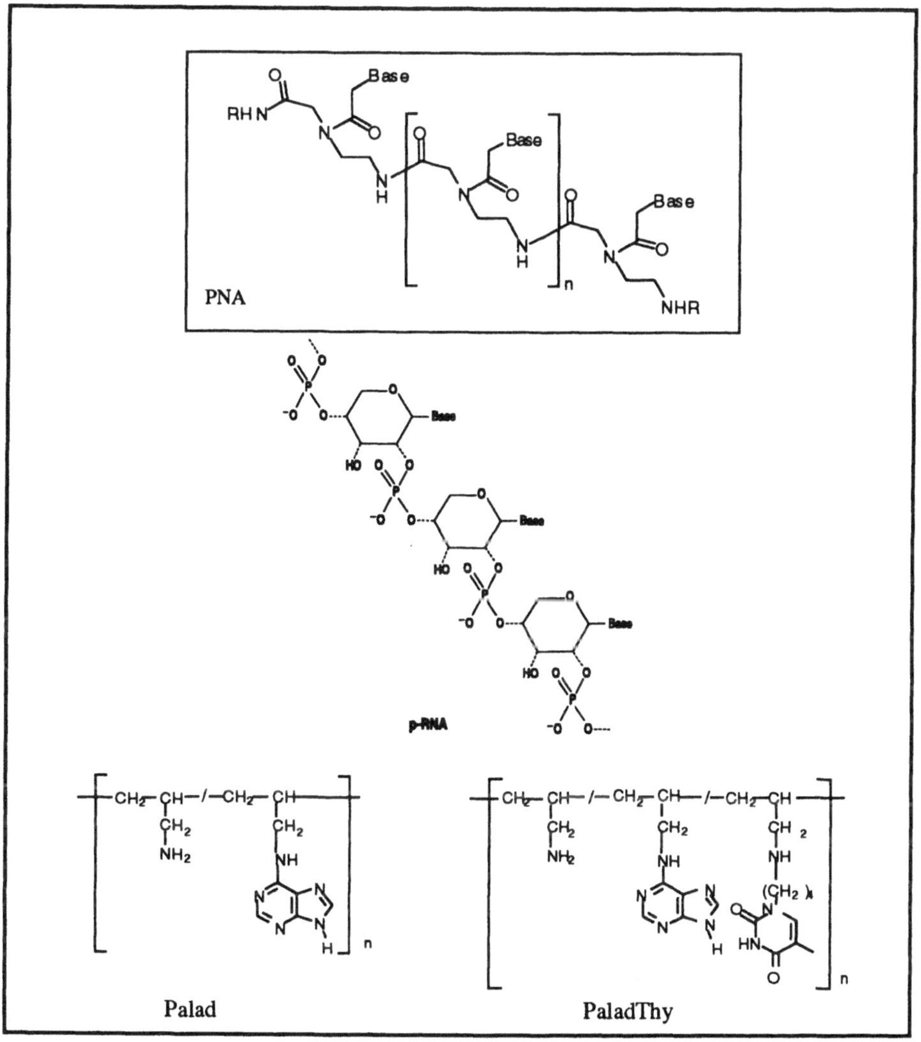

Figure 15. Structure of peptide nucleic acids (PNAs), p-RNA, Palad and PaladThy.

For a long time, it was thought that there was an absolute separation in the living cell between the roles of nucleic acids as information carriers on the one hand and of proteins as catalysts on the other. In the 1980s, Altman and Cech (who in 1989 were awarded the Nobel prize in chemistry for this work) showed that certain ribonucleic acids (ribozymes) have catalytic functions.

We can distinguish several properties of ribozymes (Maurel et Décout, 1998). The first is their ability to carry out self-splicing reactions, that is to say the cutting and joining end to end fragments of an RNA molecule. It is also the case in the self-hydrolysis of several viroids or when one RNA molecule acts on another, for example RNase P, an enzyme important in transfer RNA maturation. Large ribozymes require divalent metal ions such as magnesium ions to fold correctly. Divalent metal ions, such as magnesium, manganese, cobalt, calcium cations, are also essential for efficient catalysis by ribozymes (Pyle, 1993).

Today selected ribozymes are able to catalyse reactions pertaining to protein synthesis and are able to catalyse the synthesis of complementary strand of RNA (Picirilli et al, 1992; Samaha et al, 1995; Illangasekare et al, 1995; Zhang et Cech, 1997; Tarasow et al., 1997). Similar ribozyme activities may have played an important role in the "RNA world" either by expanding the limited number of chemical functional groups necessary for maintaining a complex metabolism in the absence of proteins or by the large diversity of their sequences .

The existence of these RNA catalysts poses questions concerning primitive catalysts used in the origins of life : were the first molecules related to our nucleic acids endowed with catalytic activity?

In the "RNA world" scenario of the origin of life, an ancestor of RNA, a common precursor to all forms of life, catalysed the necessary reactions to assemble and replicate itself and for life. It is thought to have a darwinian behavior.

Recently, methods have been developed for generating and screening large libraries of nucleic acid molecules differing in their sequences and their folding properties (for reviews see: Lehman and Joyce, 1993; Gold et al., 1995; Lorsch and Szostak, 1996; Jaeger, 1997; Osborne and Ellington, 1997; Breaker, 1997). A large range of ribozymes capable of assembling short oligonucleotides or performing the same peptidyl transferase reaction as the ribosome for linking amino acids during protein synthesis were selected. The success of these methods in the selection of RNAs possessing new catalytic properties and RNAs that bind strongly bio-organic molecules must be extended to RNA-like molecules capable of performing new reactions, in particular assembling monomers in an informative strand.

6. Exobiology

If life is an inevitable event when the proper conditions (physical and chemical) exist, though we can anticipate that there is elsewhere in the universe living organisms which precise assemblage would be different because so much depends upon history and evolution.

The term exobiology was introduced by Lederberg in 1960 extending the boundaries of biological investigations beyond the Earth, to other planets, comets, meteorites and space at large. The objective of exobiological research is to achieve a better understanding of the principles leading to the emergence of life from inorganics, its evolution, and its distribution on Earth and throughout the Universe. In this last case, the best way is to detect signatures of life.

Thanks to exobiological concepts we can formulate some prerequisites for the emergence of life. Above all, life is a planetary phenomenon. An habitable zone must be point out as a planetary system in orbit around a single (or twice) star. The planet must be at the right distance from the star to allow temperature and pressure conditions for liquid water.
A special mention must be done to liquid water. As about 90% of all life material today is occupied by water, the availability of liquid water is considered as one of the key prerequisites for life. Water is solvent which allows the synthesis and organization of biopolymers; it is also a reaction partner in most biochemical pathways. On the other hand, water is a powerful hydrolytic agent, and the scenario of prebiotic soup we have already discussed, underlines the difficulties encountered in presence of great volumes of water. Consequently survival and strategies for adaptation of life in extreme environments that are depleted of water must also be carefully examined.
Chemical criteria must be added including the fact that the biogenic elements C, H, O, N, S, P, must be much more abundant in the putative living organisms than in non-living matter. To serve as guidelines, we need a specific feature that is a carbon-based organic chemistry. Also, energy sources and all the prebiotic conditions driving a prebiotic chemistry are essentials.
Others requisites are of importance because they may also be considered as hallmarks of life. These are concerned with proto-biological, that is replication of informational molecules, catalysis and the ability to extract energy from the environment to create order and/or macromolecular organization (Schrödinger, 1944). Key structures that is building blocks of biological molecules are taking parts in these tasks.
Then, last but not least, living organisms evolve. This means that hereditary changes can influence the overall fitness of organisms and make them capable to survive (or not) environmental changes. Ultimately, there is life only when there is evolution. But we don't never forget that early cells were very different from modern cells, different enough that they should not be looked at as organisms (Woese, 1998).
These interdependent criteria as a whole will be used in the search for potential extraterrestrial habitats outside the solar system.
One word more to introduce thought-provoking notion. Following the French physician Bichat, life is the whole fonctions that resist to death. Of course Bichat (1800) as a vitalist was considered like a devil. But following now the German biologist Weismann (1892) (one of the most famous anti-vitalist) who linked the understanding of origin of life to the one of the death, the originality of life lies in the immortality of the germ plasm. Thanks to this material life is original and impredictible. That is the foundamental notion. Even if on the earth the basic reactions are oxydo-reductions which govern the photoreduction of carbon dioxide and the production of oxygen by the way of photosynthesis driven by light (UV), searching life elsewhere requires interrogations about what are we searching for ? Traces of

memories? Former life, extant life or a future one ? Are we prepare to take under consideration another kind of evolution of the energy conversions? What is upsetting in the search of the Origin of life and particularly in Exobiology is that we don't know how far we are prepared to accept differencies.

Acknowledgement : This work was in part supported by grants from CNES.

REFERENCES

Allen, D. A.; Wickramasinghe, D. T. "Discovery of organic grains in comet Wilson". *Nature* 1987, *329*, 615-616.

Altman, S. "Ribonuclease P: an enzyme with a catalytic RNA subunit." Advances in enzymology, F.F. Nord and A. Meister; Interscience, New York, 1989, vol. 62, 1-36.

Anders, E. "Pre-biotic organic matter from comets and asteroids". *Nature* 1989, *342*, 255-257.

Bernal, J. D. "The physical basis of life" Routeledge and Kegen Paul, London, 1951.

Bichat, X. "Recherches physiologiques sur la vie et la mort", (1800) édition intégrale, Editions Gérard & C°, Verviers, Belgique, 1973.

Björk, G. R. "Biosynthesis and function of modified nucleosides". *In* D. Söll and U. RajBandary (Eds.). tRNA: Structure, biosynthesis, and function. American Soc. for Microbiology, Washington D.C., 1995, pp 165-204.

Böhler, C.; Nielsen P. E. ; Orgel, L. E. "Template switching between PNA and RNA oligonucleotides". *Nature* 1995, *376*, 578-581.

Bolli, M.; Micura, R.; Eschenmoser, A. "Pyranosyl-RNA: chiroselective self-assembly of base sequences by ligative oligomerization of tetranucleotide-2',3'-cyclophosphates (with a commentary concerning the origin of biomolecular homochirality)". *Chem. & Biol.* 1997, *4*, 309-320.

Bonner, W. A. "The origin and amplification of biomolecular chirality". *Origins Life Evol. Biosphere* 1991, *21*, 59-111.

Braterman, P. S.; Cairns-Smith, A. G.; Sloper, R. W. "Photo-oxidation of hydrated Fe^{2+}. Significance for banded iron formations". *Nature* 1983, *303*, 163-164.

Breaker, R. R. "*In vitro* selection of catalytic polynucleotides". *Chem. Rev.* 1997, *97*, 371-390.

Butlerow, A. "Formation synthétique d'une substance sucrée". *C. R. Acad. Sci. Paris* 1861, *53*, 145.

Carle, G.; Schwartz, D.; Huntington, J. (Eds.) "Exobiology in solar system exploration. NASA", 1988.
Cech, T. R. "The chemistry of self-splicing RNA and RNA enzymes". *Science* 1987, *236*, 1532-1539.

Cech, T. R.; Zaug, A. J.; Grabowsky, P. J. "*In vitro* splicing of the ribosomal RNA precursor of Tetrahymena: involvement of a guanosine nucleotide in the excision of the intervening sequence. *Cell* 1981, *27*, 487-496.

Chyba, C. F.; Sagan, C. "Endogenous production, exogenous delivery and impact-shock synthesis of organic molecules: an inventory for the origins of life". *Nature* 1992, *355*, 125-132.

Chyba, C. F.; Thomas, P. J.; Brookshaw, L.; Sagan, C. "Cometary delivery of organic molecules to the early Earth". *Science* 1990, *249*, 366-373.

Corliss, J. B.; Baross, J. A.; Hoffman, S. E. "Submarine hydrothermal systems: a probable site for the origin of life". *Oceanol. Acta* 1981, *4*, suppl., 59-69.

Cronin, J. R.; Pizzarello, S. « Enantiomeric excess in meteoritic amino acids ». *Science* 1997, *275*, 951-955.

Decker, P.; Schweer, H.; Pohlmann, R. "BIOIDS, X: Identification of formose sugars, presumable prebiotic metabolites, using capillary gas chromatography / gas chromatography-mass spectrometry of *n*-butoxime trifluoroacetates on OV-225".
J. Chromatogr. 1982, *244*, 281-291.

Décout, J.-L.; Maurel, M.-C. "N^6-substituted adenine derivatives and RNA primitive catalysts". *Origins Life Evol. Biosphere* 1993, *23*, 298-306.

Décout, J.-L.; Vergne, J.; Maurel, M.-C. "Synthesis and catalytic activity of adenine containing polyamines". *Macromol. Chem. Phys.* 1995, *196*, 2615-2624.

de Duve, C. R. "Blueprint for a cell. The nature and origin of life". Neil Patterson Publishers, Burlington, North Carolina, 1991.

de Duve, C. R. "Vital dust. Life as a cosmic imperative". Basic Books, HarperCollins Publishers, New York, 1995.

Doudna, J. A.; Szostak, J. W. "RNA catalysed synthesis of complementary strand RNA". *Nature* 1989, *339*, 519-522.

Egholm, M.; Buchardt, O.; Nielsen P. E.; Berg, R. H. "Oligonucleotide analogues with an achiral peptide backbone". *J. Am. Chem. Soc.* 1992, *114*, 1895-1897.

Engel, M. H.; Macko, S. A. " Isotopic evidence for extraterrestrial non-racemic amino acids in the Murchison meteorite. *Nature* 1997, *389*, 265-268 (Comments Chyba C. pp. 234-235).

Eschenmoser, A.; Loewenthal, E. "Chemistry of potentially prebiological natural products". *Chem. Soc. Rev.* 1992, 1-16.

Ferris, J. P. "The chemistry of life's origin". *Chem. Eng. News* 1991, *62*, 22-35.

Ferris, J. P.; Ertem, G.; Agrawal, V. K. "The adsorption of nucleotides and polynucleotides on montmorillonite clay". *Origins Life Evol. Biosphere* 1989, *19*, 153-164.

Ferris, J. P. ; Hagan, W. J. "HCN in chemical evolution: the possible role of cyano compounds in prebiotic synthesis". *Tetrahedron* 1984, *40*, 1093-1120.

Ferris, J. P.; Hill, A. R.; Liu, R.; Orgel, L. E. "Synthesis of long prebiotic oligomers on mineral surfaces". *Nature* 1996, *381*, 59-61.

Ferris, J. P.; Orgel, L. E. "An unusual photochemical rearrangement in the synthesis of adenine from hydrogen cyanide". *J. Am. Chem. Soc.* 1966, *88*, 1074-1074.

Fredrickson, J.K; Onstott, T.C." Microbes deep inside the Earth". *Scientific American.* 1996, 68-73.

Gesteland, R. F.; Atkins, J. F. (Eds.) "The RNA World". Cold Spring Harbor Laboratory Press, 1993.

Gilbert, W. "The RNA World". *Nature* 1986, *319*, 618-618.

Gold, L.; Polisky, B.; Uhlenbeck, O.; Yarus, M. "Diversity of oligonucleotide functions". *Annu. Rev. Biochem.* 1995, *64*, 763-797.

Hartman, H. "Speculations on the origin and evolution of metabolism". *J. Mol. Evol.* 1975, *4*, 359-370.

Hochachka, P.W.; Somero, G.N. "Biochemical adaptations". Princeton University Press, Princeton, NJ. 1984.

Huber, C.; Wächterhäuser, G. "Activated acetic acid by carbon fixation on (Fe,Ni)S under primordial conditions". *Science* 1997, *276*, 245-247.

Illangasekare, M.; Sanchez, G.; Nickles, T.; Yarus, M. "Aminoacyl-RNA synthesis catalyzed by an RNA". *Science* 1995, *267*, 643-647.
Jaeger, L. "The new World of ribozymes". *Current Opinion Struct. Biol.* 1997, *7*, 324-335.

Joshi, P. C.; Pathak, H. D. "Formation of aminoacids and nucleic acid constituents under possible primitive earth conditions". *J. Brit. Interpl. Soc.* 1975, *28*, 90-96.

Joyce, G. F. "RNA evolution and the origins of life". *Nature* 1989, *338*, 217-224.

Joyce, G. F.; Schwartz, A. W.; Miller, S. L.; Orgel, L. E. "The case for an ancestral genetic system involving simple analogues of the nucleotides". *Proc. Natl. Acad. Sci. USA* 1987, *84*, 4398-4402.

Kagan, H. B.; Balavoine, G.; Moradpour, A. « Can circularly polarized light be used to obtain chiral compounds of high optical purity ». *J. Mol. Evol.* 1974, *4*, 41-48.

Kasting, J. F. "Earth's early atmosphere". *Science* 1993, *259*, 920-926.

Larralde, R.; Robertson, M. P.; Miller, S. L. "Rates of decomposition of ribose and other sugars: implications for chemical evolution". *Proc. Natl. Acad. Sci. USA* 1995, *92*, 8158-8160.

Lazcano, A.; Guerrero, R.; Margulis, L.; Orò, J. "The evolutionary transition from RNA to DNA in early cells". *J. Mol. Evol.* 1988, *27*, 283-290.

Lehman, N.; Joyce, G. F. "Evolution *in vitro*: analysis of a lineage of ribozymes". *Current Biology* 1993, *3*, 723-734.

Lipmann, F. " Attempts to map a process evolution of peptide biosynthesis" *Science* 1971, *173*, 875-884.

Liu, R.; Orgel, L. E. "Oxidative acylation using thioacids". *Nature* 1997, *389*, 52-54.

Lorsch, J. R.; Szostak, J. W. "Chance and necessity in the selection of nucleic acid catalysts". *Acc. Chem. Res.* 1996, *29*, 103-110.

Maurel, M.-C.; Ninio, J. "Catalysis by a prebiotic nucleotide analog of histidine". *Biochimie* 1987, *69*, 551-553.

Maurel, M.-C. "RNA in Evolution: A review". *J. Evol. Biol.* 1992, *5*, 173-188

Maurel, M.-C.; Convert, O. "Chemical structure of a prebiotic analog of adenosine". *Origins Life Evol. Biosphere* 1990, *20*, 43-48.

Maurel, M.-C.; Décout, J.-L. "Studies of nucleic acid-like polymers as catalysts". *J. Mol. Evol.* 1992, *35*, 190-195.

Maurel, M.-C.; Vergne, J.; Drahi, B.; Décout, J.-L. "Implications of adenine in primitive catalysis and exobiology". *Proc. Sixth Eur. Symp. on Life Sciences Research in Space, ESA SP-390*, 1997, 177-181.

Maurel M.-C.; Décout, J.-L. "Origins of life : Molecular foundations and new approaches" *Tetrahedron Rep*, in press, 1998.

Maurette, M.; Jéhanno, C.; Robin, E.; Hammer, C. "Characteristics and mass distribution of extraterrestrial dust from the Greenland ice cap". *Nature* 1987, *328*, 699-702.

Miller, S. L. "A production of amino acids under possible primitive earth conditions". *Science* 1953, *117*, 528-529.

Miller, S. L. "Production of some organic compounds under possible primitive earth conditions". *J. Am. Chem. Soc.* 1955, *77*, 2351-2361.

Mojzsis, S. J.; Arrhenius, G.; McKeegan, K. D.; Harrison, T. M.; Nutman, A. P.; Friend, C. R. L. "Evidence for life on Earth before 3,800 million years ago". *Nature* 1996, *384*, 55-59.

Morowitz, H. J. "Beginning of cellular life". Yale University Press (Ed.), New Haven, 1992.

Nielsen, P. E.; Egholm M.; Berg, R. H.; Buchardt, O. "Sequence-selective recognition of DNA by strand displacement with a thymine-substituted polyamide". *Science* 1991, *254*, 1497-1500.

Oparin, A. I. "The origin of life". McMillan Publishing, New-York, 1938.

Orgel, L. E. "Evolution of the genetic apparatus". *J. Mol. Biol.* 1968, *38*, 381-393.

Orgel, L. E. "RNA catalysis and the origins of life". *J. Theor. Biol.* 1986, *123*, 127-149.

Orgel, L. E. "Evolution of the genetic apparatus: a review". Cold Spring Harbor Symp. Quant. Biol. 1987, *52*, 9-16.

Orgel, L. E. "Was RNA the first genetic polymer?" . *In* Grunberg-Manago M.; Clark B. F. C.; Zachau H.G. (Eds.). "Evolutionary thinkering in gene expression". Plenum Press, New York, 1989, pp 215-224.
Orgel, L. E. "Molecular replication". *Nature* 1992, *358*, 203-209.

Orgel, L. E.; Lohrmann, R. "Prebiotic chemistry and nucleic acid replication". *Acc. Chem. Res.* 1974, *7*, 368-377.

Orò, J. "Synthesis of adenine from ammonium cyanide". *Biochem. Biophys. Res. Comm.* 1960, *2*, 407-412.

Orò, J. "Stages and mechanisms of prebiological organic synthesis". *In* Fox. S.W. (Ed.), "The origin of prebiological systems and their molecular matrices". Academic press., New York, 1965, pp. 137-161.

Orò, J.; Kimball, A. P. "Synthesis of purines under possible primitive earth conditions, I. Adenine from hydrogen cyanide". *Arch. Biochem. Biophys.* 1961, *94*, 217-227.

Osborne, S. E.; Ellington, A. D. "Nucleic acid selection and the challenge of combinatorial chemistry". *Chem. Rev.* 1997, *97*, 349-370.

Paecht-Horowitz, M. "Clay catalysed polymerization of amino acid adenylates and its relationship to biochemical reactions". *Origins Life Evol. Biosphere* 1978, *9*, 289-295.

Paecht-Horowitz, M.; Berger, J.; Katchalsky, A. "Prebiotic synthesis of polypeptides by heterogeneous polycondensation of aminoacids adenylates". *Nature* 1970, *228*, 636-641.

Picirilli, J. A.; Mc Connell, T. S.; Zaug, A. J.; Noller, H. F.; Cech, T. R. "Aminoacyl esterase activity of the Tetrahymena ribozyme". *Science* 1992, *256*, 1420-1424.

Pyle, A. M. "Ribozymes: a distinct class of metalloenzymes". *Science* 1993, *261*, 709-714.

Reichard, P. "From RNA to DNA, why so many ribonucleotide reductases?". *Science* 1993, 260, 1773-1777.

Ricard, J.; Vergne, J.; Décout, J.-L.; Maurel, M.-C. "The origin of kinetic co-operativity in prebiotic catalysts". *J. Mol. Evol.* 1996, *43*, 315-325.

Robertson, M. P.; Miller, S. L. "An efficient prebiotic synthesis of cytosine and uracil". *Nature* 1995, *375*, 772-774.

Samaha, R. S.; Green, R.; Noller, H. F. "A base pair between tRNA and 23S rRNA in the peptidyl transferase centre of the ribosome". *Nature* 1995, *377*, 309-314.

Schrödinger, E. "What is life?". Cambridge University Press, Cambridge, 1944.
Schopf, J. W. (Ed.) "Earth's earliest biosphere: its origin and evolution". Princeton University Press, Princeton, 1983.

Schopf, J. W.; Packer, B. M. "Early archaen (3.3 billion to 3.5 billion-year-old) microfossils from Warrawoona group, Australia". *Science* 1987, *237*, 70-73.

Spach, G. "Chiral versus chemical evolution and the appearance of life". Origins of life Evol. Biosphere 1984, *14*, 433-437.

Tarasow, T. M.; Tarasow, S. L.; Eaton, B. E. "RNA-catalysed carbon-carbon bond formation" *Nature* 1997, *389*, 54-57.

Wächtershäuser, G. "Before enzymes and templates: Theory of surface metabolism". *Microbiol. Rev.* 1988, *52*, 452-484.

Wächtershäuser, G. An all-purine precursor of nucleic acids. *Proc. Natl. Acad. Sci USA.* 1988, *85*, 1134-1135.

Walker JCG(1978) Oxygen and hydrogen in the primitive atmosphere. Pure Appl Geophys 116: 222-231

Walker JCG, Kein C, Schidlowski M, Schopf JW, Stevenson DJ and Walter MR 1983, Environmental evolution of the Archean-Proterozoic earth. In Schopf JW (ed) Earth's Earliest Biosphere, pp 260-290. Princeton Univ. Press, Princeton.

Walker, J. C. G. "Carbon dioxide on the early Earth". *Origins of Life* 1985, *16*, 117-127.

Weismann, A. "Essais sur l'hérédité et la sélection naturelle", traduction française par Henry de Varigny, Reinwald et Cie, Paris 1892.

White, D. H.; Erickson, J. C. "Catalysis of peptide bond formation by histidyl-histidine in a fluctuating clay environment". *J. Mol. Evol.* 1980, *16*, 279-290.

Wieland, T. "Sulfur in biomimetic peptide syntheses". Wieland, T. (Ed). de Gruyter W., Berlin, New York, 1988, pp 213-221.

Woese, C. "The universal ancestor". *Proc. Natl. Acad. Sci. USA.* 1998, 95, 6854-6859.

Zhang, B.; Cech, T. R. "Peptide bond formation by *in vitro* selected ribozymes". *Nature* 1997, *390*, 96-100.

THE CONTRIBUTION OF ISO TO EXOPLANETARY SYSTEMS RESEARCH AND ASTROBIOLOGY

PETER CLAES
Infrared Space Observatory and Far Infrared Space Telescope, European Space Agency - Space Science Department ESTEC, The Netherlands

Abstract. ISO, the Infrared Space Observatory, was launched on 17 November 1995. ISO was not suitable for directly detecting exoplanets due to its low angular resolution and not high enough infrared sensitivity. ISO's strength was in its high resolution spectroscopy and its capability to observe in as yet unseen wavelengths. A lot of its wealth of scientific data been returned are of significance to exoplanetary research and astrobiology. Examples are data on substellar mass objects (brown dwarfs), circumstellar discs and studies of objects in the Solar System with astrobiological importance such as Europa and Titan. Also its observation of the interstellar medium, star formation and comets contribute to understanding better the molecular reprocessing in planetary systems and formation of such a system. Already several brown dwarfs and their spectra and new circumstellar disks were observed with ISO. Most directly applicable to the field are the discovery of a Vega-disk surrounding a star (ρ^1 Cnc) around which a hot jupiter was discovered earlier. Another remarkable result is the discovery of a planet-forming disk around a dying star. Some of the findings of ISO will be of immediate importance for a life- and planet-searching project such as Darwin.
Key words: Stars: circumstellar disks, brown dwarfs, ρ^1 Cnc - ISO - Astrobiology: Darwin, icy satellites, ozone, Europa, Titan

1. Introduction to the ISO-satellite and its instruments

1.1. THE INFRARED SPACE OBSERVATORY SPACECRAFT

The Infrared Space Observatory(ISO) was a real *observatory* mission, whereas its predecessor IRAS was a *survey* mission. IRAS mapped and scanned the entire infrared sky for the first time in civilian space history. ISO was

J.-M. Mariotti and D. Alloin (eds.), Planets Outside the Solar System: Theory and Observations, 365–379.

the second civilian infrared telescope in space. ISO was made in Europe. The spacecraft was built by an industry consortium under leadership of Aerospatiale. The four instruments on board were built by universities and space institutes in four European countries. ISOCAM was built in Saclay (France). ISOPHOT originated from MPIA (Germany). SWS was built and conceived in SRON (Groningen, the Netherlands). LWS was made in the United Kingdom at RAL.

ISO was launched on the 17th of November of 1995. Depletion of its

Figure 1. ISO visualization generated by ESA's VisuLab

coolant (liquid He) occurred on the 9th of April 1998. The foreseen mission duration was 18 months. In practice ISO's actual lifetime reached **2.5 years**. This resulted in nearly complete sky coverage. ISO's telescope was a Ritchey-Chrétien **60 cm** (f/15) telescope of which the coldest parts(detectors) were cooled till 3K. The spacecraft was in a 24-hours highly elliptical orbit. Its perigee was at approximately 1000 km and its apogee at 70000 km. This led the satellite every orbit for a couple of hours through Earth's radiation belts. The spacecraft was capable to perform single pointing observations, rasters and slews. ISO observed objects from the Solar system till cosmological distances. Its instruments were 1000 times more sensitive than IRAS — its predecessor — and had a 20–30 times better angular resolution than IRAS. Each ISO-instrument had a 3 arcmin Field of View (FOV) at its disposal.

1.2. CAMERA : ISOCAM

The ISO-camera could detect sources from 2.5 till 17 μm. It had a InSb-detector for the short wavelength channel and a Si:Ga detector for the long wavelength channel. The detector had 32 x 32 pixels and the Pixel Field of View (PFOV) could have the values from 1.5, 3, 6 till 12 arcsec. The camera had polarimetric imaging capabilities and a spectral resolution R $= \lambda/\Delta\lambda = 40$ when using the *Contineous Variable Filter* (CVF). ISOCAM was also used when the other instruments were observing (in prime mode). It then "saw" a slightly different area of the sky and was used in *parallel mode*.

1.3. SHORT WAVELENGTH SPECTROMETER : SWS

SWS could work in the wavelength range from 2.4 till 45 μm. Its spectral resolution R $= \lambda/\Delta\lambda = 2000$ in grating mode. and 20000 in grating + Fabry-Pérot mode (from 15 till 35 μm). The instrument had three different apertures. Each aperture was used for the short wavelength and long wavelength section of the instrument. Various detector-types were used to obtain the wavelength coverage. Among those were blocked impurity band detectors (BIBIB-detectors),

1.4. LONG WAVELENGTH SPECTROMETER : LWS

LWS operated from 43–196.7 μm. Its spectral resolution R $= \lambda/\Delta\lambda =$ 100–300 in grating mode and 8000 till 10000 in grating plus Fabry-Pérot mode. Ge:Be detectors and Ge:Ga detectors were used plus stressed Ge:Ga detectors.

1.5. IMAGING SPECTRO-PHOTOPOLARIMETER : ISOPHOT

ISOPHOT could observe in the 2.5–240 μm - range. ISOPHOT was composed of three subsystems : PHT-S, PHT-P and PHT-C.

PHT-P was a multiband, multi-aperture photometer and polarimeter with 3 single detectors for the wavelength range from 3 till 120 μm.

PHT-S had two grating spectrophotometers, operated simultaneously, for the wavelength ranges of $\lambda\lambda = 2.5$–5 μm and 6–12 μm. The spectral resolution R $= \lambda/\Delta\lambda = 90$ (PHT-S).

PHT-C had two photometric FIR cameras arrays for $\lambda\lambda = 50$–240 μm. The PFOV was 43.5 arcsec ($\lambda\lambda = 50$-120 μm) and 89.4 arcsec ($\lambda\lambda = 120$–240 μm).

PHT-serendipity mode scanned the sky for new discoveries by using the PHOT-instrument during slews from one targeted position to another.

2. ISO-detection of the building bricks of life in the universe

Water vapour has been detected with ISO on a variety of objects. Examples in the Solar System are the stratospheres of the giant planets, the stratosphere of Titan and in comets such as comet Hale-Bopp (Crovisier *et al.*, 1997). Water-vapour also has been observed in the interstellar medium in dense clouds towards the center of the Galaxy. It has been seen in the vicinity of the aged star W Hydrae, from which an oxygen-rich wind blows into space. The bright infrared source GL 2591, surrounding a newly formed massive star, revealed to SWS hot and abundant water vapour. Jets of gas from very young stars such as HH-54 also showed water features in their spectrum.

Water plays a crucial role in the star-making process as well by radiating away excess heat which would otherwise prevent the parent gas from condensing under gravity to form a star. This is clearly seen in IRAS 16293-2422 which is a proto-stellar core. A team of American astronomers discovered a water-generator in the Orion Nebula (Glanz, 1998). The water is generated by means of shocks ejected from a young star in a star-forming region (Orion BN-KL). The shocks heat the gas till 2000 K. Water (an essential component of life) is, as demonstrated by ISO, clearly ubiquiteous in the Galaxy and when the many ISO-measurements will be put together in the future a "water-map" can be constructed of our Galaxy.

HCN is important because it is one of the building blocks of amino-acids, important biomonomers. HCN has been detected on Titan, in interstellar ices (Whittet *et al.*, 1996) and towards massive YSO's (Lahuis *et al.*, 1997).

Polycyclic Aromatic Hydrocarbons(PAHs) have been detected in an organic ring around HD 97300 (Siebenmorgen *et al.*, 1998). PAHs also have been found in more evolved stellar objects and as circumstellar material (Beintema *et al.*) as well as in molecular clouds such as M17-SouthWest and the frosty nebula HH100 IRS (Siebenmorgen *et al.*).

CH_4 has been detected in the stratosphere of Titan, on the giant planets, toward deeply embedded protostars (Boogert *et al.*, 1996) etc. Hydrocarbons (e.g. C_2H_2) have been detected on Titan (Coustenis *et al.*, 1998), the gas giants of the Solar System and towards massive YSO's (Lahuis *et al.*, 1997). ISO has detected CO_2 and CO as interstellar ices (Whittet *et al.*, 1996), in comets (Crovisier *et al.*, 1997) and CO_2 in molecular clouds as well (de Graauw *et al.*, 1996).

3. ISO-observations of interstellar matter and medium

Various studies of interstellar matter in various directions for various celestial sources have been performed by Laureijs *et al.*, Siebenmorgen *et al.* and Whittet *et al.* Gradually a better understanding is growing of the composition of "the" interstellar dust grain with two types of icy mantles (polar (H_2O) and non-polar (CO and CO_2))(de Graauw *et al.*, 1996) . Still better theoretical models are needed to understand better grain reactions (molecular reprocessing). People see a variety of interstellar "molecules and grains"-composition in function of the radiation field, temperature, density, etc. The principal ices in interstellar space as shown by ISO are H_2O, CO_2 and CO. Detected as well are CH_4, CH_3OH, HCOOH, XCN (including HCN !) and OCS. Detections of O_2, N_2 and NH_3-ices are missing. NH_3 is a major reservoir of nitrogen. A key-question is thus : how is the formation of ices influenced by low- and high mass star formation ? Studies of a comparison between interstellar and cometary ices are on-going, but first results indicate similar mixing ratios.

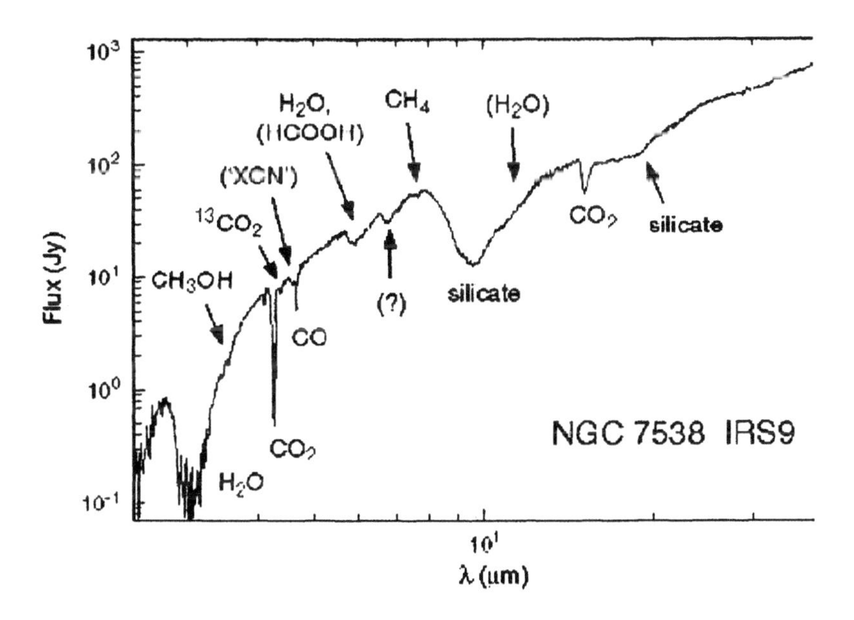

Figure 2. an ISO SWS view of interstellar ices (towards NGC 753) (Whittet *et al.*, 1996)

4. ISO-detection and characterisation of brown dwarfs

Brown dwarfs are worth studying because :

1. they are unique objects with interesting new physics.
2. they are intermediate (missing link) between objects with stellar masses and planetary masses.

3. understanding them better helps us to distinguish between formation of substellar objects in a disk and by fragmentation of a molecular cloud.
4. they represent a continuation of the IMF in the substellar domain.
5. their "non-substantial" contribution to the " missing mass"-problem (dark matter) is significant.

Comerón *et al.* observed very low mass objects with ISOCAM in the ρ Oph-cloud of which some were suspected by earlier ground-based near-IR observations, to be substellar of nature. The ρ Oph-cloud is estimated to be a few million years old. Its distance is 160 pc. The opacity of the cloud forms a natural screen against background sources. The cloud is moreover far from the Galactic equator. Photometry was combined in the R, I, J, H, K, L', ISOCAM LW1(3.6 μm), ISOCAM LW4 (6 μm) and N bands. Interstellar foreground extinction (Rieke *et al.*, 1985), possible circumstellar excess (Adams *et al.*, 1996) and photospheric spectral features-models (Allard *et al.*, 1996) were taken into account.

The purpose of the ISO-observations was to add data-points to the spectral energy distribution, as more data points allow to fit much better the observed object's spectral energy output to combined-model fits. This allows to derive the luminosity of the object provided the age is known by fitting to PMS-isochrones (as modelled by (Burrows *et al.*, 1993)). The temperature stability due to deuterium burning lasts for a time comparable to the embedded stage, so the temperature estimates should be quite robust. An independent value of the temperature was as well obtained by taking spectra from the ground for some of the objects.

The lowest value of mass this technique can yield is about 10 M_J. The best candidates and their characteristics are shown in Table 1. "A_V" is the absorption coefficient in the visible in magnitudes. "n" is the coefficient from the power law, modelling the circumstellar excess. Three objects are considered as firm detections of young still luminous brown dwarfs where 5 are transitional objects. Three objects are M-stars. Also indicated is whether a "young" or an "old" isochrone would give the best fit, where the separation between these cases is set at 2.10^6 years. This age is about half the estimated age of the ρ Oph complex.

Attempts to observe with ISOCAM Teide 1 and Calar 3 (Leech *et al.*, 1998) failed. The sources were too weak to be detected. The field brown dwarf **DENIS-P J1228.2-1547** at 210 pc (M $<$ 65 M_J) (Keck HIRES spectra exist !) was observed by ISOCAM. Accurate astrometric data exist of **Kelu-1** at 10 pc ! (M $<$ 75 M_J derived astrometrically) as well as optical spectra obtained by ESO's 3.6m telescope (using EFOSC1) at La Silla in Chile. ISOCAM photometry and fitting to atmospheric models of France Allard was performed on these objects. All these observations clearly show that the IMF does fall off but NOT abruptly below 0.08 $M_{\odot}$. This conclusion is valid both in the solar neighbourhood and further out.

TABLE 1. Best fitting parameters to the available photometry

Object	A_V	n	age	T (K)	M ($M_{\odot}$)
Oph 2317.3-1925	9	-2.2	young	2850	0.07
Oph 2317.5-1729	25	-2.7	young	3050	0.23
Oph 2320.0-1915	2	-2,1	young	2850	0.08
Oph 2320.8-1708	18	-2.9	old	3150	0.20
Oph 2320.8-1721	10	-1.6	old	2650	0.04
Oph 2321.6-1918	24	-2.6	old	2850	0.07
Oph 2322.8-1233	28	-2.8	young	2750	0.05
Oph 2331.1-1952	11	-2.3	young	2650	0.04
Oph 2349.8-2601	0	-2.1	old	2450	0.02
Oph 2351.8-2553	45	-1.1	young	2900	0.09
Oph 2404.5-2152	14	-2.1	young	2750	0.05
Oph 2408.6-2229	8	-3.0	old	2550	0.03

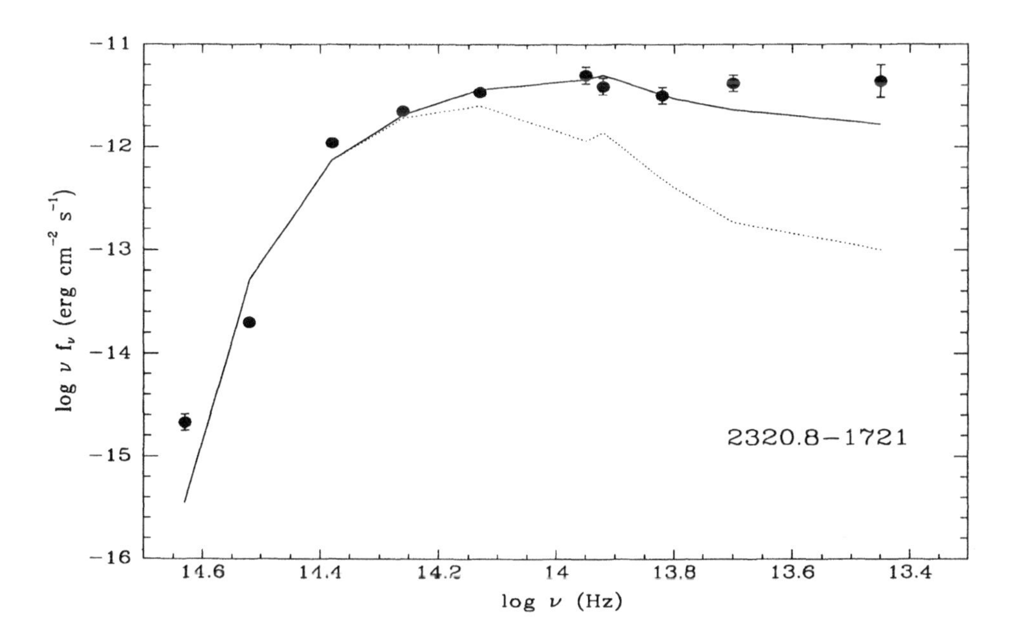

Figure 3. Photometric fitting of best brown dwarf, observed with ISO and from the ground (Comerón *et al.*, 1997)

5. ISO-observations of spectral signatures of extended gaseous envelopes of "hot jupiters"

Rauer *et al.* have perfomed SWS-observations in Fabry-Pérot mode (high resolution) of 51 Peg. They hope (data in analysis) to find spectral absorption features in the stellar spectrum, of an extended gaseous envelope around the "hot jupiter" 51 Peg b. They predict such a extended gaseous envelope for the physical reasons of atmospheric evaporation because of thermal/non-thermal escape and/or gravitational mass loss. The 51 Peg system has an inclination of nearly 90°, thus such an extended envelope could occult partially the disc of the parent star (source : proposal-abstract).

6. ISO-observations of circumstellar disks, rings and medium

A first series of observations to discuss are the preliminary results of the "Vega-proposal". Habing *et al.* selected a sample of 84 stars in total. These stars satisfy the following criteria :

1. they are single stars or very wide binaries with the companion more than one ISOPHOT diaphragm away from the parent star.
2. their distance < 25 pc.
3. their spectral type ranges from A to K.
4. they are main sequence stars.

The calibration was checked using COBE-DIRBE data which give a good value for the background flux at low resolution. The observations were performed with ISOPHOT at 60, 90, 135 and 180 μm. Preliminary results indicate that Vega-like disks are common but not ubiquitous : 10 out of 21 show IR-excess. In the case of Vega the grains have a radius of 10 μm or less. No evidence for excess was found for 47 UMa or HD 95128. The main question the proposers try to answer is whether the presence of a disk depend on the mass, age, rotation velocity or other parameters of a star.

Moneti *et al.* found evidence for circumstellar disks around post T-Tauri stars (observing with ISOCAM at 6.5 μm and 15 μm). Excess emission was found at 15 μm. This was considered as clear evidence for remnant disks around Lindroos binaries, which are wide pairs consisting of early-type MS primary with late type companions.

Waelkens *et al.* performed SWS-observations of 3 young main-sequence stars with dusty circumstellar disks. 51 Oph, HD 100546 and HD 142527 are 3 objects in evolutionary stage between Herbig Ae/Be stars and β Pic. They detected the presence of crystalline silicates which were as well found in comets, meteorites and interplanetary dust particles.

Dominik *et al.* detected a Vega-like disk associated with the planetary

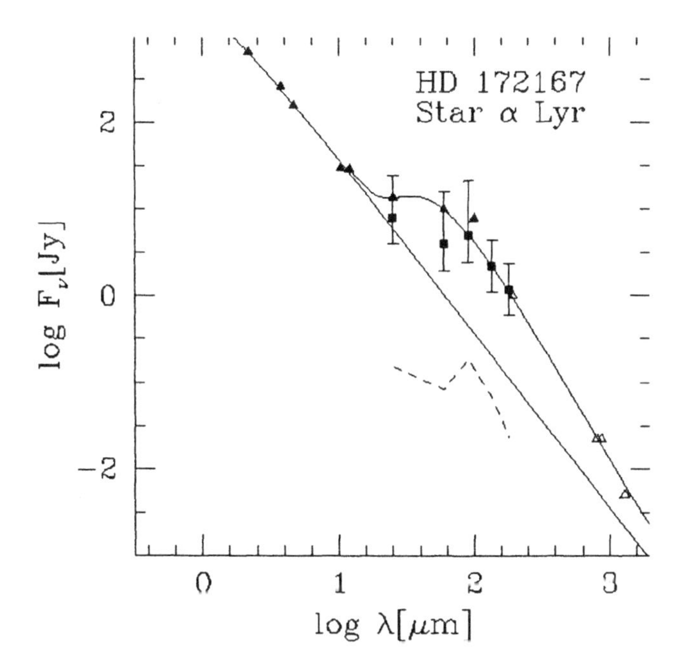

Figure 4. Infrared photometric excess of α Lyr (Habing *et al.*, 1996)

system of ρ^1 Cnc. They detected excess flux at 60 μm. They used PHOT Astronomical Observing Template (AOT) no. 3 at 25 μm and AOT P22 at 60, 90, 135 and 180 μm. ρ^1 Cnc is a G8V-star with a detected "hot jupiter" of 0.84 M_J / sin i and a possible 2nd companion of 5 M_J / sin i. A cool M5 red dwarf orbits this planetary system at a distance of 1150 AU. The model best in accordance with the observation is that of a circumstellar disk of not too large cometary dust + ice particles (grain size of 10 μm) at a distance of 50 - 60 μm. These dust grains must be replenished in some way. It has now been clearly shown that dust/icy disks and planets may coexist. Examples are the Solar Sytem (zodiacal dust), β Pic and ρ^1 Cnc. Figure 5 shows a schematic most recent view of the ρ^1 Cnc planetary system with data from radial velocity measurements and ISO combined.

Waters *et al.* detected an oxygen-rich dust disk surrounding an evolved star in the Red Rectangle. The disk ranges from 500 till 2000 AU. The name Red Rectangle refers to the morphology of an extended nebula in bipolar configuration consisting of carbon-rich dust. Deeply embedded in this nebula is a circumbinary disk consisting of oxygen-rich dust. Spectra were taken from this disk. Crystalline silicates and CO_2 spectral features were seen. The features were similar to those found around young stars with planet formation from the point of view of grain size distribution, degree of crystallization, disk dimensions and total mass, radial density distributions

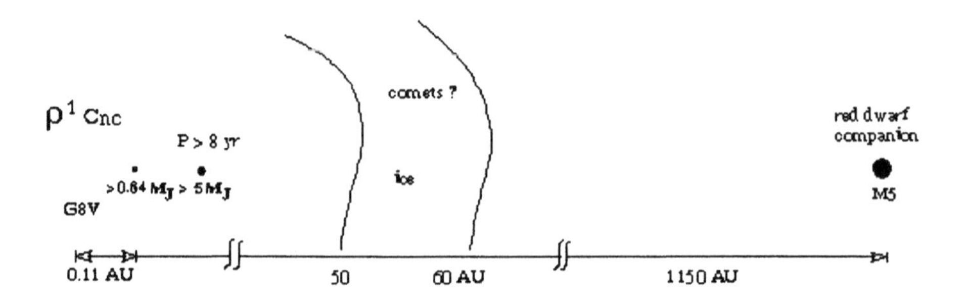

Figure 5. A schematic view of the "new" ρ^1 Cnc system, adding the ISO-result of Dominik *et al.*, (source : author)

of the grains as well as CO kinematics and depletion. The authors conclude that planet-formation may as well occur around **evolved** stars such as HD 44179 in the Red Rectangle.

Siebenmorgen *et al.* found a ring of organic molecules around HD 97300. This pre-main sequence Herbig Ae/Be star showed extended emission with CAM-CVF. The observations are consistent with a dust model of organic molecules (e.g. PAHs) as well as smaller and larger grains of silicates and graphite of a total mass of 0.03 $M\odot$. The observations indicate that the ring is made of interstellar matter. It is formed by the interaction of the stellar wind with surrounding matter. One peak of extended emission may be due to an embedded companion.

7. ISO-observations of comets

Comets are related to both interstellar matter and circumstellar matter. They are generally also considered to contribute a lot (by means of impacts) to the volatile inventory on the surfaces (oceans and ice deposits) as well as of the atmospheres of the terrestrial planets. Both supply and depletion (the last by atmospheric erosion) of volatiles may result depending on impact parameters. They are hence also linked to "early life" and present planetary atmospheric and biospheric conditions. ISO observed in particular with

PHT-S the comet Hale-Bopp (target of opportunity) and found that at 2.9 AU it was losing H_2O:CO_2:CO at a ratio of 100:22:70 (Crovisier *et al.*, 1997). Cometary volatiles have a composition similar to interstellar ices. ISO detected Mg_2SiO_4 (crystalline silicate) as dust component with SWS which was also detected in circumstellar disks, which clearly shows the link between comets and circumstellar material. This contrasts with interstellar silicates which are amorphous.

8. ISO-observations of Titan, Europa and icy satellites : astrobiology !

8.1. TITAN

Titan was observed in grating and Fabry-Pérot mode with SWS and LWS. Spectrophotometry was performed with PHT-S.

First results of SWS (Coustenis *et al.*, 1998) confirm the spectral results of the IRIS instrument of the Voyagers but give a higher precision. The grating spectra have a 5–20 times higher resolution. H_2O was detected for the first time as a weak spectral feature at 39.37 μm and at 43.89 μm. Its presence is probably due to the infall of icy micrometeorites into the atmosphere of Titan. The presence of H_2O explains very well the presence of CO and CO_2 in Titan's atmosphere as detected earlier by the Voyagers. The detection of water in the atmosphere makes Titan's fysico-chemistry even more richer, more complex than thought before. Upper limits for several other trace elements such as benzene and allene have been obtained. PHT-S revealed a signature of possible surface features. SWS revealed surface features as well in the 5–7 μm window (Courtin *et al.*, 1998). LWS suffered a lot of straylight from Saturn and the features are lost in the noise in grating mode. Fabry-Pérot results are in the process of analysis. An important result of ISO-observations will be better atmospheric models resulting from more precise determination of atmospheric mixing ratios at various heights in the atmosphere. Comparing Titan- with Saturn- and Iapetus observations will allow to determine the relative contribution of H_2O from interplanetary origin and local Saturnian origin. The contribution to the local component greatly favours Saturn(presence of nearby ring-system), whereas the interplanetary component should be equal inward and outward (apart from gravitational focusing and (for charged particles) precipitation along magnetic lines of force, both favouring Saturn).

The importance for astrobiology of better understanding Titan cannot be stressed enough. The surface has liquid and solid reservoirs of organic compounds. A complex chemistry still largely unknown dominates the atmosphere and surface. The surface is shielded from UV-rays by an organic UV-absorbing layer in the stratosphere. Liquid H_2O can persist for centuries on the surface and react with the organic compounds under certain temporary special conditions (large impacts) (Thompson *et al.*, 1992), even if the surface temperature is very low (94 K on the average). HCN has been detected in the atmosphere and is predicted to exist on the surface as well. HCN is biologically significant as it is an essential ingredient of amino-acids which are important biomonomers. Proteins which are responsible for much of the structure and most of the functions of living cells, are made from chains of amino-acids. Some of the organic compounds found in the interstellar medium or in circumstellar disks have been discovered in Titan's atmosphere as well. Especially N_2-CH_4-chemistry based products produced after radiation bombardment by energetic particles and UV are found in both environments. The difference between Titan and the interstellar environment is that

1. the liquid state exists on Titan's surface.
2. Titan has a much bigger reservoir of molecules/polymers at its surface.
3. Titan's surface is shielded from UV-radiation.
4. Titan has a complex surface with strongly varying chemical and physical conditions.

8.2. EUROPA

Interest for Europa has risen dramatically since the Voyager and Galileo-missions. This interest is explained apart from the interesting geology by the likelyhood of a subsurface water-ocean below the icy surface. Such an ocean could harbour life. HST and Galileo spectra revealed a teneous atmosphere consisting of O_2 and SO_2. A group around Thérèse Encrenaz (DESPA, Meudon) is analysing ISO SWS-spectra of Europa to find signatures of O_2, O_3, O, SO_2, OH, H, H_2O and to constrain very precise the abundances of these trace gases, resulting from interaction of the icy surface with energetic charged particles in Jupiter's radiation belts. This will lead additionnally to a better understanding of the surface-composition of this astrobiologically important object. ISO-observations look at other wavelength bands then HST or Galileo but they miss the high angular resolution of these observatories. ISO's observations are thus complementatry to these observations. Recent Galileo-observations have shown the existence of hydrated salts on the surface of Europa which may point to a brine ocean beneath (McCord *et al.*, 1998).

8.3. ICY SATELLITES

ISO made disk-averaged observations of the icy satellites Iapetus, Callisto, Ganymede and Europa with SWS. Spectra showing evidence of O_3, O_2,

CO_2, etc. are expected for some of these objects. O_3 e.g. was detected earlier on Ganymede. Spectra of icy objects could look similar to "Darwinian spectra". An exoplanetary observer could interprete at first glance these spectra as signatures of life on an exoplanet. Further analysis should show however clear differences (such as abundances, mixing ratios, etc.) between an icy object and an exo-earth. Distance to the star (Habitable Zone !) should also enable to classify these objects in one of the categories.

9. Importance of ISO for Darwin (TPF-IRSI)

ISO was a technological demonstrator of infrared cryogenic technology in space. ISO has performed studies of *zodiacal light* : ISO showed less than 15% deviation from a blackbody spectrum. The observations look like those for fine silicates. Spectra of fine silicates similar to those found in the coma of comets, collected interplanetary dust particles and the dust around β Pic. These studies are important to better understand exo-zodiacal light.

ISO has taken spectra of icy satellites in the Solar System. Some of these spectra may look like *Darwinian exoplanetary spectra*. It is important to be able to distinguish between both. Understanding Titan better is important for astrobiology not at the least because of the likelyhood of the existence of "exo-titans". ISO added more detail to the study of some exoplanetary systems. It showed the presence or absence of IR-excess in such systems (47 UMa, 51 Peg, β Pic, ρ^1 Cnc, etc.) Some ISO observations also led to an indirect better understanding of the planet-formation process.

10. Conclusions

ISO is both technically and scientifically a necesary predecessor mission for an infrared interferometer in space. ISO has paved the way politically as well for an infrared interferometric mission. ISO revealed the presence of building bricks of biomonomers everywhere in the universe. ISO-results are fully in the process for 10 years more worth of analysis and interpretation. ISO is about to show the history, characteristics of interstellar and circumstellar grains, gas and ices. ISO has given clear indications that circumstellar disks are common but not ubiquitous. ISO makes a clear evolutionary connection between interstellar materials, comets and circumstellar materials.

ISO shows the absence/presence of circumstellar disks in planetary systems. ISO showed that circumstellar disks with planet-forming characteristics can form around evolved stars and in binary systems. ISO showed the presence of young brown dwarfs in star-forming regions.

ISO found H_2O in the atmosphere of Titan which makes this object even more complex and intriging. ISO observations of icy satellites will lead to a better interpretation and analysis of Darwinian exoplanetary spectra. ISO

characterised the zodiacal light, which is important for Darwin.
New discoveries are awaiting in the data !
More information can be found on the following selected URLs :

1. http://isowww.estec.esa.nl
2. http://isowww.estec.esa.nl/science
3. http://www.ipac.caltech.edu/iso/iso.html
4. http://www.sron.rug.nl/iso (last page is a SWS-page, similar pages exist for the other instruments)

References

Adams, F.C., Lada, C.J. et al. (1987), *ApJ*, **Vol. 312**, 788

Allard, F., Hauschildt, P.H. et al. (1996), *ApJ*, **Vol. 445**, 443

Boogert, A.C.A. et al. (1996) Solid methane towards deeply embedded protostars, *Astronomy and Astrophysics : First ISO Results*, **Vol. 315 no. 2**, pp. L377-L380

Burrows, A., Hubbard, W.B. et al. (1993) , *ApJ*, **Vol. 406**, 158

Cernicharo, J., Gonzalez-Alfonso, E., Lefloch, B. (1997) Physical conditions in the interstellar and circumstellar medium from ISO observations, in K Leech, N.R. Trams, M. Perry, (eds.), *First ISO Workshop on Analytical Spectroscopy*, **ESA SP-419**, pp. 23-35

Comerón, F., Rieke G.H., Claes, P. (1998) ISO observations of candidate young brown dwarfs, *Astronomy and Astrophysics*, in press

Comerón, F., Claes, P., Rieke, G. (1997) Oph 2320.8-1721, a Young Brown Dwarf in the ρ Ophiuchi Cluster: Views from the Ground and from Space, *The ESO Messenger*, **no.89**, pp. 31-33

Comerón, F., Claes, P. (1998) ISO Observations of Brown Dwarf Candidates in ρ Ophiuchi, in R. Rebolo, E. L. Martin and M. R. Zapatero Osorio (eds.), *Brown Dwarfs and Extrasolar Planets*, A.S.P. Conference Series, San Francisco, pp. 43-50

Coustenis, A. et al. (1998) ISO observations of Titan with SWS/grating, *Astronomy and Astrophysics*, in press

Coustenis, A., Claes, P. et al. (1998) on LWS and PHT-S observations of Titan, *Astronomy and Astrophysics*, in preparation

Coustenis, A. et al. (1998) on SWS-observations of Titan, *Icarus*, in preparation

Courtin, R., Lellouch, E., Billebaud, F., Claes, P. (1998) The 5-to-7 micron spectrum of Titan, *proceedings of EGS 1998 meeting in Nice*, in press

Crovisier, J., Leech, K. et al. (1997), The Infrared Spectrum of comet Hale-Bopp, in K Leech, N.R. Trams, M. Perry, (eds.), *First ISO Workshop on Analytical Spectroscopy*, **ESA SP-419**, pp. 137-140

de Graauw, Th. et al. (1996) SWS observations of solid CO_2 in molecular clouds, *Astronomy and Astrophysics : First ISO Results*, **Vol. 315 no. 2**, pp. L345-L348

Dominik, C., Laureijs, R.J., Jourdain de Muizon, M., Habing, H.J. (1998) A Vega-like disk associated with the planetary system of ρ^1 Cnc, *Astronomy and Astrophysics*, in press

Ehrenfreund, P., Boogert, A.C.A. et al. (1997) Ices in star-forming regions: highlights from ISO, laboratory and theory, in K Leech, N.R. Trams, M. Perry, (eds.), *First ISO Workshop on Analytical Spectroscopy*, **ESA SP-419**, pp. 3-9

Glanz, J. (investigators : Neufeld, D., Harwit, M., Melnick, G., Kaufmann, M.) (1998) A Water Generator in the Orion Nebula, *Science*, **Vol. 280**, p. 378

Habing, H., J., Bouchet, P., Dominik, C. et al. (1996) First results from a photometric infrared survey for Vega-like disks around nearby main-sequence stars, *Astronomy and Astrophysics*, **Vol. 315 no. 2**, pp. L233-L236

Lahuis, F., van Dishoeck, E.F. (1997) ISO-SWS spectroscopy of gas-phase C_2H_2 and HCN towards massive YSO's: data reduction and modelling, in K Leech, N.R. Trams, M. Perry, (eds.), *First ISO Workshop on Analytical Spectroscopy*, **ESA SP-419**, pp. 275-276

Leech, K., Claes, P. et al. (1998) INFRA-RED SPACE OBSERVATORY Observations of the Pleiades Brown Dwarfs Teide 1 and Calar 3, in R. Rebolo, E. L. Martin and M. R. Zapatero Osorio (eds.), *Brown Dwarfs and Extrasolar Planets*, A.S.P. Conference Series, San Francisco, pp. 93-95

McCord, T.B. et al. (1998) Salts on Europa's Surface Detected by Galileo's Near Infrared Mapping Spectrometer, *Science*, **Vol. 280**, pp. 1242-1245

Moneti, A., Zinnecker, H. et al. (1997) Evidence for Remnant Circumstellar Disks Around Post T Tauri Stars, in Proceedings of Lisboa conference on star formation.

Rieke, G.H., Lebofsky, M.J. (1985) *ApJ*, **Vol. 228**, 618

Siebenmorgen, R., Natta, A., Krügel, E., Prusti, T. (1998) A ring of organic molecules around HD 97300, *Astronomy and Astrophysics*, in press

Thompson, W.R., Sagan, C. (1992) Organic Chemistry on Titan - surface interactions, *Symposium on Titan*, **ESA SP-338**, pp. 167-176

Waelkens, C., et al. (1998) SWS observations of young main-sequence stars with dusty circumstellar disks, *Astronomy and Astrophysics : First ISO Results*, **Vol. 315 no. 2**, pp. L245-L248

Waters, L.,B.,F.,M., Waelkens, C. et al. (1998) An oxygen-rich dust disk surrounding an evolved star in the Red Rectangle, *Nature*, **Vol. 391**, pp. 868-871

Whittet, D.C.B. et al. (1996) An ISO SWS view of interstellar ices: first results, *Astronomy and Astrophysics : First ISO Results*, **Vol. 315 no. 2**, pp. L357-L360

ARE WE ALONE IN THE COSMOS?

T.C. OWEN
Institute for Astronomy University of Hawaii
2680 Woodlawn Drive, Honolulu, HI 96822

1. THE QUEST FOR A COSMIC CONNECTION

The desire to understand our own destinies in terms of some great cosmic design is one of the strongest human tendencies. And the "design" is much more appealing if it includes some superior beings who have an interest in our destiny. We even hope this interest is strong enough to open the possibility of communication if not direct visits. We yearn for this contact: some message, some sign that we are not simply circling the sun without purpose, alone in an uncaring universe. Because of this widespread cosmic loneliness, those of us who are interested in obtaining a scientifically verifiable answer to the question posed in the title of this essay find ourselves confronting a large body of received wisdom, ranging from such authorities as the BIBLE to "they say". We must tread a narrow path, overgrown with many varieties of religious beliefs, legends, speculation and tall tales, only to find ourselves finally confronting a large sign saying "no admittance without hard evidence."

We live in an age that is hungry for myths. In ancient times, these wonderful stories of heroes, heroines, gods and monsters provided a structure our ancestors could use to make sense of the world around them. For many people, this essential function of mythology has been satisfactorily supplanted by science, a very different way of looking at the world that achieves an even better result. Despite the extraordinary successes of science, however, those marvelous mythical creatures are still with us. They have simply changed their shapes and habitats in response to modern requirements.

We can easily trace this transformation. In Homer's time, the Gods lived comfortably on Mt. Olymp while fierce monsters roamed about in parts of Grece and more distant regions of the Mediterranean world. One thinks of Scylla and Charybdis, for example, guarding the Strait of Messina between

J.-M. Mariotti and D. Alloin (eds.), Planets Outside the Solar System: Theory and Observations, 381–395.

Sicily and Italy. Four hundred years later, Periclean Athens was flourishing and there were Greek colonies in Sicily. The old mythical monsters had been replaced by new ones in northern Europe and Africa, little-known places that were far way. This retreat continued, leaving us with tales of the Yeti in the high Himalayas, Bigfoot in the dense forests of the Pacific Northwest and of course that lovable reptile undulating in the peat-filled waters of Loch Ness. The retreat has gone even farther, as the human mind has cleverly located the ultimate haven for its fantasies. During the present century, our mythical companions have left the Earth completely and are now populating planets in this system and beyond, exercising powers beyond our comprehension but never forgetting their enduring interests in the planet Earth and its human inhabitants.

In its purest form, this modern manifestation of mythology involves UFOs, malevolent Martians, and a massive re-interpretation of history that finds everything from the pyramids of Egypt to cave paintings by early humans to be strongly influenced by (or even the work of) visitors from the sky. Some modern authors have even suggested that the old classical myths were nothing more than the simple-minded responses of our ignorant ancestors who were trying to understand these apparently magical visits by galactic cosmonauts. Thus the new mythology incorporates the old, in much the same way that the invading peoples who became the classical Greeks absorbed some of the local legends and deities they found around them into their own pantheon.

In this essay, I will briefly review the UFO phenomenon and the myth of ancient astronauts, moving on to Mars and our current scientific understanding of why that planet is so different from Earth. Is the Earth unique in the universe as an inhabited planet? Many scientists don't think so, and we can see why, using the famous Drake equation to organize our ideas. If we are to find a scientific answer to the question in the title, we must do some experiments, and here the good news is that we have already started.

2. THE MYTH OF UFOs

If we are not alone in the cosmos, it might seem reasonable to expect some representatives of one or more of our nearest neighbors to pay us a visit from time to time. This apparently reasonable expectation has gained such force that the inverse argument has also been advanced: if we haven't had such visits, it means that we are in fact, alone.

What is the evidence? There have been reports of visitors coming from the sky throughout recorded history. This is only natural, as the sky is frequently chosen as a home for human deities. These deities are by definition interested in us, so it seems reasonable that they or their messengers would come and visit from time to time. In biblical times, wings or a chariot of fire might be the means of conveyance, today it is the space-ship. Tubular or saucer-shaped, these mechanical marvels swoop through our skies, flashing their lights and occasionally landing to take a closer look. Sometimes they even abduct humans, taking them for rides or conducting bizarre experiments on them. Such are the accounts that we can read almost weekly in the tabloids available in our supermarkets.

Is there any truth to all of this? It is undeniably true that people see things moving in the sky that they cannot identify. This experience happens to astronomers and physicists as well as to clergymen, police, airplane pilots, even the odd president of the United States. So the reports of UFOs are genuine, but what are these people seeing?

It usually turns out that perfectly natural objects are being mistaken for something mysterious. The planet Venus heads the list. At its brightest, Venus can cast a shadow. It is exceptionally luminous and to people unfamiliar with the sky, its appearance can be quite extraordinary. Seen through broken clouds or wind-stirred tree branches, Venus can appear to be moving rapidly and erratically. Police have chased it along country roads, jet pilots have flown after it, naval guns have been fired at it. Fortunately, Venus is far enough from us to have survived all these indignities!

Birds flying at night are also frequently "unidentified" as are clouds, meteors and artificial satellites. A UFO with flashing red and green lights is almost certainly an airplane, yet reports of many such sightings have been filed. Amidst all the natural explanations, there are problems of fraud and jokes. Pictures of flying garbage can lids, out-of-focus light bulbs, reflections of interior lights on windows showing external views have all been presented as evidence of alien spaceships. Mysterious "crop circles" cut into farmers' fields that were celebrated as the work of visiting cosmonauts turned out to be an artifice of pranksters. The sad truth is that we have no solid evidence for contemporary visits by interstellar spaceships. We have pictures of meteors carving their way through the Earth's upper atmosphere and records of airbursts produced by impacting interplanetary projectiles, fragments of rock and ice left over from the formation of the solar system. We have photographs of houses and cars damaged by meteorites and we have the stones that did the damage in our museums, along with rocks that have reached the Earth from the moon and even from Mars. We have debris from our own rockets and satellites that have crashed back into the Earth. We do not have a single fragment or picture of a spacecraft from another world.

All we have are reports. These are interesting, and many come from perfectly reputable sources, but they are not supported by any hard evidence. They are not the kind of cases that can convict a murderer, nor do they offer the kind of reliable facts one looks for in buying a used car.

We need some proof we can verify before we can accept the wonderful idea that the Earth is being visited by spacecraft. What about a conspiracy? Devout believers in alien visits often suggest that the government has the real evidence, they just won't release it to us. It is certainly true that our government keeps secrets from us, although they usually come out. In this case, however, we are talking about a world-wide, multi-government conspiracy. There is no reason to expect that alien spaceships land only in the U.S. If these spacecraft came all the way to Earth from some distant star and there are so many of them, surely they have visited many countries. Therefore we must assume that the governments of all the countries have banded together to deny their citizens this knowledge. Imagine India, Iran, Israel, Iraq, Indonesia, Ireland, Italy and Iceland, all locked in a conspiracy to prevent us from knowing about extraterrestrial visits. I think we can agree this is rather unlikely. This negative conclusion should not be discouraging. Except in clusters or associations, the distances between neighboring stars are immense, so it is not so surprising that we do not receive daily visits, even if the galaxy is teeming with life. To put matters in perspective, it is useful to realize that our fastest spacecraft, Voyagers 1 and 2, would take 100,000 years to reach the nearest star, even if they were headed directly for it, which they are not. To hop from star to star in times short by human standards, we need to move at relativistic speeds and this takes huge amounts of energy. In a beautiful example worked out by Nobelist Edward Purcell, the annihilation of the equivalent of an Empire State Building of matter and anti-matter is required for a relativistic trip to a star twelve light years away. Purcell scrupulously ignores the engineering difficulties in such an enterprise, such as the energy required to make that much anti-matter and the need to isolate the matter and anti-matter from each other until they meet in the unimaginable engine. Considering that we can already transmit messages of any length across the galaxy at the speed of light for a few dollars worth of energy, it seems reasonable to think that even highly advanced civilizations would choose this mode of communication rather than direct travel. But of course, we can't know that! We must keep open the possibility that there are some deep secrets of nature we haven't yet uncovered that might make space travel easy. Even in this case, we can do some experiments, however. We can test the idea that the Earth has been repeatedly visited since the time of its formation by extra-terrestrials who have left signs of their appearances behind them.

3. THE MYTH OF ANCIENT ASTRONAUTS

If we have no current visitors from the Great Beyond, what about visits in the past? As a first step, we can search ancient historical records to see if we discover something that is out of place (some indication of knowledge that contemporary civilizations could not possess or some artifact or accomplishment that must have required outside intervention.

The greatest champion of this approach has been Erich von Daeniken, whose book CHARIOTS OF THE GODS? sold millions of copies around the world during the seventies, and is still in print today. Von Daeniken assumes a very low level of initiative on the part of his readers, so low that they will not notice his many inconsistencies and will not check his extravagant claims for evidence of terrestrial visits by ancient astronauts. He offers a perfect example of the development of a modern myth.

There is a cave drawing at Tassili in northern Africa, he reports, that provides an accurate representation of an astronaut in a space suit. He reproduces this drawing in his book and even gives us a reference to the man, Henri Lhote, who discovered it. It is an outline drawing of human figure with no neck. A quick trip to the library reveals that the drawing is one of several reproduced by Lohte's expedition. Other drawings show similar figures that are clearly barefoot, decorated with feathers and holding bows. These are simply poorly executed representations of contemporary humans. Depictions of local animals by these same early artists are also crudely drawn. Later inhabitants of these caves produced more accurate figures, some of which are quite beautiful. These later drawings contain no figures resembling cosmonauts or their craft. Von Daeniken also writes about the discovery of a cave drawing in California that looks to him like the depiction of a slide-rule. Obviously these primitive artists would never have made slide rules of their own, so they must have seen one in the hands of some visitors from an advanced civilization. A problem with this fantasy is that most young scientists and engineers today have never seen or heard of a slide rule. This primitive device for doing calculations was the best we could do at the time von Daeniken wrote his book. It has been completely superseded even in our own rather simple civilization by hand-held electronic calculators.

A strange pattern of lines found on the dry and desolate plain of Nazca in Peru is interpreted by von Daeniken as a landing field for interstellar spaceships. Here he seems to assume that visitors from a nearby star will be bridging cosmic distances in fixed wing aircraft; interstellar bi-planes, perhaps? And so it goes, sometimes straying perilously close to the language of science. Is it really a coincidence he asks, that the height of the great pyramid of Cheops multiplied by a thousand million is nearly equal to the distance from the Earth to the sun? Obviously the ancient Egyptians didn't know this distance, so someone had to tell them, or perhaps that someone actually built those giant pyramids, a feat our author doubts the Egyptians could accomplish on their own. The pyramids are a marvel, as anyone who has seen them can attest. They have inspired tall tales for millennia, including a famous, fabulous account by Herodotus in the fifth century BC (the great pyramids were already over 1500 years old!). Yet the quarries from which the blocks were taken, the tools that were used are all still there. The Egyptians surely built these extraordinary royal tombs themselves, and it is demeaning to suggest they needed help. So what about that number? Well, if ancient astronauts were responsible, why didn't they do a better job? Why isn't the height a more precise fraction of the Earth-sun distance? After all, if you multiply the length of a common ball point pen (a "Bic Stick") by a million million, you get the distance from the Earth to the sun exactly! It appears that we are indeed confronting a coincidence.

After many pages of this kind of "evidence", von Daeniken winds up his case with an appeal to scientific authorities. Here he comes up with an interesting statement that bears some consideration. He is discussing the equation developed by astronomer Frank Drake to estimate the number N of intelligent civilizations in the galaxy. (We will discuss this equation ourselves in section 6.) To impress his readers with the seriousness of all this, von Daeniken writes (p. 141): "Fantasy and wishful thinking may be concealed in all the deliberations and suppositions, but the [Drake Equation] is a mathematical formula and thus far removed from mere speculation."

We scientists know better. There is nothing magical about an equation, and von Daeniken himself inadvertently proves this by suggesting that if one divides the height of the great pyramid into its area, one will obtain the dimensionless number π. (He should have used the perimeter, not the area.) We have to be careful (as I shall try to be in Section 6) to make this very clear.

And we can do much better with ancient records. Why stop with recorded history after all? We can search back through the rocks to a time nearly 4 billion years ago, a million times farther back than van Daeniken. We find wonderful things along this immense passage through time: diamonds, fossils, evidence of ancient microbial life, the concentrated iridium and carbon soot layer associated with the K-T extinction event, but nowhere do we find fossilized remains of ancient spaceships, or artifacts discarded by their crews. So what do we conclude from all this? First, that people are so intensely interested in the modern myth of extraterrestrial visits that they will accept almost any version of this exciting concept. Second, we must never fail to apply critical thinking when confronted by an unusual idea. "How could I test that claim?" is a good question to keep asking yourself, whether someone is trying to sell you a used car, a political candidate, an explanation for the great pyramids, or a new scientific result.

4. THE MYTH OF MALEVOLENT MARTIANS

Of all the planets in our solar system, Mars is most like the Earth. This was already apparent to astronomers over 150 years ago. They could not see the surface of Venus through their telescopes and the surface of the moon never changed. But Mars had polar caps that waxed and waned with the seasons, clouds that would come and go, including huge global dust storms, and dark areas on its surface whose intensities and outlines also exhibited both seasonal and non-repetitive changes. Here was a world like ours, smaller to be sure, and farther from the sun, but blessed with an atmosphere and seasonal cycles that might even imply the growth and death of vegetation.

This attractive idea gained stunning support around the turn of the century from the work of Percival Lowell. Lowell was a wealthy American amateur from Boston who established an observatory outside Flagstaff, Arizona to pursue his interest in Mars. Lowell saw the planet criss-crossed with lines that he interpreted as evidence of canals constructed by the planet's inhabitants to channel water from the polar caps to the desert regions. Thus was born the concept of "Martians", who have become well-established members of our mythology, more widely known than Zeus or Athena.

Lowell's ideas were attacked at the time and with increasing vigor in subsequent years. Contemporary astronomers were unable to see the "canals" and temperature measurements soon showed that Mars was far colder than Lowell had imagined. Nevertheless, a serious panic erupted in the United States on the evening of October 30, 1938, when Orson Wells broadcast his famous radioplay based on H.G. Welles' War of the Worlds. The radio broadcast purported to be an on-the-scene account of an invasion of Earth by Martians, who landed in New Jersey and were proceeding to conquer the United States with poison gas attacks. The next day the New York Times ran a front page story about the resulting "wave of mass hysteria" that "seized thousands of radio listeners throughout the nation."

Sixty years later, after several spacecraft have flown past, orbited and even landed on Mars, we have no evidence for the existence of Martians. Those wonderful canals of Percival Lowell were simply the product of wishful thinking. Not only are there no artificial structures on the planet, the Viking Landers were unable to find any evidence of microbial life or even any organic compounds in the Martian soil. Upper limits on organics were less than a part per billion, a sensitivity sufficient to find the proverbial needle in a haystack. And yet... From orbit the Viking spacecraft confirmed the presence of dry river beds and other evidence of water erosion. These days the Martian atmosphere is so thin that liquid water cannot exist on the planet's surface, but "once upon a time", Mars was wet. Evidently the planet experienced an early period in its history when the atmosphere was thicker and the surface was warmer than it is today. It is possible to determine the era when the water was flowing on Mars by counting the number of impact craters in the floor of the wider channels. The density of craters per square kilometer can then be compared with surfaces on the moon whose ages we know from rock samples returned to Earth. The result is that the time of liquid water on Mars has been established as earlier than 3.5 billion years ago. Since then, the surface of the planet has been in a state very similar to the one we find it in today (cold, dry, irradiated by solar ultraviolet light that easily penetrates the thin atmosphere, extremely hostile to the existence of life.

But in that earlier time, during the first billion years of the solar system's existence, conditions on Mars and Earth were apparently rather similar. It thus seems possible that life could have arisen on Mars at the same time it did on Earth, some 4 billion years ago, if we can demonstrate that both planets were endowed with the same starting materials. Subsequently, Mars evolved along a very different path. If life did begin there, in order to survive it would have been forced to adapt in microbial form to a deep underground environment, as some forms of life have successfully done on Earth. Otherwise, Mars would be totally lifeless today.

5. THE MYTH OF THE GARDEN OF EDEN

Why are Earth and Mars so different? Why is Earth the only one of the nine planets that harbors abundant life? What is it, exactly, that makes a planet "Earth-like"? If we can answer these questions, we will be well on our way to answering the Title's question. We want to try to establish some fundamental principles that allow us to extrapolate from our own solar system to planets anywhere in the Universe.

For our purposes, an Earth-like planet is one in a nearly circular orbit, on which both dry land and water occur, whose mass is sufficiently large that it can retain an atmosphere for 5 billion years yet sufficiently small that hydrogen can readily escape from the atmosphere into space. This is a world on which life can originate and evolve to an intelligent form with which we could communicate over interstellar distances.

In our own solar system, we have learned that Venus is too close to the sun for liquid water to be stable on its surface, while Mars is too small to retain a thick atmosphere against the effects of impact erosion. We can see the problem with Venus if we imagine moving the Earth to this planet's orbit. The increase in surface temperature that results from being closer to the sun would lead to more evaporation of the oceans, which would increase the water vapor in the atmosphere, leading to a larger greenhouse effect. The latter would in turn increase the surface temperature further, producing more evaporation, etc. etc. In other words, a positive feedback loop would be established and we would have a "runaway greenhouse". Eventually the oceans would boil and all that water would be in the atmosphere, which would be so hot that the water vapor would easily rise to altitudes where its molecules could be broken apart by solar ultraviolet light. Hydrogen could then escape into space, leaving oxygen behind to combine with the rocks. We know this actually happened on Venus because we have found a 150-fold enrichment in deuterium, the heavy isotope of hydrogen, in the remaining water vapor, compared with water in Earth. This enrichment indicates that a huge amount of hydrogen has escaped from Venus.

In the atmosphere of Mars, we also have evidence from isotopes that a large fraction of the original atmosphere has been lost. In this case it is the enrichment of argon and xenon produced by radioactive decay of parent elements that provide the clues. Forming carbonates won't cause the observed enrichment and simple thermal escape or sputtering won't touch xenon. Loss of a large fraction of the early atmosphere by some massive, non-fractionating process is required, and the intense bombardment that occurred during the first 700 million years of this small planet's history will do the trick.

What does all this tell us about the uniqueness of the Earth? Is it possible that we are living in a cosmic Garden of Eden? Perhaps our planet is so special that we are the only sentient beings in the universe! The limit of habitability set by a planet's proximity to its star is obviously fixed. No planet can be as close to its sun as Venus is to ours and have liquid water on its surface, the sine que non for life. The outer limit, being too far from the star, is harder to define. If Mars had been the size of Earth or Venus, it would have been able to sustain a thick atmosphere against the effects of impacts, as the other two planets obviously did. Such a planet could achieve and maintain a sufficient greenhouse effect to allow liquid water at its surface, at least near the equator. But where do all these gases come from in the first place? Perhaps the Earth received a unique mixture of volatiles that allowed the development of our special atmosphere, oceans and life. Present evidence suggests that this is not the case. The amounts of carbon and nitrogen in the atmosphere of Venus are very similar to those in the total volatile inventory on Earth. The pattern of Earth's noble gas abundances is close to that of Mars. As a result, most scientists think that these gases were contributed by a flux of comets and primitive meteorites in the late heavy bombardment that left the scars we still see on the surfaces of the moon, Mercury, Mars and every other ancient solid surface in the solar system. Comets will bring in water and all the necessary biogenic elements, giving each planet a sufficient endowment of starting materials to develop life. This late, heavy bombardment seems to be a natural part of planet formation, so we can expect the same starting conditions for the origin of life to exist on Earth-like planets throughout the universe.

We have just received stunning confirmation of this hypothesis through the discovery of organic compounds in an ancient rock from Mars, known as ALH 84001. These compounds, together with mono-mineralic crystals of magnetite and sulfate, carbonate globules and possible microfossils have been touted as possible evidence for ancient biological activity on Mars. Even if it does not contain evidence of life itself, this rock carries an important message. If the organic materials are not contaminants, they show that the conditions for starting life indeed existed on the surface of the only other planet in our solar system with liquid water on its surface during that critical first billion years when life began on Earth. This is strong support for the idea that life may be relatively common in the Universe.

6. THE MYTH OF MATHEMATICAL CERTAINTY

We are now ready to try to muster a scientific answer to the titular question. We can use the Drake equation as a guide. This is a formula developed by Frank Drake for estimating the number, N, of advanced civilizations in our galaxy, at a given time:

$$N = R^* f_p n_e f_l f_i f_c L$$

Here the different terms have the following meanings:

R^* = the rate of star formation in the galaxy
f_p = the fraction of stars that have planets
n_e = the number of Earth-like planets in such planetary systems
f_l = the fraction of such planets on which life develops
f_i = the fraction of life-bearing planets that produce intelligent life
f_c = the fraction of intelligent life- bearing planets that produce a civilization capable of and interested in interstellar communication
L = the average lifetime of such civilizations, measured in years.

All we have to do now is substitute a number for each term in the equation, carry out the multiplications, and we will then have the answer we seek. Solving this equation gives us the number of civilizations in the galaxy that might send us emissions, or at lease messages, ending our cosmic loneliness forever. It sounds easy, but there is nothing magical about an equation. To make a rabbit stew, you have to have a rabbit. To solve an equation like this one, you must know the numerical value of each term. Unfortunately, until just last year, we have only been able to give a number of the first term, $R^* \sim 1$. Now suddenly, thanks to the work of Mayor and Queloz in Europe and Marcy and Butler in the United States, we know for the first time that other planetary systems exist. Yet we are still not in a strong position to assign a number to f_p, as these exciting observations have not yet been extended into a general survey. Nor can we evaluate n_e. The technique that is being used by these investigators so successfully to find giant planets cannot hope to detect Earth-like planets whose masses are inadequate to perturb the motions of their central stars to an extent that would cause a detectable Doppler shift. We are therefore (still!) compelled to resort to the indirect arguments given in section 5 in order to estimate the number of earth-like planets. These arguments suggest that it is reasonable to assume one "Earth" for every 3 planetary systems. This estimate is based on the assumption that every planetary system will have at least one inner planet with a mass similar to ours, so every third system will have that planet in the "right" place. As we have seen, the restrictions on the orbital distance from the star are not so acute if the mass of the planet is sufficient.

However, we have no idea at present whether there are any other Earth-like planets that could sustain life as we know it exist! Let us first address the issue of liquid water. The popular press quickly picked up on the fact that some of the new planets were located near enough to their respective stars that their effective temperatures would be above the freezing point of water. So if liquid water could exist on these brave new worlds, why not life as well? The argument founders on the circumstances in which the liquid water would occur. In our own solar system, each of the four, giant outer planets has levels in its atmosphere where temperatures are high enough for water to exist in the liquid state. Even for Neptune, it is simply necessary to descend into the atmosphere far enough, and you are guaranteed to find a level with tropical temperatures. The problem is that none of these planets possesses a solid surface, so the liquid water only exists as cloud droplets, like the cumulus clouds we enjoy on Earth. As all the planets discovered so far are giants, most of them more massive than Jupiter, we may safely assume that even those that are close to their stars will only have liquid water in the form of clouds. Unless one is willing to entertain the idea of life existing solely in air (the equivalent of Aristophanes' Cloud Cuckoo Land) we are not talking about habitable Earth-like planets. There is another problem. If these giants that are so close to their stars came to their present locations by moving inward from more distant orbits, as many scientists suppose, they will have swept away any small terrestrial planets that were already in place, either forcing them into the stars or into themselves.

Perhaps instead of the planets themselves, we should ask about satellites as possible habitable worlds. None of the satellites of our own giant planets is massive enough to maintain an atmosphere at the Earth's distance from the sun. The gases would rapidly escape. We must therefore assume that these newly discovered giants have satellites with masses close to that of Earth, if we expect these moons to become Earth-like planets. Perhaps they do, especially those planets with masses even greater than that of Jupiter. If so, we suddenly have a new set of environments on which life might arise and flourish. Unfortunately, we must leave the subject in this speculative vein. Returning to our equation, we must recognize that there is no certainty about the numerical values for the other terms. We must simply use the best scientific evidence and logic at our command to estimate these numbers. For example, we can suggest that the product (f_p n_e) will be less than 0.3, but perhaps greater than 0.03. Until we detect Earth-size planets, we won't know. Moving to f_l, we can see how important that Mars rock is. If it really turns out that life began on Mars, we would feel some confidence that $f_l \sim 1$. Without that knowledge, we just don't know. f_l could be 10^{-9} ! The next term f_i is even tougher. Biologists seem to be divided about the inevitability of the evolution of intelligence, once life gets going. Suppose there had been no K-T extinction event? The dinosaurs had been around for 150 million years and didn't seem to be getting much smarter. If they had prevented the development of mammals, the Earth might still be shaking under their mighty but rather stupid perambulations. Lacking another example of life on another planet, we can't say.

We are perhaps on firmer ground in guessing that f_c is near unity. As we have seen in the first three sections of this essay, curiosity about extraterrestrial beings is a very deep-seated human attribute. Once an intelligent species has emerged, it seems inevitable that attempts will be made to communicate with hypothetical soul-mates elsewhere in the galaxy.

Putting all these terms together, various authors find values of N ranging from L 10^{-10} to L/10. We can see why those who try to solve the Drake equation look longingly at L, the average lifetime of an advanced civilization, to compensate for possible low values among the f's. L might be a very large number, even larger than the 150 million achieved by the dinosaurs. Or it might not. What lessons can we learn from ourselves?

7. THE MYTH OF MANIFEST DESTINY

Our own civilization first achieved the ability to participate in interstellar communication about 50 years ago. This represents our value of L, so far! We have obviously been around a lot longer than that, long enough to see some of the problems we must solve if we want to make L a truly large number. To the extent that we represent an average example of an advanced civilization, (an assumption that Joseph Shklovskii and Carl Sagan have called the "principle of mediocrity") we can assume that the same problems will be confronted by civilizations elsewhere in the galaxy.

We can consider two general classes of difficulties, those that are caused by external agents and those we cause ourselves. In the first category, we think of natural catastrophes such as a giant impact by an asteroid or a comet, like the one that did away with the dinosaurs. We can only imagine a massive change in the Earth's climate brought about by some as yet unknown characteristic of the sun, or the outbreak of a deadly disease. Remarkably, we have achieved a level of scientific and technological maturity that allows us to cope with such events. We are steadily improving our ability to detect incoming asteroids and comets at distances sufficient to allow us to intercept and deflect or destroy them. The sun may yet surprise us, but 4.5 billion years of experience (plus observations of many other solar-type stars and a lot of astrophysical theory) suggest that the sun's luminosity will stay very constant for another 5 billion years. Despite the recent concern over AIDS, it appears that modern medicine is capable of coping with the danger of epidemics.

All this capability carries its own threat, however, leading to the problems we cause ourselves. To reach this level of maturity, we also have achieved the ability to destroy ourselves totally in a thermonuclear holocaust. While we seem to be reducing that particular danger these days, we are interacting with our environment in many other self-destructive ways. The solar luminosity may not change, but by pumping more CO2, CH4 and fluorocarbons into the atmosphere, we are increasing the greenhouse effect, leading inevitably to global climate changes. And this exposes perhaps the most serious threat we offer ourselves: the rampant increase in the human population. To quote Charles Darwin: "There is no exception to the rule that every organic being naturally increases at so high a rate that, if not destroyed, the earth would soon be covered by the progeny of a single pair."

By steadily reducing the threats to our existence, we have become the cancer that is destroying our planet. As of the 1990 census, we are multiplying with a doubling time of only 40 years. We have to learn to regulate (and that really means to diminish) our population if we want our species to survive with a high level of technology for many more millennia. Otherwise we face the danger of being overwhelmed by social problems that could limit our own value of L to about 100 years. Humans will continue to survive, but the problems of survival will become so acute that spending money to try to communicate with extraterrestrial civilizations will seem an almost criminal extravagance. We hold our future in our own hands; there is nothing "manifest" about our destiny. This means we are able at last to transform the mythology about superior cosmic beings into a scientific enterprise: we can do the experiment necessary to search for them. Certainly we will also continue to amass the data needed to supply values for the various terms in the Drake equation. But wouldn't it be wonderful to leap frog that effort by making direct contact? This is a concept that humanity has been notoriously unwilling to accept. Each tribe, each ethnic group, each nation, each religion seems to think it is the best, the truest, the one with special privileges. Yet we must learn to accept complete responsibility for our own actions if we can ever hope to make L a number that challenges the 150 million years achieved by the dinosaurs.

8. "THE GREATEST ADVENTURE LEFT TO HUMANKIND"

Let's end by looking at the bright side of our new-found abilities. After centuries of speculation about extraterrestrial life, we have finally developed the technology that will allow us to find out whether or not there are other advanced civilizations out there among the stars. During the 50 years we have had this capability, there have been only a few deliberate efforts to receive or transmit signals over cosmic distances. Television, FM and radar transmissions have all been steadily leaving the planet, however, heading out into the depths of interstellar space at the speed of light. Inhabitants of hypothetical planets around many of the bright stars we see in our skies each night have already been exposed to our signals. The number and sensitivity of our radio telescopes has continued to increase, and new discoveries of transmissions from atoms and molecules in our galaxy and its nearer neighbors continue to pour in. We have detected signals from molecules of formaldehyde, ammonia, ethyl alcohol, and several tens of other molecules, in giant interstellar clouds. We have discovered radio beacons emanating from stars smaller than the Earth that are composed only of neutrons. At least one of these tiny remnants of supernova explosions has its own system of planetary bodies, revealed by their gravitational effects on the star's rotation, encoded in the radio beacons. In all this richness of received information, however, there is not yet a single shred of evidence for an artificial signal, something that would indicate the presence of another advanced civilization.

This may turn out to be true, but we are in no position to reach any conclusion yet. We can understand non-detection of radio signals as easily as we understand the absence of visitors to our planet. Sensitive as they are, our giant telescopes and their receivers can only search an exceedingly tiny fraction of the radio spectrum in a given direction in space at a given time. We do not know at what frequencies our hypothetical friends are broadcasting, nor do we know where they are! Specialists in this field have begun to overcome these problems by building receivers that can monitor 10's of millions of frequencies simultaneously, feeding these signals through computers that can test them for evidence of alien intelligence. Coupling such receivers to dedicated radio telescopes would allow surveys of the entire sky, as well as directed searches focused on the stars considered most likely to have planets like Earth.

Regrettably, intelligence on our own planet is often not as high as we might hope. A NASA program to embark on what Frank Drake has rightly called "the greatest adventure left to humankind", was voted down amid hoots of laughter from our fellow Americans in the U.S. Senate. Undaunted, scientists have used private money to start over again, The Planetary Society, a 100,000 member organization founded by Carl Sagan, has sponsored a project called BETA (Billion-channel Extraterrestrial Assay). It has a receiver that can detect a quarter of a billion frequencies simultaneously to search for signals, using a 26-meter radio telescope belonging to Harvard University. The famous film director Steven Spielberg has contributed funds for this search. The Planetary Society has also lent its support to another SETI project called SERENDIP (Search for Extraterrestrial Radio Emissions from Nearby Developed Intelligent Populations) at the University of California, Berkeley. The doomed NASA effort has been revised by a group of scientists at the newly organized, privately funded SETI Institute. They have dubbed their enterprise "Project Phoenix."

Are we alone in the cosmos? We won't know until we do the experiments that can tell us. Let us hope the Phoenicians and their colleagues will be able to marshall the necessary resources to allow this grand adventure to continue until we have the answer.

Adapted from a chapter in EVOLUTION! FACTS AND FALLACIES, ed. J.W. Schopf, Academic Press (New York) (1998).

STRATEGIES FOR REMOTE DETECTION OF LIFE

A. LÉGER
Institut d'Astrophysique Spatiale, CNRS, Bât. 121
Université Paris-Sud - F-91405 Orsay, France
e-mail: leger@ias.fr

1. Introduction

The history of Astronomy has taught us not to be anthropocentric. When searching for extrasolar lives, we have to keep in mind this lesson, not trying to search for our twins. Specifically, we should base our quests on laws and properties that are expected to apply everywhere in the Universe.

An implication of this statement is that, before searching for life, we must give a definition of life as universal as we can.

Then, we shall consider whether there are tracers of such a phenomenon that can be detected by remote sensing - a non trivial question -

2. What could be alien life?

Biophysicists and biochemists agree to define a living being as a system that:

(1) contains *information* (negentropy),

(2) is able to *replicate* itself,

(3) undergoes few random changes in its information package that allow its
evolution by a Darwinian selection of the most performing.

Searching for such beings, several light-years away from us, seems *a priori* a hopeless task. We show thereafter that, very fortunately, this is not the case.

3. Life signatures on Exoplanets

Due to the fundamental aspect of this question, proposals have been made for several decades.

J.-M. Mariotti and D. Alloin (eds.), Planets Outside the Solar System: Theory and Observations, 397–412.

3.1. STATE OUT OF EQUILIBRIUM

Lovelock (1965, 1975) has noticed that the gases of the Earth atmosphere, which are far from equilibrium, have a biological origin. A typical example is the simultaneous presence of CH_4 and O_2. He has concluded that the presence of a large departure from equilibrium is a criterion for the presence of a biological activity.

It must be note that there is *no definitive physical basis* for this criterion. Earth receives photons from a hot source, the Sun (5780 K), and emits photons towards a cold source, Space ($\simeq 3K$). This is exactly what is required for a system to be able maintaining a state out of equilibrium but this does not imply that the system is a living one...

However, it is true that, on our planet, the only processes that actually produce large quantities of gases out of equilibrium in the atmosphere are biological ones. The criterion could be considered as valid if atmospheric models were to fail when trying to produce the advocated mixture of gases. In other words, the criterion "presence of large amounts of out of equilibrium gases in a planetary atmosphere" should not be rejected but it has to be *qualified* in each of its application.

Sagan et *al.* (1993) have considered the detection of CH_4, N_2O and O_3 by the Galileo spacecraft when flying by the Earth as indications of life on our planet. But they did so only after having shown that abiotic processes are unable to produce the observed amounts of gas.

3.2. SIMULTANEOUS PRESENCE OF CH_4 AND H_2O

J. Kasting, at this Summer School, has shown that on the Primitive Earth, abiotic processes could produce a volumic mixing ratio of methane, $[CH_4]$, of 10^{-4} but to reach the level of 10^{-2} he has to imply biological processes. Therefore, the presence of large amounts of CH_4 ($[CH_4] \geq 10^{-2}$), in presence of H_2O and UV produced OH radicals which are a strong sink for methane, could be an indication of life. This statement has to be qualified by extensive atmospheric modellings that investigate whether abiotic processes can produce methane in such quantities as well.

3.3. A SPECIAL CASE FOR LIFE SUPPORT: ORGANIC CHEMISTRY IN SOLUTION IN WATER

What could be the supports of life?

Can we derive a criterion for the presence of life from the generic definition we have adopted?

Trying to avoid being biased by the nature of life on Earth, one can consider different possibilities for the support of its information. Organised physical structures as those of the magneto-optic or semiconductor memories of our computers could be thought of, but building such systems by natural processes appears unlikely.

A more attractive possibility is a chemical support: the coding by a sequence of chemical entities (chain cells) in a *linear macromolecule* which behaves as a message written with letters. A replication process can be imagined provided that cells have homo or hetero affinities. Of course, we have in mind the example of DNA but this can happen with other macromolecules. During this Summer School, M.C. Maurel has given very interesting examples of macromolecules which are not DNA but can carry similar information (p-RNA, PNA, PALADTHY).

Carbon chemistry...

If one considers that linear macromolecules are a favorable case for storing the information of living beings, *carbon* based chemistry, organic chemistry, appears to be the most powerful. Carbon can be easily oxidised (CO_2) or reduced (CH_4) which allows the production of a great variety of chemical species. This property is unique. Even Si, the element most similar to C in Mendeleev classification, makes much stronger bonds with O than with H and has a poorer chemistry.

This richness of C chemistry is confirmed by the study of a medium where physical conditions (pressure, temperature...) are quite different from those in our laboratories, the Interstellar one: among 112 species identified in 1996, 84 contain carbon and only 8 silicon.

... in solution in Water

The requirement that a living being can evolve implies that it is a place where many chemical reactions take place. Macromolecules can react rapidly, only if they are in *solution* . Solid state reactions are too slow and large molecules cannot stand the temperatures required for them to be in gas phase.

As pointed out by A. Brack (1993), among the different possible solvents, liquid water is a special one. It has the highest dielectric constant, $\epsilon = 80$, that allows salt ionization and, most important, it has the capability of building H-bonds with dissolved molecules. The latter property allows specific conformation of macromolecules when in solution in water thanks to the solvent attraction of their hydrophilic groups (OH, CO, $COOH$...) and repulsion of their hydrophobic ones (CH, CH_3 , aliphatic chains...). These macromolecule conformations allow very specific chemical reactions, as "key-keyhole" ones, which are valuable in building reproducible complex structures with a rich information content. Alternative solvents as liquid hydrocarbons, alcohols, liquid NH_3 are less favorable because of a lower ϵ or/and lack of H-bonding.

Chemically, H_2O has some activity, i.e. hydrolysis, that are important to select between different chemical pathways, destroying the products of many of them (A. Brack 1993). However, its activity remains moderate as opposed, for instance, to liquid NH_3 that attacks essentially any organic compounds and would destroy all the products of prebiotic chemistry, impeding the appearing of life.

Last but not least, T. Owen (1980) has pointed out that water is also appropriated for a subtle reason: it is indirectly more resistant to UV because some of its photolysis products, O_2 and O_3, protect it from further attack.

Consequently, restricting our quest for lives to those based on carbon chemistry in solution in water is probably not so severe.

The reader could be some what disappointed: initially, we have strongly stated that we should not look for a twin of life on Earth, and now we are reducing our search to organic chemistry in solution in water, the very basis of life on our planet... However, the similarity between exo-lives that we are searching for and the terrestrial one stops there. For instance, by no means we assume that information should be stored in DNA, using the Genetic Code for transcription to proteins which perform the chemical actions!

In other words, we have reached the conclusion that it is *not mere chance* if life on Earth is based on carbon and water. This system is a very appropriate one to fulfil the requirements of the general concept of life as it has been defined.

T. Owen (1980) concluded that our present understanding of life requirements and planetary conditions provides us with some real support for *carbon-water* chauvinism. We are not just reproducing home.

3.4. IMPLICATION 1 : DEFINITION OF THE HABITABLE ZONE AROUND A STAR

The definition of the Habitable Zone (HZ) around a star results: it is the region where a planet can sustain liquid water at its surface, at least during some time of its local year.

J. Kasting et *al.* (1993) have further proposed the concept of Continuously Habitable Zone (CHZ) requiring that such a situation lasts long "enough" for life to evolve towards elaborated structures. This excludes massive stars which have a short life.

The case of low mass stars, MV ones, is of special interest. On one hand they are very attractive because they have a low luminosity implying, for a given size and temperature of a planet, a more favourable star/planet luminosity ratio and an easier direct detection. On the other hand, the habitable planet candidates are closer to them than around other stars. A 300 K planet, without Greenhouse effect, is at a heliocentric distance: $a(300K) \simeq 1AU\ (L/L_{\odot})^{1/2}$. Now, tidal forces are proportional to a^{-3}, the lower L the stronger they are and they can phase lock the planetary spinning and orbital rotation, if resonance phenomena do not prevent the process (Mercury should be phase locked around the Sun by tidal forces but is not, due to resonances).

This locking could make the planetary (always) sunny side torrid and the dark one frozen, which would be unfavourable for life development... However, such a fate is not for sure because the resulting difference of temperature will pump up winds that efficiently reduce the temperature gradient on the planet, if the atmosphere is thick enough. The case of Venus is illustrative: although it is spinning very slowly (local day is 240 terrestrial days), its dark face is only few degrees lower than its sunny one. Using a 3D atmospheric model, Joshi et *al.* (1997) concluded that over a large range of conditions, planets around MV stars are very likely to be habitable

3.5. IMPLICATION 2 : A POSSIBLE CRITERION FOR DETECTING LIFE

It is thought that carbon is mostly fully oxidised (CO_2) in the primitive atmosphere of terrestrial planets (D. Gautier, 1992). If a carbon based life has developed at a large scale on one of them, it requires a large amount of organic molecules and has to make them from the abundant raw material, CO_2. This implies the reduction of the latter by a process as:

$$CO_2 + H_2O + energy \rightarrow (CH_2O) + O_2^{\nearrow},$$

where the energy can have different origins: stellar photons, planetary internal heat... In any case, there is a release of *free oxygen.* A *burrial* of synthesized organic carbon (CH_2O) is also necessary to allow the accumulation of O_2, otherwise, the oxygen output would be exhausted by subsequent oxidation of organic matter. This requires the presence of *plate tectonics* on the planet. Under such conditions, the extensive development of life on a planet can produce an oxygen rich atmosphere.

Now, oxygen is a very reactive gas that can oxidises several volcanic output gases and iron or sulfurs contained in rocks. If not continuously replenished, it would disappear rapidly from an atmosphere provided plate tectonics or volcanism are present. The later bury surface materials and bring up fresh reducing ones. Another active trap for free oxygen is the water weathering of rocks that bring fresh material in contact with atmospheric O_2 and favours its oxidation. On present Earth, if photosynthetic production of O_2 was stopped, this gas would disappear in 4 10^6 yr (J. Kasting, private communication; H. Holland, 1984).

As a consequence, the massive presence of free O_2 (P_{O_2} > 10 mbar) in an exo-planet atmosphere appears as an indication of the presence of an efficient production of this gas. Thereafter, we discuss whether it can be considered as a realiable indication of the presence of a C based life.

4. Qualifying the H_2O - O_2 criterion

As any criterion, the H_2O - O_2 criterion has to be qualified, searching whether some abiotic processes cannot produce these gases as well. If abiotic processes cannot do so but under well identified conditions, the criterion would still be useful.

4.1. EARTH-LIKE PLANETS

Photodissociation of CO_2 has been considered as a source of abiotic O_2 (J. Rosenqvist and E. Chassefière, 1995). These authors have found that only a small amount of O_2 (P_{O_2} < 5 mbar) can be produced under the most favourable conditions.

A more plausible possibility is the photodissociation of H_2O by the UV photons of the star, followed by H escape from the planetary atmosphere.

On Earth-like planets, i.e. rocky ones with ground temperature close to 300K, this process is strongly reduced by the presence of a *cold trap* at the tropopause which blocks most of the ascending water vapor and makes it return to the planet surface as rain or snow. Then, H_2O is a minority gas in the upper atmosphere, the UV rich one. In that region, photolysis of water produces H and OH. If H efficiently escapes from the atmosphere the end result would be the production of O_2. However, when H is a minority species, its escape is severely limited by the diffusion through the majority gas and it can recombine with OH or another oxidising species, prior to escaping. These processes have been studied in detail by E. Chassefière (1996) who concluded to a low production of abiotic O_2 under Earth-like conditions.

As a confirmation, 99.9999% of breathable oxygen on Earth is produced by photosynthesis. Only 1 ppm comes from abiotic H_2O photodissociation (J. Walker, 1977) although the latter species is quite abundant at its surface.

Earth is large enough a planet for sustaining volcanism for billions of years after its formation (uranium and potassium radioactive decays are volumic heating processes whereas radiative cooling is a surface one). Volcanoes produce reducing gases (H_2, CO, SO_2...) and spread reducing rocks (with Fe^{++}) at the surface of the planet and both react with atmospheric O_2. Therfore, volcanism provides a powerful sink for free O_2 on Earth size planets.

As already mentioned, water weathering of rocks provides another efficient O_2 sink. It is active on planets with emerged continents and water rains.

As a result, on a large terrestrial planet, or one with continents and H_2O rains, sources of O_2 can maintain a non vanishing amount of this gas in the atmosphere *only* if their production rate overcomes the fixation rate by these sinks.

Now, we review the presently known situations where oygen can be abundant, but *abiotic*, in the atmosphere of a planet and where the H_2O - O_2 criterion does not apply. They are those where the preceding mechanisms do not work. Namely:

4.2. PLANETS WITH MOIST RUNAWAY

For a planet in the CHZ of its star, the CO_2/carbonate cycle is a mechanism that stabilizes the surface temperature thanks to a negative feedback. An increase of the temperature produces more water evaporation, enhanced dissolution of CO_2 in water rains, enhanced fixation of CO_2 into carbonates and therefore a weaker Greenhouse effect of this gas that cools back the planetary surface (see J. Kasting's lecture) .

If a planet is close enough to its star, this mechanism can fail. This will happen if the surface temperature reaches values high enough for dissociating carbonates into CO_2 and oxides. The negative feedback will be blocked as atmospheric CO_2 is no more fixed into the ground. A massive input of H_2O into the stratosphere will occur and the UV photolysis of water becomes a powerful source of O_2 as diffusion does not reduce the H escape any more when water is a major constituent. An atmosphere rich in abiotic O_2 can build.

Kasting et *al.* (1993) have made a conservative estimate of the distance to the star above which this moist runaway cannot occur (the inner limit of the CHZ). They have found $(a)_{min} = 0.95(L/L_{\odot})^{1/2}\,AU$. If the planet is still closer to its star, the Greenhouse effect of the evaporated water can produce a hot runaway of the temperature, but this is not necessary for leading to the disappearance of the whole content of free water at the surface of the planet.

During a moist runaway, the successive phases are:
(1) H_2O, O_2 and inert gases M (e.g. N_2, CO_2...) are abundant in the whole atmosphere. This is the active H_2O photolysis episode. Such a phase is

relatively short (0.1-0.3 Gyr);
(2) O_2 and M are still abundant but H_2O is exhausted. The duration of that phase will depend upon the rate at which O_2 sinks can capture oxygen. It can last longer (1-3 Gyr) than the first phase because the O_2 sinks can be temporary saturated (finite volcanic outputs...);
(3) M are the only species left. This is the present situation of Venus.

Consequently, several features allow the *identification* of this abiotic production of O_2:
(i) the planet is out of the CHZ i.e. at a renormalized distance, $a_r = a(L/L_{\odot})^{-1/2}$, smaller than 0.95 AU,
(ii) in most cases, O_2 is present but H_2O is absent.

This is why a better criterion for the presence of a photosynthetic activity is: "the *simultaneous* presence of O_2 and H_2O ".

4.3. ICY PLANET WITHOUT VOLCANISM

Another situation can lead to an atmosphere rich in abiotic O_2: if oxygen sinks are inactive and allow the accumulation of the low production of O_2 by UV photolysis (J. Kasting, 1997).

It is expected when a planet is small enough. Its volcanism, with the associated oxygen sinks, can disappear soon after the planet formation. Quantitative calculations are necessary, but there is probably room between the Earth's size and Mars' one for a planet to be large enough to retain a thick atmosphere (0.1-10 bar) but small enough for volcanism to vanish after say 2 Gyr.

The sink due to water weathering will be missing too if the planet is cold enough for having very low water evaporation, scarce snowfalls and no liquid water rains.

Consequently, a planet with size in between Earth and Mars, located farther than the CHZ, can have an atmosphere rich in abiotic O_2 .

There are features allowing *the identification of this situation* too:
(i) its has a small size;
(ii) whatever is the local season, the temperature of its surface is lower than the H_2O freezing point (the planet is located outer the CHZ);
(iii) the strength of the water vapor bands in the planet spectrum is weak.

Points (i) and (iii) should be made quantitative.

4.4. O_2 BROUGHT BY COMETS ?

Recently, Noll et *al.* (1997) observed O_3 in the spectrum of icy moons of Jupiter and interpreted it as the result of the chemical synthesis of O_2 during the bombardment of H_2O ice by H^+ and subsequent formation of O_3 by solar UV irradiation. They have inferred that if pre- cometary grains are proton irradiated in a pre-planetary nebula, comets would be rich in O_2 and bring these species to the terrestrial planets of their solar system, providing an abiotic source of O_2 for their atmosphere.

However, it has been objected (Léger et *al.*, 1998) that pre-cometary grains are *not* significantly irradiated during the pre-planetary period as shown by calculations using recent models for protoplanetary disks and confirmed by independent observations:
(i) terrestrial geological records show that O_2 was not present on the Young Earth during the heavy bombardment period when the cometary input was orders of magnitude larger than now;
(ii) no O_2 has been detected in Halley comet by the Giotto mass spectrometer (upper limit: $[O_2]/[H_2O] < 0.5\,\%$).

Noll et *al.*'s hypothesis appears therefore with no theoretical nor observational support and cannot be considered as a falsification of the proposed criterion.

In conclusion, the presently known situations that can lead to a planet with a massive content in abiotic O_2 in their atmosphere *can be well identified.* The H_2O - O_2 criterion does apply to identify inhabited terrestrial exoplanets when the later situations can be ruled out.

4.5. FALSE NEGATIVES ?

Are there possibilities that life exist on a planet and that the H_2O - O_2 criterion is not fulfilled (false negative)? The answer is definitely *yes* .

This is the case if life on a planet has remained located in small niches and, more generally, if its production of O_2 by photosynthesis is not able to overcome the oxygen sinks. This had been the situation on Earth up to about 1 Gyr ago, if the threshold is set at $P_{O_2} = 10$ mbar (Fig. 1).

The implication is that missions able to detect H_2O and O_2 are able to show evidence of exolives - a major scientific and human discovery - but that negative answers on a small number of solar systems would not prove that life does exist only on Earth. However, if the solar system sample is large enough and that no detection is made it would be an indication that life is probably not a frequent phenomenon in the Universe - a discovery with important implications -.

5. Ozone, a better tracer than Oxygen

When T. Owen (1980) proposed the O_2 criterion, he also proposed to search for the spectroscopic signature of this gas in the visible (A and B band at 760 and 720 nm, respectively). In practice, this proposal is not operational because in this wavelength range the star is much brighter than the planet ($5\,10^9$ for Sun/Earth) and the spectrum of the latter would be extremely difficult to obtain.

Searching for the O_2 signature in the radio domain ($\lambda = 5mm$) is not operational either because of the dilution factor of the planetary solid angle with respect to the antenna lobe of any radio telescope that can be thought of, in the near to mid-future.

R. Bracewell (1978, 1979) pointed out that the star/planet contrast is more favourable in the IR than in the visible ($7\,10^6$ at $10\mu m$ for Sun/Earth) and R. Angel et *al.* (1986) showed that the mid IR region is of special interest because it contains *many* informative spectroscopic signatures:
- 15 μ m CO_2 band whose presence (or absence) would indicate a major similarity (or difference) with the atmospheres of solar terrestrial planets,
- 6-8 μ m and 16-20 μ m H_2O depressions, telling whether it is habitable or not,
- 9.6 μ m O_3 band telling whether O_2 is present, implying a photosynthetic activity when the planet fulfills well defined conditions.

The IR spectrum of the solar terrestrial planets having an atmosphere illustrates the informative power of this part of the spectrum (Fig. 2). Observing these spectra would indicate that Venus, the Earth and Mars atmospheres contain CO_2 but the Earth alone is habitable (H_2O) and inhabited (O_3).

The O_3 band has another interesting property: it is a very sensitive tracer of O_2. The dependence of the former gas abundance upon the latter is not linear but *logarithmic* (Paetzold 1962, Kasting et *al.* 1985, Léger et *al.* 1993a). On Earth, a small amount of O_2 (10 mbar) would lead to a O_3 abundance only twice smaller than the present one (Fig. 1). The dependence of the band intensities upon the abundance of O_2 and O_3 require a detailed calculation which is in progress, but it can already be foreseen that, in the preceding situation, the O_2 bands will be much more reduced than the O_3 ones, making the detection of the latter species easier than that of the former. As an illustration, if the band intensities were proportional to the abundance of the species, an outer observer of the Earth, with the capability of measuring these bands at their present level with a signal to noise ratio of 12, could have detected life on Earth (at 6 σ) for 1 Gyr if observing the O_3 band, but only for 0.5 Gyr if observing the O_2 one.

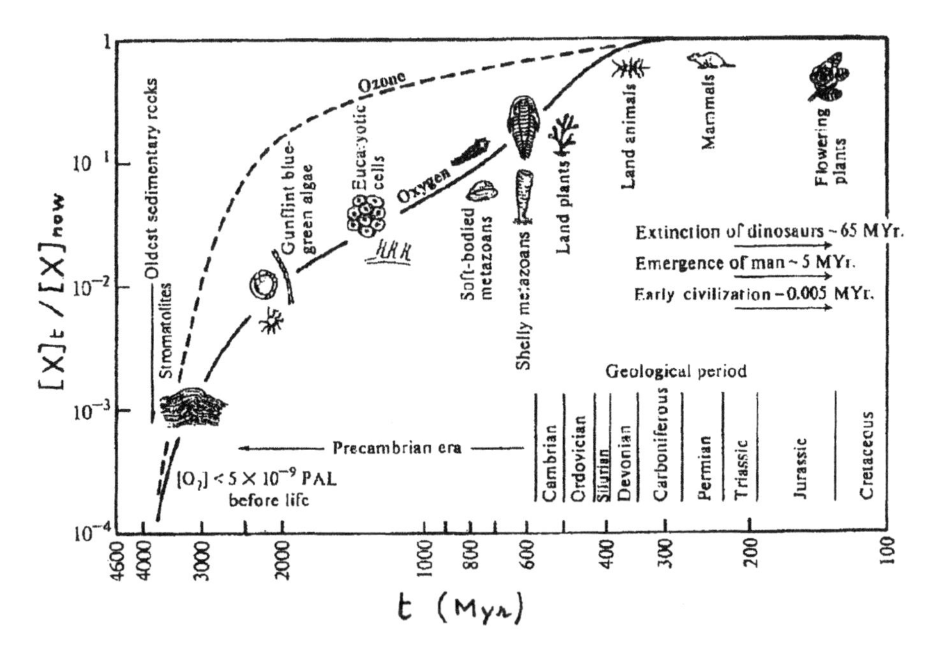

Figure 1. A possible history of the O_2 and O_3 content of Earth's atmosphere as quoted by G.Blake (1995)

Now, the statement becomes: "the simultaneous detection of H_2O and O_3 that corresponds to a significant amount of O_2 ($P > 10$ mbar) is a strong indication of on-going biological activity on the planet".

It applies (see also Kasting 1997) when:
- the planet is located in the CHZ of its star, conservatively estimated to $a_r = 0.95 - 1.15\,AU$ (Kasting et *al.* 1993);
- it is large enough for volcanism to be still active.

The presence of a *photosynthetic* activity implies that of *living beings* for a subtle reason. An atmosphere containing O_2 and O_3 blocks UV photons with $\lambda < 300$ nm. Therefore, the photosynthesis of organic compounds from CO_2 requires the successive absorption of two ground photons, with a storage of energy in between. A system able to perform this function has to be so complex that likely it is the result of a biological evolution.

Considering the implication of the confidence in this criterion, it is highly desirable that the scientific community keeps on trying to falsify/qualify it. At present, *it stands* but in situations that can be clearly identified.

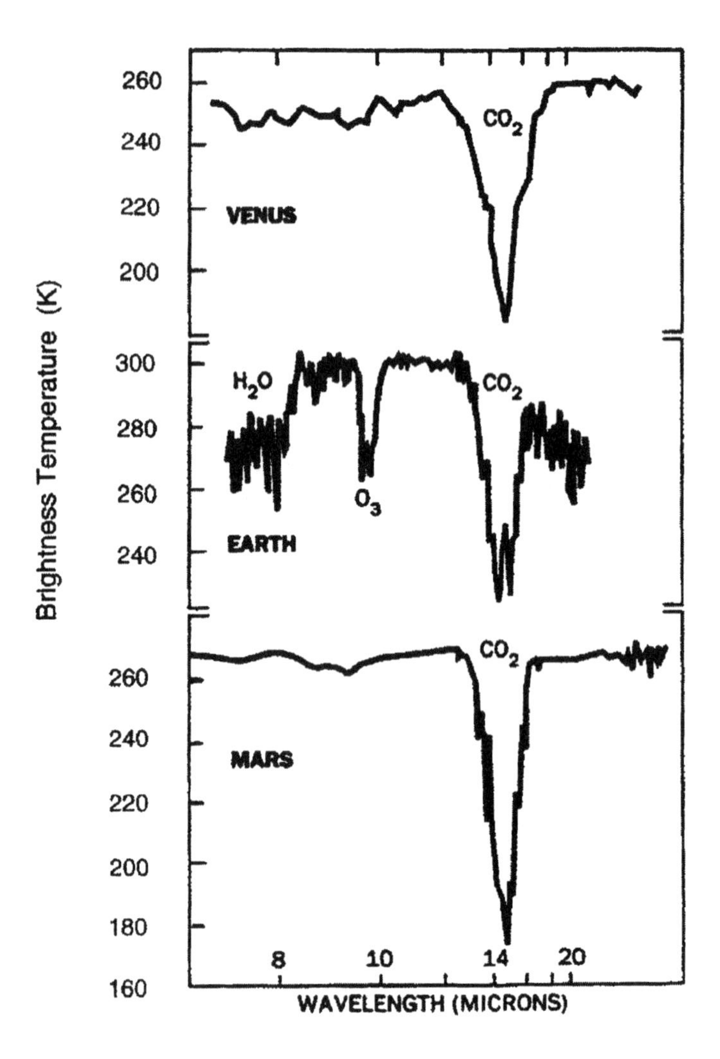

Figure 2. IR emission of the 3 telluric planets of the Solar System that have an atmosphere, expressed in brightness temperature. Note the strong CO_2 band at $15\mu m$, which is present in the spectra of all them, the structure shorter than $8\mu m$ due to H_2O and the O_3 band at $9.6\mu m$. The latter two spectral features are specific to the Earth and reveal that the planet is habitable (H_2O) and has a photosynthetic activity, (O_3) tracing the presence of a carbon based life. (Adapted from Hanel et *al.* 1992).

6. Concepts for IR Interferometers

The possibility for an infrared space interferometer to detect Earth-like planets around nearby stars, and to perform low resolution spectroscopy of their atmosphere has been seriously considered, following the pioneering works of Bracewell and Mac Phie (1979) and Angel et *al.* (1986). These efforts have led to two similar proposals, DARWIN-IRSI (Léger et *al.* 1993b, 1996) and OASES- TPF (Angel and Woolf 1996, 1997; Beichman 1996) now under consideration at ESA and NASA, respectively.

J.-M. Mariotti has given a somewhat detailed discussion of these missions at the Summer School. Here, we outline only few basic features that are common to both proposals:
- they favour the infrared range, because of the reduced contrast ratio between the star and the planet, and because of the existence of the spectral signatures of CO_2, H_2O, O_3, and possibly CH_4 (vibrational modes of molecules are in that spectral range);
- they propose passively cooled spacecrafts in 5 AU or 1 AU aphelia orbits;
- they are based on interferometric concepts, since, at 10 pc, a 1 AU separation between the star and the planet corresponds to an angular separation of 0.1 arcsec when projection conditions are optimum, i.e. the resolution of a 24 m telescope operating at 10 μm...

The numbers of solar systems that can be detected by such an interferometer are given in Fig. 3

Fig. 4 shows the recovered spectrum of the Earth in a simulated observation of the Solar System from distance, in the interferometer nulling mode,as compared with the exact one at the same spectral resolution. The absorption features of water, ozone and carbon dioxide would *be detected.*

7. Conclusion

Space projects have been recently proposed for observing a few hundreds of nearby single stars searching for terrestrial planets, and performing a low resolution spectroscopic analysis of their infrared emission in order to reveal the presence, or absence, of CO_2, H_2O and O_3 in their atmosphere.

The *simultaneous* detection of H_2O and O_3 corresponding to a massive presence of free O_2 ($P_{O_2} > 10$ mbar) in the atmosphere of a terrestrial exoplanet located in the *Continuously Habitable Zone* of its star, i.e. at distances renormalized by $(L_\star/L_\odot)^{1/2}$ between 0.95 and 1.15 AU, possibly between 0.95 and 2.4 AU (Forget and Pierrehumbert, 1997) appears to be a good criterion for the presence of a *photosynthetic activity* at the surface of the planet and therefore to that of a C based life (see also Kasting 1997).

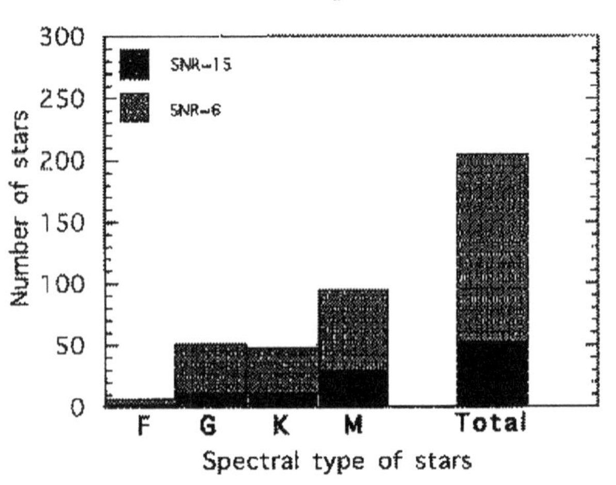

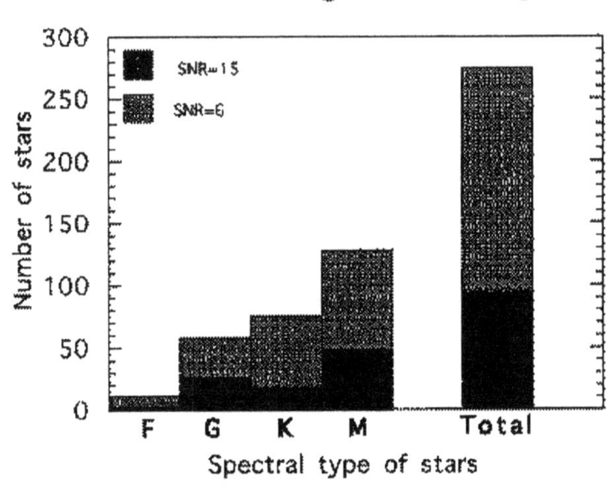

Figure 3. Histograms of the number of stars studied

Although ambitious, these projects could be realized using space technology either already available or in development for others missions. They could be built and launched during the first decades of the next century if there is the will to do so.

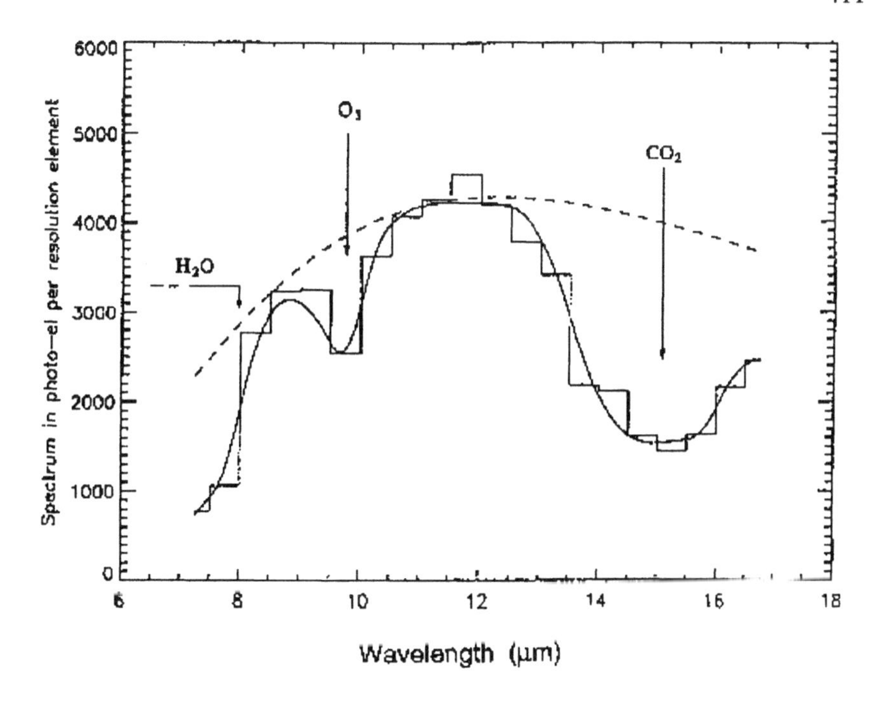

Figure 4. Simulation of the spectrum recovery of an Earth twin planet located around a G2V star at 10 pc, by the proposed DARWIN mission; $\Delta\lambda = 0.5\ \mu$m, integration time = 700h. The solid line is the exact Earth spectrum at this resolution and the dash line, the emission from a blackbody at 300K. Deviations from the latter point out the spectral features. Absorption of water ($\leq 8\ \mu$m), ozone (9.6 μm) and carbon dioxide (15 μm) are clearly detected. They would indicate that the exo-planet has an atmosphere containing a gas common to all the Solar telluric with an atmosphere (CO_2). It is habitable (H_2O) and is a location where a photosynthetic activity takes place (H_2O and O_3 simultaneously). The latter point would be a major discovery (Courtesy of B. Mennesson).

DARWIN-IRSI or OASES-TPF are not dead ends but beginnings of a program. The next missions should have higher spectral resolution providing richer information (compare Fig.2 and Fig.4) and higher sensitivity providing larger samples of solar systems and better statistics.

Their goal, detection of Earth-like planets and possibly extra-terrestrial life, justifies a vigorous action. Likely, the quest for extra-terrestrial life will be a major issue of Science during the XXIst Century.

References

Angel J.R., Cheng A.Y. and Woolf N.J., 1986: *Nature* 322, 341-343

Angel J.R. and Woolf N.J., 1996: *Scientific American* 274, 46 (April 1996)

Angel J.R. and Woolf N.J., 1997: *Astrophys. J.* 475, 373

Beichman, C. A.,1996: "a Road Map for Exploring Planetary Systems" JPL Publication 96-22

Blake G., 1995: proceedings of the ExNPS Kick off conference, 18-19 April 1995, ASEPS Doc. 710-95-004, JPL, Pasadena, CA, USA

Bracewell R.N., 1978: *Nature* 274, 780-781

Bracewell R.N. and McPhie R.H., 1979: *Icarus* 38, 136-147

Brack A., 1993: *Origin of Life and Evol. of the Biosphere*, 23, 3

Forget F. and Pierrhumber R.T.,1997: *Science* 278, 1273

Gautier D., 1992: Primitive planetary atmospheres: Origin and evolution, in *Frontiers of Life* , Trân Thanh Vân J.&K. et *al.* (eds), Editions Frontières, 91192 Gif (France)

Kasting et *al.*, 1985: *J. Geophys. Res.* 90, 10497-10510

Kasting J.F., 1997: *Origin of Life and evolution of the Biosphere* 27, 291-307

Hanel R.A., Conrath B.J., Jennings D.E. & Samuelson R.E., 1992: *Exploration of the Solar System by IR remote sensing* , Cambridge Univ. Press, USA

Joshi M.M., Haberle R.M. & Reynolds, 1997: *Icarus* 129, 450-465

Kasting J.F., Whitmire, D.P. and Reynolds R.T., 1993: *Icarus* 101, 108

Kasting J.F., 1997: *Origin of Life and evolution of the Biosphere* 27, 291-307

Léger A., Pirre M. & Marceau F., 1993a: *Astr. Astroph.* 277, 309 - 313

Léger, A. et *al.* 1993b: *The DARWIN Mission Concept*, proposal to the ESA "Horizon 2000 Plus" planning process

Léger A., Ollivier M., Altwegg K., Woolf N. J., 1998: A&A (in press)

Léger A., Mariotti J.-M., Mennesson B., Ollivier M., Puget J.-L., Rouan D., Schneider J., 1996: *Icarus* 123, 249-255

Lovelock J.E., 1965: *Nature* 207, 568-570

Lovelock J.E., 1975: *Proc. R. Soc. Lond. B* 189, 167

Noll K.S., Roush T.L., Cruikshank D.P., Johnson R.E., Pendleton Y.J., 1997: *Nature* 388, 45-47

Owen, T., 1980: in *Strategies for the Search for Life in the Universe*, Papagiannis ed., Reidel, p177

Paetzold H.K., 1962: "La Physique des Planètes" vol. 7, *Mem. Soc. Roy. Sci., Liège*, p 452

Rosenqvist J. and Chassefière E., 1995: *Planet. Space Sci.*, 43, 3

Sagan C., Thompson W.R., Carlson R., Gurnett D. & Hord C. 1993: *Nature* 365, 715-721

Walker J.C., 1977: *Evolution of the Atmosphere,* Macmillan, N.Y

THE SCENARY OF EXTRA-SOLAR PLANET SEARCH, FROM AN EXTRAGALACTIC VIEWPOINT

D.ALLOIN
European Southern Observatory
Alonso de Cordova 3107, Vitacura,
Santiago 19, Chile

1. Introduction

For an astronomer used to navigate in the extragalactic world, it is more than a challenge to provide the conclusions of an Advanced Study Institute dedicated to the search of extra-solar planets! Let me handle this task with the eyes of someone new to the field and just curious and amazed about it: I shall provide some of the remarks I drew after the presentations and a number of messages I picked up all along this 10 days of exciting discussions.

We heard numerous, extensive and well-documented lectures of which I have selected the following topics to give some conclusions or remarks: (a) the formation of planets, (b) the detection of planets, (c) the discrimination between brown dwarfs and planets, (d) the solar system as a bench-mark, (d) the planets habitability.

Some of the suggestions by lecturers and participants will then be presented, and, finally, I shall give the usual ending words.

2. The formation of planets

An essential question here is to understand the links between star and planet formations. Indeed, the processes involved in star formation, such as the fragmentation in the parent molecular cloud, the initial mass function of the stars formed within the cloud, the occurence of multiplicity..., should all be understood as they set the stage to the formation of planets. As an output of star formation modeling, we should get the initial conditions for models of protoplanetary discs. As well, our observational knowledge on Young Stellar Objects (YSO) can be used to put constraints on proto-

J.-M. Mariotti and D. Alloin (eds.), Planets Outside the Solar System: Theory and Observations, 413–418.

planetary discs, as regards to their size, structure, composition, gas-to-dust content and lifetime.

Let me review hereafter, for the successive steps in star formation, the status of theory/modelisation, and some essential parameters like disc size and lifetime:

- class 0 and class I objects, the parent molecular cloud gives birth to a protostar: the main question here is the evacuation of the angular momentum. Regarding modeling, one deals with the fragmentation process, with the gravitational collapse and the subsequent formation of a viscous accretion disc: the interplay accretion/outflow has to be understood, the role played by magnetic fields might be quite important (hence the need for performing an MHD treatment of the problem) and the type of resulting accretion disc is still under discussion. However, thick massive discs (around one solar mass) with follow-up fragmentation would evolve too rapidly and seem to be unlikely. Conversely, thin and light discs (one hundredth solar mass) appear to be more suitable because of their subsequent gentle evolution. It should be reminded here that the disc size at this stage is about 1000 AU and the phase lifetime is about 500000 years.

- class II and class III objects, which will lead at the end to the formation of planetesimals (around 10^{16}g): important questions here are the dust-to-gas ratio and the growth of grains. Modeling has to use fluid/particle SPH codes and has to predict both the radial and vertical distribution of particles. The lifetime of this phase is typically of 10^{6}years.

- then comes the aggregation of planetesimals, the building up of protoplanets (the isolation mass at 1 AU is about 10^{25}g) and the birth of planets. The modeling of the evolution from the previous phase to and along this one depends strongly on whether turbulence is taken into account or not. Without turbulence, dust is expected to settle down in the disc mid-plane. While if turbulence is considered, one rather deals with collisions and sticking of particles (ice-glued) and a progressive building up of larger bodies, all throughout the disc. N-body codes are used in the modeling of the late part of this phase. The evolution time along this sequence varies a lot, whether one assumes that there exist "seeds" (10^{4}years) or not (10^{8}years). Obviously, another critical parameter is the initial radial distribution of planetesimals in the disc.

Close comparisons between observational results (YSO) and model predictions could improve the understanding of these phases. On the side of theory, the identification of the dominant physical processes is still on the way and a consistent picture of the full evolution through the successive phases should be worked out. The impact of some parameters, such as the stellar mass or the stellar multiplicity, has to be tested with respect to the stability and lifetime of the protoplanetary disc. More generally, the dynamics of the systems may play a major role and feature out the planets/disc interactions or the migration of planets. On the side of observations, there is a gap in radius between the knowledge we have of YSO discs (known at a distance larger than 10 AU from the central star) and the searches for planets (searched for at a distance less than 10 AU from the star). This gap slows down our progress in understanding the complete evolution from star formation to planet formation, but there is hope that future instruments and improved modeling will sort out the question.

3. The detection of planets

When one comes to the point of trying to detect planets, the first obvious words are: planets are small,
planets are faint,
planets are bloomed out by their parent star!
This turns their detection into a real observational challenge and astronomers have shown creativity and persistence in trying to overcome these difficulties.

Planets can be probed through their gravitational pull: this property is used in various techniques which have been developped to detect planets. One should mention the radial velocity search, the astrometric search, the milli-second pulsar search,and the micro-lensing search.

Planets may be singled out through an image or a spectral signature: there is hope that the many ambitious experiments in progress will allow to perform this type of detections. Let us recall the transit method, the use of adaptive optics in a coronographic mode, the dark speckle approach and the nulling interferometer, the search for reflected light from the planet, and also the detection of some by-products of a planetary system like polarized light and zodiacal light.

Each of these approaches probes part of the parameter space (planet mass, planet period, distance to central star). In that sense they are quite complementary approaches. Most of them, if not all, involve monitoring: then one has to think of dedicated telescopes and databases. As well, nearly all of these techniques are pushing the observing capabilities to their limits: in radial velocity precision, in sensitivity, in contrast range... A whole set of instruments dedicated to the search for extra-solar planets are being built or planned, either for groundbased or spaceborn telescopes. A major concern regards the instruments stability. One is simply amazed by the inventivity of astronomers and engineers in this field, either for the instruments principles or for the technical advances. It is likely that the necessary funds to build these experiments can be raised, thanks to important cost-savings in building new generation telescopes (for a given mirror size).

4. The brown-dwarf/planet discrimination

The following question comes to mind: how are we going to discriminate between planets and brown-dwarfs?

On the side of theory they can be separated by their origin. Indeed, brown-dwarfs should be the result of fragmentation (then, what is the lower mass limit of brown-dwarfs?), while planets are the result of aggregation (then, what is the upper mass limit to planets?).

On the side of modeling, although they have a similar radius, they can be distinguished through their machinery (structure and source of energy) , and their expected spectrum.

Searches for both types of objects are very active, although the number of objects found in each class is still small. Much progress is expected with the development of deep surveys and a new generation of large telescopes.

5. The Solar System as a bench-mark

Time has come to look at the Solar System from some distance and to see it not only as a collection of planets and small bodies but also as a single planetary system. It is then possible to feature its evolution from icy planetesimals to earth-like planets or cores of giant planets and then to giant planets which subsequently captured gas to build up their atmosphere. The interplay between planets and minor bodies, the changes in the orbits of planets are also interesting questions to analyse in the Solar System.

Some more questions come immediately to mind: how unique is the Earth? What has been the delivery of material from other bodies like dust aggregats or meteorites? How could we interpret the spread by a factor 10 in the D/H value within the Solar System? How can we explain the formation of water oceans on the Earth? Is it related to a large rate of comet impacts? The study of the content in noble gases is quite important as they are clues to the exchanges between the planet and its environment (indeed they don't escape and they are chemically inert).

In conclusion, one should also ask the following question: how confident can we be about the understanding we have today of the history/evolution of the Solar System?

6. The habitability of a planet, the possibility of life?

In order to better identify the parameters which allowed life to appear on Earth, one has to perform some archeological investigations about the Earth's physical conditions. What have been the favourable factors which made it possible for life to start on Earth? I picked up three at least in the course of various lectures: (i) the rapid outgasing of rocks, (ii) CO and

CH_4 oxidized into CO_2 and (iii) the motion of continental platforms. One has to monitor as well the evolution of the planet energy balance and more particularly the evolution of its atmosphere. Major questions regarding the primitive atmosphere of the Earth are the evolution of its composition, the enrichment from the volcanic and mineral buffers, the hydrogen budget...All these problems raise a vivid interest within the concerned community.

Another important question is to define the limits of the habitable zone within a planetary system. For a sun-like star, the range is from 0.95 AU (limit of water loss) to 1.4 UA (limit set by green-house effects). Of course, these limits will change with the effective temperature of the central star.

How did life start on Earth? We know that the interstellar material contains a wealth of complex molecules. Researches are even persued to identify some amino acids in the interstellar matter, although this is a very difficult task because of the complexity of their spectral signature. Anyhow, there is a tremendous jump from complex molecules to living systems.

Two characteristics of the life-phenomenon are: (i) its capacity to *replicate*, which implies that there exists a mechanism to transmit information (DNA) but which also calls for the existence of a protection system (kind of a close configuration), and (ii) its capacity to *improve*, which implies that changes should be possible through certain routes (kind of an opening system). The needs in energy are covered through exchanges with the outside (photo-cells, glucose degradation..). The interplay close/open is an interesting problem to study. As well, it is important to understand the building up of complexity which certainly plays a major role in the occurence of life.

Finally, let's summarize the critical timescales involved in the complete process from planet formation to life-phenomenon occurence: (i) star formation: about 10^6years, (ii) planet formation: 10^8 to 10^9years, (iii) from protocells to eukaryotes: around $2x10^9$years, and (iv) from eukaryotes to multicell systems: about 10^9years.

As to how we shall search for other living systems in the Universe, various life-signatures have been discussed: O_2, O_3, H_2O... Passionate discussions were held under the olive trees touching upon the definition of life, upon human responsibility in the exploration of other "worlds"...

7. Suggestions and final words

At the end of the ASI, two interesting suggestions were made. First, it has been proposed to set up an european TMR (Training Mobility Research programme of the EC) on the topic of exo-planets research. Second, it has been suggested to set up an email Letter on planet/YSO researches, so that the community remains in contact and exchanges rapidly various progresses in the field.

At the end of this ASI, I should say that I have been quite impressed by the youth and vigor of research activities in the field, by its multidisciplinarity aspects as indeed one has to span a fairly broad range of knowledge in order to link the differents steps involved in the occurence of life. Lastly, let's comment on the simple formula which was given during the ASI and which provides the probable number of advanced civilizations in the Universe:

$$N = R_{*}.f_{p}.n_{e}.f_{l}.f_{i}.f_{c}.L.$$

R_* is the rate of star formation, f_p is the fraction of stars having planetary systems and n_e is the number of Earth-like planets within a planetary system: deriving figures for these quantities is some work for the astronomers and they are on the way.

The terms, f_l, which represents the fraction of Earth-like planets which have developped life, and f_i, which is the fraction of life-bearing planets that produced intelligent life, are to be derived by biologists and it is hard to know where they stand in this quest because the parameters are numerous and complex.

The last two terms, f_c, which is the fraction of Earth-bearing planets with a civilization interested in communication and able to do it, and L, which is the average life-time of such a civilization, have to be discussed rather by sociologists and philosophers...Progress on such interrogations is still ahead of us.

This means at the end that we are far from having a figure for N and it leaves room for many other future meetings on the subjects related to the life-phenomenon in the Universe!

Index

CPSIA information can be obtained
at www.ICGtesting.com
Printed in the USA
LVHW081620080320
649318LV00016B/724